Systems Biology

Series Editor
John M. Walker
School of Life Sciences
University of Hertfordshire
Hatfield, Hertfordshire, AL10 9AB, UK

For other titles published in this series, go to
www.springer.com/series/7651

METHODS IN MOLECULAR BIOLOGY™

Systems Biology

Edited by

Ivan V. Maly

Department of Computational Biology, School of Medicine,
University of Pittsburgh, Pittsburgh, PA, USA

Humana Press

Editor
Ivan V. Maly
Department of Computational Biology
School of Medicine
University of Pittsburgh
Pittsburgh, PA
USA

ISBN: 978-1-62703-949-9
ISSN: 1064-3745
DOI: 10.1007/978-1-59745-525-1
ISBN: 978-1-59745-525-1 (eBook)
ISSN: 1940-6029 (eBook)

Softcover re-print of the Hardcover 1st edition 2009

Printed on acid-free paper

springer.com

Preface

The rapidly developing methods of systems biology can help investigators in various areas of modern biomedical research to make inference and predictions from their data that intuition alone would not discern. Many of these methods, however, are commonly perceived as esoteric and inaccessible to biomedical researchers: Even evaluating their applicability to the problem at hand seems to require from the biologist a broad knowledge of mathematics or engineering. This book is written by scientists who do possess such knowledge, who have successfully applied it to biological problems in various contexts, and who found that their experience can be crystallized in a form very similar to a typical biological laboratory protocol.

Learning a new laboratory procedure may at first appear formidable, and the interested researchers may be unsure whether their problem falls within the area of applicability of the new technique. The researchers will rely on the experience of others who have condensed it into a methods paper, with the theory behind the method, its step-by-step implementation, and the pitfalls explained thoroughly and from the practical angle. It is the intention of the authors of this book to make the methods of systems biology widely understood by biomedical researchers by explaining them in the same proven format of a protocol article.

It is recognized that, in comparison to the systems biology methods, many of the laboratory methods are much better established and their theory may be understood to a greater depth by interested researchers with a biomedical background. We intend, however, this volume to shatter the perceived insurmountable barrier between the laboratory and systems-biological research techniques. We hope that many laboratory researchers will find a method in it that they will recognize as applicable to their field, and that the practical usefulness of the basic techniques described here will stimulate interest in their further development and adaptation to diverse areas of biomedical research.

Pittsburgh, PA ***Ivan V. Maly***

Contents

PART IV: METHODS FOR LARGER-SCALE SYSTEMS ANALYSIS

Contributors

EIVIND ALMAAS • *Biosciences and Biotechnology Division, Lawrence Livermore National Laboratory, Livermore, CA, USA*

GARY AN • *Division of Trauma/Critical Care, Department of Surgery, Northwestern University Feinberg School of Medicine, Chicago, IL, USA*

ARIEL I. BALTER • *Biocomplexity Institute and Department of Physics, Indiana University, Bloomington, IN, USA*

MICHAEL L. BLINOV • *Richard Berlin Center for Cell Analysis and Modeling, University of Connecticut Health Center, Farmington, CT, USA*

JOHN H. CARSON • *Richard Berlin Center for Cell Analysis and Modeling, University of Connecticut Health Center, Farmington, CT, USA*

RICHARD BERLIN • *Center for Cell Analysis and Modeling*

MICHAL ČERVEŇANSKÝ • *Faculty of Mathematics, Physics and Informatics, Comenius University, Bratislava, Slovak Republic*

ARVIND K. CHAVALI • *Department of Biomedical Engineering, University of Virginia, Charlottesville, VA, USA*

JACOB CZECH • *National Resource for Biomedical Supercomputing, Pittsburgh Supercomputing Center, Carnegie Mellon University, Pittsburgh, PA, USA*

JOSEPH DADA • *School of Computer Science and Manchester Centre for Integrative Systems Biology, University of Manchester, Manchester, UK*

MARKUS DITTRICH • *National Resource for Biomedical Supercomputing, Pittsburgh Supercomputing Center, Carnegie Mellon University, Pittsburgh, PA, USA*

JAMES R. FAEDER • *Department of Computational Biology, University of Pittsburgh School of Medicine, Pittsburgh, PA, USA*

ANDREW T. FENLEY • *Biosciences and Biotechnology Division, Lawrence Livermore National Laboratory, Livermore, CA, USA*

RIMA GANDLIN • *Department of Mathematical Sciences, Carnegie Mellon University, Pittsburgh, PA, USA*

RALPH GAUGES • *Department of Modeling of Biological Processes, Institute for Zoology/BIOQUANT, University of Heidelberg, Heidelberg, Germany*

CHEOL-MIN GHIM • *Biosciences and Biotechnology Division, Lawrence Livermore National Laboratory, Livermore, CA, USA*

JAMES A. GLAZIER • *Biocomplexity Institute and Department of Physics, Indiana University, Bloomington, IN, USA*

RANDY W. HEILAND • *Biocomplexity Institute and Department of Physics, Indiana University, Bloomington, IN, USA*

SUSAN D. HESTER • *Biocomplexity Institute and Department of Physics, Indiana University, Bloomington, IN, USA*

MANJU M. HINGORANI • *Molecular Biology and Biochemistry Department, Wesleyan University, Middletown, CT, USA*

WILLIAM S. HLAVACEK • *Theoretical Division and Center for Nonlinear Studies, Los Alamos National Laboratory, Los Alamos, NM, USA and Department of Biology, University of New Mexico, Albuquerque, NM, USA*
RAQUELL M. HOLMES • *Richard Berlin Center for Cell Analysis and Modeling, University of Connecticut Health Center, Farmington, CT, USA*
STEFAN HOOPS • *Virginia Bioinformatics Institute, Virginia Polytechnic Institute and State University, Blacksburg, VA, USA*
JUCHANG HUA • *Department of Biological Sciences and Center for Bioimage Informatics, Carnegie Mellon University, Pittsburgh, PA, USA*
DANIEL T. KAMEI • *Department of Bioengineering, University of California, Los Angeles, CA, USA*
KERRY J. KIM • *Center for Cell Dynamics, University of Washington Friday Harbor Laboratories, Friday Harbor, WA, USA*
URSULA KUMMER • *Department of Modeling of Biological Processes, Institute for Zoology/BIOQUANT, University of Heidelberg, Heidelberg, Germany*
BERT J. LAO • *Department of Bioengineering, University of California, Los Angeles, CA, USA*
SUNGMIN LEE • *Biosciences and Biotechnology Division, Lawrence Livermore National Laboratory, Livermore, CA, USA*
ANDRE LEVCHENKO • *Whitaker Institute for Biomedical Engineering, Institute for Cell Engineering, Department of Biomedical Engineering, Johns Hopkins University, School of Medicine, Baltimore, MD, USA*
MIKHAIL K. LEVIN • *Richard Berlin Center for Cell Analysis and Modeling, University of Connecticut Health Center, Farmington, CT, USA*
IVAN V. MALY • *Department of Computational Biology, University of Pittsburgh School of Medicine, Pittsburgh, PA, USA*
PEDRO MENDES • *School of Computer Science and Manchester Centre for Integrative Systems Biology, University of Manchester, Manchester, UK and Virginia Bioinformatics Institute, Virginia Polytechnic Institute and State University, Blacksburg, VA, USA*
ROBERT F. MURPHY • *Center for Bioimage Informatics, Lane Center for Computational Biology and Departments of Biomedical Engineering, Biological Sciences and Machine Learning, Carnegie Mellon University, Pittsburgh, PA, USA*
ALI NAVID • *Biosciences and Biotechnology Division, Lawrence Livermore National Laboratory, Livermore, CA, USA*
JUSTIN NEWBERG • *Department of Biomedical Engineering and Center for Bioimage Informatics, Carnegie Mellon University, Pittsburgh, PA, USA*
MARTA NOVOTOVÁ • *Institute of Molecular Physiology and Genetics, Slovak Academy of Sciences, Bratislava, Slovak Republic*
MATTHEW A. OBERHARDT • *Department of Biomedical Engineering, University of Virginia, Charlottesville, VA, USA*
JASON A. PAPIN • *Department of Biomedical Engineering, University of Virginia, Charlottesville, VA, USA*

JÚLIUS PARULEK • *Institute of Molecular Physiology and Genetics, Slovak Academy of Sciences, Bratislava, Slovak Republic and Faculty of Mathematics, Physics and Informatics, Comenius University, Bratislava, Slovak Republic*
SMITA S. PATEL • *Department of Biochemistry, Robert Wood Johnson Medical School, Piscataway, NJ, USA*
RANJIT RANDHAWA • *Department of Computer Science, Virginia Polytechnic Institute and State University, Blacksburg, VA, USA*
SVEN SAHLE • *Department of Modeling of Biological Processes, Institute for Zoology/ BIOQUANT, University of Heidelberg, Heidelberg, Germany*
CLIFFORD A. SHAFFER • *Department of Computer Science, Virginia Polytechnic Institute and State University, Blacksburg, VA, USA*
MILOŠ ŠRÁMEK • *Faculty of Mathematics, Physics and Informatics, Comenius University, Bratislava, Slovak Republic and Austrian Academy of Sciences, Vienna, Austria*
JOEL R. STILES • *National Resource for Biomedical Supercomputing, Pittsburgh Supercomputing Center, Carnegie Mellon University, Pittsburgh, PA, USA*
MACIEJ H. SWAT • *Biocomplexity Institute and Department of Physics, Indiana University, Bloomington, IN, USA*
SHLOMO TA'ASAN • *Department of Mathematical Sciences, Carnegie Mellon University, Pittsburgh, PA, USA*
JOHN J. TYSON • *Department of Biological Sciences, Virginia Polytechnic Institute and State University, Blacksburg, VA, USA*
C. JOANNE WANG • *Whitaker Institute for Biomedical Engineering, Institute for Cell Engineering, Department of Biomedical Engineering, Johns Hopkins University, School of Medicine, Baltimore, MD, USA*
SOOYEON YOON • *Biosciences and Biotechnology Division, Lawrence Livermore National Laboratory, Livermore, CA, USA*
IVAN ZAHRADNÍK • *Institute of Molecular Physiology and Genetics, Slovak Academy of Sciences, Bratislava, Slovak Republic*
BENJAMIN L. ZAITLEN • *Biocomplexity Institute and Department of Physics, Indiana University, Bloomington, IN, USA*
JASON W. ZWOLAK • *Department of Biological Sciences, Virginia Polytechnic Institute and State University, Blacksburg, VA, USA*

Part I

Introduction

Chapter 1

Introduction: A Practical Guide to the Systems Approach in Biology

Ivan V. Maly

Summary

This essay provides an informal review of the modern systems-centric biological methodology for the practical researcher. The systems approach is defined, and a generic recipe for employing it in biomedical research is offered. General caveats are discussed that pertain to biological complexity, to explanation in molecular terms, and to bottom-up investigation. An outlook on the development of systems biology is also given.

Key words: Systems biology, Methodology, Mathematical modeling, Complexity, Physiology.

1. Introduction

What is systems biology? Over the past 40 years, practicing systems biologists delimited their field in a great number of ways. At times the definition was restricted to applications of the formal systems theory in biology; more recently, the tendency has been to focus on biomolecular interactions or on multivariate analysis as systems biology's proper subject and method *(1–5)*. This essay is written from a conservative biologist's perspective and takes a less specific view of the topic. First, what should we call a system, in the context of the scientific way to parse the world into concepts? We recognize a system in a certain number of different and interacting objects. Noninteracting objects do not form a system. Also, it is hardly useful to see any substantial number of interacting, but identical objects as a system. Science has powerful methods to study aggregate behavior of identical objects. Notably, such methods tend to disregard the corpuscular nature of the

Ivan V. Maly (ed.) *Methods in Molecular Biology, Systems Biology, vol. 500*

DOI: 10.1007/978-1-59745-525-1_1

individual objects and take a view of their collections as continua. In this case, biology can freely borrow methodologically from established areas of physics. In contrast, studying behavior of collections of interacting nonidentical objects remains a methodological challenge. We will, therefore, restrict the meaning of "system" to a system of interacting nonidentical parts. Study of a living object by discerning so-defined systems in it will then be called systems biology. Its status as a distinct discipline should engender no jealousy: The definition limits the subject of systems biology to what the scientific method is currently handling perhaps least confidently.

The difficulty appears to stem from the limitation of the human mind itself *(6, 7)*: we are nearly incapable of considering more than a few things at a time. Psychophysical experiments suggest the limit of about seven, which corresponds well to the number of the nonidentical, interacting elements that deserve to be called a system, as commonly perceived in the systems biology practice. It is important to observe that the limitation is not just to our intuition, but to rational reasoning as well. We can consider larger systems of course, but the effects resulting from interactions of more than a few elements at a time will likely be missed. To reason about systems whose complexity is beyond our immediate grasp, we must extend our mind with formal deduction, under the general rubric or mathematics. As applied to the natural world, it is termed mathematical modeling. Involving mathematics in nonsystems biological research can be necessitated by a desire of quantitative precision in understanding; in systems biology, it is indispensable for any progress whatsoever, because even the crudest qualitative effects are liable to be overlooked by the unaided mind that has evolved for rather different purposes.

That quantitative precision is rarely sought in modern systems biology is important to recognize, so as not to confuse the nature of the mathematics employed with the goals of the investigation. And certainly, those who are just considering employing systems analysis especially should not decide against it, if it is qualitative inference about their subject that they are after. We owe the quantitative, continuous-variable flavor of our most widely applied mathematics to its original development for the purposes of celestial mechanics and similarly particular problems of quantitative precision. As a consequence, the modern systems-biological modeling has to be done most commonly in terms of dynamics of continuous quantities first, to an exceeding precision ("phosphorylation goes up by 73%"), and then the results are reinterpreted in terms of qualitative statements about discrete events ("this genotype permits cell division"), to arrive at the kind of knowledge that is actually being sought. This is no different from how experimental measurements are most commonly employed.

2. The Basic Protocol

So how should we conduct a mechanistic systems analysis of the object of our investigation? The following may be suggested:

1. On the basis of the existing empirical knowledge, make a choice of the model elements (system parts) that is efficient for specifying the system (e.g., mitogens and their activators in the system of cell proliferation control). The efficient choice is not unique and requires a great deal of creativity on the part of the biologist. Some of the most general caveats are discussed in the next section of this chapter.
2. Identify those properties of the elements that would reflect the relevant interactions between the elements (e.g., mitogen phosphorylation levels).
3. Express the element properties, and interactions that affect them, in the structure of a mathematical model. (e.g., rate of phosphorylation equals activator concentration times the rate constant, and so on.) Practice shows that it is not very important what exact kinetic laws, etc., to use to begin – this may be "ironed out" later to the extent that it matters for the actual goal of the investigation. It is highly advisable, however, from the beginning to take pains to be consistent with dimensions of all quantities introduced (not to add apples to oranges in your formulae).
4. Determine the following by means of numerical analysis of the model, guided closely by biological thinking (this is a great deal of work if done systematically, but only a thorough analysis is worth the trouble):
   (a) What known features of the system can be objectively (mathematically) derived from the known features of its elements and interactions (e.g., the mitogen activity can indeed be "predicted" to peak at intervals, as with cell division.). Determining this gives us confidence in our understanding of the system in terms of the selected elements and interactions. Simultaneously, it codifies our knowledge completely and unambiguously in the form of the model that "by itself can do what the system does the way we think it does." Only for the simplest "systems" found seldom in biology is this outcome not worth the effort of the formal analysis.
   (b) What elements and interactions implicated in the system at the outset prove to be dispensable for the explanation achieved. (e.g., the peaks are predicted whether activator C is available or not.) These elements are relieved of duty, and in a sense the rigorous explanation becomes simpler than the intuitive one was. This outcome can direct further

experimentation, although it predicts a "negative result." (Similarly it may be worth checking why we think we know that behavior of the system which turns out impossible to derive formally from what we know about the parts.)

(c) What previously unknown features of the system can be deduced from the previously known interactions and elements. Determining this directs further experimental study of the system as a whole. It corresponds to the most conventional notion of "prediction" and of the role of modeling.

(d) What previously unknown elements or interactions, or what features of them (e.g., how much of what protein species, how fast a reaction) must be assumed to explain the known features of the system. Determining this directs further experimental investigation into the system elements. This is perhaps the most valuable outcome in the framework of top-down investigation, which is advocated below.

The author contends that attempting analysis along these lines can benefit nearly every line of biological investigation, where appreciable empirical knowledge has been accumulated, i.e., most any line of investigation in modern biology. Chapters in this volume describe specific approaches and their caveats at the level of detail that should give a head start even to a complete novice. To those accustomed to the use of mathematical modeling in experimental research (physicists) or to systems-centric thinking (engineers), the generic "protocol" outlined above may look even trivial. There are, however, certain problems about applying these methodological ideas to modern biology. It is unlikely that these problems should be resolved in the same systems biology framework, but recognizing them might be of help in setting up a systems approach to the subject of the investigation more effectively and in interpreting its results more consciously.

3. Methodological Caveats

3.1. The Molecular Approach and Complexity

Successful explanation of the whole in terms of the parts depends critically on the investigator's choice of the parts to implicate in the explanation. Challenging problems in modern biomedicine concern explanation of behavior of cells, these apparently simplest physicochemical systems which are alive in every sense of the word. It has proved a challenging objective to explain cells in terms of the biomolecular species that are found in them. Systems biology is called upon to facilitate this task with its methods, not replace the objective. Singular examples of molecular self-organization into life-like assemblies *(8)* are certainly fascinating, enough so to consider the general approach of explaining cells in terms of molecules, with the help of the systems biology methods.

Complexity of explanation of cells in terms of molecules is widely acknowledged by biologists and their engineer and physicist colleagues. The latter especially are familiar with the problem of carefully selecting parts in terms of which to explain the whole. A popular quip among physicists is "modeling bulldozers with quarks," i.e. subsubatomic particles – meaning an absurdly bad methodological choice, which would result in clearly insurmountable difficulties of complexity. Instead, it is of course recommended to explain the workings of a bulldozer in terms of its obvious functional parts: wheels, gears, levers. Modern biological knowledge suggests that biomolecules, proteins especially, are such functional units within the cells. The analogy appears powerful enough to motivate applications of the systems analysis, proven to be adequate for the bulldozers and gears. The greater complexity of the cell as a biomolecular system is seen only as a worthy challenge, not the absurd one from the quip about quarks. Indeed, the absurdity was meant to reside in the number of parts, and it seems clear that there are "more quarks" in a bulldozer than there are biomolecules in a cell. The reader is invited to "do the math" using the estimates he favors. We alluded in the beginning to the fact that it is the number of different parts that challenges the scientific method, whereas the count of identical parts can be of little concern. There are perhaps only a few quark types, elementary particle types, subatomic particle types, and atomic types in a bulldozer, which matter is arranged rather uniformly into the relatively few types of mechanical parts, such as the wheels, gears, and levers. On the whole, there are very few *different* parts in a bulldozer, even counting quarks, compared with the most conservative estimates of the number of chemically distinct polypeptides alone in a human cell. From the systems standpoint, then, it should appear more challenging to attempt explaining cells in terms of their constituent biomolecules than bulldozers from quarks.

The appealingly straightforward approach of explaining cells in terms of molecules, with the help of the systems biology methods, is thus faced with a complexity problem of proportions that are considered plainly insurmountable in another context. Our analysis was based, however, on the strict chemical definition of the component unit, such as the polypeptide chain of unique sequence. In modern biology, this is rarely the definition of a protein as the functional unit. Rather, proteins are identified by their function itself. As was noted early by Albrecht-Buehler *(9)*, referring to a biomolecular species in the modern molecular-biological discourse almost never means referring to any of its actual physical properties. The chemical structure, to a degree, is still part of the definition – for example, chemically distinct entities with the same function may be recognized as separate isoenzymes. The multisubunit nature of the typical protein of molecular cell biology, its variability arising from existence of isoforms of each component, and the extreme variability of posttranslational modifications may also at times be acknowledged. However, the actual usage of the term protein as a unit of cell-biological explanation is hardly encumbered by these complications. It is even unimportant in the modern usage of "protein," which exact polypeptide subunits it is meant to include, and which are its auxiliary adaptors, etc. Molecular motor dynein is one well-known example of this methodologically powerful definition of a "protein," which has traded structural clarity for functional one. From the systems standpoint, this has the potential of reducing complexity enormously and making the task of explaining cells in terms of the so-defined "molecules" feasible again.

However, in a certain alarming sense, some of us may find that our subject has left with the complexity, for functions are not things. Yes, useful reasoning about cells in terms of the functionally defined "molecules" is possible. Yet this activity is very different from the straightforward (if impossible) explanation of cells from the actual, chemically-defined molecules. Taking a less "objectivist" approach, we may find explanation of the system in terms of functions of its parts, even without any attempt to otherwise define the parts, acceptable and perhaps even more efficient than explanation in terms of any material parts. Explanation in terms of functions is appealing logically, because in the logical framework, we would expect to be able to proceed from assuming the known functions of the parts. In the situation of any natural-scientific investigation, however, the elementary functions are always more or less unknown. Is this fact a minor annoyance that ought not to stand in the way of the straightforward research strategy, or should it affect our methodological outlook? Shrager *(10)* suggests that it should, because functions that we assign to the (biomolecular) parts depend, in the actual research practice, on the explanation that we are constructing for the system. Should tautology slip in through this way of defining the parts in terms of which to study the system, no ordinary amount of objective experimentation may be able to remove it. At the very least, this dependence of the parts on the whole – introduced entirely by the investigator! – means that the investigation and explanation are no longer from the parts to the whole – which, it might appear, was our intended goal and method. Let us abstract now from the problematic use of functionally-defined biomolecules as system parts, and discuss other ways in which the bottom-up character of the explanation can be degraded.

3.2. External Constraints and Bottom-Up Explanation

In most realistic systems analyses, some of the model elements are not strictly internal to the system. Mathematically, they are boundary conditions or other types of constraints. Biologically, they account for actual physical boundaries to the motion of the truly internal system components, for the fact that something outside the metabolic system under study limits the availability of a nutrient, and so on. As noted by Bradley *(11)*, in reality there are no external constraints because there are no real boundaries of the system: what constitutes the system is the investigator's choice. It does not present any insurmountable logical or mathematical problem to delimit the system arbitrarily and take into account the external constraints on it that will correspond to the particular chosen division between the interior and exterior of the system. How useful or, alternatively, misleading, the use of external constraints will be, will, however, depend on such soft matter as how well circumscribed the system is in reality – which in the context of the natural-scientific investigation is unknown. If the system deals with reaction and diffusion of intracellular molecules, it is relatively reasonable to impose boundary conditions on their fluxes to account for the bounding nature of the cell membrane (for example, no-flux boundary conditions, to describe impenetrability). In contrast, if we concern ourselves with the molecular dynamics of components of a lipid raft within the cell membrane, it is comparatively difficult to constrain their motion in a way preserving much of the bottom-up predictive power of our model, because the lipid raft perhaps "imposes constraints" on the membrane it is in about equally with the membrane imposing constraints on it – as observed in his contemporary context by Bradley *(11)*. System analyses that do not explicitly make reference to any external constraints are methodologically suspect too, for they should appear to have taken the arbitrary delimitation of the system literally, as the system being isolated, which should usually be quite unrealistic in the biological context. Thus, the decision to invoke certain external constraints on the system in the explanation or modeling should strike a consciously determined balance between how "realistic" and how "bottom-up" the investigator wants his model or explanation to be.

There is no doubt at least that in reality, biomolecular components of cells are under constraints of the cell structure. Harold *(12)* convincingly argues in the modern, molecular-centric context that at least some of the basic structural features of cells, such as the cell boundary itself, are not in any useful sense fully derivative from the component molecules, even though they of course consist of them. Such basic structures are templates for their own propagation between cell generations, which they achieve by directing collective organization of their molecular components. This fact should favor the methodological choice of realistic over "bottom-up" system models in the above dilemma. The modern cell and systems biology is nonetheless faced with an enormous number of molecular species, which have been chemically identified and whose collective organization would need to be directed in this fashion by the cell structures. In comparison to this number, there are few cell structures, which have been identified by the more traditional cell biology, and which may act as the organizers. We should not compare apples to oranges; yet, logic notwithstanding, in the actual research practice, the sheer numerical discrepancy of our knowledge of the cell structures and of their molecular components appears to play a role in methodological decisions. This discrepancy seems to suggest that there is complexity in the cell, which stems from the number of chemically distinct molecular components, and which should give rise to cellular behavior that will become known to us when we derive it from the molecular interactions that must be taking place. It is far less often that the same discrepancy is seen as an indication that there may be relative simplicity in the cell and its behavior, which will become known to us better when we study them directly further. Whether one ascends from the molecular chaos and expects it to self-organize on the cellular level, or descends from the exquisitely structured organism body (multi or unicellular) and expects the functional structures to continue into the subcellular domain, depends on one's scientific background. In leading education systems today, future researchers majoring in nonbiological quantitative disciplines receive no exposure to organismal biology, while biology curricula lack both the quantitative component and a rigorous organism-level component. It should then come as no surprise, and be seen as a circumstance largely external to the scientific development proper that in the realm of modern quantitative biology research there is a severe imbalance of representation of the above two views on what we should be looking for on the subcellular level. Correcting this by broadening the curricula or by recognizing boundaries of individual expertise in interdisciplinary areas will be about equally difficult.

Bottom-up explanation may retain its desirability even in the face of the trade-off with the model realism, when the incorporation of the system environment and of the external constraints has been carefully considered. However, it has long been known that in the systems-biological context in particular, properties of the parts differ within the system from their properties in isolation *(13)*. Thus, even if we studiously avoid defining parts by their function in the system, any fully successful investigation still would not be able to proceed from the bottom up – by first studying the properties of the parts and then deducing from them the system behavior. The statement that the properties of parts in the system are different from their properties in isolation may sound like lazy logic, something coming from a lack of effort to properly define the parts, the properties, and the system. The author concedes that it may be possible to define these things so that this dictum does not hold logically, and the properties of the parts are the same in the system as in isolation. This kind of logical dissection should be possible given a complete knowledge of the system. In a real study, the (complete) knowledge of the system is not available, and is exactly what we are after. Therefore, from the standpoint of the research practice in natural science, parts will indeed have different properties in the system than in isolation and, as observed by Yates et al., this circumstance should be of special concern to the systems biologist *(13)*.

Explanation from the bottom up is, nonetheless, central to our very notion of understanding a system. We must conclude that the way to explain (formulate and convey understanding) must, in systems biology in particular, be different from the way to study (obtain new understanding). All problems with the bottom-up approach then seem to arise merely from substantiation of the explanatory deduction – a fallacy that is the mirror image of teleology. Designing the model and the entire study to explain cellular functions in the organism, selecting the system parts based explicitly on their function in the system and irrespective of their molecular or nonmolecular nature, and eventually presenting the obtained understanding of the system functions in the deductive manner may not have the straightforward appeal, yet applying the systems method consciously in this fashion should not suffer from the discussed difficulties.

4. Outlook

We have discussed some general methodological choices pertaining to the systems method. The most fundamental question, however, is whether to apply the systems method at all. The purpose of this book is to help a biomedical investigator to give an affirmative answer to this question. Can, however, any lesson be derived from the fact that the answer has been, for great many, in the negative for a long time? One comparatively comfortable explanation of this fact posits that the reason why experimentalists rejected modeling in the past was the repulsive "spherical cells (cows)" that theorists of the time liked to play with. (The quip recurred at a seminal, historically recent meeting, *14)*. Although the author has no first-hand knowledge of why most experimentalists shunned biological theory, he can find no evidence in the literature of the prevalence of the "spherical" theories. In fact, theories characteristic of the 1960s and 1970s appear rather exquisite by our modern standards.

Theories of that era often were not theories of biological systems modeled as consisting of interacting molecules. Instead they might deal with physiological processes such as material fluxes through the kidney *(13)*, or the respiratory cycle and its modulation *(15)*. Their conceptual depth and quantitative connection to the physiological experiments did not suffer at all from the nonmolecular nature of both the theory and experiment. At the same time, theories that dealt with metabolite-mediated interactions of enzymes in the elaborate framework of classical biochemistry *(16)* were laying the foundation for much of today's work, and the kinetic theory of the cytoskeletal structures *(17)* was taking essentially the form these models have now.

The new science of molecular cell biology took off on its own, leaving the rather exquisitely measured and modeled world of organismal physiology (and, arguably, of metabolism and self-assembly) behind. Why did the system modelers not catch up? It should be only understandable if the detailed physiological and biochemical knowledge of the time looked far more attractive as an object to model. Qualitative and fragmentary to the extreme in the individual examples sought out to support the bold, system-denying interpretation of the "one gene, one protein, one function" paradigm, the results of this new type of biological studies could not be expected to attract systems thinkers. A student of molecular biology in the 1990s will remember how exceptions to its intentionally fragmentary nature, such as the far earlier *lac* operon study, stood out in the course material as if they were some alien science. Should we call this intervening period the Dark Ages for systems biology? After all, it did appear to be merely a lull following a period when the quantitative systems method enjoyed an appropriate place among the research tools of biology.

The natural shift of interest to the new, molecular cell biology circa 1980 did not have to lead to the systems method being relegated to "crank science," if remembered at all, by an entire scientific generation engaged in the most active subfield of biology. It may be argued that this only happened because in this case the normal, productive shift of the mainstream interest coincided with what Wiley called destruction of "biodiversity" in biological research itself *(18)*: A major, near-monopolist funding source on which much of biology had come to depend happened to be shrinking, and an understandable decision was to concentrate support on what was new and therefore most promising. In a science that benefits from its status of a popular occupation, such events and decisions may have a stronger effect than the natural evolution of scientific interests: They may rescue the science from stagnation, or they may nip development of a healthy methodology.

Is systems biology in its Renaissance? Molecular cell biology and molecular genetics can be said to have matured in the above sense, making some of the most active subfields in current biology receptive as well as amenable to the systems method again. The new generation of systems biologists is determined to ask truly systems-centric questions about biological objects and to conquer complexity through the use of computers and interdisciplinary collaboration. A wider attribution of these aspirations to our scientific predecessors *(1, 19)* will probably come. Which of their intellectual threads will be picked up, and especially whether the productive aspects of the "pre-molecular" worldview will be rediscovered by postgenomic biology, remains to be seen. It will also be a matter of individual decisions how much impact the factors external to science will have on the development of biological methodology in our time. The subjective retrospection and methodological outlook that were given above together argue that it is imperative that the results of our current efforts become known to the enthusiastic systems biologists of 2050: Freed from dogmatic constraints, the systems approach does not promise a quick and sure solution to the currently recognized problems as much as it opens for exploration a particularly challenging face of a diversity of biological phenomena.

References

1. Mesarović, M. D. (1968) *Systems theory and biology–view of a theoretician, in Systems Theory and Biology* (Mesarović, M. D., ed.) Springer, New York, pp. 59–87.
2. Kitano, H. (2002) Systems biology: A brief overview. *Science* 295, 1662–1664.
3. Hood, L., Heath, J. R., Phelps, M. E., and Lin, B. (2004) Systems biology and new technologies enable predictive and preventative medicine. *Science* 306, 640–643.
4. Westerhoff, H. V. and Alberghina, L. (2005) *Systems biology: did we know it all along? in Systems Biology: Definitions and Perspectives* (Alberghina, L. and Westerhoff, H. V., eds.) Springer, Berlin, pp. 3–9.
5. Ideker, T., Winslow, L. R., and Lauffenburger, D. A. (2006) Bioengineering and systems biology. *Ann. Biomed. Eng.* 34, 257–264.
6. Garfinkel, D. (1980) Computer modeling, complex biological systems, and their simplifications. *Am. J. Physiol.* 239, Rl–R6.
7. Miller, G. A. (1956) The magical number seven, plus or minus two. Some limits on our capacity for processing information. *Psychol. Rev.* 63, 81–97.
8. Kirschner, M., Gerhart, J., and Mitchison, T. (2000) Molecular "vitalism". *Cell* 100, 79–88.
9. Albrecht-Buehler, G. (1990) In defense of "nonmolecular" cell biology. *Int. Rev. Cytol.* 120, 191–241.
10. Shrager, J. (2003) The fiction of function. *Bioinformatics* 19, 1934–1936.
11. Bradley, D. F. (1968) *Multilevel systems and biology–view of a submolecular biologist, in Systems Theory and Biology* (Mesarović, M. D., ed.) Springer, New York, pp. 38–58.
12. Harold, F. M. (2001) *The Way of the Cell.* Oxford, New York.
13. Yates, F. E., Brennan, R. D., Urquhart, J., Dallman, M. F., Li, C. C., and Halperin, W. (1968) *A continuous system model of adrenocortical function, in Systems Theory and Biology* (Mesarović, M. D., ed.) Springer, New York, pp. 141–184.
14. Doyle, J. (2001) Computational biology: Beyond the spherical cow. *Nature* 411, 151–152.
15. Yamamoto, W. S. and Walton, E. S. (1975) On the evolution of the physiological model. *Annu. Rev. Biophys. Bioeng.* 4, 81–102.
16. Heinrich, R. and Rapoport, T. A. (1974) A linear steady-state treatment of enzymatic chains. General properties, control and effector strength. *Eur. J. Biochem.* 42, 89–95.
17. Oosawa, F. and Asakura, S. (1975) *Thermodynamics of the Polymerization of Protein.* Academic Press, New York.
18. Wiley, H. S. (2008) Think like a cockroach. *The Scientist* 22, 29.
19. Weaver, W. (1948) Science and complexity. *Am. Scientist* 36, 536.

Part II

Methods for Analyzing Biomolecular Systems

Chapter 2

Computational Modeling of Biochemical Networks Using COPASI

Pedro Mendes, Stefan Hoops, Sven Sahle, Ralph Gauges, Joseph Dada, and Ursula Kummer

Summary

Computational modeling and simulation of biochemical networks is at the core of systems biology and this includes many types of analyses that can aid understanding of how these systems work. COPASI is a generic software package for modeling and simulation of biochemical networks which provides many of these analyses in convenient ways that do not require the user to program or to have deep knowledge of the numerical algorithms. Here we provide a description of how these modeling techniques can be applied to biochemical models using COPASI. The focus is both on practical aspects of software usage as well as on the utility of these analyses in aiding biological understanding. Practical examples are described for steady-state and time-course simulations, stoichiometric analyses, parameter scanning, sensitivity analysis (including metabolic control analysis), global optimization, parameter estimation, and stochastic simulation. The examples used are all published models that are available in the BioModels database in SBML format.

Key words: Simulation, Modeling, Systems biology, Optimization, Stochastic simulation, Sensitivity analysis, Parameter estimation, SBML, Stoichiometric analysis.

1. Introduction

Biochemical networks are intrinsically complex, not only because they encompass a large number of interacting components, but also because those interactions are nonlinear. Like many other nonlinear phenomena in nature, their behavior is often unintuitive and thus quantitative models are needed to describe and understand their function. While the concept of biochemical networks arose from the reductionist process of biochemistry, where the focus was on studying isolated enzymatic reactions, it is now better

Ivan V. Maly (ed.) *Methods in Molecular Biology, Systems Biology, vol. 500*

DOI: 10.1007/978-1-59745-525-1_2

understood in the framework of *systems biology*, where the focus is on the behavior of the whole system, or at least several reactions, and particularly on what results from the interactions of its parts. Computational modeling is thus a technique of systems biology as important as its experimental counterparts. This chapter covers the definition and analysis of computational models of biochemical networks using the popular software COPASI. It provides essentially a practical view of the utility of several computational analyses, using established models as examples. All software and models discussed here are freely available on the Internet.

From a modeling perspective, biochemical networks are a set of chemical species that can be converted into each other through chemical reactions. The focus of biochemical network models is usually on the levels of the chemical species and this usually requires explicit mathematical expressions for the velocity at which the reactions proceed. The most popular representation for these models uses ordinary differential equations (ODEs) to describe the change in the concentrations of the chemical species. Another representation that is gaining popularity in systems biology uses probability distribution functions to estimate when single reaction events happen and therefore track the number of particles of the chemical species. As a general rule, the latter approach, known as stochastic simulation, is preferred where the numbers of particles of a chemical species is small; the ODE approach is required when the number of particles is large because the stochastic approach would be computationally intractable.

1.1. ODE-Based Models

Each chemical species in the network is represented by an ODE that describes the rate of change of that species along time. The ODE is composed by an algebraic sum of terms that represent the rates of the reactions that affect the chemical species. For a chemical species X:

$$\frac{\mathrm{d}X}{\mathrm{d}t} = \sum_{\text{all reactions } i} s_i \cdot v_i, \qquad (1)$$

where s_i is a stoichiometry coefficient that is the number of molecules of X consumed or produced in one cycle of reaction i, with a positive sign if it is produced or negative if consumed, and v_i is the velocity of reaction i. Obviously, for reactions that do not produce or consume X the corresponding s_i is zero.

The velocity of each reaction is described by a *rate law* that depends on the concentrations of the reaction substrates, products, and modifiers (*see* **Note 1**). Rate laws are the subject of chemical and enzyme kinetics and are generally nonlinear (except the case of first-order mass action kinetics). Often these rate laws are saturable functions, i.e., have finite limits for high concentrations of substrates, products, and also for many modifiers (*see* **Note 2**).

An example of a rate law is depicted in **Eq. 2**, which represents a rate law of reaction with one substrate (S), one product (P), and a competitive inhibitor (I)

$$v = \frac{\frac{E \cdot k_{\text{cat}}}{K_{mS}}\left(S - \frac{P}{K_{\text{eq}}}\right)}{1 + \frac{S}{K_{mS}} + \frac{P}{K_{mP}} + \frac{I}{K_I}}. \tag{2}$$

In **Eq. 2**, the limiting rate of reaction ("$V_{\max}$") is directly represented as a product of the concentration of the enzyme and the turnover number ($E \cdot k_{\text{cat}}$). It is usually good practice to make this product explicit, since it is then possible to have the enzyme concentration be a variable of the model too. This is important if the model includes protein synthesis, degradation, or protein–protein interactions.

These ODE models can be used to simulate the dynamics of the concentrations of the chemical species along time given their initial values. This is achieved by numerical integration of the system of ODE which can be carried out with well-established algorithms (for example *(1, 2)* but *see* **Note 3**). It is also useful to find steady states of the system, which are conditions when the concentrations of the chemical species do not change. If the steady state is such that the fluxes are also zero, then the system is in chemical equilibrium, otherwise the fluxes are finite meaning that the concentrations do not change because the rates of synthesis balance with the rates of degradation for every chemical species. Steady states can be found using the Newton–Raphson method which finds the roots of the right-hand side of the ODE (which must be zero by the definition of steady state). Alternatively steady states can also be found by integration of the ODE. COPASI can use either one of these strategies or a combination of the two (*see* **Note 4**).

Other model analyses can be carried out but a description of their theory in any detail is beyond the scope of this article. Some of them are described at a high level in **Subheading 3**, whenever they are used.

1.2. Stochastic Models

When analyzing a biochemical system which contains small numbers of particles of each reactant, the assumption of continuous concentrations fails and consequently the underlying basis of the ODE representation also fails. Moreover, in such conditions, stochastic effects become more pronounced and may lead to dynamics that differ significantly from those that would result from the ODE approach. In the conditions described above, one should then use a stochastic discrete approach for the simulation of the system dynamics.

Stochastic models represent the number of particles of each chemical species and use a reaction probability density function (PDF) to describe the timing of reaction events. Gillespie developed a Monte Carlo simulation algorithm, known as the stochastic simulation algorithm (SSA) first reaction method, that simulates the stochastic dynamics of the system by sampling this PDF *(3, 4)*. The theoretical derivation of this method is too involved to be described here, and the reader is directed to the original publications *(3, 4)* or a recent review *(5)*. It is important to stress that one simulation run according to this approach is only one realization of a probabilistic representation, and thus provides limited amount of information on its own. In the stochastic formalism, it is very important that simulations are repeated for a sufficient number of times in order to reveal the entire range of behavior presented by such a system (i.e., to estimate a distribution for each chemical species and its dynamic evolution).

2. Materials

2.1. Copasi

The software COPASI *(6)* will be used throughout this chapter. COPASI is freely available for noncommercial use (*see* **Note 5**) and executable versions are provided for the most popular operating systems: Microsoft Windows, Apple Mac OS X, Linux, and Sun Solaris. The source code of COPASI is also available under an open source license and so it can be compiled for other architectures.

New versions of COPASI are released often and there is a distinction between *stable* and *development* versions. Development versions are those where new features are introduced; stable versions have no new features and differ from the previous development release only by having bug fixes. While testing is more intense in stable releases, the reader is encouraged to download the latest development release. Irrespective of being stable or development releases, COPASI releases are labeled with a *build number*, which is sequential (*see also* **Note 6**).

2.1.1. Installing COPASI

Instructions for installation of COPASI depend on the operating system version, but all start with downloading the appropriate binary from the project's Web page http://www.copasi.org. Choose the *download non-commercial* option from the site's menu and then select the appropriate version for your platform. Download will proceed after selecting the nearest server and accepting the license terms. Once the file has finished downloading, the installation instructions are different for each platform.

For Microsoft Windows, the downloaded file, Copasi-XX-WIN32.msi, is an installation program which you should run by double clicking it. For Apple OS X, the downloaded file, Copasi-XX-Darwin.dmg, is a disk image containing a folder named "copasi." You can either start COPASI directly from the disk image or drop the folder into your applications folder and start it from there.

Finally, for Linux or Solaris, you need to unpack the archive where you want to install it (it can be in a system-wide location like /usr/local/copasi, or in a user home, such as ~/copasi). For optimal performance you should set the environment variable COPASIDIR to /usr/local/copasi (or wherever you have installed it).

2.2. BioModels Database

"BioModels" is a database that archives biochemical models that have previously been published in peer-reviewed journals *(7)*. Some examples in this chapter use models that are available there and so to avoid entering those models manually it is best to download them from BioModels.

Models in this database are primarily distinguished by their identifier, which is in the form BIOMDxxxxxxxxxx where the x's represent a number. To download a specific model, point your Web browser to http://www.ebi.ac.uk/biomodels/, select the *search* option, type the model ID in the search box, and then click on the link for that model's page (*see* **Note 7**). Once in the model's page you can examine the model's characteristics, including the citation of the original publication, who was the author of the model, etc. To download the model in a format that COPASI (and most other systems biology software) can read select the link entitled **SBML L2 V1** (*see* **Note 8**) at the very top of the page, and download it to your local computer (*see* **Note 9**). This will be a file entitled BIOMDxxxxxxxxxx.xml which COPASI can import.

3. Methods

3.1. Model Construction and Basic Simulation

3.1.1. Model Specification

The COPASI user interface is composed of two main areas: a hierarchical organization of functions on the left (a *tree*), and a larger area on the right which contains the controls related with the function selected on the tree. All features related with model specification reside on the first main entry of the tree, appropriately named *Model*. When this entry is selected on the left, the right displays the basic information about the model, such as its name, the units used, and a large field for comments (*see* **Note 10** and **11**). Expanding the *Model* subtree reveals two other entries: *Biochemical* and *Mathematical*, these are different views of the model. Specification of a new model is done in the *Biochemical* part as the other is only for examining equations and matrices.

The most practical way to enter a model is to start by adding its component reactions. Select *Reactions* and then double click the first empty row of the table on the right. This will change to display the detailed reaction window. Enter a name for the reaction and type its chemical equation, for example "NAD + ethanol = acetaldehyde + NADH" (*see* **Note 12**). The equals sign has quite a specific meaning, not only does it separate substrates from products, but it also means the reaction is considered kinetically reversible. If you want the reaction to be irreversible then you should use instead the combination of characters "->" (*dash* and *right angle bracket*).

After entering the reaction equation, you should select the appropriate rate law for this reaction. COPASI only allows selecting rate laws that match the characteristics of the reaction entered: same number of substrates and products and reversibility. You can chose a rate law from the menu; if the appropriate one is not available, then you can add one yourself by pressing the *New Rate Law* button. Type the rate law in the box, for example: $V/K*(A*B-P*Q)/(K + A + B + P + Q)$. Next select the appropriate type of each symbol in the equation (*see* **Note 13**). You should also mark the reaction as *reversible* or *irreversible* (*see* **Note 14**). **Figure 1** shows this window when entering the above rate law. When you finish press *Commit* and go back to *Model Reactions* where you can now select this rate law for the reaction (assuming it was reversible with two substrates and two

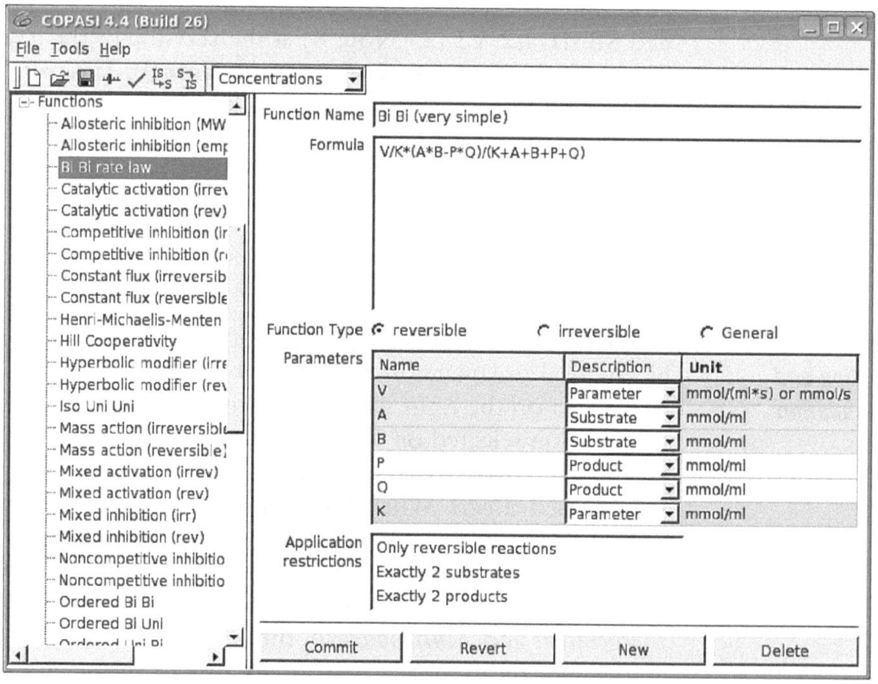

Fig. 1. Definition of a kinetic rate law.

products). At that point you will be able to enter the values of the parameters (in this case V and K).

You can define more compartments and change their sizes (volumes) on the entry named *Compartments*. By default there is always a compartment called simply "compartment" of unit volume. You can define any number of compartments of any positive size.

After entering all reactions you can examine the chemical species by selecting *Species* on the tree which will show a summary table with all species included in the reactions. You can change their initial concentrations, set their compartment and change their type. The species type determines how its concentration (actually its particle number) is calculated in the models: it can be set by *reactions*, which means that its concentration will be determined by the ODE generated from the reactions or by the SSA; it can be *fixed*, meaning that it becomes a parameter of the model; set by *assignments*, which are algebraic expressions (see below); or set by arbitrary *ODE* (i.e., entered directly by the user). You can add extra species directly at the end of the table. If you double click any row, you will then see a more detailed page for that species alone, which additionally lists all reactions where the species is involved in.

The entry marked *Global Quantities* in the tree is to add explicit mathematical expressions that are to be calculated in the model (unlike the ODE that are defined implicitly from the reaction stoichiometry and rate laws). There are three types of global quantity (1) *fixed*, which are arbitrary constants; (2) *assignment*, which are new variables that have their value calculated by algebraic expressions; or (3) *ODEs*, which are new variables that are determined by an explicit ODE. These global quantities are useful to expand the model to include features that are not directly linked with the biochemistry. As an example, suppose you would want to calculate the ratio of NADH/(NAD + NADH) at all times in your model, either because you just want to monitor it, or maybe you want to make it affect something else. Then you should double click the list of global quantities to enter a detailed form. There you should enter its name, select the type as *assignment*, then enter the expression in the larger box. Note, however, that you are not allowed to type "NADH" or "NAD," since these are variables of the model (species) and you will instead have to select them from a dialog box: press the small button that has the COPASI logo (*see* **Note 15**), then select *Species*, *Transient Concentrations*, and select "[NADH](t)"; the division sign and the brackets are typed directly. ODEs are set the same way, except that the expression is now the right-hand side of the differential equation that will be integrated in the model simulations.

3.1.2. Importing and Exporting SBML

The popular SBML format is used as a means of sharing models between systems biology software, so an important feature of

COPASI is that it can indeed read and write models in this format. However, it is important to realize that COPASI can represent a small number of features that are not possible to be specified in SBML and those will be lost on export to SBML. On the other hand there are features of SBML that are not yet implemented in COPASI – when loading files with such features a warning is produced such that you are aware of this fact. When importing SBML there are also often warnings about issues with the files that are either not like the specifications require, or because they follow bad practices. In any case, even with warnings, COPASI almost always succeeds in importing the model if not totally, at least partially.

3.1.3. Time Course

Once a new model has been entered or loaded it is ready to be used for simulation. There are two basic types of simulation: *Time Course* and *Steady State* which are entries under the *Tasks* branch. Let us use model number 10 of BioModels, a MAP kinase model *(8)*, to illustrate these basic tasks.

To run time course, select the appropriate entry on the tree on the left and the time-course control window will appear on the right. You have to decide for how long you want to run the time course (in model time, not real time) and enter that value in the box labeled *Duration* (enter 1,000 for this example); you also need to decide how many *Intervals* in the time course you will want to sample, or alternatively the *Interval size* (when you set one, the other one updates automatically). Below are several control variables of the numerical methods, which are outside the scope of this article. Simply press the *Run* button to carry out the simulation, which will take place very fast. Expand the *Time Course* entry on the tree to reveal *Results* and select it. This displays a table with the numerical values of the time series, which can be saved to a file (button *Save data to file*).

It is also very useful to visualize the results of a time series simulation in a plot. To create it press the button *Output Assistant* (located at the level of *Time Course*) and select the first line entitled "Concentrations, Volumes, and Global Quantity Values," then press the button *Create!*. This creates a plot definition that will plot all variables, however, the plot is constructed while the simulation runs, and thus you must run it again to make the plot appear. The legend of the plot is composed of buttons, one for each curve, and by pressing them you select/deselect that curve from being displayed.

3.1.4. Steady-State Simulations

Another important task is the calculation of a steady state of the model. Select the *Steady State* entry under *Tasks*. The control variables of the steady state deserve some attention, mainly the *Steady-state resolution*, which is the smallest value of a change in species concentration that is the smallest distinguishable from zero. This value will be used to decide when to stop iterations, but also to recognize a steady state. Smaller values of this parameter lead to more accurate solutions. Another important set of control variables are named *Use Newton, Use Integration, Use Back Integration*, which are related to the strategy used to find the steady state. These variables take the values 1 or 0 meaning to use them or not, respectively. The Newton method is a solver for nonlinear algebraic equations which is very fast. However, the Newton method is not guaranteed to converge, therefore the integration method can be used to help. Integration is the method used to calculate a time course – this method attempts to find a steady state as it goes along a time course. The back-integration method is a fail safe device that is used when the other two cannot converge and may be able to find an *unstable* steady state. Pressing the *Run* button will trigger all calculations. The results are also in a branch of the tree, below the *Steady state* and can also be saved to a file like the time course results.

3.1.5. Scanning and Sampling Parameters

One of the most frequent aims of using models to study biochemical networks is to find out how certain parameters affect several aspects of the system. Thus it is likely that one needs to carry out several steady-state and/or time-course simulations at different values of the parameters of interest. COPASI supports this activity by providing a flexible scheme for changing parameter values with associated simulations, which is termed *Parameter Scan* and is under the tree branch *Multiple Task* (*see* **Note 16**). The *Parameter scan* window (**Fig. 2**) is an interface that allows us to specify a series of hierarchical changes in model parameters which culminate with the execution of a task (e.g., time-course or a steady-state simulation).

We will use here the model of the branch point of threonine/methionine biosynthesis of Curien et al. *(9)*, which is number 68 in BioModels. This is a very simple model of the branch point with only one variable chemical species that has one input and two output fluxes. Curien et al. study the effect of the Cysteine (Cys) and S-Adenosylmethionine (AdoMet) on the partition of the output fluxes. One issue that you may wonder about is that while AdoMet is a chemical species, the authors of the SBML file decided to represent it as a constant in the kinetic rate law of the enzyme threonine synthase (TS). While this is not incorrect, it would have been clearer to define it as a chemical species with fixed concentration.

We will first investigate the effect of AdoMet on the partition of fluxes. In order to be able to visualize the results, we must first define a plot where the flux of the TS and CGS are plotted as a function of AdoMet. Plots are defined under the main hierarchy *Output* and then under *Plots*. A list of plots is displayed (currently empty) and there you should double click the last empty

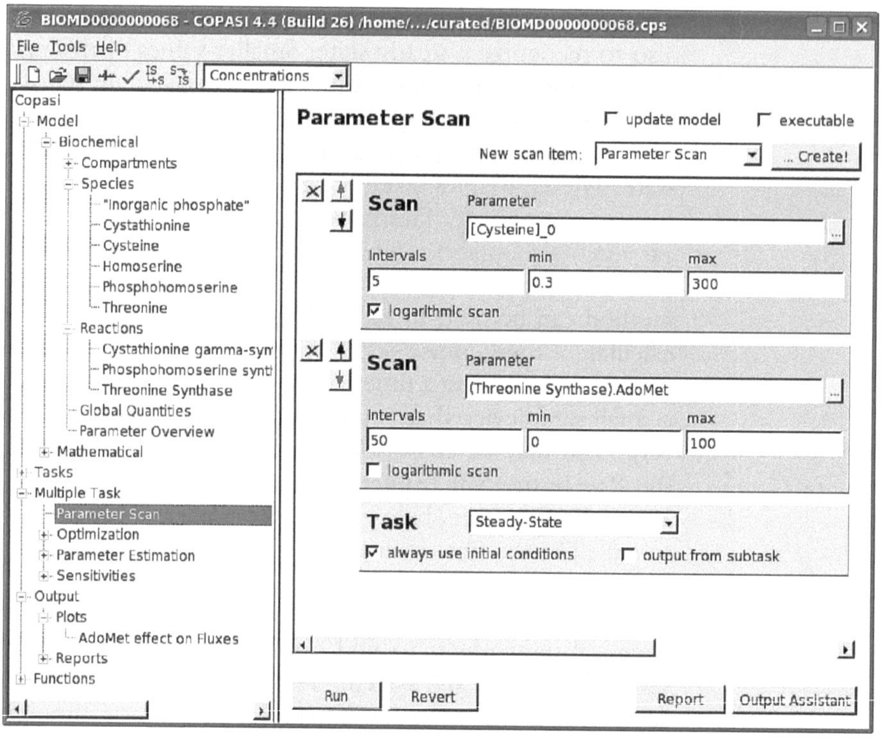

Fig. 2. Parameter scan window. The structure on the center right is a stack of operations that are carried out in order. Thus the example shown is for changing the initial concentration of Cysteine between 0.3 and 300 in five intervals spaced logarithmically, then for each of those change the parameter AdoMet between 0 and 100 in 50 equally spaced intervals, and finally to run a steady-state calculation for each value of the above.

row which will generate the plot definition window. There you should define the name of the plot ("AdoMet effect on Fluxes" is suggested) and then create a new curve (press *New curve*) where you select the parameter AdoMet in the *X*-axis (under *Reactions-Reaction Parameters-Threonine synthase-AdoMet*) and flux(Cystathionine gamma-synthase) for *Y*-axis (under *Reactions-Concentration fluxes-flux(Cystathionine gamma-synthase)*). Because you want to plot both fluxes in the same plot, you should then create a new curve again and select the same item for the *X*-axis, but then flux (*Threonine Synthase*) for the *Y*-axis. At this point each curve in the plot is represented under a different tab which have long titles; it is advisable to make the titles of each curve smaller strings for esthetic reasons, rename the first one to J(CGS) and the second to J(TS). The plot definition is ready; you can go back to *Parameter Scan*.

The task to be carried out in this case is *Steady state* and we want to scan the parameter AdoMet, so in *New scan item* at the top select *Parameter scan* and then press...*Create!* A new entry will appear in the stack above the steady-state task. There you need to select the parameter to change, press the button marked "..." and select *Reactions, Reaction Parameters, Threonine synthase, AdoMet*. The minimum and maximum values that this parameter will be changed also have to be entered, for example 0 and 100 (as in **Fig. 2** of **ref.** *9*), and finally the number of intervals desired, a value of 50 will produce a smooth curve. At the bottom of the stack, in the Tasks slot (*see* **Fig. 2**), you should disable the check *output from subtask* since we only want the final estimate of the steady-state calculation (*see* **Note 17**). At this point you can press *Run* and see the result appear as a plot in a new window (**Fig. 3**). The results plotted can be saved by selecting the menu entry *Save data*. You can also switch on or off each of the curves, simply by pressing its entry in the legend. This plot shows that increasing values of AdoMet push the flux toward the CGS reaction, as shown in the original *(9)*.

Let us now ask whether the behavior changes with different values of Cys. To do this we simply add this model entity to the scan and therefore perform a two-dimensional scan. In the *Parameter Scan* window, add another scan item by pressing ...*Create!* once again. A new slot appears where you need to select *Species, Initial concentrations, [Cysteine](t = 0)*. Set it to vary between 0.3 and 300 with five intervals and tick *logarithmic scan* (*see* **Note 18**). You can also move this slot to the top by pressing the up arrow on the left (*see* **Note 19**). Press *Run* again, and now you will see several curves for each of the fluxes plotted. Each of them is for different values of Cys. As you can see, Cys acts by affecting the initial flux partition and also by moving the value of AdoMet where both fluxes are equal.

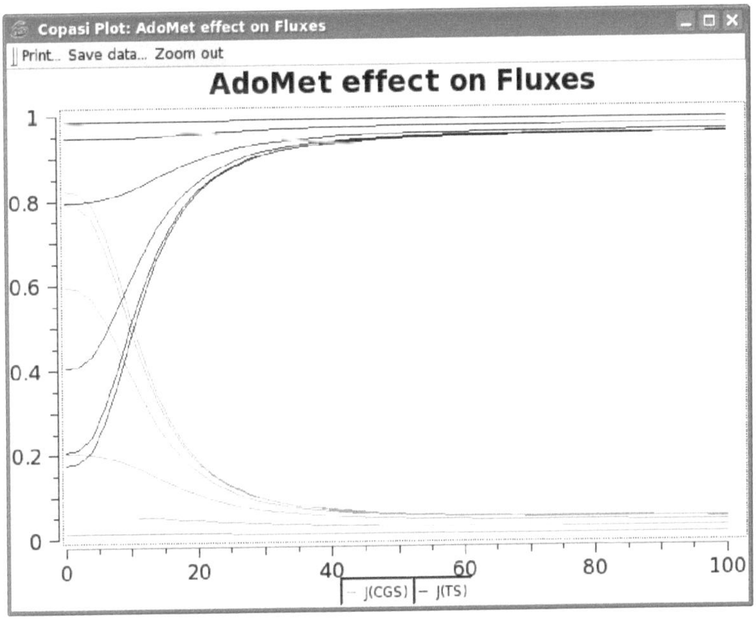

Fig. 3. Results of a two-dimensional parameter scan.

Finally, we will see how to do a random sample, rather than a scan. We shall probe 10,000 random values of the two parameters in the same range. For this remove the two slots of AdoMet and Cys by pressing the button marked X. Now select *Repeat* in the *New scan item* at the top and create a new slot; set the number of iterations to 10,000. Next select *Random distribution* in the *New scan item* and select the parameter AdoMet as above, and set it to the same limits as above. Repeat the same for Cys, also with the same limits. You can have both sampled from a uniform distribution. Note that the stack of operations should be read from the top and it means: repeat 10,000 times a random value of AdoMet and a random value of Cys and calculate the steady state. To visualize these results it is best to just plot symbols and not connect them with lines, so you have to go back to the plot definition and change *Type* to *symbols* for each of the curves (it was *lines*). Go back to *Parameter Scan* and press *Run*. The run now takes some more time (there are 10,000 simulations, after all) and finally you should obtain a plot as in **Fig. 4**. Each point plotted represents the steady-state flux for a pair of values of AdoMet and Cys. It is very easy to add more parameters to this sampling, by adding more slots associated with those parameters and moving it below the *Repeat* slot. It is even possible to combine a sequence of repeats below scans below other repeats, etc. This feature of COPASI is a very powerful means to program complex simulations.

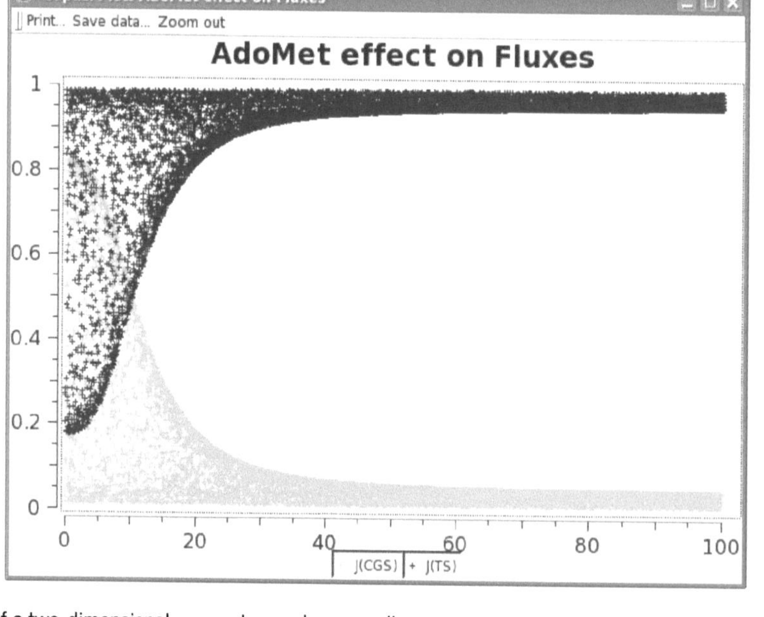

Fig. 4. Results of a two-dimensional parameter random sampling.

3.2. Stoichiometric Analyses

While the analysis of the dynamics of biochemical models is seen as the ultimate goal of these models, some properties of the model reveal themselves even without considering the kinetics of the reactions involved. These properties are sometimes known as *structural properties* because they depend only on the structure of the network, or as *stoichiometric properties* because they depend only on the stoichiometric coefficients of **Eq. 1**. COPASI provides two stoichiometric analyses (1) identification of elementary flux modes *(10, 11)* and (2) identification of mass conservation relations *(12)*.

3.2.1. Elementary Flux Modes

Elementary flux modes are the minimal subsets of reactions that would still be able to maintain a steady state if isolated from the rest of the network *(10)*. They are the basic components of flux and any observable flux is a linear combination of these. They can also be seen as "functions" that the network fulfills because they represent parts of the network that could still operate even when the rest of the network had been removed. They can be useful to identify what functions would be lost by removing a specific reaction from the network (e.g., by a gene knockout) and also to calculate maximal yields of a certain end product that can be obtained from some substrate *(11)*.

Let us use a model of erythrocyte metabolism by Holzhütter *(13)*, which is model 70 in the BioModels database. Download the corresponding SBML file and import it into COPASI as described before. You can examine the model by inspecting the various categories under the *Model* section, where you will find that it is composed of 38 reactions. The *Elementary Modes* task is under *Tasks-Stoichiometry*. To run this task simply press *Run* button on the bottom left of the right pane as there are no other choices to make. The table on the right pane should now be filled and at the top there is an indication that the model can be decomposed into 105 elementary modes. The table, depicted in **Fig. 5**, lists each of the modes in detail. The first column indicates whether the mode is reversible or irreversible (*see* **Note 20**); the second column lists the reactions that compose the mode and the third column lists the actual reaction equations. Note that in the second column, the name of the reaction is preceded by a number which is a multiplier for that flux, and if the number is negative then the flux of that reaction goes in the reverse direction in this elementary mode. In the mode depicted in **Fig. 5**, Glucose transport operates in the reverse direction (glucose is exported) while Bisphosphoglycerate mutase operates in the forward direction threefold faster than Glucose transport.

Elementary flux modes can either link a source to a sink (external substrates and products of the model) or be cyclic. The overall chemical reaction of the mode depicted in **Fig. 5** is 2*External Lactate + PRPP = 2*External Pyruvate + 3 External Phosphate + Glucose outside (*see* **Note 21**). This erythrocyte

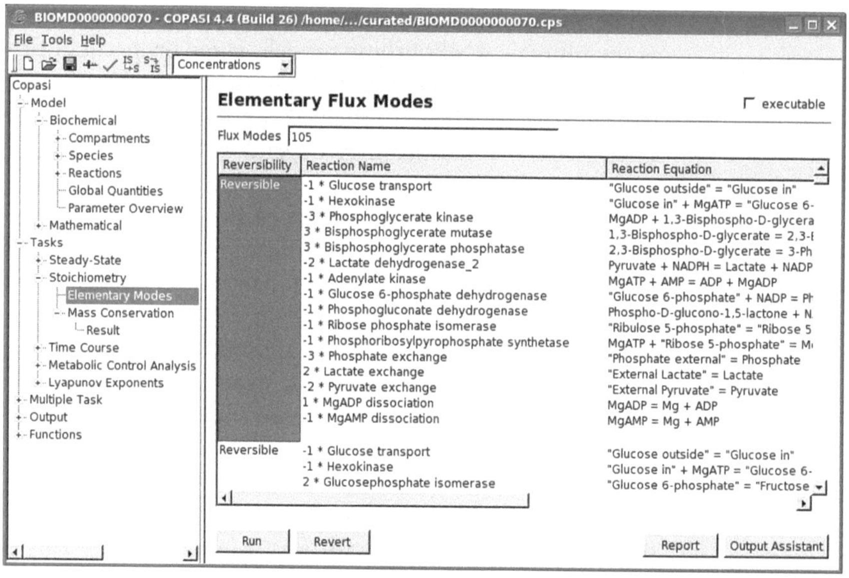

Fig. 5. Elementary flux modes.

model also contains cyclic flux modes. In the list of 105 there is, for example, a reversible mode composed of Phosphoglycerate kinase, Bisphosphoglycerate mutase, Bisphosphoglycerate phosphatase, and ATPase. Cyclic modes have no net production or consumption of any metabolite (thus they are sometimes called *futile cycles*).

To save the results of the elementary mode analysis you first need to set a report file: press the button labeled *Report* on the bottom right, then press *Browse* and enter a filename for your report in the desired folder. You will then have to press the *Run* button again in order to create the report. The report is a tab-delimited text file that contains a table with the same information as displayed in the front-end. You can read this file with a plain ASCII text editor, such as "wordpad" in Windows; you can also import this file into a spreadsheet program like "Excel."

3.2.2. Mass Conservation Relations

Mass conservation relations are algebraic sums of amounts of chemical species that are constant in any state of the model. These algebraic sums imply that the amounts of some chemical species are constrained, such that one of them can be directly calculated from the others using the algebraic expression. A special case of mass conservation relations is when there is conservation of a chemical moiety (*see* **Note 22**).

Let us continue with the erythrocyte model, and examine the mass conservation relations that it contains. The *Mass Conservation* task is also under *Tasks-Stoichiometry* and is also run by pressing the *Run* button on the bottom left of the right pane.

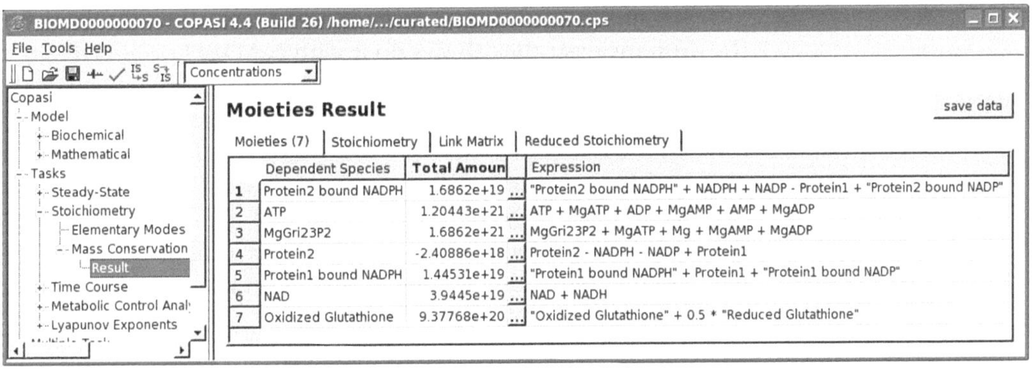

Fig. 6. Mass conservation relations.

The results of this analysis will appear in a new entry marked *Results* that appears below *Mass Conservation* (in the tree on the left), which you have to select to inspect the results.

The erythrocyte model has seven mass conservation relations as shown in **Fig. 6**. The results are listed in a table where the first column identifies the chemical species that COPASI will calculate from the mass conservation (the *dependent* species, *see also* **Note 23**). The second column lists the total number of particles of this conservation relation. The third column contains a button labeled "..." which creates a new global quantity that mirrors the total number of particles. Finally, the fourth column contains the actual expression which is constant. In the erythrocyte model, the first of these relations reads: "Protein2 bound NADPH" + NADPH + NADP − Protein1 + "Protein2 bound NADP" = 1.6862×10^{19}. That means that adding the number of particles of all the species with a positive sign and subtracting those with a negative sign adds up to 1.6862×10^{19} particles. This algebraic expression is constant throughout any condition of this system, except when the initial amounts of any of the chemical species involved change (in which case the total would be different). In particular, this expression is always true during any time course and thus does not depend on the dynamics of the system. The reader may recognize the second relation in **Fig. 6** to be the conservation of the Adenine moiety, the third is conservation of Mg, the fifth is conservation of Protein 1, the sixth is conservation of the NAD moiety, and the seventh conservation of the glutathione moiety. Together, relations 1 and 4 represent the conservation of the Protein 2 moiety (when summing the two, NADP, NADPH, and Protein 1 cancel out, leaving just the Protein 2 forms). Together, relations 1, 4, and 5 represent the conservation of the NADP moiety and also that its total is inversely related with the total of the free protein forms (expressed by the result of computing relation 1 − relation 4 + relation 5). This last complex relation appears due to the fact that whenever the free forms of the proteins react they always do it with NADPH or NADP – the three moieties (NAD, Protein 1, and Protein 2) are intertwined.

Note that there are other results of this task, which can be inspected by selecting the tabs *Stiochiometry*, *Link Matrix*, and *Reduced Stoichiometry*. These are the matrices that are used to calculate these conservation relations and are described in the theoretical derivations of Reder *(12)*. To save all of the results of this task, just press the *Save data* button on the top right corner, which creates a tab-delimited ASCII file.

3.3. Sensitivity Analyses

As discussed in the context of parameter scans, it is frequently desirable to investigate the behavior of a model systematically. In addition, every model contains a number of parameters (kinetic constants, initial concentrations, and so on) whose values are not all known exactly. Changing the values of the parameters will of course change the behavior of the model, so it is interesting to find how much the model depends on parameters. Sensitivity analysis describes how much does a specific parameter change the behavior of the model. This is useful for several reasons:

- In many cases the value of a parameter is unknown. For example, while K_m values of enzymes can be measured relatively easily in vitro, often the enzyme concentrations in vivo are not well known. In this situation, sensitivity analysis can tell us if it is important to know a specific parameter value. If a parameter is found not to affect the system very much, a rough guess for its value may be sufficient. If, on the other hand, a parameter influences the behavior of the model significantly, steps must be taken to find out its value more accurately, either by executing more experiments or by literature searches.
- Sometimes the aim of research is to change the behavior of the system. Perhaps we want to increase the yield of some biotechnological production process, or to find a drug that inhibits a metabolic pathway. Sensitivity analysis can give hints about which parameters should be changed to achieve a specific effect.
- Robustness with respect to external influences is an important property of biological systems. Living organisms need to be able to function under a wide range of environmental conditions. This means some biological processes need to be rather insensitive to parameter changes. On the other hand an organism needs to react to its environment, so other processes need to be very sensitive to external influences. Therefore robustness (or the lack of robustness) is an interesting property of biological systems, and sensitivity analysis is a way to determine this.

One should note that sensitivities as they will be described below are only able to provide a local description of robustness. This means that its results are only valid for a given parameter set (set of environmental conditions). If several parameters were to change at the same time then the individual sensitivity coefficients would also be expected to change. COPASI contains two frameworks for doing sensitivity analysis: metabolic control analysis (MCA) and generic sensitivities.

3.3.1. Metabolic Control Analysis

MCA is a concept developed by Kacser and Burns *(14)* and Heinrich and Rapoport *(15)*. Its most practical formulation deals only with steady states (*see* **Note 24**) and provides means to quantify how much the rates of the various reactions of a network affect the concentrations and fluxes at the steady state. A deeper description of the theory does not fit this text and the reader is directed to specialized reviews and books *(16–19)*.

As an example we use a model of sucrose accumulation in sugar cane *(20)*, model 23 in BioModels. After importing the SBML file into COPASI select *Tasks* and *Metabolic Control Analysis* in the tree on the left (*see* **Note 25**). Then simply click the *Run* button on the right. The *Results* window then presents a screen with three tabs labeled *Elasticities* (**Fig. 7**), *Flux Control Coefficients* (**Fig. 8**), and *Concentration Control Coefficients.*

The *elasticity coefficients* (or simply elasticities) quantify the amount of change of a reaction rate with the change in concentration

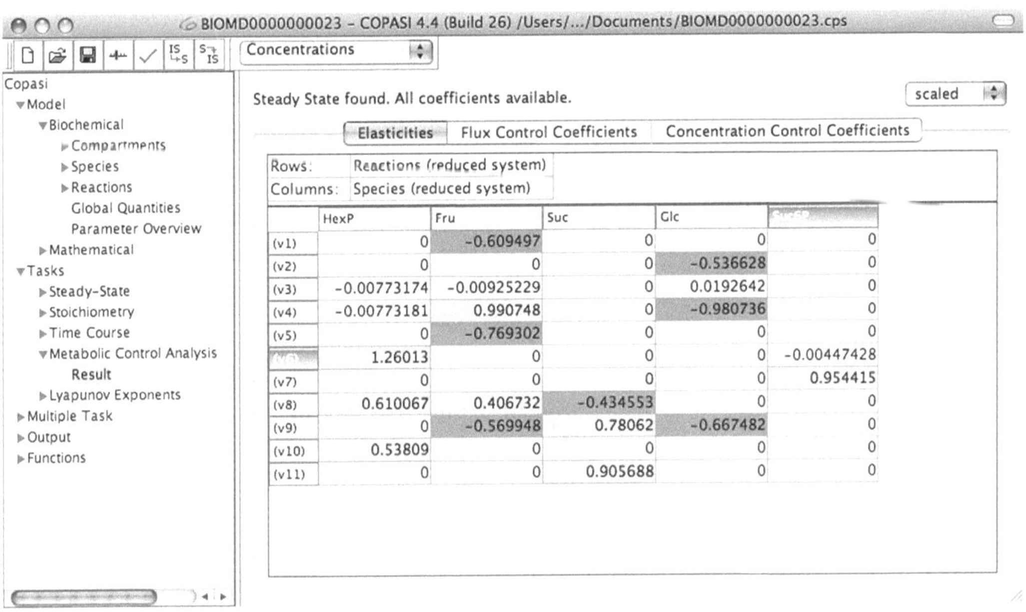

Fig. 7. Display of elasticity coefficients. Note that the cells of the matrix are colored according to the magnitude of the values, green for positive values and red for negative (colors not shown in this figure).

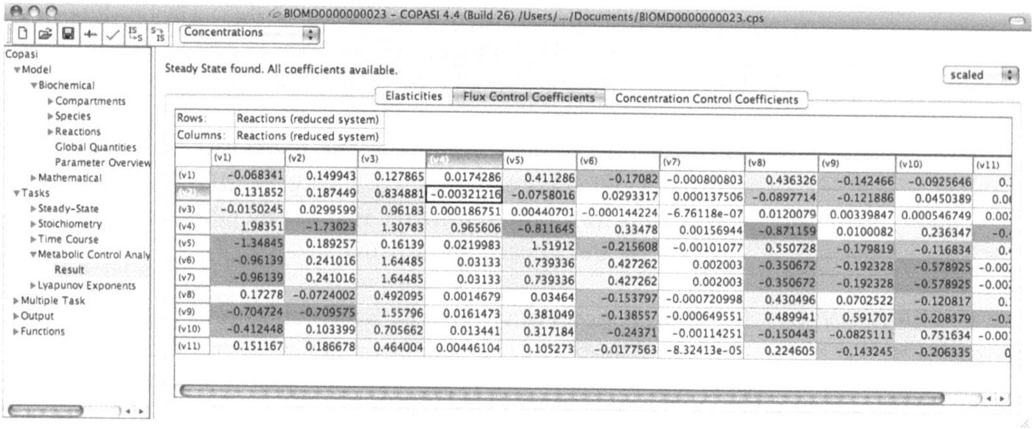

Fig. 8. Display of flux control coefficients.

of a certain chemical species. The elasticities of all the reactions with respect to all the species in the model are calculated by COPASI and displayed in a table where the columns correspond to the species and the rows to the reactions. Consider the line labeled "(v8)" (**Fig. 7**): the numbers in this line describe how the flux of reaction *v8* (HexP + Fru = Suc + UDP) changes with changes of the concentrations of the different species. Notice positive values for "HexP" and "Fru," which are the substrates of the reaction. This means an increase of 1% in Fructose concentration would increase the speed of the reaction by 0.61% (*see* **Note 26**). Correspondingly the elasticity with respect to the product (*Suc*) is negative – an increase in product concentration would lead to a lower flux. Another case, in line "(v4)" the negative value for *Glc* indicates that glucose is an inhibitor for this reaction. An elasticity equal to zero means that the metabolite concentration has no influence on the reaction rate (*see* **Note 27**).

The elasticities are properties strictly of a single reaction and are independent of the rest of the system (the elasticity of reaction A toward species B does not depend on reactions C, D, etc.). The calculation of elasticities is carried out only from the kinetic rate law of the respective reaction. Likewise, in an experiment the elasticity could be measured in vitro using the purified enzyme, so long as the concentrations of its substrates and products are set to their physiological value (and the enzyme properties remain the same after purification).

Note that in COPASI all sensitivities (i.e., MCA and generic sensitivities) can be displayed with either scaled or unscaled values. The scaled values describe relative changes, e.g., a scaled sensitivity of 0.5 means that if the parameter is increased by 10% the target value will increase by 5% (0.5 times 10%). The unscaled sensitivities describe absolute changes, e.g., an unscaled elasticity of 0.5 could mean that increasing the substrate concentration by

1 μM will result in an increase of the reaction flux by 0.5 μM/s (if those are the units that are used in the model). The scaled sensitivities are the ones most discussed in literature, particularly for MCA (but *see* **ref.** *12*).

The next tab shows the *Flux Control Coefficients* (**Fig. 8**). Unlike the elasticities, control coefficients are global properties that depend on the whole system. They quantify the extent of change of the steady-state flux of one reaction when another reaction is made slower or faster. For the MCA formalism it does not matter *how* the reaction is made faster or slower, but in practice changing the enzyme concentration is the most practical solution. Imagine a system in a steady state in which at some point we increase the concentration of one of the enzymes by 1%. After some time a new steady state will be reached, potentially all the concentrations and fluxes in the system will have changed slightly. The relative change of one of the reaction fluxes is the flux control coefficient of this reaction with respect to the reaction with the changed enzyme concentration. Like in the case of the elasticities, all combinations of flux control coefficients are calculated by COPASI and displayed in a table where the column indicate the rate of reaction that is changed and the row indicates the flux of the reaction that has been affected. The fact that the table contains almost no zeros already indicates that these are global properties of system: a change in one reaction changes the steady-state fluxes of all reactions.

In the example of the sucrose accumulation model, one thing that stands out immediately is that two lines ("(v6)" and "(v7)") are identical. This is very common in flux control coefficients and comes from the fact that the two reactions (HexP→UDP + Suc6P and Suc6P→Suc + P) form a chain without any branches in between, so that their steady-state flux is always the same. The original publication of the model discusses the accumulation of sucrose in sugar cane (reaction *v11*) vs. the hydrolysis of sucrose (reaction *v9*), arguing that the sucrose accumulation is most effective when the flux of *v11* is large and that of *v9* is small. Inspection of the last row of the table calculated in COPASI indicates the control that each reaction has over the flux of *v11* and it is interesting that the largest coefficient (0.464) corresponds to *v3*. This means that with an overexpression of hexokinase (the enzyme that catalyzes *v3*) by 10% we expect an increase in the rate of sucrose accumulation of about 4.6%. However, reaction *v3* also has a high control over the flux of *v9*, in fact much larger: 1.558; thus *v3* is not a good candidate for manipulation because while it would stimulate sucrose accumulation it would lead to a much larger increase in the hydrolysis of sucrose actually decreasing the overall efficiency. Rohwer et al. argue that the fructose/glucose transporters (*v1* and *v2*, the first two columns) are better candidates for this purpose since increasing their rates causes a simultaneous increase in sucrose accumulation and decrease in hydrolysis (as indicated by a negative control coefficient).

From this example of a relatively simple model with only five variables, it becomes evident that there is no intuitive way to reason about the response of the system to perturbations from the network structure alone. In even moderately complicated models it is impossible to predict which enzymes control the fluxes without performing actual sensitivity analysis calculations.

Another issue that may be obvious to some readers from **Fig. 8** is that the values of each row in the table sum up to 1. This is a reflection of the flux control summation theorem *(14)*, which allows us to reason about the system. For example, if the value of a flux control coefficient is known to be 0.3 then one can be sure that also other reactions will control that specific flux (since the coefficients have to add up to 1).

The third tab of the results window contains the *concentration control coefficients.* These are similar to their flux counterparts, and describe how the steady-state concentrations change depending on the changes in specific reaction rates. The main difference between these and the flux control coefficients is that they add up to 0 rather than 1.

An important thing to keep in mind about both the *elasticities* and the *control coefficients* is that they provide information only about small changes to the model. So while you can in many cases reliably predict from the control coefficients what the effect of a 5% increase in the expression of one enzyme will be, it is not possible to predict the effect of a tenfold increase or decrease. This type of information, however, could be obtained from parameter scans, i.e., by direct numerical simulation, however, that would be for a single parameter at a time (*see* **Note 28**).

MCA is a powerful concept, and the way it is implemented in COPASI is numerically robust *(12)*. Basically whenever COPASI is able to find a steady state, the MCA calculations will also provide reliable results.

3.3.2. Generic Sensitivities

Control coefficients are concepts geared toward an interpretation that is dominated by changes in enzyme concentrations (derived from gene expression), as they only measure the effects of changing the overall rate of reactions. It is also interesting to study how other parameters, such as K_m, affect the model behavior. In MCA, these generic sensitivities are known as *response coefficients* and measure the change in a system property effected by any system parameter (*see* **Note 29**), however, these have no known special summation theorems. COPASI can also calculate these generic sensitivities and to access this feature we select *Multiple Task and Sensitivities* in the tree on the left; the corresponding sensitivities window is depicted in **Fig. 9**. Basically these generic sensitivities (response coefficients in the vocabulary of MCA) are for arbitrary

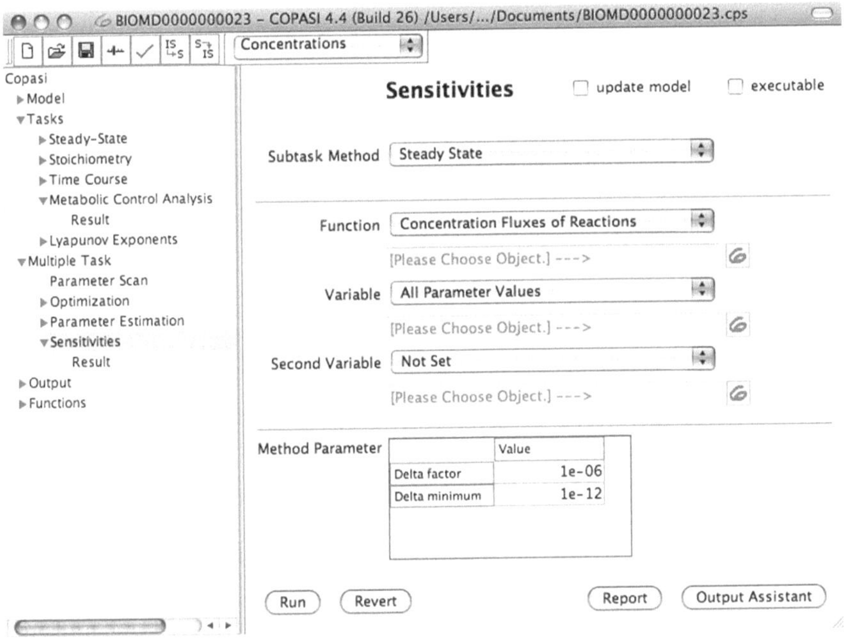

Fig. 9. Generic sensitivities window.

values in the model (*Functions* in **Fig. 9**) with respect to arbitrary parameters (*Variables* in **Fig. 9**) and are calculated numerically using finite differences (i.e., not using a matrix method from elasticities, *see* **Note 30**).

Generic sensitivities can also be calculated for time courses, but we will start with a steady-state example. Make sure that *Subtask method* is set to Steady State. In the *Function* select *Concentration Fluxes of Reactions*, meaning that we want to calculate how the steady-state reactions fluxes (measured in concentration units) are affected by parameter changes. Next the parameters of interest need to be selected in *Variables*. For this example, select *All Parameter Values* that will calculate the sensitivities with respect to all kinetic parameters in the model. After pressing the *Run* button results will appear in its window, and we shall discuss the *Scaled* tab (**Fig. 10**). Once again, the rows correspond to the reactions (as in the flux control coefficients table) and the columns correspond to the kinetic parameters of the model. Since there are usually several parameters for each reaction, this table does not fit entirely on the screen and the scroll bar needs to be used.

A comparison of this table with that of **Fig. 8** reveals that columns 3 and 6 here are identical to columns 1 and 2 of **Fig. 8**. This is expected because the sensitivities of fluxes toward v_{max} parameters can be shown to be the same as flux the control coefficients (unless there are enzyme–enzyme interactions). However, we can see from the sensitivities that some of the inhibition constants (e.g., column 1) also strongly affect the fluxes.

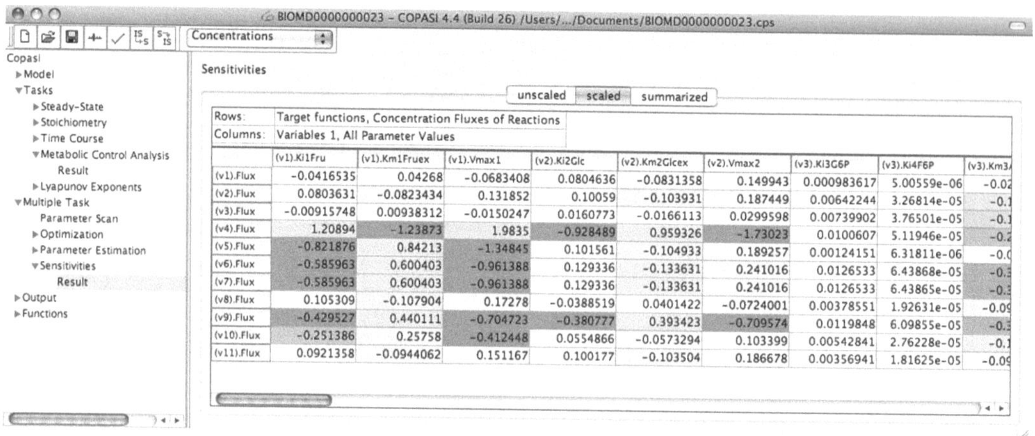

Fig. 10. Results of generic sensitivity analysis.

The generic sensitivities feature allows the calculation of many other kinds of sensitivities as well. For example, the sensitivity of a simulation result with respect to the initial concentrations could also be calculated. It is also possible to calculate second-order sensitivities (*sensitivities of sensitivities, see* **ref.** *21)* which can help determining whether sensitivity analysis results are valid over a larger parameter range.

3.4. Tuning Models with Optimization Methods

Optimization is the search for maximum or minimum values of some function (the *objective function; see* **Note 31**). In biochemical modeling, optimization can be used to find conditions in which the model behaves in some desired way *(13, 22)*. Because biochemical models are composed of nonlinear functions, their variables may have several minima or maxima, thus the problem is usually of *global optimization* where one wants to find the largest of all maxima or the smallest of all minima. Global optimization problems are hard to solve and it is well known that no single algorithm is best for all problems *(23)*. Thus COPASI is equipped with a diversity of optimization algorithms that follow very different strategies (*see* **Table 1**), and in general one should search the best solution with more than one algorithm (and at least one should be a global optimizer).

To demonstrate an application of optimization we will continue analyzing the model of sucrose accumulation in sugar cane *(20)*, which is model 23 in BioModels. Remember that accumulation of sucrose is measured by the steady-state flux of reaction *v11* but there is also a certain amount of sucrose hydrolysis, reaction *v9*, that decreases the efficiency of accumulation. So one important question is what conditions lead to a low proportion of sucrose hydrolysis relative to accumulation. This can be seen as a typical optimization problem, where we are interested in minimizing the ratio of fluxes J_{v9}/J_{v11} – our objective function. In all

Table 1
Optimization algorithms available in COPASI Version 4.4 (Build 26)

Algorithm	Strategy	Type	References
Evolutionary programming	Evolutionary algorithm with adaptive mutation rate without recombination	Global	(*24*)
Evolution strategy (SRES)	Evolutionary algorithm with numerical recombination, selection by stochastic ranking	Global	(*25*)
Genetic algorithm	Evolutionary algorithm with floating-point encoding and tournament selection	Global	(*26*)
Genetic algorithm SR	Variant of Genetic algorithm where selection is by stochastic ranking	Global	(*25, 26*)
Hooke and Jeeves	Direct search algorithm based on pattern search	Local	(*27*)
Levenberg–Marquardt	Gradient-based, adaptive combination of steepest descent and Newton method	Local	(*28–30*)
Nelder–Mead	Direct search method based on geometric heuristics	Local	(*31*)
Particle swarm	Inspired on social insect search strategies; works with population of candidate solutions like evolutionary algorithms	Global	(*32*)
Praxis	Direct search method based on the alternate direction (minimize one dimension at each time)	Local	(*33*)
Random search	Random search with uniform distribution (a shotgun approach)	Global	
Simulated annealing	Monte Carlo method that mimics the process of crystal formation (biased random search with Boltzmann distribution)	Global	(*34*)
Steepest descent	Gradient method based on first derivatives (estimated by finite differences)	Local	
Truncated Newton	Based on Newton method (uses second derivatives)	Local	(*35*)

optimization problems, it must also be specified which parameters of the model are allowed to change in order to meet the objective. In this particular example, let us imagine that we could manipulate the steady-state level of the enzymes of reactions *v1*, *v2*, *v3*, *v4*, and *v5* (e.g., by overexpression or by interfering with the upstream regulatory sequences of their genes). The question then becomes what would be the best combination of the levels of these enzymes to achieve the lowest possible ratio J_{v9}/J_{v11}. The parameters that are allowed to change are then the *Vmax* of the five reactions.

In COPASI, the optimization task is found under *Multiple Tasks* and then *Optimization* in the tree on the left. The application of optimization to biochemical modeling consists typically of three parts (1) the objective function, (2) the adjustable parameters, and (3) the search algorithm. This is mirrored in COPASI's interface as seen on **Fig. 11**. First, the objective function must be set by entering the mathematical expression J_{v9}/J_{v11}, this is done by selecting the required model entities from a menu that is activated by pressing the small button with the COPASI icon (at the right) (*see* **Note 15**). J_{v9} appears as <(v9).Flux>, then you have to enter the division sign from the keyboard, and finally select J_{v9} which appears as <(v11).Flux> (*see* **Note 32**). If you wanted to instead maximize this expression you should precede it by a minus sign (so that you minimize its symmetric).

Next you need to select the adjustable parameters, i.e., those that are allowed to change. To add one parameter to the list press the *New* button (the one with a blank page) and then the button with the COPASI icon to select the actual parameter. COPASI provides a shortcut to add all parameters: select the first

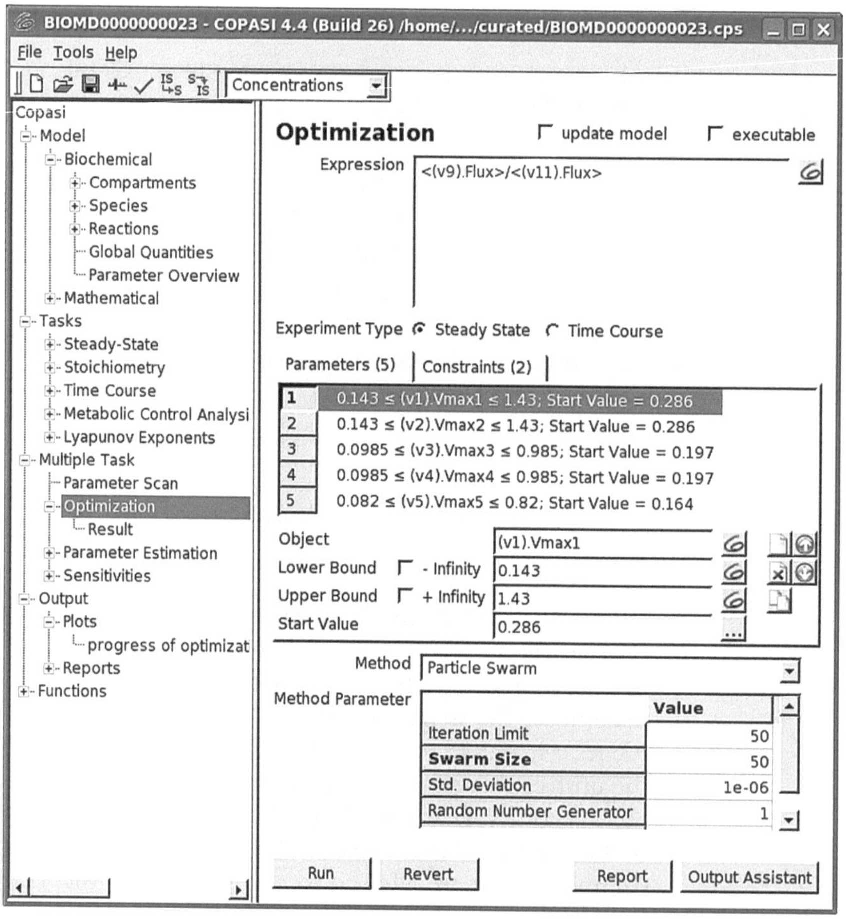

Fig. 11. Optimization window, with an objective function definition at the top, a list of adjustable parameters at the center, and an optimization algorithm at the bottom.

one by expanding *Reactions*, then *Reaction parameters* and then *v1* where you select *Vmax1*; but then rather than just accepting that, expand the other reactions of interest one by one, and while pressing the CTRL key (or the APPLE key on Macs) also select also *Vmax2, Vmax3, Vmax4*, and *Vmax5*. When you finish, all five parameters will be listed. You will realize that they are listed inside the interval between –infinity and +infinity, which is quite large indeed, but in general we want tighter limits. Let us say that it is feasible to downregulate the enzyme concentrations to 50% and to overexpress it by 400% in this example. To change all of the boundaries together select all five rows, then remove the check on –inf and type –50% on the box; similarly, remove the check on +inf and type +400% on the box below, after changing the cursor to another field, the limits of each parameter will have changed to the appropriate values (*see* **Note 33**). The start values are by default those that are specified in the model section but they could be changed; yet we shall leave them as they are now (*see* **Note 34**). Please note that if the start value of a parameter is outside the boundaries specified, COPASI will force it to the nearest boundary during the optimization.

Finally, one needs to select the method of optimization desired. For our first attempt let us use the *Truncated Newton* method and press *Run* which will quickly finish. Then move on to the *Results* section (on the left tree, below Optimization). This will show that the objective function value obtained was 0.000593843 (*see* **Note 35**), and below you will see listed the values of *Vmax* for each of the reactions. You will realize that *Vmax1* and *Vmax2* are close to the upper limit specified (indeed as argued in **ref.***20)*, and *Vmax3, Vmax4*, and *Vmax5* are near the minimum specified. This means that we would need to overexpress the first two enzymes and downregulate the remaining three. The reader may wonder about this solution, particularly if compared with the network diagram in **Fig. 1** of **ref.** *20*, one clue is given by the concentrations achieved in this solution, which you can inspect if you switch to the tab named *Species* (at the top of the right pane). Both Fructose and Glucose are very highly concentrated (almost 1 molar for Fructose) – it seems that the best way to minimize hydrolysis of sucrose and maximize its storage is to maintain a very high concentration of the products of the hydrolysis. However, this solution may not be achievable in practice due to the high concentrations of the intermediates.

After considering the results of the previous analysis, it becomes interesting to ask the same question but now not allowing the concentrations of Glucose and Fructose go above 100 mM. This is a new set of requirements of the method named *constraints* as they attempt to force the solution to a more restricted domain. To enter constraints, return to the *Optimization* page, and select the tab named *Constraints* in the center of the page.

Then let us add the constraints like we added the adjustable parameters, pressing *New* and then the COPASI icon, and then expand *Species* and *Transient concentrations* and select *Fru* and *Glc*. Set the lower limit to 0 and the upper to 100. Now press *Run* again and inspect the result, which is now a ratio of fluxes of 0.0679853 (about 100× higher than the previous solution), and the concentration of Fru is 89 and Glc is 99.

Now select a different method of optimization, for example *Particle swarm* and set the *Iteration Limit* to 50 (the default of 2,000 is way too long for this problem) and run again. This method takes longer, and you will see a window appear with a progress dialog, which shows the number of function evaluations and the current value of the objective function. At the end it is possible that a window appear with several warnings, if so please *see* **Note 36**. In the end, it will show an objective function value of 0.0584978 or somewhere close to that. Run this a few times and note that the result differs each time; this is because the algorithm is stochastic and it does not always necessarily converge to the same value (*see* **Note 37**). Note that now the concentrations of Fru and Glc are within 0.1% of the upper limit of 100. This shows the great utility of optimization methods in biochemical modeling, and the MCA/sensitivity approach would never be able to answer this constrained problem. With optimization we can solve practical problems with realistic constraints (not just calculations based on infinitesimal changes). It is also very reassuring to realize that the modeler is entirely driving the process *by the definition of objective functions and constrains*, which are a means of directing the computations to solve specific problems. Optimization is an excellent way to explore the space of behavior of complex multidimensional models, such as those of biological systems.

3.5. Parameter Estimation

Biochemical models depend on many parameters, but quite frequently the values of these parameters are unknown and have to be estimated from some data. Parameter estimation is a special case of an optimization problem, in which one attempts to find values for a set of model parameters that minimize the distance between the model behavior (simulation results) and the data. COPASI provides specific parameter estimation functionality that is based on the optimization methods described in **Subheading 3.4**.

COPASI measures the distance between model and data using an expression that is derived from a least-squares approach *(36)*. The objective function used is:

$$O(p) = \sum_i \sum_j \sum_k \omega_{k,i} \left(X_{k,i,j} - Y_{k,i,j}(\boldsymbol{p}) \right)^2, \qquad (3)$$

where $X_{i,j,k}$ is the experimental value of variable i at measurement j within experiment k and the corresponding simulated data point is given by $Y_{k,i,j}(\boldsymbol{p})$ where $\boldsymbol{p}$ is the vector of parameter values used for the simulation. It is important that the data for the different variables be of comparable magnitudes so each group of values for each variable in each experiment is multiplied by a weight $\omega_{k,i}$ (*see* **Note 38**).

3.5.1. Example

To illustrate parameter estimation we shall use the MAP kinase cascade model of Kholodenko *(8)* which is model 10 in BioModels. You will also need some experimental data, and a file (MAPKdata.txt) is provided at http://www.comp-sys-bio.org/tiki-index.php?page=CopasiModels. You must download this file and store it in the same folder where you have put the SBML file with the model that was downloaded from BioModels. The data contained in this file are for "measurements" of the single-phosphorylated form of MAPK and of the phosphorylated MAPKK at various time points (*see* **Note 39**). The problem then consists of adjusting the V_{max} parameters of a few reactions in order for the model to be as close to the data as possible.

3.5.2. Experimental Data

As implied in the objective function above (**Eq. 3**), COPASI allows fitting the model to multiple experiments simultaneously. The software also allows using steady-state and time-course data, which can even be used together (i.e., some experiments be time courses while others are steady-state observations). The experimental data must be provided in ASCII data files with columns of data delimited by tabs or commas; each column will be mapped to a model entity. Since COPASI knows nothing about your data files, there is a necessary step of creating a mapping between the data columns and model entities. To make this mapping easier we suggest that the data file should include a row of column headings. Additionally, if there are several experiments in a single file, these experiments should be separated by an empty line (allowing COPASI to detect the beginning and end of each experiment's data automatically). Each column of an experiment data file must be classified as one of the types listed in **Table 2**. Even if some columns are not needed, they must be classified as *ignored*. It is important that all columns of type *independent* and *dependent* are actually mapped to the actual model entities they correspond to.

At this point it is best to proceed with the MAPK example and you should examine the structure of the data file with a plain text editor, for example Notepad on Windows (a spreadsheet will also work, as long as you do not overwrite the file). Then import the SBML file in COPASI and select *Multiple Tasks* and *Parameter Estimation*. The complete specification of the data file format is done in a dialog box (**Fig. 12**) that is invoked with the *Experimental Data* button. To add the data file press the *New* button (blank page) that is above the box named *File*. Select the MAPKdata.txt file that you have previously downloaded, and then COPASI will automatically recognize that there is one experiment in this file

Table 2
Classification of data types for mapping experimental data to the model entities

Data type	Meaning
Independent	Independent model items are those which need to be set before the experiment takes place. Possible model elements are initial concentrations but could also be kinetic parameters. Note that in time-course experiments only the first row of independent data columns is used (since it refers to the initial state of the system). Columns of this type must be associated with elements of the model
Time	This column type is only available for time-course experiments and is a special case of an independent model item. Obviously one and only one column of this type may exist in each time course experiment. COPASI will attempt to automatically identify this column if there are column headers but it may fail and in such a case you must set this type for the appropriate column
Dependent	The dependent data are those that were measured in the experiment and are entities in the model that are variables (i.e., determined from the solution of equations rather than set by the modeler). These are the target data that COPASI attempts to match, and are the data specified in the objective function (**Eq.** 3). Columns of this type must be associated with the actual model elements that they correspond to
Ignored	These are columns of data that the user does not want to include in the problem. Columns marked in this way are not taken into account in the parameter fitting task. This is useful to ignore potential irrelevant columns of data files. This setting is also useful to "switch-off" using one data column when desired

that it names *Experiment* (you can change this if you like) and that it goes from line 1 to line 11, including the header in line 1. You must indicate that these data are from a time course, so select the appropriate check box on the *Experiment Type*. The table at the bottom of this dialog box indicates the columns found in the data file for the current experiment, and under the heading *Column Name* it reproduces the titles in the header (line 1 of the file). The first column has been identified as type *Time* because of its title, the remaining two are set to the default type *ignored*. Since these columns contain the measurements of the concentrations of MAPKKK-P and MAPK-P you have to set their type to *dependent*. When you do that, a new dialog appears for you to point to the actual model entity that this column represents, select *Species* and then *Transient Concentrations* and chose the appropriate one (*see* **Note 40**). Repeat the process with the other column; when you finish the dialog box should look the same as **Fig. 12**.

Note that COPASI has already determined values for the weights ($\omega_{k,i}$ in **Eq.**); the brackets indicate that they were calculated rather than set by you. However, you are free to change any weight by editing them and removing the brackets (but note that they should always be positive numbers smaller or equal to 1).

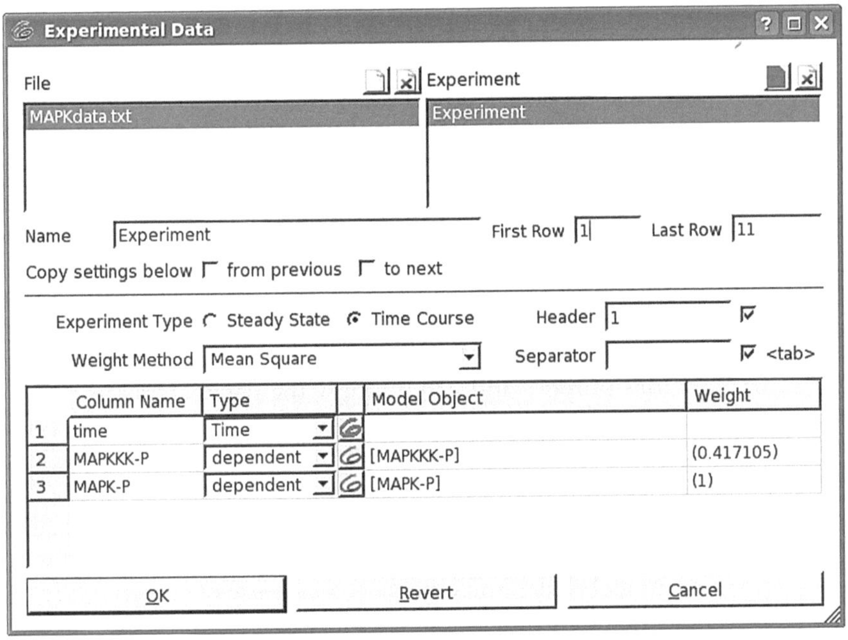

Fig. 12. Experimental data definition window.

It is important to realize that changing the weights affects the ability of the software to perform the fit, and particularly bad choices might entirely prevent success of the fit. COPASI contains three different methods to calculate these weights (*mean*, *mean square*, and *standard deviation*, respectively), as depicted in **Eqs. 4–6**:

$$\omega_{j,k} = 1 / \left|\left\langle X_{j,k} \right\rangle\right|, \tag{4}$$

$$\omega_{j,k} = 1 / \sqrt{\left\langle X_{j,k}^2 \right\rangle}, \tag{5}$$

$$\omega_{j,k} = 1 / \left(\left\langle X_{j,k}^2 \right\rangle - \left\langle X_{j,k} \right\rangle \left\langle X_{j,k} \right\rangle\right). \tag{6}$$

The *mean* and *mean square* methods (**Eqs. 4** and **5**) assure that data columns with small values contribute in the same order of magnitude to the objective function as columns containing large values. The *standard deviation* method (**Eq. 6**) sets larger weight to columns that have little fluctuations.

3.5.3. Estimated Parameters and Constraints

Obviously the exercise of parameter estimation requires one to select the parameters that are to be estimated. Typically these are initial values (concentrations, volumes, etc.) or parameters of the kinetic functions of the reactions (or arbitrary ODE if there are any in the model). The selection of these parameters and their boundaries is specified in exactly the same way as for optimization (*see* **Subheading 3.5.2**).

Sometimes it is necessary to estimate a parameter differently for each experiment, meaning that the software should estimate one value per experiment rather than a single value that best fits *all* experiments (which is the default). For example this is needed when one has executed replicate experiments but where one is not confident that the initial concentration of a chemical species is the same in all experiments. COPASI is able to deal with this, allowing the user to restrict the effect of a parameter to a subset of the experiments listed (obviously this only matters when there are several experiments, but this not the case in the present example). The button labeled *Duplicate for each experiment* is there for this purpose and will multiply the parameters selected when it is pressed to as many new parameters as there are experiments.

It is also possible to define constraints, just like in the optimization task. But beware that adding any arbitrary constraints may well render a problem unsolvable if the constraints cannot be fulfilled. Remember that the main constraints you want for the model is that it fits the data, so the use of constraints in parameter estimation should be taken with care or avoided if possible.

For the present example of the MAPK model, select the reaction limiting rates *V1*, *V2*, *V5*, *V6*, *V9*, and *V10* and set their limits to be -90% and +90% of their original values, in the same way as in the optimization example above.

3.5.4. Fitting the Data

At this point the parameter estimation problem has been completely specified and the actual fitting task can proceed using any of the optimization methods available (*see* **Table 1**). Before running the task it is advisable to define a plot to monitor the progress of the fit and another one to examine results. It is also important to save the file in COPASI format (in case you want to come back to it later, since the SBML file does not contain instructions for the parameter estimation). To create the plots mentioned press the button *Output Assistant* which lists a series of plots and reports that are commonly useful. You can select the plot named *Progress of Fit* and press *Create*, which will generate a plot of the values of **Eq. 3** vs. the number of function evaluations (*see* **Note 41**). The second plot of interest is called *Parameter Estimation Results per Experiment* and it consists of the experimental values of the variables (i.e., contained in the data file) plotted against the values of their corresponding simulated value. The plot also contains the weighted residuals of each data point (i.e., the terms calculated inside the summation in **Eq. 3**).

After defining the plots, save the file, select an optimization method, and press *Run*. For this example select the Levenberg–Marquardt method and run it with the default values. After a short while the method will have finished and you will have plots like those of **Fig. 13**. You should also examine the results by selecting the *Results* page (below *Parameter Estimation* on the left). There you will see the statistics for the sum of squares, though be aware

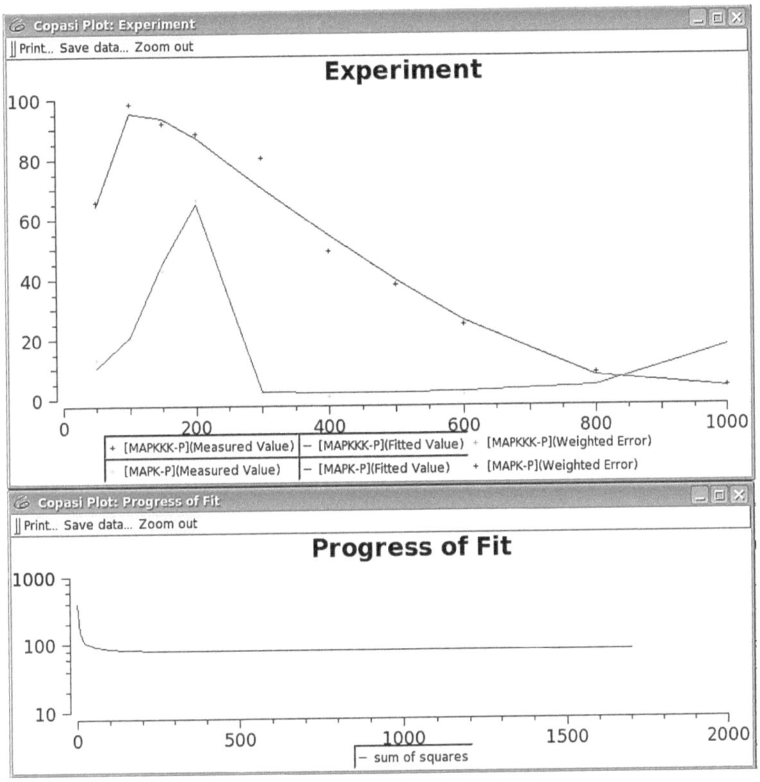

Fig. 13. Results of fitting model parameters to a data set. The plot on the top overlays the experimental data (*crosses*) over the model behavior after fitting (*lines*). The plot at the bottom displays the progress of the sum of squares (**Eq. 3**) as the optimization algorithm progressed (note the logarithmic scale of the *Y*-axis).

that these are problem dependent and you should not compare sums of squares between different problems (not even the same problem with different data sets). More useful are the statistics for the estimated parameters on the second tab, where you will likely see that the coefficients of variation of the estimated parameter values are smaller than 35%, which is very good given the presence of noise in the data. You can also examine the parameter correlation matrix that provides information about dependencies between parameter estimates.

3.6. Stochastic Simulation

Along with the traditional ODE approach, COPASI is also equipped to carry out stochastic simulations based on the theoretical framework derived by Gillespie *(4)*. The *Time Course* task can easily be executed with the algorithm of Gibson and Bruck *(37)* (*see* **Note 42**) and this is as simple as selecting the Gibson–Bruck method from a pull-down menu (**Fig. 14**). This is particularly appealing to those who normally carry out simulations with the ODE approach but sometimes have a need to switch to the stochastic approach. Of course, this also means that COPASI is equally useful for modelers who mostly use the stochastic approach.

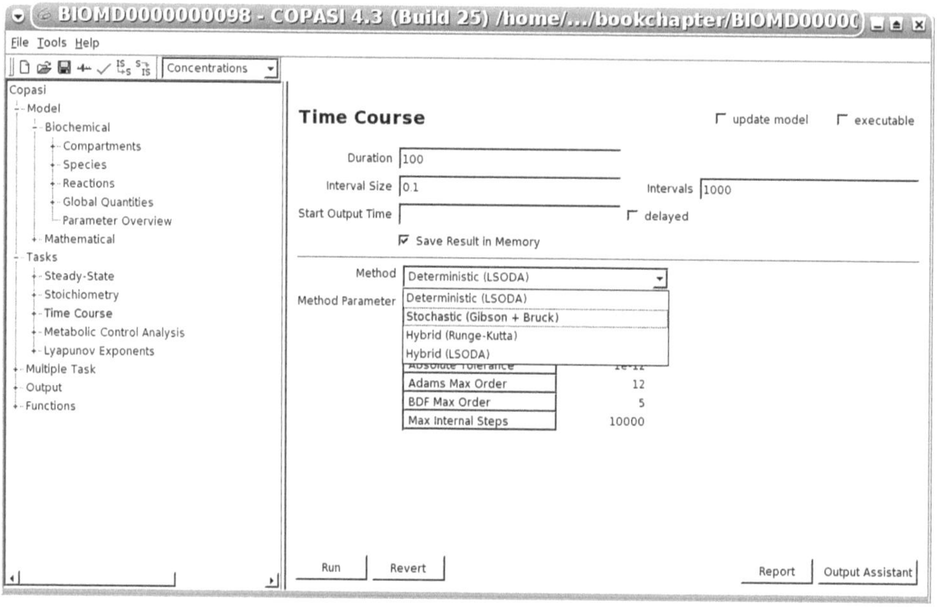

Fig. 14. Switching to a stochastic simulation approach in the *Time Course* window.

Let us consider an example using a model of calcium oscillations by Goldbeter *(38)*, which is model 98 in BioModels. After importing the SBML, go to the *Time Course* task. It is useful to define a trajectory plot of the number of particles against time, which can be done via the *Output Assistant*: chose either the second option (*Particle Numbers, Volumes, and Global Quantity Values*) which will have a scale of numbers of particles, or the first option which will output the corresponding concentrations to the computed particle numbers in the course of the simulation. **Figure 15** shows the outcome of a stochastic simulation for the calcium model.

There are several issues that have to be considered to carry out successful stochastic simulations. The first consideration is that in this approach reversible reactions must be handled as two separate irreversible reactions (the forward and reverse directions). In ODE-based simulations, the forward and backward reaction rates are usually aggregated and thus can cancel each other out (resulting in a null rate); in stochastic simulations each single reaction event has to be considered separately and even if there is no net rate, the actual cycling rate will be explicitly represented. In order to facilitate the conversion of ODE-based models to the stochastic representation, COPASI provides a feature that, at the modeler's request, converts all reversible reactions to the corresponding individual forward and backward reactions (**Fig. 16**). This useful tool adjusts the model automatically – the reaction scheme and the kinetics – and is able to work for a wide range of kinetic rate laws, such as mass action and standard enzymatic kinetics. Nevertheless, there are certain cases when it is not able to dissect rate laws into two separate irreversible kinetic functions. These cases can be very

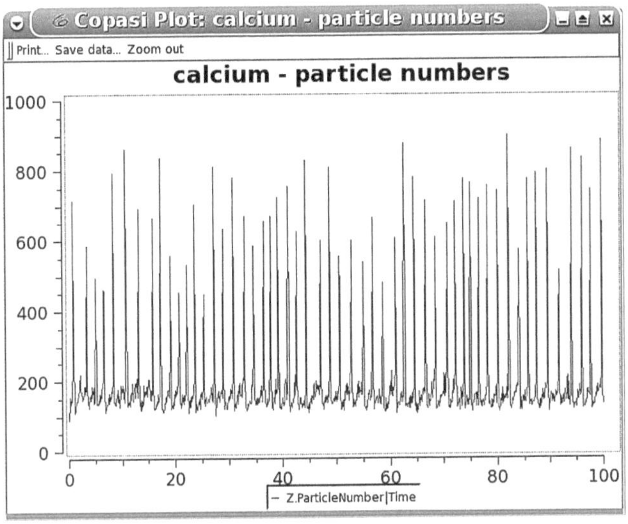

Fig. 15. Trajectory of calcium oscillations using the stochastic simulation algorithm.

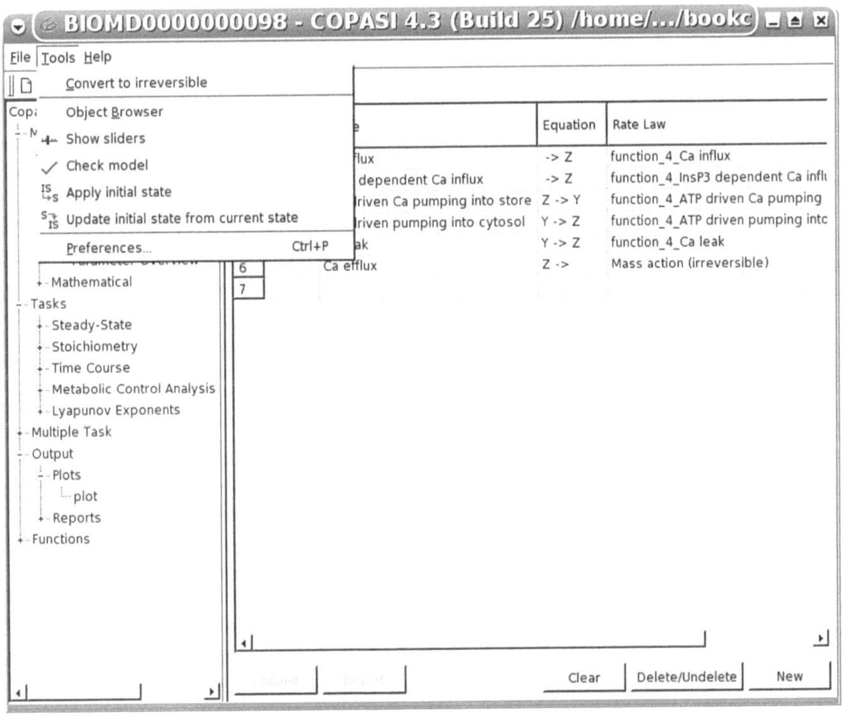

Fig. 16. Menu option to convert a model to be composed only of irreversible reactions.

complex rate laws or rate laws that are actually not appropriate (e.g., an expression that is never negative, thus that is not really reversible). When COPASI cannot automatically convert all reactions, the user will have to adjust the model her/himself.

Often, models are specified without considering the specific volume of the compartment. But for stochastic simulations the volume of the systems is crucial: the volume should not be too big so that the computed particle numbers are not too high and within numerical possibilities of a computer (*see* **Note 43**). This should pose no problem, since it is the purpose of these stochastic simulations to deal with systems that have relatively low particle number. Thus, it is important that the volume of the system be defined in the compartment description in such a way that the particle numbers are not too high.

Another consideration is whether or not the assumptions implied in the rate law of a specific reaction still holds in the presence of low particle numbers. Thus, when stochastically simulating a reaction network which has been described by a set of ODEs all reaction rates have to be converted to a corresponding reaction probability. This is rather simple and straightforward in the case of mass action kinetics *(3)*. However, enzyme kinetic rate laws represent the overall rate of a series of elementary mass action reactions (binding of substrate to enzyme, isomerizations, etc.). An important question is then whether it is justifiable to use such a rate expression in stochastic simulations. Several authors *(39, 40)* have shown that as long as the initial assumptions for the assumed kinetics hold (e.g., excess substrate, fast reversible enzyme–substrate complex formation, etc.), it is indeed justifiable to assume the enzymatic reaction to constitute one single step with a corresponding rate law. The modeler must then ensure that the initial assumptions still hold.

Stochastic simulations are computationally expensive. If a large system is considered which contains some species with high particle numbers and some others with low particle numbers then the use of a hybrid method should be taken into consideration. In COPASI there are currently (version 4.4 Build 26) two hybrid methods implemented. These methods dynamically divide the system into two subsystems: one of them contains reactions with participants that occur in large quantities and is simulated by numeric integrations of ODE; the other one contains reactions that have no participants in large quantities and is stochastically simulated (*see* **Note 44**). In many cases, this approach will speed up the simulation. The two hybrid methods differ only in their numerical integration algorithm – one uses Runge–Kutta, the other uses LSODA.

Since repeated runs of the stochastic simulation will differ considerably, as long as the stochastic influence is noticeable, it is advisable to execute many runs in order to sample a distribution. This can be easily done by using the *Repeat* function of the *Parameter scan* task in COPASI already discussed in a previous section (**Fig. 17**). If a plot of particle numbers over time has been defined, this repeated run will result in multiple time courses being overlaid in a single plot. However, this is not very useful when the dynamics is complex as in the example of calcium oscillations. In these cases, it is best to define a histogram (**Fig. 18**)

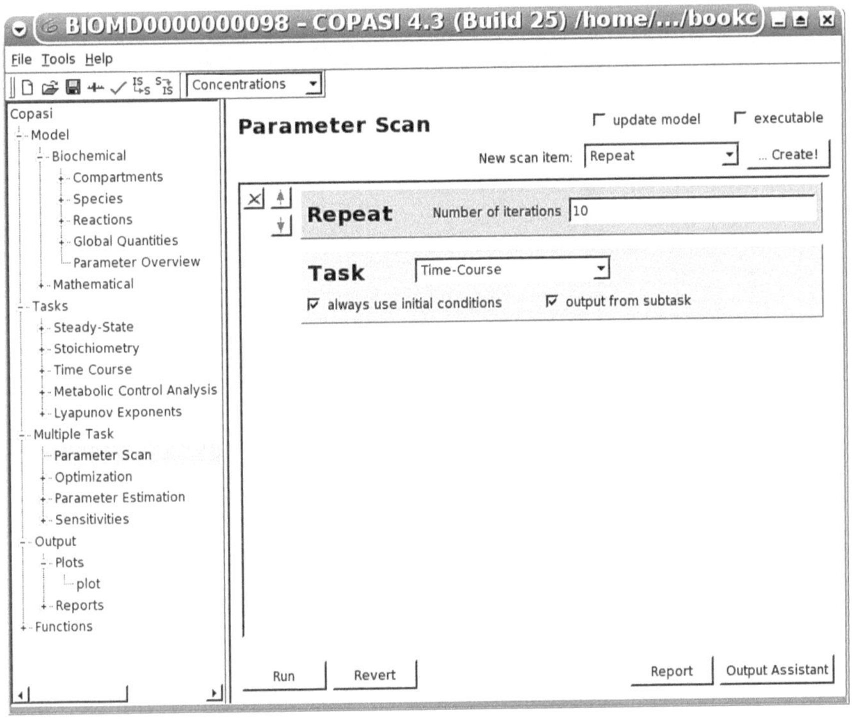

Fig. 17. Using the *Parameter scan* window to repeat the same stochastic trajectory several times.

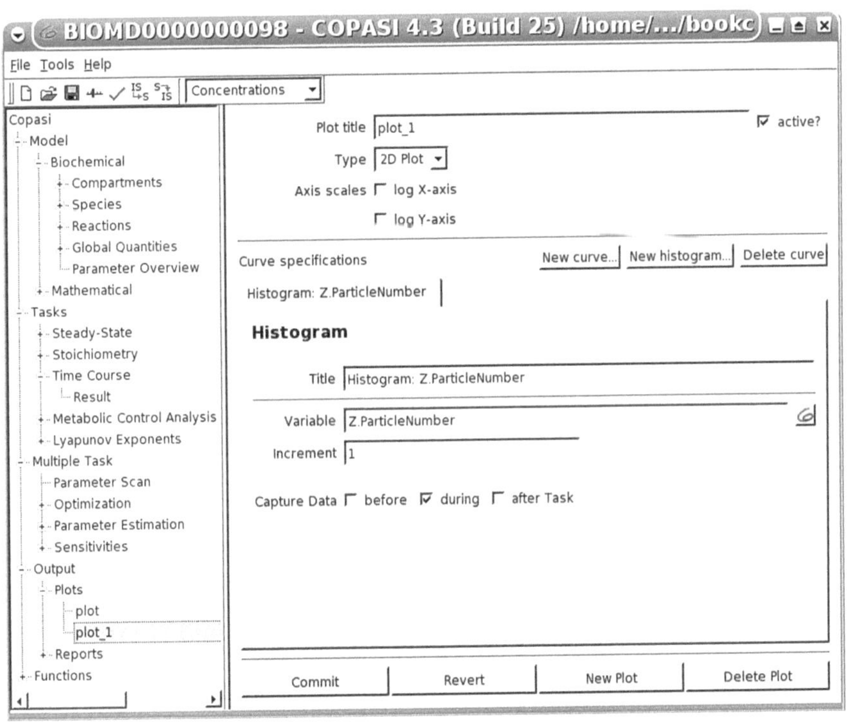

Fig. 18. Defining a histogram plot of a species concentration.

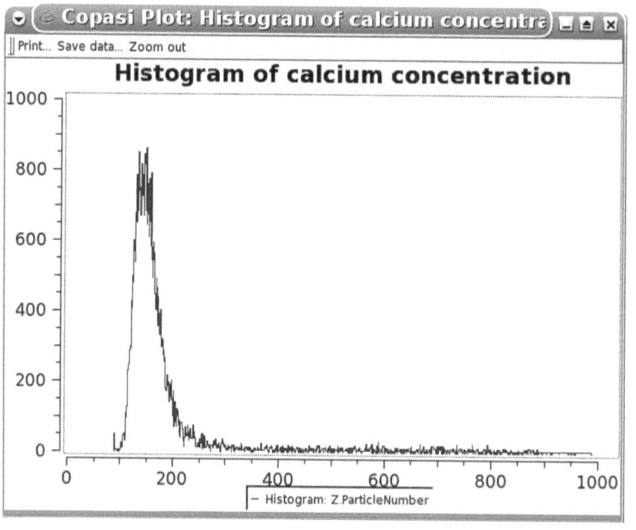

Fig. 19. Histogram of calcium concentration for ten runs of the stochastic simulation algorithm.

to display the cumulative concentration distribution, which is a better way to summarize stochastic simulations (**Fig. 19**).

4. Notes

1. A modifier is a chemical species that affects the rate of reaction but which, unlike substrates or products, is not transformed by the reaction itself. A special case of modifier is the enzyme that catalyzes the reaction, but this class also includes inhibitors and activators.
2. But often not for enzymes, for which the rates usually depend linearly on their concentration – the exception is when there are enzyme–enzyme interactions.
3. It is well known that these equations are often stiff, meaning that they contain very fast and very slow components and this poses a significant numerical problem. Beware of software that does not include ODE integrators (or solvers) that are able to cope with stiff ODEs. Methods such as forward Euler or Runge–Kutta are *not* appropriate when stiffness is present in the equations and can lead to completely spurious solutions because they accumulate truncation error. COPASI uses the LSODA method which is adaptive and is stable under stiff conditions.

4. COPASI can use any of the following three strategies: Newton method, ODE integration forward in time or integration backward in time. If all three are chosen, it first tries the Newton method and if this does not converge, it then integrates in time for a while and then tries the Newton method again – this is repeated ten times, each time integrating even further ahead (10× what was done earlier). If at the end a steady state is not found it will then go back to the original starting point and apply the same strategy but now integrating backward in time. Backward integration, if successful, will find an unstable steady state. The user has control over this strategy by selecting the parameters "Use Newton," "Use Integration," and "Use Back Integration."
5. A commercial license is also available for purchase allowing use of COPASI for applications that are for commercial profit. Go to http://www.copasi.org/commercial for further details.
6. Distribution filenames are in the format: Copasi-XX-YYYYYY.ZZZ where XX is the build number, YYYYYY is a reference to the operating system (*WIN32* for Windows, *Darwin* for OS X, *Linux*, *SunOS* for Solaris, and src for the source code), and ZZZ is the appropriate extension for the type of file, which depends on the operating system.
7. Alternatively you can *browse* the database and find the model that way. However, such a method will become essentially unworkable as the database grows.
8. This means the model is in SBML level 2 version 1; in the future the BioModels database may supply the model in another level/version of SBML so the title of this link may become something like SBML Lx Vy for level x and version y.
9. To download this file you should right click the link and then select an option that allows saving the link to disk (like *save link as...* in Firefox). If you simply click the link your browser will likely show a blank page with some sentences and then a (long) list of parameter names. This is actually part of the model and appears because the browser is trying to interpret the XML encoding as if it was HTML.
10. Additionally there is also an initial value for time, this is only important in the case when some rate equations reference time explicitly (nonautonomous models). In that case, the value of time at the start of the simulation is important and the modeler may need it to be some value different from zero.
11. There is also a selection for the interpretation of rate equations, as there are differences between the ODE and the stochastic approaches. Note that this selection only indicates whether the kinetics used are *meant* for one or the other

approach, not that the approach will be used. In fact, this feature exists so that COPASI can automatically adapt the rate equations to the required approach.

12. There must be one space between each chemical species name, otherwise COPASI will interpret the whole string as one species name. This is because the character "+" is allowed in species names, thus the space is needed to delimit species names from symbols that are not part of the species name.
13. *Substrate* and *product* are obvious; *modifier* is any chemical species that is not transformed by the reaction (inhibitors, activators, and the enzyme if represented explicitly); *volume* is the volume of any compartment; *time* is obvious; and *parameter* is anything else that does not fit any of the other categories.
14. You should not mark as reversible a rate law that can only produce positive values; to be reversible a rate law must be able to take negative values (i.e., flux in the opposite direction). Conversely, an irreversible rate law should not be able to produce negative values.
15. In COPASI, the buttons that are marked with the program's icon are always used to select model entities.
16. *Multiple task* groups a set of computational analyses that require running multiple simulations at each time.
17. This *output from subtask* button would need to be checked if the task was a time course and we wanted the whole time course to be plotted rather than just the final value (although there are also circumstances where that could be desirable thus the choice given to the user).
18. It is best to scan in logarithmic space when the parameter varies by more than one order of magnitude, otherwise most of the samples will lie in the upper order of magnitude.
19. The order of the scan items in the stack is important for the way in which the plot is constructed, but otherwise produces the same results since it generates a regular grid and executes the task at each grid position. The order of the stacks only affects the order in which the grid positions are visited.
20. Obviously a flux mode can only be reversible if *all* reactions that compose it are also reversible.
21. An astute biochemist will realize that there is a carbon and two oxygens missing on the substrate side of this equation, and obviously there should be a CO_2 in that side of the equation. This is missing because the modeler made the decision of not including CO_2 in the model (it should be in the Phosphogluconate dehydrogenase reaction). Since the erythrocyte is not known for fixating CO_2 then the mode must

operate in the reverse direction, i.e., production of PRPP from glucose. In this case, since the mode is reversible it means that one would not know this fact from the stoichiometry alone.

22. A chemical moiety is a set of atoms bound in a fixed structure which are part of molecules, which in a chemical context are referred to as "chemical groups."
23. This means that the dependent species is not calculated from a differential equation, but rather from this mass conservation relation. Thus each mass conservation relation reduces the number of ODE by one.
24. Formalisms of MCA have also been derived for time-dependent states *(41, 42)* but they are rather complicated and some of the coefficients therein are hard to conceptualize, so it is not usually applied.
25. Since COPASI only uses the MCA steady-state formalism, the software first needs to find a steady state before doing the MCA calculations. It is a good idea to investigate the steady state(s) of a model before running MCA, especially regarding the stability of a steady state. While it is technically possible to calculate the MCA for an unstable steady state it is of little practical value.
26. The value of 1% change is here used only for illustration as a "small" change, the coefficients are actually defined only for infinitesimal changes and all the theory is based on that.
27. Since the framework of MCA is based on linearizations and reaction kinetics are generally nonlinear, the values of the elasticities depend on the actual concentrations of the chemical species, so they have to be calculated for specific cases.
28. While it is possible in theory to carry out a large multidimensional scan, the computational time of that exercise would be prohibitive and is beyond simple improvements in computer efficiency (it is an *NP*-complete problem) and thus is essentially impossible for models larger than four or five variables.
29. Control coefficients are actually a special case of response coefficients that have unit elasticity.
30. This in practice consists of finding the steady state, then changing one of the parameter values slightly, and then calculating the new steady state and using ratios to estimate the differentials (the change applied is very small).
31. Maximizing a function is the same as minimizing the symmetric function.
32. There are two types of fluxes in COPASI which only differ by scale: "concentration flux" is expressed in concentration per unit time, while "particle flux" is expressed in numbers

of particles per unit time. In this case you should select "concentration flux." However, what is important is that both be of the same type, since this is a ratio.

33. It is also possible to chose another parameter for the upper or lower bounds, in which case we just need to specify which one with the usual button with the COPASI icon (to the left of the text field). In fact, it is even possible to choose another *estimated* parameter (i.e., one on the list to adjust) as long as that parameter appears in the list before it is used as a boundary value.
34. You may manually override the initial value by highlighting the parameter and then entering a number in the box labeled Start Value or use the tool button labeled as "..." to chose other options, such as random values within the interval.
35. Your numbers may be slightly different due to different precision of different computer architectures, but it should be a number in this range.
36. Possibly there were several warnings of the type "CTrajectoryMethod (12): Internal step limit exceeded," which mean that for some parameter values COPASI could have failed to find a steady state through integration of the ODEs (due to the equations being too stiff). This is not a problem since it may have solved the steady state with the Newton–Raphson method. Even if it indeed failed completely to find a steady state for some parameter combinations, the method will have still converged, as you can judge by the final result. This is one advantage of population-based algorithms: they still work even when the objective function is not continuous (which is what it would look like if the numerical solution could not be obtained).
37. The likelihood that it gets to the same result increases with the length that the algorithm is left running; if you run it with the default 2,000 iterations, it will likely always converge to the same value, but it will run for a much longer time of course.
38. These weights are scaling factors; they are not dependent on the quality of the experimental measurement like a standard deviation.
39. These data were actually created with a slightly modified version of the model where some parameters were changed, a time course was simulated and then noise was added to the values of MAPK-P and MAPKKK-P.
40. This is why it is useful to have column headers because COPASI displays them and you can remember what this column is. This is particularly important if you have a data file with many columns.

41. A function evaluation is the complete calculation needed to simulate the data that needs to match the experimental data. Therefore it consists of calculating all time courses and steady states corresponding to each experiment.
42. Gibson and Bruck's next reaction method *(37)* is a more efficient version of the original Gillespie first reaction method. It achieves better performance by an intelligent use of data structures. For example it stores dependencies between the reactions in dependency graphs and this avoids redundant recalculations of the reaction propensities.
43. Stochastic simulations determine the time interval between reactions, and this time is dependent on the number of particles. If there are too many particles the interval between any two reactions is extremely small, meaning that it would just take too long to simulate any time interval of interest (i.e., at least milliseconds).
44. The division between the subsystems is done with respect to the participating particle numbers and there is a control variable that corresponds to this threshold value which can be adjusted by the user.

Acknowledgments

We thank the users of COPASI whose feedback on the software is crucial for its continued improvement. We also thank *all* developers who have actively contributed to COPASI. COPASI development is supported financially by generous funding from the US National Institute for General Medical Sciences (GM080219), the Virginia Bioinformatics Institute, the Klaus Tschira Foundation, the German Ministry of Education and Research (BMBF), and the UK BBSRC/EPSRC through The Manchester Centre for Integrative Systems Biology.

References

1. Garfinkel, D., Marbach, C. B., and Shapiro, N. Z. (1977) Stiff differential equations. *Ann. Rev. Biophys. Bioeng.* 6, 525–542.
2. Petzold, L. (1983) Automatic selection of methods for solving stiff and nonstiff systems of ordinary differential equations. *SIAM J. Sci. Stat. Comput.* 4, 136–148.
3. Gillespie, D. T. (1976) A general method for numerically simulating the stochastic time evolution of coupled chemical reactions. *J. Comput. Phys.* 22, 403–434.
4. Gillespie, D. T. (1977) Exact stochastic simulation of coupled chemical reactions. *J. Phys. Chem.* 81, 2340–2361.

5. Gillespie, D. T. (2007) Stochastic simulation of chemical kinetics. *Ann. Rev. Phys. Chem.* 58, 35–55.
6. Hoops, S., Sahle, S., Gauges, R., Lee, C., Pahle, J., Simus, N., Singhal, M., Xu, L., Mendes, P., and Kummer, U. (2006) COPASI – a COmplex PAthway SImulator. *Bioinformatics* 22, 3067–3074.
7. Le Novere, N., Bornstein, B., Broicher, A., Courtot, M., Donizelli, M., Dharuri, H., Li, L., Sauro, H., Schilstra, M., Shapiro, B., Snoep, J. L., and Hucka, M. (2006) BioModels Database: a free, centralized database of curated, published, quantitative kinetic models of biochemical and cellular systems. *Nucleic Acids Res.* 34, D689–D691.
8. Kholodenko, B. N. (2000) Negative feedback and ultrasensitivity can bring about oscillations in the mitogen-activated protein kinase cascades. *Eur. J. Biochem.* 267, 1583–1588.
9. Curien, G., Ravanel, S., and Dumas, R. (2003) A kinetic model of the branch-point between the methionine and threonine biosynthesis pathways in Arabidopsis thaliana. *Eur. J. Biochem.* 270, 4615–4627.
10. Schuster, S. and Hilgetag, C. (1994) On elementary flux modes in biochemical reaction systems at steady state. *J. Biol. Syst.* 2, 165–182.
11. Schuster, S., Fell, D. A., and Dandekar, T. (2000) A general definition of metabolic pathways useful for systematic organization and analysis of complex metabolic networks. *Nat. Biotechnol.* 18, 326–332.
12. Reder, C. (1988) Metabolic control theory. A structural approach. *J. Theor. Biol.* 135, 175–201.
13. Holzhütter, H. G. (2004) The principle of flux minimization and its application to estimate stationary fluxes in metabolic networks. *Eur. J. Biochem.* 271, 2905–2922.
14. Kacser, H. and Burns, J. A. (1973) The control of flux. *Symp. Soc. Exp. Biol.* 27, 65–104.
15. Heinrich, R. and Rapoport, T. A. (1974) A linear steady-state treatment of enzymatic chains. General properties, control and effector strength. *Eur. J. Biochem.* 42, 89–95.
16. Fell, D. A. (1992) Metabolic control analysis – a survey of its theoretical and experimental development. *Biochem. J.* 286, 313–330.
17. Heinrich, R. and Schuster, S. (1996) *The Regulation of Cellular Systems.* Chapman & Hall, New York, NY.
18. Fell, D. A. (1996) *Understanding the Control of Metabolism.* Portland Press, London.
19. Cascante, M., Boros, L. G., Comin-Anduix, B., de Atauri, P., Centelles, J. J., and Lee, P. W. (2002) Metabolic control analysis in drug discovery and disease. *Nat. Biotechnol.* 20, 243–249.
20. Rohwer, J. M. and Botha, F. C. (2001) Analysis of sucrose accumulation in the sugar cane culm on the basis of in vitro kinetic data. *Biochem. J.* 358, 437–445.
21. Höfer, T. and Heinrich, R. (1993) A second-order approach to metabolic control analysis. *J. Theor. Biol.* 164, 85–102.
22. Mendes, P. and Kell, D. B. (1998) Non-linear optimization of biochemical pathways: applications to metabolic engineering and parameter estimation. *Bioinformatics* 14, 869–883.
23. Wolpert, D. H. and Macready, W. G. (1997) No free lunch theorems for optimization. *IEEE Trans. Evolut. Comput.* 1, 67–82.
24. Fogel, D. B., Fogel, L. J., and Atmar, J. W. (1992) Meta-evolutionary programming, in *25th Asilomar Conference on Signals, Systems & Computers* (Chen, R. R., ed.). IEEE Computer Society, Asilomar, CA, pp. 540–545.
25. Runarsson, T. and Yao, X. (2000) Stochastic ranking for constrained evolutionary optimization. *IEEE Trans. Evolut. Comput.* 4, 284–294.
26. Michalewicz, Z. (1994) *Genetic Algorithms + Data Structures = Evolution Programs.* Springer, Berlin.
27. Hooke, R. and Jeeves, T. A. (1961) "Direct search" solution of numerical and statistical problems. *J. ACM* 8, 212–229.
28. Levenberg, K. (1944) A method for the solution of certain nonlinear problems in least squares. *Quart. Appl. Math.* 2, 164–168.
29. Goldfeld, S. M., Quant, R. E., and Trotter, H. F. (1966) Maximisation by quadratic hill-climbing. *Econometrica* 34, 541–555.
30. Marquardt, D. W. (1963) An algorithm for least squares estimation of nonlinear parameters. *SIAM J.* 11, 431–441.
31. Nelder, J. A. and Mead, R. (1965) A simplex method for function minimization. *Comput. J.* 7, 308–313.
32. Kennedy, J. and Eberhart, R. (1995) Particle swarm optimization. *Proc. IEEE Int. Conf. Neural Netw.* 4, 1942–1948.
33. Brent, P. R. (1973) A new algorithm for minimizing a function of several variables without calculating derivatives, in *Algorithms for Minimization Without Derivatives* (Brent, P. R., ed.). Prentice-Hall, Englewood Cliffs, NJ, pp. 117–167.
34. Corana, A., Marchesi, M., Martini, C., and Ridella, S. (1987) Minimizing multimodal functions of continuous variables with the "simulated annealing" algorithm. *ACM Trans. Math. Softw.* 13, 262–280.

35. Nash, S. G. (1984) Newton-type minimization via the Lanczos method. *SIAM J. Numer. Anal.* 21, 770–788.
36. Johnson, M. L. and Faunt, L. M. (1992) Parameter estimation by least-squares methods. *Methods Enzymol.* 210, 1–37.
37. Gibson, M. A. and Bruck, J. (2000) Efficient exact stochastic simulation of chemical systems with many species and many channels. *J. Phys. Chem. A* 104, 1876–1889.
38. Goldbeter, A., Dupont, G., and Berridge, M. J. (1990) Minimal model for signal-induced Ca2+ oscillations and for their frequency encoding through protein phosphorylation. *Proc. Natl Acad. Sci. USA* 87, 1461–1465.
39. Rao, C. V. and Arkin, A. P. (2003) Stochastic chemical kinetics and the quasi-steady-state assumption: application to the Gillespie algorithm. *J. Chem. Phys.* 118, 4999–5010.
40. Cao, Y., Gillespie, D., and Petzold, L. (2005) Multiscale stochastic simulation algorithm with stochastic partial equilibrium assumption for chemically reacting systems. *J. Comput. Phys.* 206, 395–411.
41. Acerenza, L., Sauro, H. M., and Kacser, H. (1989) Control analysis of time dependent metabolic systems. *J. Theor. Biol.* 137, 423–444.
42. Ingalls, B. P. and Sauro, H. M. (2003) Sensitivity analysis of stoichiometric networks: an extension of metabolic control analysis to non-steady state trajectories. *J. Theor. Biol.* 222, 23–36.

Chapter 3

Flux Balance Analysis: Interrogating Genome-Scale Metabolic Networks

Matthew A. Oberhardt, Arvind K. Chavali, and Jason A. Papin

Summary

Flux balance analysis (FBA) is a computational method to analyze reconstructions of biochemical networks. FBA requires the formulation of a biochemical network in a precise mathematical framework called a stoichiometric matrix. An objective function is defined (e.g., growth rate) toward which the system is assumed to be optimized. In this chapter, we present the methodology, theory, and common pitfalls of the application of FBA.

Key words: Systems biology, Metabolic reconstruction, Flux balance analysis, Metabolomics, Constraint-based modeling, Genome-scale network.

1. Introduction

The availability of sequenced and annotated genomes, coupled with a tremendous knowledge base in scientific literature, has facilitated the construction of genome-scale models of metabolism for a wide variety of organisms *(1–5)*. Metabolic network reconstructions include stoichiometric detail for the set of known reactions enzymatically catalyzed in a particular organism. These metabolic reconstructions can be built from the bottom-up, i.e. from the level of the gene to whole pathways acting in concert, and can include thousands of reactions and genes. Because of their enormous size, computational methods are required to quantitatively analyze large-scale biochemical networks. Although traditional ordinary differential equation (ODE)-based models of metabolism allow for characterization of dynamic cell states, full-scale dynamic modeling is often difficult for networks consisting

Ivan V. Maly (ed.) *Methods in Molecular Biology, Systems Biology, vol. 500*

DOI: 10.1007/978-1-59745-525-1_3

of thousands of reactions because of a paucity of necessary parameter values *(6)*. Therefore, methods are needed in which kinetic parameters are less critical for prediction of cell phenotype. Flux balance analysis (FBA) is one such technique for analysis of large-scale biochemical systems under conditions where kinetic parameters do not need to be defined: namely, at steady state (*see* **ref.** *7* for historical perspective on FBA).

In this chapter, we discuss how to perform FBA and provide details for troubleshooting mistakes. FBA is a constraint-based method; first a space of possible phenotypes is defined by imposing constraints on a biochemical system, and finally an objective function is optimized within that space to determine the system's most likely phenotypic state. The state space of an FBA problem consists of steady-state (i.e., constant growth rate/exponential phase) fluxes through all reactions in the biochemical network. Predictions of values for these fluxes are obtained by optimizing for an objective (e.g., maximizing growth rate, minimizing energy use, maximizing end-product production), while simultaneously satisfying constraint specifications *(8)*. The set of constraints can be grouped into any one of four categories: (a) physicochemical (e.g., conservation of mass), (b) topological (e.g., compartmentation and spatial restrictions associated with metabolites/enzymes), (c) environmental (e.g., media composition, pH, temperature), and (d) thermodynamic (e.g., reaction reversibility) *(8, 9)*.

FBA employs a mathematical formalism derived from the mass action expression:

$$\frac{\mathrm{d}C}{\mathrm{d}t} = Sv$$

in which C, and v are vectors of metabolite concentrations and reaction fluxes, respectively, and t is time. S is a stoichiometric matrix composed of rows corresponding to metabolites and columns corresponding to reactions for a given metabolic network (*see* **Subheading 3** for more details). FBA is performed on metabolic networks at steady-state since intracellular metabolic kinetics are much faster than changes in cellular phenotype (e.g. growth rate of a cell), and therefore the phenotype of a metabolic network quickly stabilizes to a steady solution *(8)*.

By definition, the change in concentration of metabolites over time is equal to zero when a system is at steady-state:

$$\frac{\mathrm{d}C}{\mathrm{d}t} = 0$$

Therefore, the set of possible steady-state flux distributions through the metabolic network can be represented as the vector v in the equation:

$$Sv = 0$$

This equation constitutes the main constraint set representing a biochemical network in FBA, as elaborated in **Subheading 3.2** and **Fig. 1A–C**.

Numerous extensions to FBA have been recently developed. Following is a brief overview of some of these extensions:

- *DFBA*: An abbreviation for dynamic flux balance analysis. Two independent approaches were proposed to address diauxic growth (changes in growth rate as one carbon source is depleted and the cell switches to utilizing another carbon source) in *Escherichia coli*, namely dynamic optimization (DOA) and static optimization (SOA) *(10)*. DOA considers the full time course of bacterial growth via the formulation of a nonlinear programming (NLP) problem (*see* **ref.** *8* for differences in mathematical programming strategies), while SOA entails solving multiple linear programming (LP) problems over several discretized time steps *(10)*. The SOA was similar to a previous method *(11)* except for the inclusion of constraints regarding the rate-of-change of metabolite fluxes in DFBA. For large-scale biochemical networks, SOA would be more applicable due to the computational limitations of the NLP aspect of DOA *(10)*.
- *EBA*: A more rigorous addition of thermodynamic constraints to FBA via energy balance analysis was proposed *(12)*. Using this method, the flux space calculated by FBA is further constrained and flux distributions that are thermodynamically infeasible are removed. The *E. coli* metabolic network was simulated under glucose-minimal medium, and in addition to the flux distribution, chemical potentials and conductances were also predicted for every reaction in the network. Information on reaction conductance was used to characterize enzyme regulation in the switch between aerobic and anaerobic conditions. Furthermore, some gene knockouts predicted incorrectly by FBA were correctly predicted using EBA *(12)*.
- *rFBA*: Regulated FBA considers the effects of transcriptional regulation on metabolism. The FBA flux space is further constrained by accounting for regulatory elements (e.g., if a particular transcription factor is present, then a given reaction does not occur). By incorporating Boolean rules, defining enzymatic regulation at discretized time steps and iteratively simulating the fluxes and concentrations in the network, rFBA was applied to a prototypic network *(13)*. The use of rFBA on a large-scale integrated metabolic and regulatory network of *Saccharomyces cerevisiae* consisting of 805 genes and 775 regulatory interactions has also been demonstrated *(14)*. Gene expression profiles were simulated under genetic and environmental perturbations, and growth phenotypes were characterized for various transcription factor knockout strains (subject

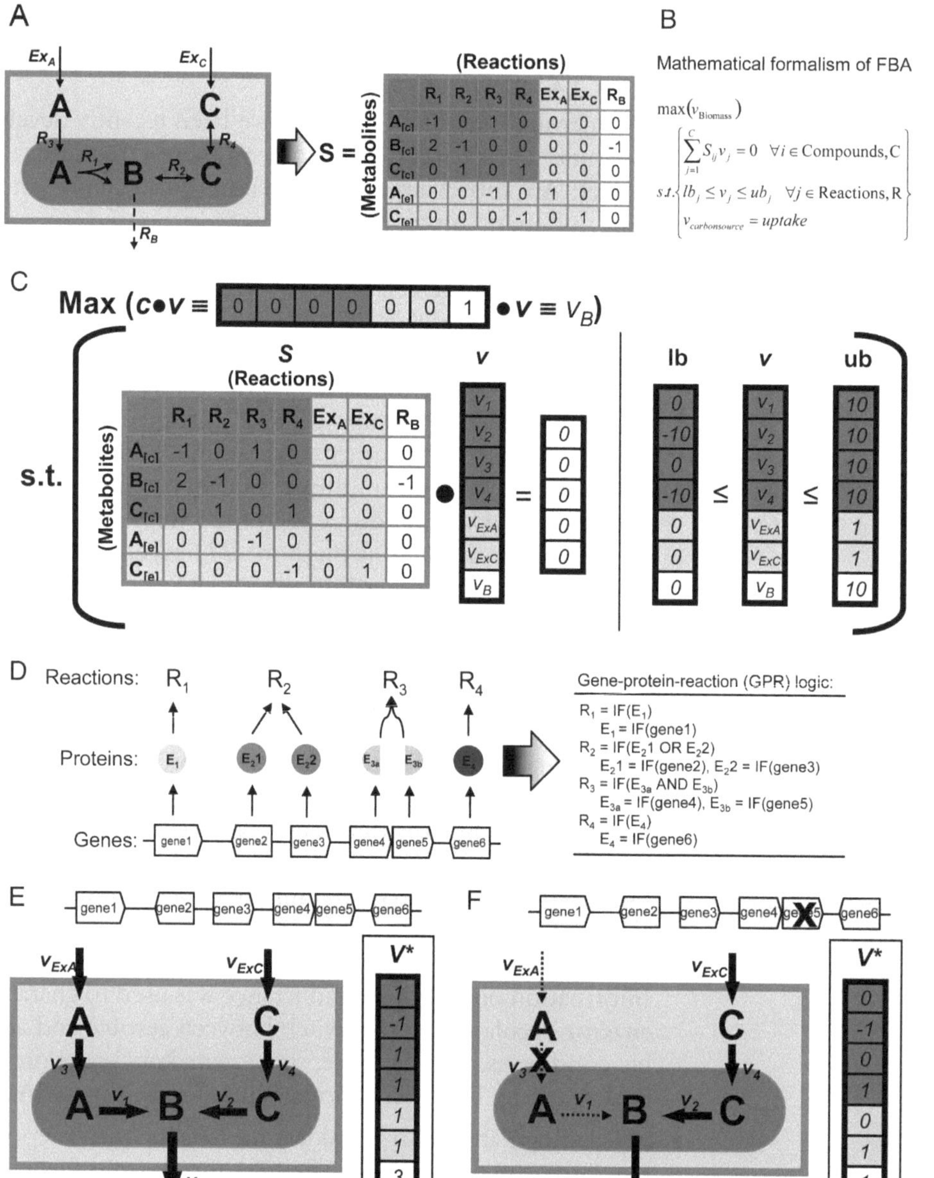

Fig. 1. The formulation of an FBA problem is demonstrated. (**A**) A simple prototypic metabolic network is converted into an ***S*** matrix. Different shades of elements in the ***S*** matrix represent the location of the corresponding reaction/metabolite in the system. R_B represents the objective reaction. (**B**) The mathematical formalism of FBA where ***"S"*** is the stoichiometric matrix composed of rows corresponding to metabolites and columns corresponding to reactions, "*v*" is the vector of fluxes through the associated reactions, ***"lb"*** and ***"ub"*** are the lower and upper bounds on the fluxes, "v_{Biomass}" (or v_B) is the objective flux and "$v_{\text{carbonsource}}$" is the uptake flux. The abbreviation s.t. stands for "subject to." (**C**) The mathematical formalism is illustrated in matrix format. (**D**) Genes encoding proteins that catalyze reactions in the prototypic system are shown on the left. This gene–protein-reaction (GPR) network is then modeled with Boolean logic, as shown on the right. (**E**) The solution to the FBA problem introduced in panel (**B**) is presented in flux vector *v**. (**F**) The solution to an FBA performed after knocking out *gene5* is similarly presented in *v**.

to different carbon source requirements) of *S. cerevisiae (14)*. The recent development of the R-matrix formalism for transcriptional regulatory systems also facilitates the integration of metabolism and regulation and serves to increase the applicability of rFBA *(15)*.

Further extensions to FBA are continually being developed. For example, efforts are currently underway to implement an integrated, dynamic FBA (idFBA) on a whole-cell model of metabolism, signaling and regulation *(17)*. An approach called minimization of metabolic adjustment (MOMA) was developed to analyze the effect of gene knockouts on metabolic network phenotype. In MOMA, the suboptimal solution of a mutant strain (following a gene knockout) is generated, which represents the smallest Euclidean distance to the optimal FBA solution of a wild-type strain *(18)*.

In this chapter, we highlight available resources to reconstruct the metabolic network of a particular organism and present computational tools that are available to perform FBA. In addition, with the help of a prototypic metabolic network, the process of setting up an FBA problem is described in detail. The steps involved in formulating GPR relationships, defining an objective function and troubleshooting problems that commonly occur during FBA simulations are also presented. FBA is a powerful tool to quantitatively analyze metabolic networks and has been extensively applied to genome-scale models of *E. coli (5)*, *Saccharomyces cerevisiae (3)*, and many other organisms *(2, 4, 19–22)*.

2. Materials

2.1. Tools for Metabolic Reconstruction

Metabolic network reconstructions, representing species-specific large-scale models of metabolism, are commonly built using information from biological databases and literature sources. A variety of online public resources are available to aid in generating a metabolic network reconstruction, including:

Genomic resources:

GeneDB (http://www.genedb.org/) and *TIGR* (http://www.tigr.org/) provide genomic information and functional annotations for a wide variety of organisms.

Enzyme resources:

ENZYME (http://ca.expasy.org/enzyme/) is an enzyme nomenclature database. *BRENDA* (http://www.brenda-enzymes.info/) provides enzyme, associated reaction, and pathway information. *BRENDA* also includes organism-specific information on localization and links to experimental literature references.

Species or organism-specific resources:

There are several resources dedicated to particular organisms or species. Examples include Pseudomonas Genome Database (http://www.pseudomonas.com/), EcoCyc (http://ecocyc.org/), and Xenbase (http://www.xenbase.org/common/).

Other resources:

KEGG (http://www.genome.jp/kegg/), *MetaCyc* (http://metacyc.org/), and *NCBI* (http://www.ncbi.nlm.nih.gov/) include gene, protein, and reaction information for several organisms. *KEGG* also contains a pathway module with detailed maps that are useful when reconstructing a metabolic network. In addition to gene and molecular databases, *NCBI* also includes literature databases providing expansive collection of prior research relating to the organism of interest.

The resources highlighted above are only a sampling of all the databases that are available for metabolic network reconstructions. The above list is certainly not comprehensive; therefore, the reader is directed to reviews that list other publicly and commercially available online databases *(23, 24)*.

2.2. Software Packages

Performing FBA does not require the use of specialized computing hardware. It can be executed on a standard desktop platform, using one of a variety of software tools. Generally, FBA is performed using an optimization package such as LINDO (Lindo Systems Inc., Chicago, IL) or GAMS (GAMS Development Corporation, Washington, DC), using MATLAB (The MathWorks Inc., Natick, MA), or using any other software package that allows for fast and efficient LP computation. Specifically in MATLAB, FBA can be carried out using the Optimization Toolbox. Via the Optimization Toolbox, the *linprog* function can be used for LP calculations (*see* **Subheading 3.6**). Several dedicated tools have also been created that perform FBA and other systems-biology related tasks in MATLAB. A few of these tools are highlighted below:

- The COBRA Toolbox *(25)* can execute FBA and many other systems biology applications. The COBRA Toolbox performs operations on models presented in systems biology markup language (SBML). Therefore, the SBML Toolbox must be installed in MATLAB for the COBRA Toolbox to function. For LP optimization, the COBRA Toolbox supports at least five LP solvers that are all available online: lp_solve (free), glpk (free), LINDO, CPLEX, and Mosek.
- FluxAnalyser *(26)* is a tool that provides a graphical user interface in MATLAB through which FBA and other systems biology related analyses can be performed.
- Metabologic *(27)* is a tool that executes FBA and is designed to aid in setting up optimal ^{13}C-NMR experiments to determine the proper flux distribution through a system.

A toolbox called SNA also exists for performing FBA and other network-related analyses in the Mathematica platform *(28)*. In addition, several other standalone tools capable of performing FBA are available. MetaFluxNet *(29)* is one such standalone software package that performs FBA and other analysis techniques. It also uses lp_solve as the default LP solver, but can support others, including CPLEX, LINDO, GAMS, AMPL, and MATLAB LP. Commercial software packages have also been developed for large-scale FBA and systems analyses of metabolic networks, including Simpheny (Genomatica Inc., San Diego, CA) and Discovery (INSILICO Inc., Stuttgart, Germany).

3. Methods

3.1. Prerequisites

To perform FBA, components of a biochemical network must be defined and represented in precise, mathematical forms. Since FBA is geared toward analysis of networks at steady state (i.e., nonvarying fluxes over a defined time window), kinetic parameters are not required. However, FBA requires a stoichiometric matrix, known bounds on reaction fluxes and an objective function. To illustrate the steps in formulating an FBA problem, a prototypic system is presented in **Fig. 1A** and the associated rigorous mathematical formalism in **Fig. 1B**. This prototypic metabolic system will be referenced throughout this section.

Note: For FBA of large-scale metabolic networks, the biomass reaction is often – but not necessarily – chosen as the objective function. In the following sections, the flux of the objective function is referred to as v_B (when in reference to the prototypic network presented in **Fig. 1**), $v_{Biomass}$ (when describing typical FBA of large-scale networks where the objective is the biomass function), or v_{obj} (when the objective function is left unspecified). Flux values denoted with an asterisk (e.g., $v_{biomass}$*) denote the optimal value as determined by FBA.

3.2. Building the Matrices

1. *S-matrix*: The stoichiometric matrix, or *S*, is a matrix composed of the stoichiometric coefficients for all reactions in a biochemical network. By convention, columns in *S* represent reactions, and rows represent species (e.g., metabolites) participating in the reactions. Each substrate and product of a reaction must be assigned a stoichiometric coefficient, $s_{i,j}$, dictating how many moles of that compound are consumed or produced in the reaction. Metabolites not participating in a reaction gain a coefficient of zero for that particular reaction. **Figure 1A** shows the conversion of a sample metabolic network into an *S* matrix. Note that substrate coefficients are

represented by negative numbers, while product coefficients are represented by positive numbers. Therefore, R_1 [(1) A → (2) B] becomes the first column of S in **Fig. 1A**, with a "−1" coefficient denoting the "loss" of 1 units of metabolite $A_{[c]}$ and a "+2" coefficient denoting the "gain" of 2 units of metabolite $B_{[c]}$ (where [c] represents an intracellular cytosolic metabolite).

2. *Exchange reactions*: Exchange reactions involve metabolites in the surrounding environment that are allowed to enter and/or leave the system. The constraints on the exchange reactions will dictate what resources are available in the *in silico* cell culture. Exchange reactions are represented in the S matrix as column vectors with a +1 coefficient for the metabolite being exchanged, but zeros for all other metabolites (*see* columns Ex_A and Ex_C in the S matrix in **Fig. 1A**).
3. *Lower and upper bounds*: The standard constraint set for FBA includes a lower bound (lb) and an upper bound (ub) for every reaction in the system. These bounds are represented by column vectors, with coefficients representing minimum and maximum fluxes for the corresponding reaction (*see* **Figs. 1B, C**). In effect, reaction reversibility rules are formulated. As an example, for an irreversible reaction, the lower bound would be set to *0*. Lower and upper bounds for the prototypic system are shown in **Fig. 1C** as column vectors *lb* and *ub*. Bounds for exchange reactions represent flow of nutrients into and out of the biochemical system, while bounds for transport reactions (occurring across cell and subcellular compartment membranes) and intracellular reactions (occurring within the confines of the cell membrane) represent physicochemical constraints on reaction rates, due to thermodynamics or maximal uptake rates.
4. *Additional components*: The mathematical formulation of a FBA problem is shown in **Fig. 1C**. The objective of the FBA problem in the top of **Fig. 1C** is the dot product $c \cdot v$, where c is a vector containing objective coefficients for each reaction in the system, and v is the vector of reaction fluxes in the system. Since a typical FBA problem involves the optimization of only one reaction, c will typically contain all zeros except for a "1" corresponding to the reaction flux being optimized. In the prototypic network presented in **Fig. 1A**, reaction R_B is the objective reaction, and the corresponding flux is denoted as v_B, an abbreviation for $v_{Biomass}$ (*see* **Figs. 1B, C**). The cv expression is a more formal way of representing the objective rather than as flux v_B. A zero vector representing the right side of the steady-state expression $S \cdot v = 0$ is also required.

3.3. Setting up the Optimization Problem

1. Typically, the number of metabolites in a biochemical system is less than the number of fluxes, and thus an S matrix represents an under-determined system for which there exists a range of possible solutions at steady state *(8)*. To choose the solution v^* that is most optimal for a given objective, FBA employs a standard optimization technique called Linear Programming (LP), denoting an optimization wherein the constraints and the objective function are all linear with respect to the instrument variables (the fluxes, v). The LP formalism for the prototypic system is shown in **Fig. 1C**. A typical mathematical expression for the standard FBA optimization is also provided. The fluxes in vector v are constrained between their lower and upper bounds, and the expression $c \cdot v$ is maximized as explained in **Subheading 3.2.4**.
2. An uptake constraint ($v_{\text{carbonsource}} = uptake$) simulating a controlled and steady flux of carbon source into the growth medium can also be included within the optimization problem, as indicated in the mathematical setup in **Fig. 1B**. This constraint aids in normalizing the fluxes in the system to a particular chosen value. As an example, in determining growth yield of a bacterium growing on glucose as a sole carbon source, $v_{ExGLUCOSE}$ and v_{Biomass} serve as the uptake and objective fluxes, respectively. These are similar to v_{ExA} and v_{B} in the prototypic network. Maximizing biomass is a common objective function in large-scale metabolic networks (*see* **Subheading 3.7**). By ensuring that 1 unit of biomass drains exactly 1.0 g of metabolite mass (e.g., the total amount of protein, DNA, RNA, carbohydrate, lipid, and other components in biomass – *see* **Subheading 3.7**) from the system, the resulting biomass flux will equal the growth rate (in units of hour^{-1}), and the yield can be calculated as:

$$\text{Yield} = \frac{v_{\text{Biomass}}}{v_{\text{ExGLUCOSE}}} = \frac{[1/\text{h}]}{\left[\dfrac{(\text{mmol glucose})}{(\text{grams dry weight biomass} \times \text{h})}\right]} = \left[\frac{\text{grams dry weight biomass}}{\text{mmol glucose}}\right]$$

Note: The units of flux for all intracellular reactions are mmol metabolite/gram dry weight/h. The only exception is the biomass reaction with units of flux equal to 1/h (assuming the biomass drain is scaled to 1.0 g as described below).

When exact experimental measurements are not available in the literature, but relative amounts are known (as is usually the case), the biomass drain should be set to 1.0 g of metabolite mass by scaling the biomass reaction coefficients (keeping their ratios constant with respect to each other) to satisfy the equation:

$\text{Mass}^{D}_{\text{biomass}} = \sum_{i \in D} s_{i,\text{biomass}} \cdot M_i = 1g$, where M_i is the mass of metabolite i and D is the set of biomass components drained from the system. Note that the uptake constraint is implicitly expressed in **Fig. 1C** as the constraints on v_{ExA} and v_{ExC}.

3.4. Interpreting Results

1. The results of the FBA problem for the prototypic system are shown in **Fig. 1E**. Fluxes are denoted by arrows on the network map, and the optimal flux vector v^* is shown in the inset panel. The v^* vector can be interpreted as a set of fluxes through all of the reactions in the metabolic network that will lead to an optimal value of v_B (the objective flux), which in this case has a value of 3 (*see* **Fig. 1E**). Note, the v^* vector obtained is not necessarily *the only* solution that will result in the optimal value of the objective function; it is merely *an* optimal solution. Flux variability analysis can help identify the range of possible v^* solutions for a particular FBA problem *(30)*.
2. A flux in v^* is interpreted with the reaction column in the S matrix to which it corresponds. For instance, suppose that the R_1 column in **Fig. 1C** were modified to:

$$R_1^{\text{new}} = 10 \cdot R_1 = \begin{bmatrix} -10 \\ 20 \\ 0 \\ 0 \\ 0 \end{bmatrix}$$

The resulting flux through R_1^{new} would be $v_1^{\text{new}*} = 0.1$ as opposed to $v_1^* = 1.0$, but otherwise the FBA problem would be unchanged. Importantly, the same amount of mass would be passed through R_1 per unit time per gram of cellular dry weight, as determined by the equation: $(v_1^{\text{new}*})(R_1^{\text{new}}) = (v_1^*)(R_1)$ Therefore, the flux v_1^* is scaled according to the stoichiometric coefficients of metabolites participating in reaction R_1.

3.5. Formulating GPR Relationships

The inclusion of the GPR relationships facilitates the characterization of the genotype to phenotype relationship. Reactions in the S matrix are linked to genes that encode the associated enzymes. This association allows for the analysis of the effects of gene knockouts and transcriptional regulation of metabolism. A sample set of GPR rules are provided for the prototypic metabolic network (*see* **Fig. 1D**). GPR relationships can be expressed with Boolean logic (e.g., AND/OR associations). Isozymes (enzymes E_21 and E_22 in the prototypic network in **Fig. 1D**) have an OR relationship as they can independently catalyze R_2. Subunits of a protein complex (enzymes E_{3a} and E_{3b}) can be associated with an AND relationship as both subunits are required to catalyze R_3.

The Boolean logic for GPRs is indicated in the table on the right of **Fig. 1D**.

Performing an *in silico* gene knockout experiment involves identifying the target gene in the reconstruction and removing the associated reaction(s) from the metabolic network, according to the GPR Boolean logic. Mathematically, this is achieved by constraining the upper and lower bounds on the corresponding reaction fluxes to zero. An example gene knockout simulation is presented in **Fig. 1F**. Here, *gene5* is "knocked out." According to the GPR Boolean logic in **Fig. 1D**, *gene5* codes for E_{3b}, a subunit protein required for R_3. Therefore, a *gene5* knockout will eliminate the activity of R_3. Without an active R_3, the resulting flux of the objective function is reduced from 3 to 1 (*see* **Fig. 1E, F**, respectively).

3.6. Sample MATLAB Code

Sample MATLAB code performing FBA on the metabolic network presented in **Fig. 1A** is provided in **Fig. 2**. This code utilizes the *linprog* function in MATLAB, which is available through the Optimization Toolbox.

```
%% Generate S-matrix:
S = [-1   0   1   0   0   0   0
      2  -1   0   0   0   0  -1
      0   1   0   1   0   0   0
      0   0  -1   0   1   0   0
      0   0   0  -1   0   1   0];

%% Generate additional matrices:
ub = [10;10;10;10;1;1;10];
lb = [0;-10;0;-10;0;0;0];
RS = [0;0;0;0;0];
c = [0 0 0 0 0 0 1];

%% Perform FBA optimization (Figure 1e):
v_e = linprog(-c,[],[],S,RS,lb,ub)

%% Remove v3 and re-run FBA (Figure 1f):
S(:,3) = zeros(size(S(:,3))); % removes v3
v_f = linprog(-c,[],[],S,RS,lb,ub)
```

Fig. 2. Sample MATLAB code for performing FBA illustrated in **Fig. 1e, f**. The stoichiometric matrix of the prototypic network is presented along with the additional matrices introduced in **Fig. 1**. The *linprog* command (via the Optimization Toolbox) is implemented to solve the FBA problem.

3.7. Defining the Biomass Reaction

A common objective function that is maximized when analyzing a metabolic network is the cellular growth rate. To simulate growth, a biomass demand reaction is formulated wherein all essential metabolites are drained in the ratio required for the subsequent production of cellular components. To generate a biomass reaction, the dry weight cellular composition of the organism of interest needs to be obtained from experimental literature or estimated using data from highly related organisms. The cellular composition describes the percentage of proteins, RNA, DNA, carbohydrate, lipid, polyamines, and other constituents of a given cell. Below, each of the various cellular components is analyzed in further detail.

1. *Analyzing protein content*: The percent prevalence of each of the 20 amino acids in protein needs to be calculated. If data are unavailable in existing literature, this percent prevalence can be determined by downloading the amino acid sequence corresponding to every ORF in the genome of the organism of interest and implementing a character count. Using molecular weight information from online databases (e.g., KEGG), the percentage by weight of the 20 amino acids can be calculated. As an example, suppose that the percent dry weight composition of protein in a cell is 50% and the percent by weight of alanine is 10%. To calculate the coefficient for alanine in the biomass reaction in units of mmol per gram of dry cell weight: {Alanine mmol/gDW}=

$$\frac{0.1\ \text{gm of Alanine}}{1\ \text{gm of protein}} \times \frac{0.5\ \text{gms of protein}}{1\ \text{gDW}} \times \frac{1\ \text{mol Alanine}}{89.05\ \text{gm of Alanine}} \times \frac{1000\ \text{mmol}}{1\ \text{mol}} = 0.5615$$

   Similar calculations are carried out for all 20 amino acids.

2. *Analyzing RNA and DNA content*: The same method applied to protein content is applicable to analyzing RNA and DNA content in biomass. For RNA monomers, the percent prevalence of A, C, G, and U needs to be determined from the genome. For DNA content, the G + C statistic can be used. Subsequently, the percent by weight calculations of nucleotides (ATP, CTP, GTP, and UTP) and deoxynucleotides (dATP, dCTP, dGTP, and dTTP) can be carried out by obtaining relevant molecular weight information. Using the percent dry weights of RNA and DNA in a cell, the coefficients of nucleotides and deoxynucleotides in biomass can be calculated as illustrated above for the amino acid alanine.

3. *Analyzing carbohydrate and lipid content*: Percentage by weight data for carbohydrate and lipid (neutral and polar) contents needs to be obtained from the literature. For determining the

molecular weight of lipids, information on the fatty acid composition in the organism of interest is also required. Using the fatty acid composition, the molecular weight of an average length fatty acid chain can be computed. This calculation will aid in the molecular weight calculations of neutral and polar lipids.

4. *Other components*: Depending on the organism, other components such as polyamines and cofactors can also be included in the biomass reaction. Also, growth and non-growth associated ATP demands will need to be determined from experimental anaerobic chemostat cultivation.

Published metabolic reconstructions have previously described in detail the calculations involved in formulating a biomass reaction *(21, 31)*.

4. Troubleshooting and Special Cases

4.1. Troubleshooting an FBA Problem: In Silico Organism Does not "Grow"

Suppose that a FBA problem is set up as described in **Subheading 3** but the linear optimization yields $v_{\mathrm{Biomass}}{}^{*}= 0$ or an all-zero v^{*} vector. Suppose further that these results were obtained under environmental conditions where experimental literature indicates the organism is able to grow. This type of error, which is quite common during the model-building process, can be troubleshooted in several ways as described below:

1. Check the exchange reactions that allow metabolites to be exchanged into or out of the metabolic network. A lack of some essential exchange reactions or the existence of exchange reactions with incorrect directionality (i.e., exchange reactions that output an essential metabolite rather than inputting it from the surrounding medium) can result in the organism not growing *in silico*. To evaluate these scenarios, ensure that every extracellular metabolite (as per the surrounding medium specifications) has an exchange reaction. All of the extracellular metabolites can also be allowed to exchange freely in a reversible manner by relaxing the upper and lower bounds for their corresponding exchange reactions to allow a nonzero flux in either direction (e.g., by setting $ub = 100$ and $lb = -100$ for each exchange reaction).
   a. If $v_{\mathrm{Biomass}}{}^{*}$ is still equal to *0*, the error might be internal to the network (*see* **Subheading 4.2**).
   b. If a nonzero flux for $v_{\mathrm{Biomass}}{}^{*}$ is obtained, the source of the error can be narrowed down to inaccuracies with exchange fluxes. The lower and upper bounds on all the exchange fluxes should be reset to their original values one-by-one, and the value of $v_{\mathrm{Biomass}}{}^{*}$ should be reassessed as each bound

is reset. The constraints for exchange fluxes are determined by the surrounding medium specifications (e.g., if FBA is performed on *E. coli* metabolic network growing in a glucose minimal medium, the exchange reaction for glucose will be constrained for input; however, the exchange reactions for all other carbon sources will be fixed to allow output only). If adjusting the lower or upper bounds on a particular exchange reaction reduces $v_{Biomass}{}^{*}$ to *0*, an error might be present in the intracellular metabolic pathway associated with that exchange reaction.

c. If the model grows with all exchanges unconstrained but not with the specific exchange reaction bounds that are relevant to the *in silico* medium of interest, there may be gaps present in the associated pathways that are preventing certain essential metabolites from being produced for biomass (*see* **Subheading 4.2**). In FBA of genome-scale models of metabolism, sources of hydrogen, oxygen, nitrogen, sulfur, and sometimes other elements (e.g., phosphorus) must be allowed to exchange into the system in addition to the carbon source(s) the system grows on. Further, certain organisms cannot synthesize all 20 amino acids that are included in the biomass reaction. Exchanges should be provided for those amino acids that the organism uptakes from the surrounding medium.

2. If optimizing for the biomass reaction fails under experimentally observed growth conditions, performing FBA optimizations for each component of biomass individually helps to identify a particular component that cannot be produced by the metabolic network. First, a demand reaction should be created for each biomass component. A demand reaction is defined purely for modeling purposes to produce a drain on the essential metabolite of interest. After creating demand reactions for every biomass component, each biomass-component-demand reaction should be optimized in turn. Demand reactions yielding no flux indicate that the associated biomass metabolite cannot be produced. Subsequently, the pathways where the particular component is participating need to be investigated for the presence of gaps, incorrect reaction thermodynamics, etc. For cofactors in the biomass reaction, *see* **Subheading 4.3**.

3. Shadow prices, solutions to the dual problem in the LP optimization, provide additional insight for troubleshooting the application of FBA. Most LP solvers allow a user to view the dual variables, of which there exists one per constraint in the optimization. These shadow prices or dual variables represent the quantity by which the optimal value of the

objective function would improve (i.e., increase in the case of maximization) if the corresponding constraint were relaxed by one unit. Shadow prices (represented by the variable "*u*") exist for every metabolite in the network since each metabolite represents a separate constraint in the expression $S \cdot v = 0$. The concept of the dual variable is illustrated in **Fig. 3**. The objective function in this figure is represented by the grayscale gradient from v_1 to v_2, which indicates how altering v_1 and v_2 will affect the optimal value of the objective flux, v_{obj}. In this example, increasing the upper bound of v_2 (ub_2) by 1 increases v_{obj}* by 1, whereas increasing the upper bound of v_1 (ub_1) by 1 increases v_{obj}* by 2. Therefore, the dual variable associated with the ub_2 constraint is $u_{ub2} = 1$, and the dual variable associated with the ub_1 constraint $u_{ub1} = 2$. Hence, the dual variables give insight into which reactions are most limiting on the objective flux, as high values of dual variables indicate that the associated upper bound, lower bound, or metabolite in the *S* matrix is strongly limiting to the objective. For instance, if the shadow price associated with a certain metabolite in the

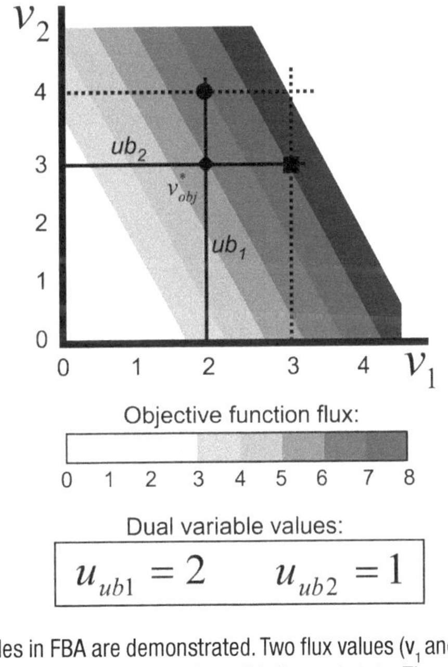

Fig. 3. Dual variables in FBA are demonstrated. Two flux values (v_1 and v_2) are plotted on the *x* and *y* axes along with their upper bound (*ub*) constraints. The shaded gradient on the plot represents values of the objective function, v_{obj}. The optimal solution v_{obj}* occurs at the intersection of ub_1 and ub_2, and is denoted with a diamond (◆). A circle (●) represents v_{obj}* after relaxing the ub_2 constraint by 1 flux unit, and a square (■) represents v_{obj}* after relaxing the ub_1 constraint by 1 flux unit. Therefore, the dual variables for ub_1 and ub_2 are 2 and 1, respectively, as shown at the bottom of the figure.

S matrix has a higher value than any other shadow price in the system, this could indicate that there is a gap in consumption or production of that metabolite. In this way, dual variables of FBA can aid in building and troubleshooting problems in the metabolic network during reconstruction.

4.2. Gap Analysis

If certain pathways in the intracellular metabolic network are incomplete and as a result the model cannot grow, gap analysis can be used to identify portions of the network that need to be further reconstructed. A gap occurs when a metabolite in the model is either consumed only or produced only. Different types of gaps exist in a metabolic network as illustrated by dotted reactions in **Fig. 4**. Here, four different types of gaps are shown, namely: (1) a direct gap in the pathway, (2, 3) gaps in cofactor

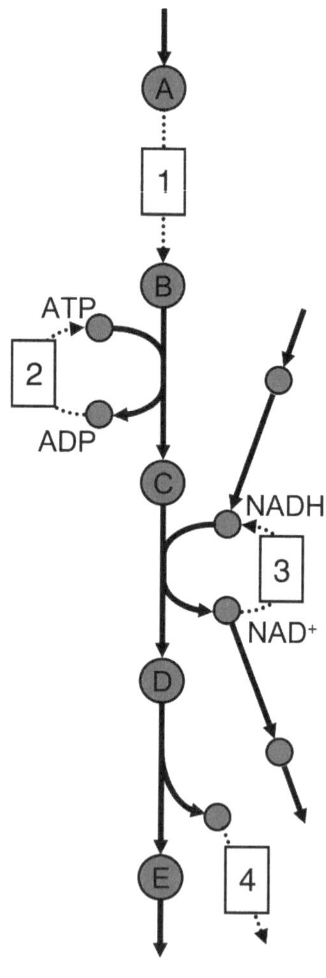

Fig. 4. Different types of network gaps are illustrated. Four network gaps are shown in a metabolic pathway: (1) a direct gap in the pathway, (2) a cofactor cycling gap, (3) a cofactor cycling gap where cofactors are both consumed and produced, and (4) a gap in the processing of a secondary byproduct of the pathway.

cycles necessary to the pathway, and (4) a gap in a peripheral pathway crucial for processing a byproduct of the main pathway. Gaps (2) and (3) differ in that while (2) is a true gap wherein ATP is consumed only and ADP is produced only, (3) is not a true gap, since both NADH and NAD+ can be produced and consumed within the network by other reactions, as illustrated in **Fig. 4**. Gap (3) represents a gap in the cofactor cycle of NADH but not in the synthesis pathway of NADH. Such gaps can be difficult to identify but are sometimes responsible for no-growth results in FBA.

4.3. Dealing with Cofactors

1. Cofactors differ from other metabolites in that they are not generally produced from basic building blocks in the formation of biomass or some other cellular product. A sample cofactor cycle is shown in **Fig. 5**, wherein ATP is hydrolyzed to ADP by one of the many metabolic reactions that use ATP, and then ADP is energy-charged back to the ATP form by the addition of a phosphate (Pi). Note that ATP is neither consumed nor produced during one full cycle of this process. Therefore, it is not necessary to synthesize from basic building blocks cofactors that are cycled in FBA (unless cofactor synthesis is accounted for in the reconstruction as well, in which case some portion of cofactors will be synthesized and some will be cycled, as explained in **Subheading 4.3.3**).
2. A common problem in the application of FBA to metabolic networks is the production of "free" energy through cofactor cycles, which are allowed to run backwards and violate thermodynamics (*see* **Fig. 5** illustrating the ATP cofactor cycle). To check for free energy loops, all exchange fluxes in the system need to be closed. Subsequently, FBA needs

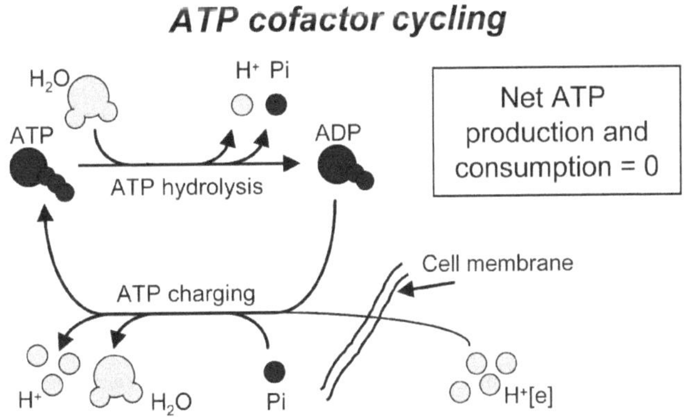

Fig. 5. The process of cofactor cycling is presented. ATP is cycled between hydrolyzing reactions and ATP charging reactions, with a net ATP production and consumption of zero. This illustrates how cofactors can be cycled but not be consumed or produced during FBA.

to be performed wherein the ATP cycling demand reaction ($ATP+H_2O \rightarrow ADP+Pi+H^+$) is maximized. If $v_{ATP\ cycling}^*$ is nonzero when metabolites are not allowed to enter or leave the system, some combination of reactions with nonzero flux must be participating in a free energy loop and erroneously fueling the production of ATP (*see* also **ref.** *32)*.

3. Of special note for the biomass reaction: cofactors can be both consumed and cycled in the biomass reaction. For instance, some ATP will be consumed for an ATP synthesis demand, while a portion of ATP will be cycled in an ATP energy demand in the same biomass reaction *(2, 33)*.

5. Conclusion

In this chapter, we have provided an overview of the theory and details for the application of FBA. We presented an overview of some recent extensions to this method. We also highlighted resources that are available to reconstruct a metabolic network and perform analysis using FBA. In addition to detailing the steps involved in setting up an FBA problem, we described the methods involved in formulating GPR relationships and defining an objective function (biomass reaction). Finally, we detailed techniques in troubleshooting problems that commonly occur in FBA analysis of metabolic networks. FBA is increasingly applied to the analysis of biochemical systems, driving experimental design, and providing insight into fundamental biology.

Acknowledgements

The authors wish to thank the National Science Foundation CAREER program (grant# 0643548 to JP) for financial support. MAO and AKC contributed equally to this work.

References

1. Sun, J., Lu, X., Rinas, U., and Zeng, A. P. (2007) Metabolic peculiarities of *Aspergillus niger* disclosed by comparative metabolic genomics. *Genome Biol.* 8, R182.
2. Heinemann, M., Kummel, A., Ruinatscha, R. and Panke, S. (2005) In silico genome-scale reconstruction and validation of the *Staphylococcus aureus* metabolic network. *Biotechnol. Bioeng.* 92, 850–864.
3. Duarte, N.C., Herrgard, M.J., and Palsson, B.O. (2004) Reconstruction and validation of *Saccharomyces cerevisiae* iND750, a fully compartmentalized genome-scale metabolic model. *Genome Res.* 14, 1298–1309.

4. Duarte, N. C., Becker, S. A., Jamshidi, N., Thiele, I., Mo, M. L., Vo, T. D., Srivas, R., and Palsson, B. O. (2007) Global reconstruction of the human metabolic network based on genomic and bibliomic data. *Proc. Natl. Acad. Sci. USA* 104, 1777–1782.
5. Feist, A. M., Henry, C. S., Reed, J. L., Krummenacker, M., Joyce, A. R., Karp, P. D., Broadbelt, L. J., Hatzimanikatis, V., and Palsson, B. O. (2007) A genome-scale metabolic reconstruction for *Escherichia coli* K-12 MG1655 that accounts for 1260 ORFs and thermodynamic information. *Mol. Syst. Biol.* 3, 121.
6. Stelling, J., Klamt, S., Bettenbrock, K., Schuster, S., and Gilles, E. D. (2002) Metabolic network structure determines key aspects of functionality and regulation. *Nature* 420, 190–193.
7. Kauffman, K. J., Prakash, P., and Edwards, J. S. (2003) Advances in flux balance analysis. *Curr. Opin. Biotechnol.* 14, 491–496.
8. Lee, J. M., Gianchandani, E. P., and Papin, J. A. (2006) Flux balance analysis in the era of metabolomics. *Brief. Bioinform.* 7, 140–150.
9. Covert, M. W., Famili, I., and Palsson, B. O. (2003) Identifying constraints that govern cell behavior: a key to converting conceptual to computational models in biology? *Biotechnol. Bioeng.* 84, 763–772.
10. Mahadevan, R., Edwards, J. S., and Doyle, F. J. 3rd (2002) Dynamic flux balance analysis of diauxic growth in *Escherichia coli*. *Biophys. J.* 83, 1331–1340.
11. Varma, A. and Palsson, B. O. (1994) Stoichiometric flux balance models quantitatively predict growth and metabolic by-product secretion in wild-type *Escherichia coli* W3110. *Appl. Environ. Microbiol.* 60, 3724–3731.
12. Beard, D. A., Liang, S. D., and Qian, H. (2002) Energy balance for analysis of complex metabolic networks. *Biophys. J.* 83, 79–86.
13. Covert, M. W., Schilling, C. H., and Palsson, B. (2001) Regulation of gene expression in flux balance models of metabolism. *J. Theor. Biol.* 213, 73–88.
14. Herrgard, M. J., Lee, B. S., Portnoy, V., and Palsson, B. O. (2006) Integrated analysis of regulatory and metabolic networks reveals novel regulatory mechanisms in *Saccharomyces cerevisiae*. *Genome Res.* 16, 627–635.
15. Gianchandani, E. P., Papin, J. A., Price, N. D., Joyce, A. R., and Palsson, B. O. (2006) Matrix formalism to describe functional states of transcriptional regulatory systems. *PLoS Comput. Biol.* 2, e101.
16. Shlomi, T., Eisenberg, Y., Sharan, R., and Ruppin, E. (2007) A genome-scale computational study of the interplay between transcriptional regulation and metabolism. *Mol. Syst. Biol.* 3, 101.
17. Lee, J. M., Gianchandani, E. P., Eddy, J. A., and Papin, J.A. (2008) Dynamic analysis of integrated signaling, metabolic, and regulatory networks. *PLoS Comput. Biol.* 4, e1000086.
18. Segre, D., Vitkup, D., and Church, G. M. (2002) Analysis of optimality in natural and perturbed metabolic networks. *Proc. Natl. Acad. Sci. USA* 99, 15112–15117.
19. Jamshidi, N. and Palsson, B. O. (2007) Investigating the metabolic capabilities of *Mycobacterium tuberculosis* H37Rv using the in silico strain iNJ661 and proposing alternative drug targets. *BMC Syst. Biol.* 1, 26.
20. Thiele, I., Vo, T. D., Price, N. D., and Palsson, B. O. (2005) Expanded metabolic reconstruction of *Helicobacter pylori* (iIT341 GSM/GPR): an in silico genome-scale characterization of single- and double-deletion mutants. *J. Bacteriol.* 187, 5818–5830.
21. Oh, Y. K., Palsson, B. O., Park, S. M., Schilling, C. H., and Mahadevan, R. (2007) Genome-scale reconstruction of metabolic network in *Bacillus subtilis* based on high-throughput phenotyping and gene essentiality data. *J. Biol. Chem.* 282, 28791–28799.
22. Kim, T. Y., Kim, H. U., Park, J. M., Song, H., Kim, J. S., and Lee, S. Y. (2007) Genome-scale analysis of *Mannheimia succiniciproducens* metabolism. *Biotechnol. Bioeng.* 97, 657–671.
23. Francke, C., Siezen, R. J., and Teusink, B. (2005) Reconstructing the metabolic network of a bacterium from its genome. *Trends Microbiol.* 13, 550–558.
24. Reed, J. L., Famili, I., Thiele, I., and Palsson, B. O. (2006) Towards multidimensional genome annotation. *Nat. Rev. Genet.* 7, 130–141.
25. Becker, S. A., Feist, A. M., Mo, M. L., Hannum, G., Palsson, B. O., and Herrgard, M. J. (2007) Quantitative prediction of cellular metabolism with constraint-based models: the COBRA Toolbox. *Nat. Protoc.* 2, 727–738.
26. Klamt, S., Stelling, J., Ginkel, M., and Gilles, E. D. (2003) FluxAnalyzer: exploring structure, pathways, and flux distributions in metabolic networks on interactive flux maps. *Bioinformatics* 19, 261–269.
27. Zhu, T., Phalakornkule, C., Ghosh, S., Grossmann, I. E., Koepsel, R. R., Ataai, M. M., and Domach, M. M. (2003) A metabolic network analysis & NMR experiment design tool with user interface-driven model construction for depth-first search analysis. *Metab. Eng.* 5, 74–85.
28. Urbanczik, R. (2006) SNA–a toolbox for the stoichiometric analysis of metabolic networks. *BMC Bioinformatics* 7, 129.

29. Lee, D. Y., Yun, H., Park, S., and Lee, S. Y. (2003) MetaFluxNet: the management of metabolic reaction information and quantitative metabolic flux analysis. *Bioinformatics* 19, 2144–2146.

30. Mahadevan, R. and Schilling, C. H. (2003) The effects of alternate optimal solutions in constraint-based genome-scale metabolic models. *Metab. Eng.* 5, 264–276.

31. Forster, J., Famili, I., Fu, P., Palsson, B. O., and Nielsen, J. (2003) Genome-scale reconstruction of the *Saccharomyces cerevisiae* metabolic network. *Genome Res.* 13, 244–253.

32. Price, N. D., Famili, I., Beard, D. A., and Palsson, B. O. (2002) Extreme pathways and Kirchhoff's second law. *Biophys. J.* 83, 2879–2882.

33. Varma, A. and Palsson, B. O. (1993) Metabolic capabilities of *Escherichia coli*. 2. Optimal-growth patterns. *J. Theor. Biol.* 165, 503–522.

Chapter 4

Modeling Molecular Regulatory Networks with JigCell and PET

Clifford A. Shaffer, Jason W. Zwolak, Ranjit Randhawa, and John J. Tyson

Summary

We demonstrate how to model macromolecular regulatory networks with JigCell and the Parameter Estimation Toolkit (PET). These software tools are designed specifically to support the process typically used by systems biologists to model complex regulatory circuits. A detailed example illustrates how a model of the cell cycle in frog eggs is created and then refined through comparison of simulation output with experimental data. We show how parameter estimation tools automatically generate rate constants that fit a model to experimental data.

Key words: Systems biology, Parameter estimation, Model validation.

1. Introduction

Mathematical models of gene-protein regulatory networks play key roles in archiving and advancing our understanding of the molecular basis of cell physiology. Models provide rigorous connections between the physiological properties of a cell and the molecular wiring diagrams of its control systems. A simple example is the set of reactions controlling the activity of MPF (mitosis promoting factor) in *Xenopus* oocytes *(1)*, which we refer to herein as the frog egg model. In the diagram of this network (**Fig. 1**), vertices represent substrates and products (collectively referred to as species), solid directed edges represent biochemical reactions, and dashed directed edges represent regulatory signals.

Ivan V. Maly (ed.) *Methods in Molecular Biology, Systems Biology, vol. 500*

DOI: 10.1007/978-1-59745-525-1_4

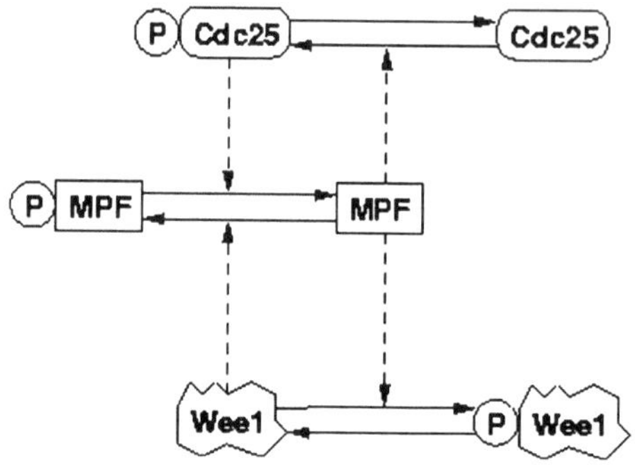

Species	Description	Phosphorylated
M_a	Active MPF	no
M_i	Inactive MPF	yes
C_a	Active Cdc25	yes
C_i	Inactive Cdc25	no
W_a	Active Wee1	no
W_i	Inactive Wee1	yes

$$\frac{dM_a}{dt} = (v_c' \cdot C_i + v_c'' \cdot C_a) \cdot M_i - (v_w' \cdot W_i + v_w'' \cdot W_a) \cdot M_a$$

$$\frac{dC_a}{dt} = \frac{V_c \cdot M_a \cdot C_i}{K_{mc} + C_i} - \frac{v_c''' \cdot v_c \cdot C_a}{K_{mcr} + C_a}$$

$$\frac{dW_a}{dt} = -\frac{v_w \cdot M_a \cdot W_a}{K_{mw} + W_a} + \frac{v_w''' \cdot v_w \cdot W_i}{K_{mwr} + W_i}$$

Fig. 1. Network diagram, mapping of species names, and the corresponding set of ordinary differential equations for a model of the mitotic regulatory system in frog eggs. The regulation of MPF (mitosis promoting factor) by Wee1 (kinase) and Cdc25 (phosphatase) controls when the cell enters mitosis. Notice the two positive feedback loops whereby MPF activates Cdc25 (MPF's activator) and inactivates Wee1 (MPF's inactivator). The active forms (M_a, C_a, and W_a) have associated differential equations. The total amounts of MPF (M_T), Wee1 (W_T), and Cdc25 (C_T) are conserved (i.e., remain constant throughout the process). $M_i + M_a = M_T$, $W_i + W_a = W_T$, and $C_i + C_i = C_T$. Therefore, the inactive forms (M_i, C_i, and W_i) do not have differential equations because they can be calculated from these conservation relationships.

Collectively, these biochemical reactions cause the concentrations of the chemical species (S_i) to change in time according to a set of differential equations (one for each species)

$$\frac{dS_i}{dt} = \sum_{j=1}^{R} b_{ij} v_j, i = 1, \ldots, N,$$

where R is the number of reactions, N is the number of species, v_j is the velocity of the jth reaction in the network, and b_{ij} is the stoichiometric coefficient of species i in reaction j ($b_{ij} < 0$ for substrates, $b_{ij} > 0$ for products, $b_{ij} = 0$ if species i takes no part in reaction j). **Fig. 1** shows differential equations derived from the reactions in the network diagram. The set of rate equations and associated parameter values is a mathematical representation of the temporal behavior of the regulatory network.

Since the purpose of these models is to codify a systems-level understanding of the control of some aspect of cell physiology, it is necessary to validate a proposed model against observed behavior of the reference system. In most cases, it is essential to model the behavior not only of the wild-type form of the organism, but also of many mutant forms (where each mutant form typically represents one or two variations in the genetic specification of the control system). For example, if we are modeling the cell cycle of an organism, then we would wish to know features such as the cell size at division, the time required for various phases of the cell cycle (G1, S, G2, M), and the viability or point of failure for each mutation. Measurements of the amounts for various control species within the cell over time would also be valuable information. In the case of a thoroughly studied organism such as *Saccharomyces cerevisiae* (budding yeast), a model can be compared against many dozens of mutants defective in the regulatory network.

A realistic model of the budding yeast cell cycle consists of over 30 differential equations and 100 rate constants and is tested against the phenotypes of over 150 mutants *(2)*. A model of this complexity represents the upper limit of what a dedicated modeler can produce "by hand" with nothing but a good numerical integrator like LSODE *(3)*. Beyond this size, we begin to lose our ability even to meaningfully display the wiring diagram that represents the model, let alone comprehend the information it contains, or determine suitable rate constants in the corresponding high-dimensional space. To adequately describe fundamental physiological processes (such as the control of cell division) in mammalian cells will require models of 100-1000 equations. To handle this next generation of dynamical models will require sophisticated software to automate the modeling process: network specification, equation generation, simulation and data management, and parameter estimation.

There are a number of distinct approaches to simulation. Deterministic models usually represent the system of chemical reactions with ordinary differential equations *(4-6)*. In some cases, partial differential equations are used to account for spatial effects *(7)*. Stochastic modeling is in its infancy, and most often is done by some variation of Gillespie's algorithm *(8-10)*. For the remainder of our discussion, we will consider only deterministic simulation by ordinary differential equations (ODEs).

Creating a model that mimics the observed behavior of a living organism is a difficult task. This process involves a combination of biological insight, persistence, and support by good modeling tools. In the following sections, we will describe the model development process that we employ and the software tools that we have developed to construct and test models. We then provide a detailed example of how the tools can be used to create a simple model of the frog egg cell cycle and to estimate the associated rate constants.

2. The Modeling Process

Successful modeling of macromolecular regulatory networks can be aided by software tools based on a well-defined modeling process. Such tools should support the line of thought followed by modelers as they approach a problem. Mid-sized models of macromolecular regulatory networks track reactions among tens of species and are tested against hundreds of experimental observations. Thus, modelers need tools that help to organize the relevant information and automate as many steps of the process as possible. **Figure 2** shows our conception of the modeling process. The modeler starts with an idea about an organism and a regulatory system to model. Next, the modeler gathers information (from the literature and from their own experiments) related to the regulatory system of the organism. During the literature search, the modeler builds a hypothesis from information already published, continuously checking the hypothesis against the existing literature. Once the modeler has a testable hypothesis about the regulatory system, the hypothesis can be codified into four types of technical information:

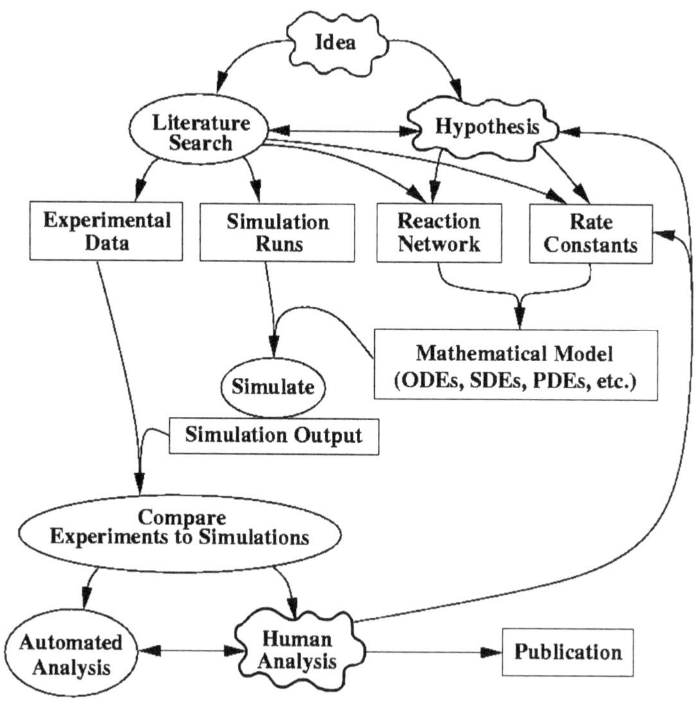

Fig. 2. The modeling process. Once the modeler has generated a testable hypothesis about the organism, he or she must assemble the four necessary collections of information (experimental data, simulation runs, reaction network, and rate constants). This defines both the mathematical model and the behavior that the model must reproduce. The modeler then will repeatedly simulate and update the model, perhaps with the aid of automated analysis tools, until an acceptable result is obtained.

- Experimental data: The information that will be used to validate the model. This information might come as time series data of the concentrations of certain regulatory chemical species, as other observables such as the average size of cells at division, or as qualitative properties such as the viability or inviability of a mutant.
- Simulation runs: Specifications for the simulations that will be made to model the experimental data. For example, each simulation might relate to a specific mutation of the organism. The specification will define the distinct conditions necessary to simulate that mutation, such as differences in rate constants from the wild-type values.
- Reaction network: The chemical equations that describe the regulatory processes.
- Rate constants: The parameters that govern the reaction rates.

Typically, the experimental data and simulation run descriptions are part of the problem definition and are not subject to frequent modifications. Nor are they considered to be "right" or "wrong" in the same way as the reaction network and rate constant values typically will be. The network and rate constants together define the mathematical model that will be simulated, compared with experimental observations, and judged "acceptable" or "unacceptable."

One simulation run of an ODE model takes only a fraction of a second on a typical desktop computer in 2007. As described above, a complete model actually involves a large collection of simulations, to be compared against a collection of experimental results. This entire set of runs might take a second or so for a smaller model such as our frog egg example on a desktop computer for one choice of rate constants, and about a minute or two for a larger model needed to describe the budding yeast cell cycle.

Once an initial specification of these four types of information has been made, the next phase of the process begins. This is a simulation-compare-update loop, whereby simulation results are compared with the experimental data. In some way, either a human or a computer will make a judgment as to the quality of the relationship between the two. At that point, since the model is typically judged unsatisfactory, the modeler will make adjustments and repeat the cycle. We prefer to view this as a double loop, in that changes to rate constant values are made much more frequently than changes to the reaction network. That is, the modeler will typically "twiddle" the rate constants so long as progress is being made in matching simulation output to experimental data. When changes to the rate constants appear no longer to improve the match, then the modeler will attempt to improve the model by changing the reaction network, which in turn will trigger another round of changes to the rate constants. The process is continued until the model is judged satisfactory or totally hopeless.

Modelers often try to assign values to rate constants by a time-consuming process of "parameter twiddling" and visual comparison of simulation results to experimental data. A better approach is automated parameter estimation (once the modeler is confident that the basic structure of the reaction network is sound enough). To fit a model to experimental data by automated optimization algorithms requires thousands to millions of repetitions of the full calculations.

The process of comparing real-world observation (experimental data) with the mathematical model (time-series output from a simulation) is called model validation. Model validation is closely related to automated (or manual) parameter estimation, because both require that some measure of the quality of the model can be made. In the case of automated parameter estimation, we need a way to take the experimental data and the output from a simulation run, and create a single number as a measure of the quality of the fit.

This can be extremely difficult. First, the simulation data (usually in the form of time series plots) might not be similar to the form of the experimental data (often qualitative information such as whether a cell is viable or not). In general, some complex computation must be done to relate the two. The function that does this computation is called a *transform* and is discussed further in **Subheading 3.3.1**. Second, although it might be a simple judgment to measure the goodness of fit between one simulation and one experiment, it is often difficult to judge the goodness of fit of an entire ensemble of runs, where improvements in matching some experiments might come at the cost of worse fits for others. The function that balances these fits is called the objective function and is discussed in **Subheading 3.3.2**.

3. Software Tools

Before the current generation of modeling tools for systems biology was developed, many stages in the modeling cycle described in **Subheading 2** were done by hand. This presents two problems. First, it takes a great deal of time and effort to convert the original intuitive concept of a model into a suitable set of reaction equations and simulations. Second, there are many opportunities for errors, especially at the (essentially mechanical) step of converting a reaction mechanism into differential equations.

A wiring diagram, like **Fig. 1**, nicely represents the topology of a reaction network (reactants, products, enzymes). But it is not a good representation for specifying the kinetics of the network (the reaction rate laws, v_j). A large reaction network can become so complex that even its topological features are obscured by a large number of intersecting lines. Obscurity is increased by the fact that there is no standard format for drawing such graphs. Without precise notational conventions, it is impossible to convert a wiring diagram unambiguously into a model, either by hand or by machine.

Another approach for deriving a model is to explicitly write out the chemical reactions. This loses some of the intuitive appeal of the diagrammatic approach, but allows for a more compact definition of a reaction network. Normally, the modeler has already made a hand or CAD-drawn version of the network in graphical form, showing the interactions in a qualitative sense but without the quantitative information of the rate equations or the rate constant values.

Models often include concepts not captured by the differential equations alone. Conservation relations are defined by linear combinations of species concentrations that remain constant throughout a simulation: $T \equiv \sum_{i=1}^{N} a_i S_i(t)$, where a_i is a constant and T is constant. Such constraints arise from linear dependencies in the stoichiometry matrix: $\sum_{i=1}^{N} a_i b_{ij} = 0$. Events are special actions that trigger in the model under given conditions. For example, cellular division could be represented by a halving of cell mass, and might occur when a given function involving some number of chemical species reaches a threshold during a simulation.

The key to successfully creating and managing such complex models is to use software tools that organize the information in a coherent way and catch inconsistencies and errors early in the process. In this section, we will describe the JigCell Model Builder *(11,12)*, which is used to define the reaction equations and rate constants of the model. We then present the JigCell Run Manager *(13)*, which is used to define a series of simulation runs that will generate output to validate the model. Finally, we describe the parameter estimation tool (PET) *(14)*, which supports exploration of the parameter space and automated parameter estimation with the goal of selecting rate constant values that best fit the simulation output to the experimental data.

Underlying any such software tool is a representation scheme for describing a model, that is, a language for expressing the model in a complete and formal sense. The systems biology markup language (SBML) *(15,16)* has now become the standard reference language for reaction network modeling. SBML describes all necessary features pertaining to the reaction network, conservation relations, events, and rate constants. SBML does not describe all data necessary for modeling, including information describing the simulation runs and experimental data from **Fig. 2**, which must be stored in separate files. SBML also is not a suitable language for human comprehension. Thus, software tools are needed to provide an interface between the user and SBML.

3.1. The JigCell Model Builder

The JigCell Model Builder (referred to herein as the "Model Builder") is used to define the components that make up an SBML model. The Model Builder uses a spreadsheet interface, allowing a large amount of data to be displayed in an organized manner.

The Model Builder provides functionalities for both first-time users and expert modelers. The Model Builder supports the definition of events and user-defined units. An event, such as cell division, can be defined by specifying a condition that must be met to trigger the event, and the changes that result due to the event. A major goal of the Model Builder is to minimize the time and errors associated with translating a regulatory network to a set of equations. As the user enters reaction equations, rate laws, and functions into their cells in the main spreadsheet, several other spreadsheets are updated to track the various entities that make up a model. After the user has finished defining a model using the Model Builder, this model can be used with other SBML-compliant software to simulate the response of the model to given conditions.

The Model Builder's interface is broken into 10 spreadsheets, all accessible by clicking on the appropriate tab. **Figure 3** shows the "Reactions" spreadsheet. There is a spreadsheet for each of the eight SBML components in a model (reactions, functions, rules, compartments, species, parameters, units and events). There is one spreadsheet for conservation relations and one spreadsheet for the equations (including both ODEs and rule equations).

The "Reactions" spreadsheet is the primary tool used to create the reaction network of a model. The other spreadsheets are either partially or completely filled by the Model Builder from the "Reactions" spreadsheet. A reaction represents any chemical transformation, transport, or binding process that can change the amount of one or more species. Each row defines a single chemical reaction. **Figure 3** shows the "Reactions" spreadsheet loaded with the frog egg model. The three main columns in this spreadsheet are: "Reaction," "Type," and "Equation."

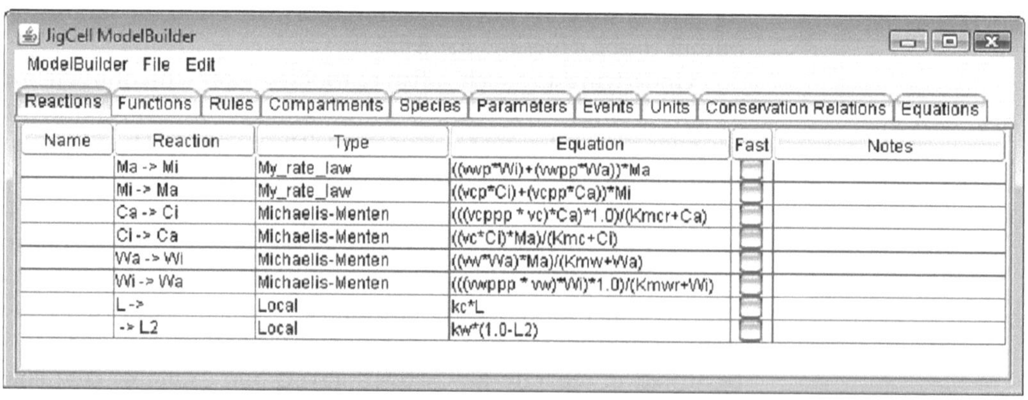

Fig. 3. The "Reactions" spreadsheet.

1. The "Reaction" column defines the species (reactants and products) and their stoichiometries. A list of substrates separated by "+" signs is entered first. An arrow (→) is then entered, and is followed by a list of products, also separated by '+' signs. Substrate and product names can contain any combination of letters, numbers, underscores, and apostrophes. There is no limit to the number of species that can be entered as substrates or products. The stoichiometry of a reaction is defined by placing a number and an '*' character in front of the species (e.g., $3 * M_a$).
2. By picking a rate law from a drop down list in the "Type" column, the user can specify the kinetics of the reaction being defined. The Model Builder provides three built-in rate laws (mass action, Michaelis Menten, local) and also allows users to define their own rate laws in the "Functions" spreadsheet. For all rate laws other than local, the Model Builder will enter the associated rate law in the "Equation" field. The local type allows the user to define the reaction rate of a single reaction without creating a new rate law. If the user selects local, the equation field will remain empty until the user defines the equation for the reaction rate. Local rate laws may contain algebraic expressions with parameters and species.
3. The "Equation" column specifies the equation for the rate of the reaction. If the reaction type is not Local, the "Equation" column displays the unsubstituted equation of the selected rate law until the user edits the rate law equation by clicking on the cells in this column. Clicking on one of these cells displays the "Parameters/Modifiers" Editor (**Fig. 4**), where the user assigns "interpretations" to the rate constants and modifiers. The "interpretations" can be numeric constants, expressions, species, or species-related expressions. The Model Builder partially fills the "Parameters/Modifiers" Editor when built-in rate laws are used (e.g., S1 becomes Ci in **Fig. 4** automatically because the user defined the reaction Ci → Ca). The Model

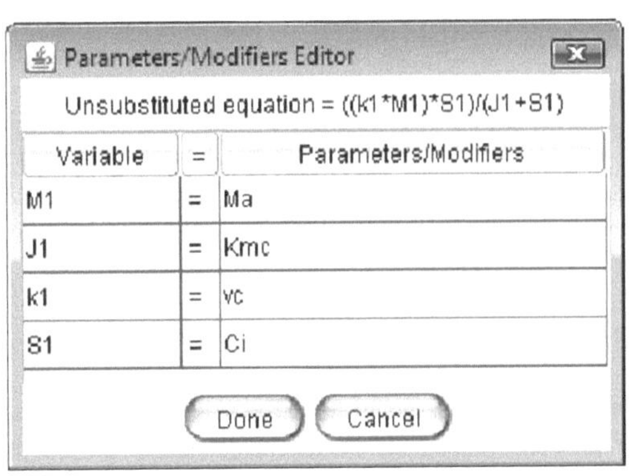

Fig. 4. The "Parameters/Modifiers" Editor.

Builder will substitute the user's interpretations (entered via the "Parameters/Modifiers" Editor) into the equation field of the "Reactions" spreadsheet so that the user can see the final rate law used to govern the reaction. Expressions are evaluated to numerical values when the model is simulated.

The "Functions" spreadsheet (**Fig. 5**) is used to create and edit function definitions. A function definition is a named mathematical function that may be used throughout the rest of a model. For example, user-defined rate laws are created as function definitions. Checking the box in the "Rate Law" column causes the newly created rate law to be included in the drop-down list of rate laws in the "Type" column of the "Reactions" spreadsheet. Functions are defined with place holders for arguments of the form A#, where # is some number. The function *My_rate_law* in (**Fig. 5**) contains five arguments A1–A5. These arguments can be assigned in the "Parameters/Modifiers" Editor (**Fig. 4**) when the function is selected as the rate law for a reaction. Otherwise, to use this function it may be called like this: *My_rate_law(vwp, Wi, vwpp, Wa, Ma)*. Any of the function arguments can be a parameter, species, or algebraic expression.

The "Rules" spreadsheet (**Fig. 6**) serves two purposes. First, it displays algebraic rules, which are the conservation relations in

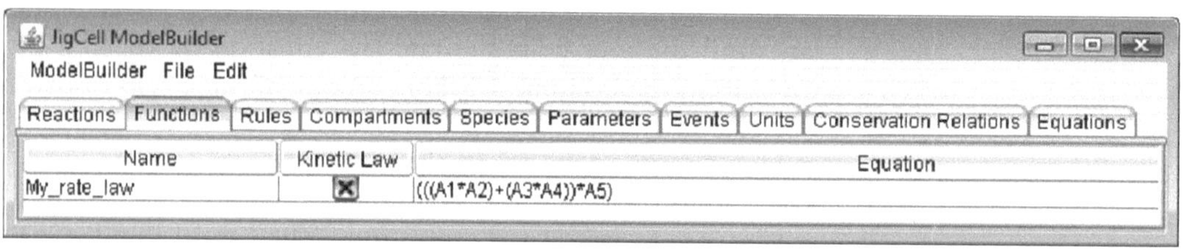

Fig. 5. The "Functions" spreadsheet.

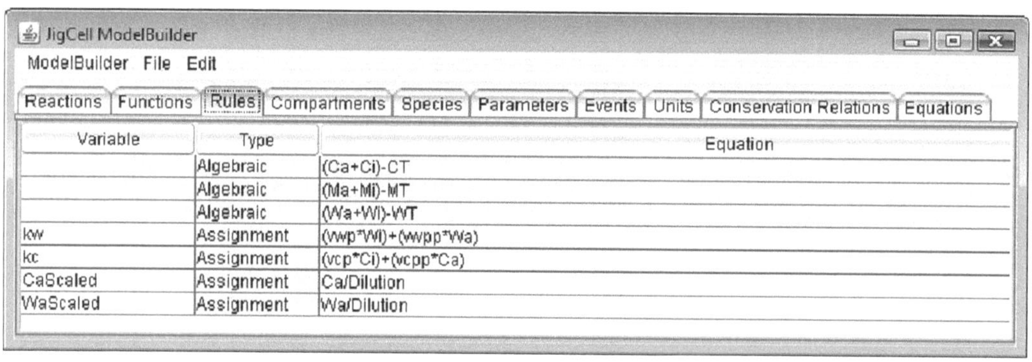

Fig. 6. The Model Builder "Rules" spreadsheet. The algebraic rules are automatically created by the Model Builder from the conservation relations. The lines for kw and kc define the rates for the reactions of L2 and L, respectively. CaScaled and WaScaled scale the concentrations of Ca and Wa to 1 after they have been diluted by Dilution. See Subheading 4.1 for more about dilution.

the model. The program deduces these relations from the stoichiometric matrix of the model and displays each conservation relation in the form $(a_1 S_1 + a_2 S_2 + \ldots) - T = 0$, where T is the conserved quantity and a_1, a_2, ... are constants calculated from the stoichiometry matrix. The user cannot edit an algebraic rule on this spreadsheet but may specify how the Model Builder uses the rule on the "Conservation Relation" spreadsheet. The second purpose of the "Rules" spreadsheet is to create and edit assignment rules. Assignment rules are used to express equations that set the value of variables. The "Variable" field in the assignment rule can be a species, parameter or compartment. In the case of species the "Equation" field sets the quantity to be determined (either concentration or substance amount), in the case of compartments the "Equation" field sets the compartment's size, and in the case of parameters the "Equation" field sets the parameter's value. The value calculated by the assignment rule's "Equation" field overrides the value assigned in the "Compartments," "Species," or "Parameters" spreadsheet.

The next three tabs are used to define compartments, species, and parameters. A compartment represents a bounded space in which species can be located. Spatial relationships between different compartments can be specified. Modelers are not required to enter compartment information when defining a model, as a single compartment called "cell" is created by default. The "Species" spreadsheet (**Fig. 7**) provides a list of all species that are part of a chemical reaction or defined in a Rule. The list of species is generated automatically by the Model Builder, though a user can add, delete, and modify species. There are several editable attributes associated with each species. The "Parameter" spreadsheet (**Fig. 8**) is used by the Model Builder to manage all parameters and their values associated with a model. A parameter is used to declare a value for use in mathematical formulae. The Model Builder recognizes as a parameter any name on the "Reactions" spreadsheet that is not defined as a species.

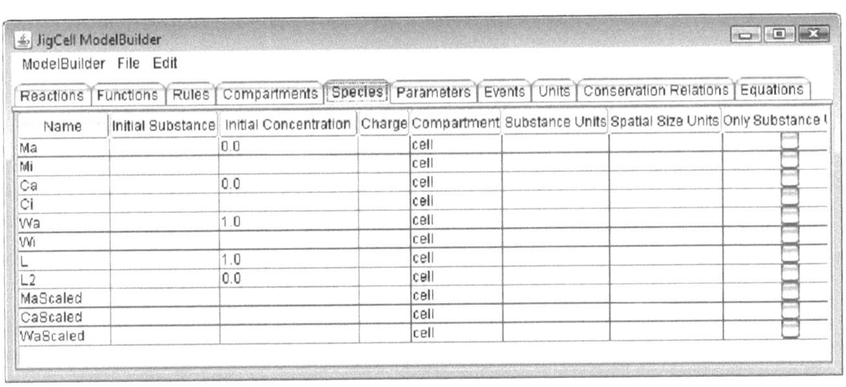

Name	Initial Substance	Initial Concentration	Charge	Compartment	Substance Units	Spatial Size Units	Only Substance U
Ma		0.0		cell			
Mi				cell			
Ca		0.0		cell			
Ci				cell			
Wa		1.0		cell			
Wi				cell			
L		1.0		cell			
L2		0.0		cell			
MaScaled				cell			
CaScaled				cell			
WaScaled				cell			

Fig. 7. The Model Builder "Species" spreadsheet.

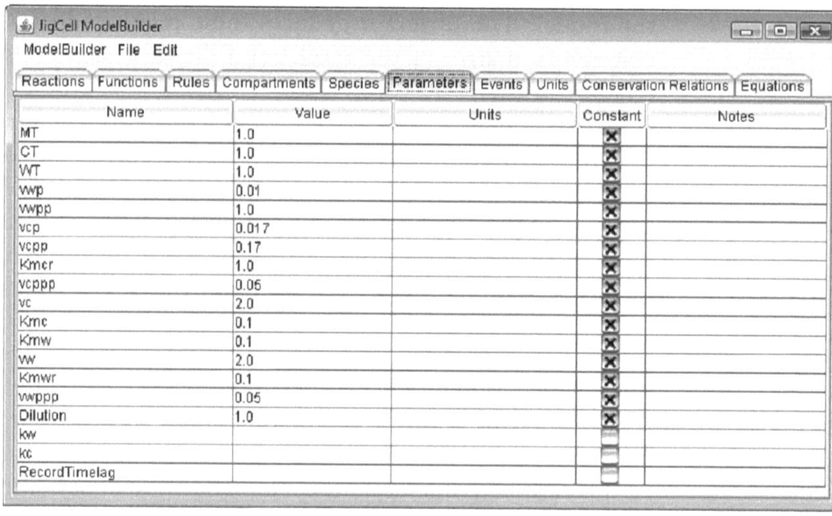

Fig. 8. The "Parameter" spreadsheet.

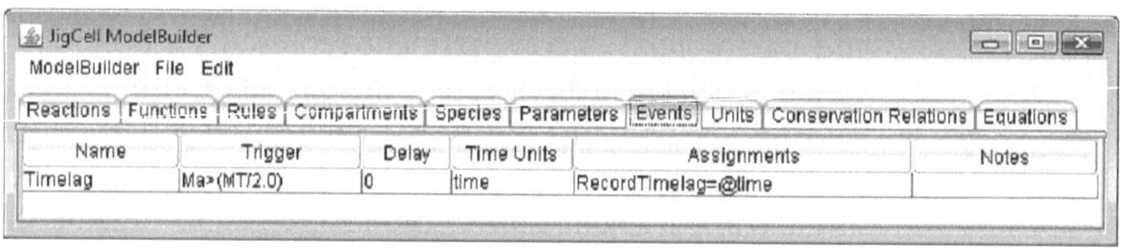

Fig. 9. The "Events" spreadsheet. The symbol "@time" represents time in the system of differential equations. This event sets "RecordTimelag" to the value of time when the "Trigger" becomes true and is used to get the time for active MPF (Ma) to reach half the total MPF concentration (MT). This is provided as an example of how events are defined, but it is not used in the later modeling example.

The "Events" spreadsheet (**Fig. 9**) allows the user to define actions associated with a model. For example, when modeling the cell cycle, some trigger for cell division must be defined and the consequences of that division must be specified. The "Name" column provides an (optional) identifier for an event. The "Trigger" column defines the conditions under which the event takes place. The format of this entry allows the user to specify an equality relationship. Whenever the relationship entered in the "Trigger" column is satisfied, the actions specified in the "Assignments" column will occur. The "Event Assignment Editor" lets the user define the changes that will occur when an event is executed.

The "Units" Spreadsheet lists all unit types used in the model, along with their definitions. A unit definition provides a name for a unit that can then be used when expressing quantities in a model. The Model Builder has a number of basic units and 5 built-in unit definitions (area, length, time, substance, and volume). Complex unit definitions such as *meter/second*^{2} can be created.

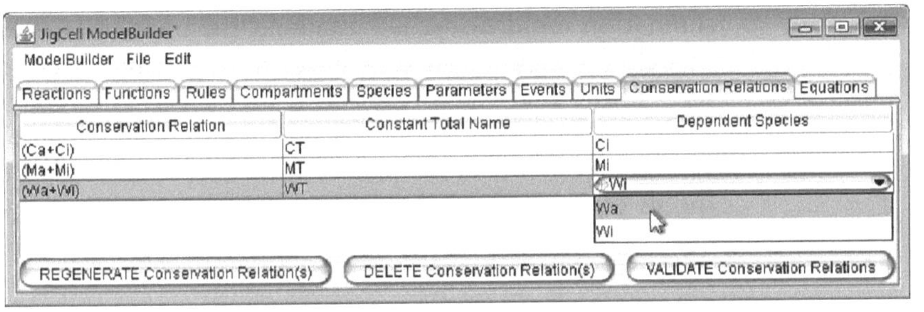

Fig. 10. The "Conservation Relations" spreadsheet.

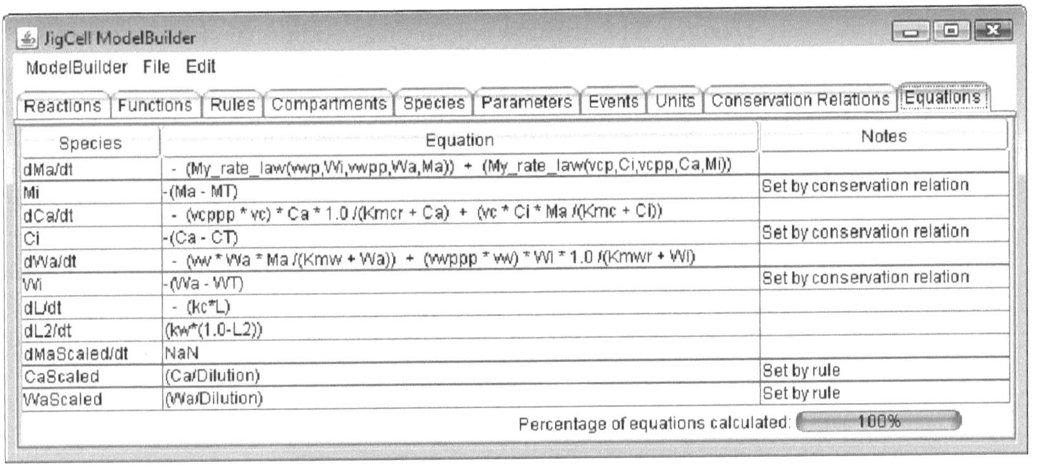

Fig. 11. The "Equations" spreadsheet.

The "Conservation Relations" spreadsheet (**Fig. 10**) is used to view a list of all conservation relationships that exist between species in the model. The list of conservation equations is generated automatically, using Reder's method *(17)*.

The "Equations" spreadsheet (**Fig. 11**) allows the modeler to see a list of the different types of equations that define the model. The user does not edit equations here, as they are created automatically from data entered on other spreadsheets. The "Equation" column displays differential equations, assignment rule equations, conservation relation equations, or the condition set on the species when no equation exists.

3.2. The JigCell Run Manager

The JigCell Run Manager (referred to herein as the "Run Manager") lets users define specifications for an ensemble of simulation runs. Hierarchies of simulations can be built up, whereby a given simulation inherits parameter changes from a "basal" run definition. This hierarchical organization of simulations is useful because models are often validated against a collection of experimental protocols, each one of which requires only slightly different simulation conditions. For example, the budding yeast cell cycle model must capture the differences among many dozens of mutations of the wild type organism. If the "basal" run represents the wild-type organism, then the hierarchy can define unambiguously and compactly the deviations from wild-type that are necessary to specify each mutant type.

Users input the description of ensembles using five spreadsheets: Runs, Basal Parameters, Basal Initial Conditions, Simulator Settings, and Plotter Settings. The "Runs" spreadsheet (**Fig. 12**) specifies how to simulate a certain experiment. The name column can (optionally) be used to identify the experiment being simulated. The parents column lists all runs from which the row inherits changes. The changes column lists additional changes to parameters, initial conditions, simulator settings, and plotter settings that are needed for this run. These changes are specified using the "Changes" editor (**Fig. 13**), which opens when clicking on the changes cell for a particular run. The changes for a particular run override the changes inherited from any parents, and these changes propagate to its children. Color is used to reflect where the changes are made: Blue is used to indicate changes made in the current run (locally) and green to indicate changes inherited from a parent run (or some previous ancestor). This information is also indicated in the "Parents" column of **Fig. 13**, which indicates either the name of the ancestor that caused that parameter's setting to change, or states "local" if the change was explicitly made by the user for this run. **Figure 12** shows a "Runs" spreadsheet for simulating some experiments done on frog egg extracts to characterize the activation of MPF.

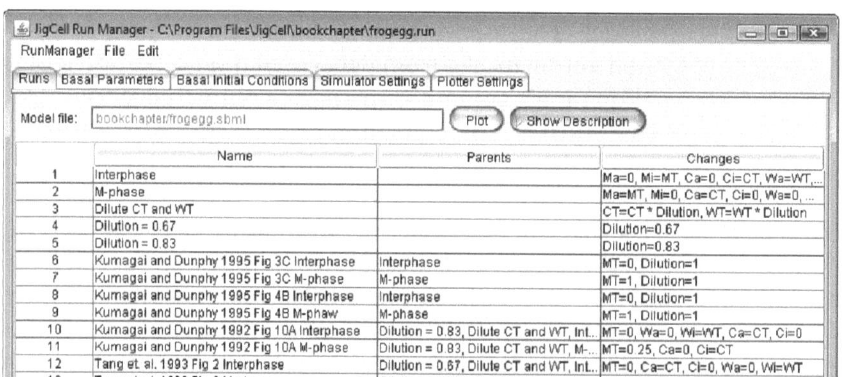

Fig. 12. The "Runs" spreadsheet.

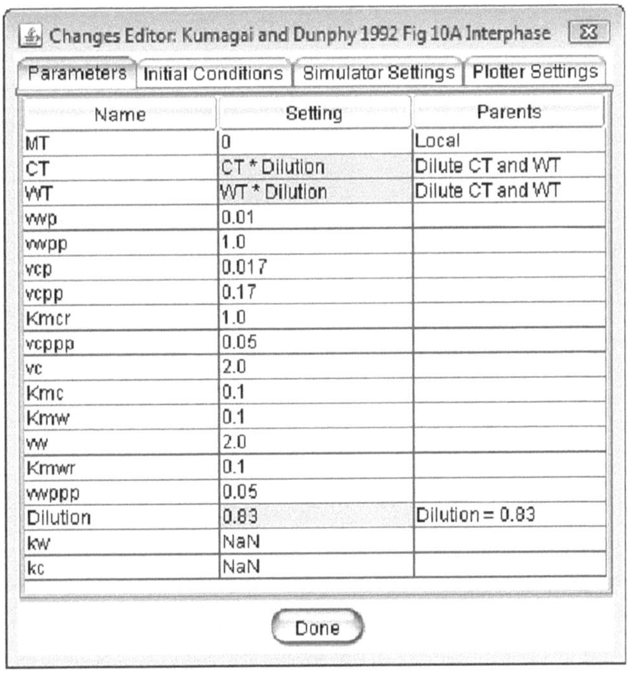

Changes Editor: Kumagai and Dunphy 1992 Fig 10A Interphase

Parameters | Initial Conditions | Simulator Settings | Plotter Settings

Name	Setting	Parents
MT	0	Local
CT	CT * Dilution	Dilute CT and WT
WT	WT * Dilution	Dilute CT and WT
vwp	0.01	
vwpp	1.0	
vcp	0.017	
vcpp	0.17	
Kmcr	1.0	
vcppp	0.05	
vc	2.0	
Kmc	0.1	
Kmw	0.1	
vw	2.0	
Kmwr	0.1	
vwppp	0.05	
Dilution	0.83	Dilution = 0.83
kw	NaN	
kc	NaN	

Done

Fig. 13. The "Changes" Editor for a particular run. In the "Setting" column of MT the cell would be colored blue to represent a local change. In the "Setting" column for CT, WT, and Dilution, the cell would be colored green to represent changes inherited from a parent run.

Each row corresponds to a separate experiment. The run named "Interphase" (on row 1) describes changes to the initial model to simulate an extract starting in interphase. This run is then set as a parent to the run named "Kumagai and Dunphy 1995 **Fig. 3C** Interphase" on row 6. The run on row 6 inherits all its parent's changes and represents an experiment to measure the phosphorylation of MPF by Wee1 during interphase. The "Changes" column displays changes made by the current run but not changes inherited from the parents.

The Run Manager provides a "Plot" button on the "Runs" spreadsheet that generates an immediate simulation for a specified row and then plots the results.

The "Simulator Settings" spreadsheet (**Fig. 14**) specifies the simulator to be used and appropriate values for the simulator's configuration parameters, such as total time of integration, tolerances, output interval, etc. In this case, the simulator chosen is XPP *(18)*. Other simulators are also provided, such as StochKit *(19)* (for stochastic simulation) and Oscill8 *(20)*.

The "Plotter Settings" spreadsheet (**Fig. 15**) enables the user to specify the variables to be plotted from a simulation run's output. The "Plotter Settings" spreadsheet also contains options to customize the plot by selecting colors, mark styles, whether to connect points, etc.

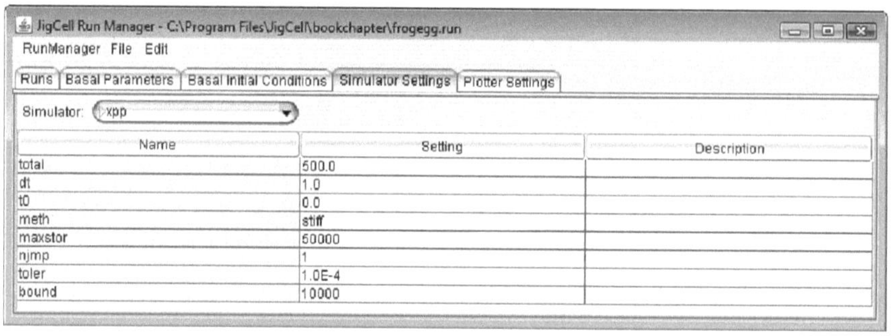

Fig. 14. The "Simulator Settings" spreadsheet.

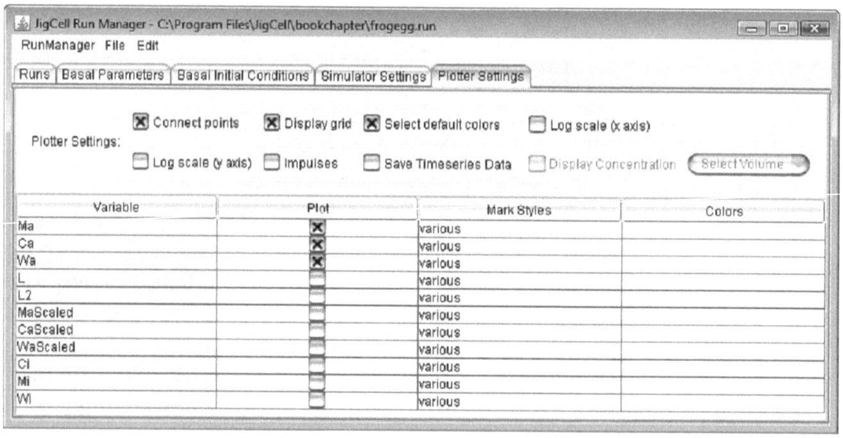

Fig. 15. The "Plotter Settings" spreadsheet.

3.3. (PET) Parameter Estimation Toolkit:

The Parameter Estimation Toolkit (PET) is designed to help users explore parameter space and fit simulation output (e.g., time course simulations) to experimental data. Typical use of PET follows the modeling process discussed in **Subheading 2**:

1. The user imports an SBML file created by the Model Builder or some other SBML editor.
2. A basal parameter set is created directly from the SBML file or imported from the Run Manager's basal file.
3. Simulation runs are defined in PET or imported from a run file created by the Run Manager.
4. At this point the user may simulate the model, even though experimental data have not yet been defined.
5. Experimental data are defined and transforms set up for the simulation runs.

6. Experimental data and model output are compared by the user (Human Analysis) or by the parameter estimator (Automated Analysis). Parameters are adjusted to seek a better fit of the model to the data.

PET supports cut and paste of experimental data into and from applications, such as Microsoft Excel, copying of plots into presentations or other documents, and generation of PDF files containing plots. PET supports undo and redo of most operations (including all delete operations), semantic checks of user input, and color coding (e.g., of parameters changed by the user in the "Edit Basals" spreadsheet).

The following subsections detail some general features of PET. Specific examples of these features are provided in **Subheading 4**.

3.3.1. Experimental Data and Transforms

Users enter experimental data and define what transforms to use on the model output in the "Edit Data" screen (**Fig. 16**). Transforms

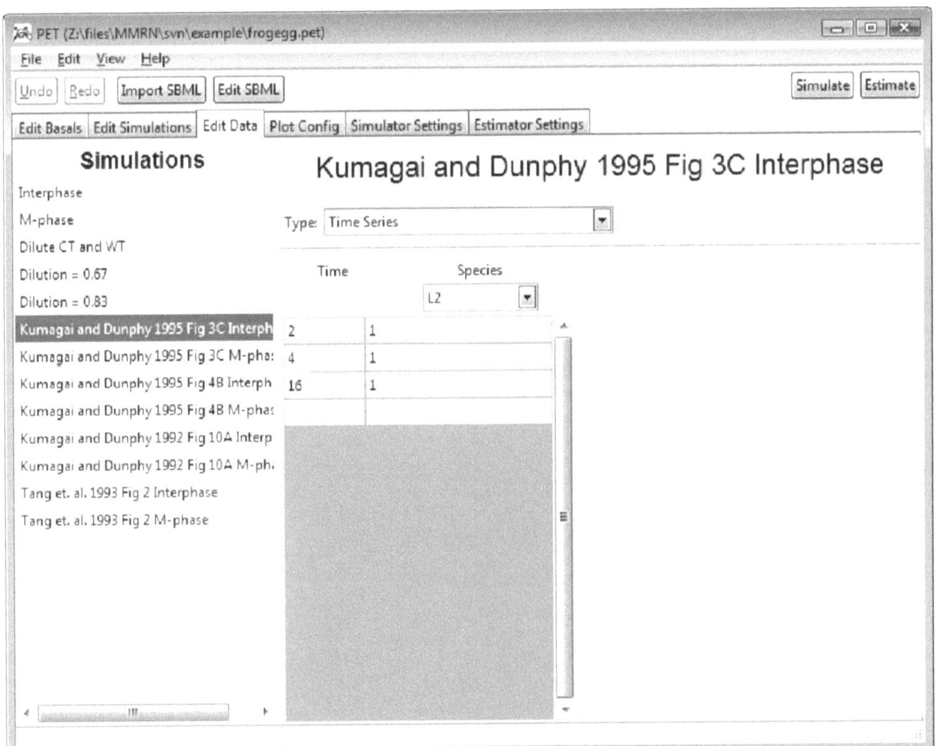

Fig. 16. The "Edit Data" screen shows experimental data and the set up for transforms. This figure shows a list of numbers for the time series concentration of L2. The "Time Series" transform is selected for the type of experimental data.

convert the time series data generated by a simulation into a form comparable to the experimental data. For example, experimental data might measure the time it takes for a specific event to happen (timelag) or how much of a species must be added to a system to change a steady state (threshold), or the viability of a mutant. In these cases, the computer simulation must produce a number comparable to the experimental datum (i.e., measuring the same observable). Automated parameter estimation routines then take the difference between the experimental observation and the transformed output of the model, and attempt to minimize this difference by adjusting parameter values. A transform might be quite sophisticated. For example, it might need to analyze the time series output for some measurement (such as cell size) to deduce that an oscillation is taking place, and its period. Transforms are implemented as FORTRAN functions.

The name of every simulation run defined in the "Edit Simulations" screen (**Fig. 17**) appears in the "Edit Data" screen (**Fig. 16**).

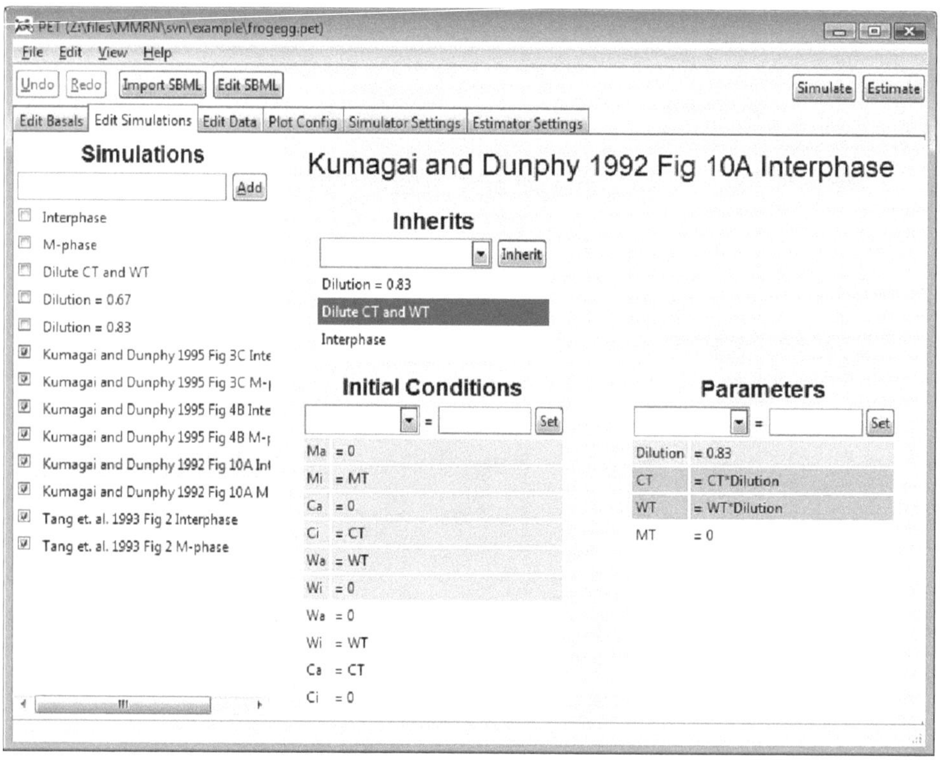

Fig. 17. The "Edit Simulations" screen showing parameter and initial condition values. PET highlights inherited changes in gray. When a parent is selected in the "Inherits" list, the changes inherited from that parent are highlighted in a pastel purple (also shown in gray in this figure).

In the "Edit Data" screen, the user can select the name of a simulation run and define experimental data and a transform. Note that some run specifications might not define either experimental data or a transform. These specifications might be inherited by other runs (e.g., the "M-phase" and "Interphase" runs in the example in **Subheading 4**), or the modeler might wish to store these specifications for another purpose.

3.3.2. Parameter Exploration and Estimation

A user can explore parameter space by setting parameter values (**Fig. 18**), clicking the "Simulate" button, and view the results (**Fig. 19**). This will generate time course plots of selected species (**Fig. 19**). Changes in basal parameters and initial conditions can be made in the "Edit Basals" screen. The user might wish to keep track of multiple basal sets, which are all displayed in the "Edit Basals" screen. When the user clicks the "Simulate" button a simulation is run for each basal set checked in the "Edit Basals"

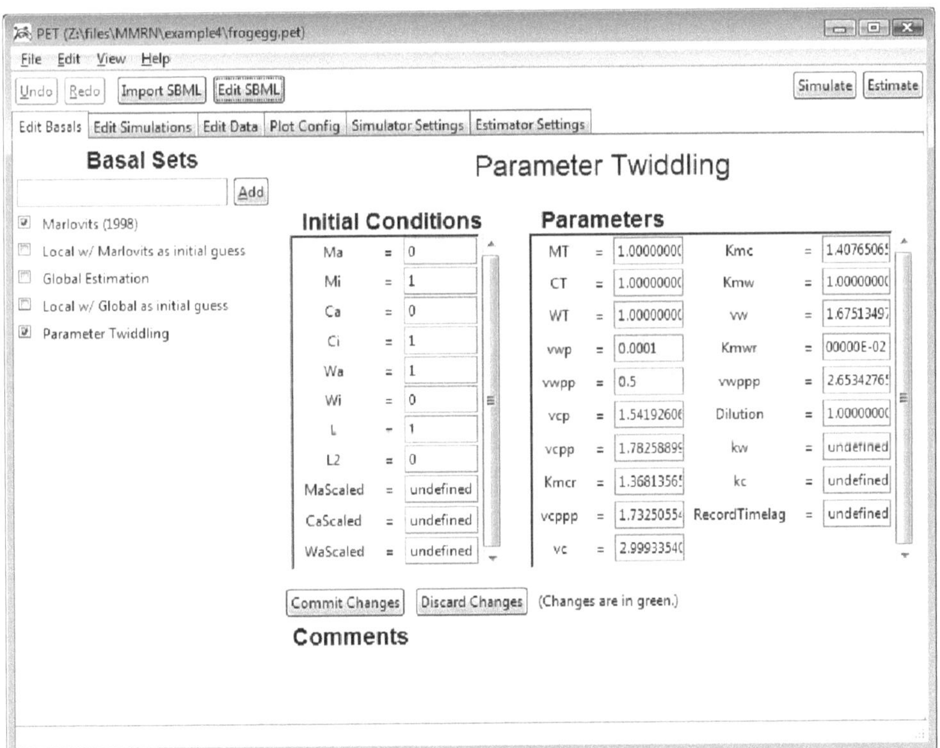

Fig. 18. The "Edit Basals" screen lets users define basal sets of initial conditions and parameters. Changes made to parameters and initial conditions are highlighted in green (parameters vwp and vwpp in this figure). The "Commit Changes" button saves changes and removes the highlight colors. Alternatively, the "Discard Changes" button will restore all changed values to the last commit or the original basal set, whichever is more recent.

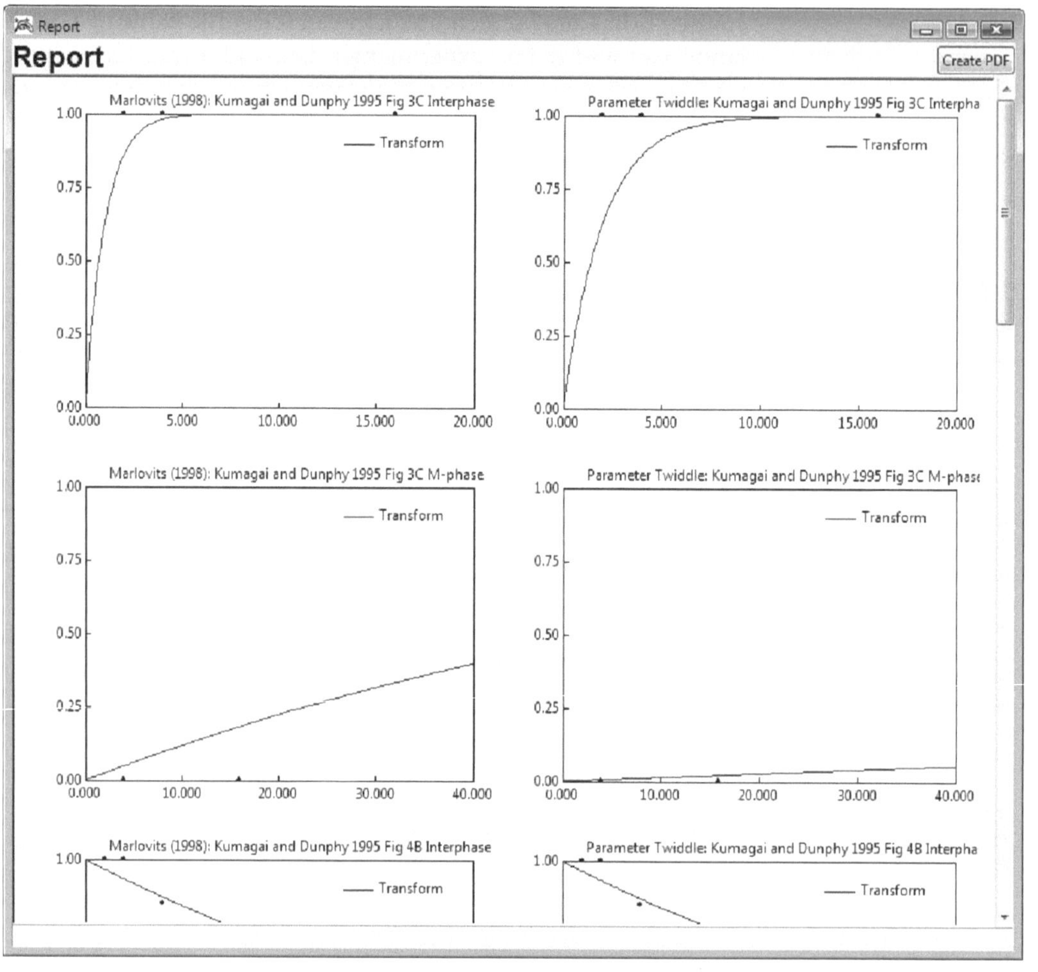

Fig. 19. The PET report window shows the plots using the basal set named "Marlovits (1998)" (*left column*) side-by-side with plots using the basal set shown in Fig. 18 (*right column*). Each simulation run takes a row in the grid of plots. The simulation run "Kumagai and Dunphy 1995 Figure 3C Interphase" is on the first row, "Kumagai and Dunphy 1995 Figure 3C M-phase" is on the second row, and so forth. As many simulation runs and basal sets will be simulated as the user checks in the "Edit Simulations" (Fig. 17) and "Edit Basals" (Fig. 18) screens in PET. This feature of PET allows the user to quickly compare multiple basal sets to experiments and assess which basal set best fits experimental data.

screen, paired with each simulation checked in the "Edit Simulations" screen. For the example, in **Figs. 17** and **18,** sixteen simulations are performed: the eight simulations checked in **Fig. 17** are run for each of the two basal parameter sets checked in **Fig. 18**. Every simulation performed generates a plot and the appropriate experimental data from the "Edit Data" screen is plotted with the model simulation points. This allows the user to quickly compare model simulations with experimental data.

By manually changing parameters, running simulations, and viewing plots, a user might discover parameter values that bring the simulations into acceptable agreement with the experimental data. But this manual process is time consuming. PET also provides automated parameter estimation, which searches for parameter values that best fit a model to experiments. Automated parameter estimation can be configured through the "Estimator Settings" screen (**Fig. 20**) and then run with the "Estimate" button.

Two algorithms are currently available in PET for automated parameter estimation: ODRPACK95 *(21, 22)* and VTDirect *(23)*. Both minimize an objective function defined as the weighted sum of squares of the differences between the model and experimental data:

$$E(\beta) = \sum_{i=1}^{n} w_{\varepsilon_i}\varepsilon_i^2 + w_{\delta_i}\delta_i^2, \tag{1}$$

$$f_i(x_i + \delta_i; \beta) = y_i + \varepsilon_i, \quad \text{for} \quad i = 1 \ldots n, \tag{2}$$

where β is the parameter vector (referred to as a parameter set in this chapter), each f_i is a function of the model (e.g., a time course simulation) and could be different for each i, x_i is the

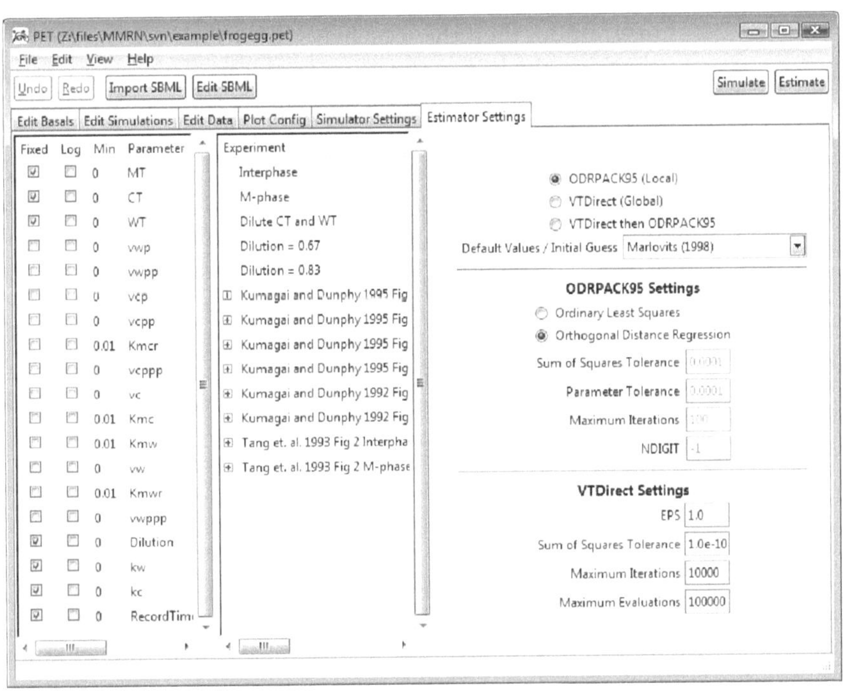

Fig. 20. The "Estimator Settings" screen shows the parameters to be estimated and ranges on those parameters (*left*), experimental data weights (*center*), and algorithm settings (*right*).

i^{th} independent experimental datum (e.g., time), y_i is the ith dependent experimental datum (e.g., species concentration), δ and ε are the respective errors attributed to the independent and dependent experimental data, and w_δ and w_ε are the weights for δ and ε supplied by the user (PET automatically calculates default values for these). The algorithms search for a δ and β to minimize **Eq. 1** (note that ε can be calculated from **Eq. 2** once δ and β are chosen). Zwolak et al. *(21)* and Boggs et al. *(24, 22)* explain this objective function in more detail. ODRPACK95 is a local optimization algorithm based on Levenberg-Marquardt. VTDirect is a global optimization algorithm based on the "DIViding RECTangles" algorithm of Jones *(23)*.

When estimating parameters automatically, the user can select which experiments are to be fit by checking them in the "Edit Simulations" screen (**Fig. 17**). For a particular "estimation," the user might allow only certain parameters to be varied by PET. The fixed parameters might be part of a conserved quantity, have a known value, or are not well constrained by the current data. Such parameters are selected as "fixed" by checking the box in the "Fixed" column of the "Estimator Settings" screen (**Fig. 20**).

Ranges on each parameter can also be defined (and *must* be defined for global optimization with VTDirect). **Figure 20** shows the "Estimator Settings" screen in PET where the ranges can be edited. When the parameter range extends over multiple orders of magnitude, then the user may wish to use a logarithmic scale by checking the box in the "Log" column. This feature is only available for global estimation and affects the way VTDirect searches parameter space. For example, for a linear scale with a range of 0.01–1000 for some parameter p_1, VTDirect might select values of approximately 200, 400, 600, and 800. If a logarithmic scale is selected, the equivalent points selected by VTDirect would be 0.1, 1, 10, and 100. In the linear case, small values of p_1 are never explored, which might not be desirable.

Weights can be assigned to the experimental data to reflect relative confidence in the data in the "Estimator Settings" screen (**Fig. 20**). These are the weights appearing in **Eq. 1**. PET assigns default values for the weights of

$$w_{\delta_i} = \frac{1}{x_i^2 + 1}, \quad w_{\varepsilon_i} = \frac{1}{y_i^2 + 1}.$$

These weights can reflect error bounds on the data determined by repeats of the experiment, if available. Larger values for the weight can be assigned for data with small error bounds. Similarly, smaller values for the weight can be assigned for data with large error bounds.

4. A Modeling Example

We now provide a detailed example of how our tools are used to build a model, based on the modeling process described in **Subheading 2**. The model used here was derived from Marlovits et al. *(1)* and Zwolak et al. *(25, 26)* and can be seen in **Fig. 1**. It models the regulation of entry into mitosis in frog egg extracts by MPF, Cdc25, and Weel. Experimental data from Kumagai and Dunphy *(27, 28)* and Tang et al. *(29)* are fit using local and global optimization. We discuss an alternative model motivated by the parameter set returned from the global optimizer. Readers interested in pursuing the example further might consider implementing this alternative model as an exercise.

4.1. Entering the Molecular Network

We begin by entering the molecular network from **Fig. 1** into the Model Builder. Each reaction appears as a line in the reaction spreadsheet (**Fig. 3**). Michaelis-Menten kinetics are used for the forward and reverse reactions of Cdc25 and Weel. A user defined rate law (My_rate_law) is used to define MPF phosphorylation and dephosphorylation by the active forms of Cdc25 (C_a) and Weel (W_a) as well as a small residual activity of the inactive forms of Cdc25 (C_i) and Weel (W_i). Two species, L and L2, are added to the model for comparison to measurements of labeled MPF. L is used to measure the rate at which Cdc25 removes the phosphate group from MPF (Kumagai and Dunphy *(28)* Figure 3C). L2 is used to measure the rate at which Weel adds the phosphate group to MPF (Kumagai and Dunphy *(28)* Figure 4B). The map of names used in the model to the biological names can be seen in **Fig. 1**.

The Marlovits *(1)* parameter set ($\beta_{Marlovits}$ in **Table 1**) is entered into the Model Builder via the "Parameters" spreadsheet, and exported to a basal file for later use with the Run Manager and PET. Initial conditions for the species are defined in the "Species" spreadsheet for interphase (**Table 2**). Interphase is defined as a state of low MPF and Cdc25 activity and high Weel activity.

In some experiments, a buffer is added to an extract, thereby diluting the endogenous concentrations of proteins in the extract. The dilution factor is set to 1 for the experiments from Kumagai and Dunphy Figures 3C and 4B *(28)*. For the other experiments we use a dilution factor ("Dilution") relative to the Kumagai and Dunphy *(28)* experiments. The dilution of species in the model is handled in the Run Manager, as discussed in **Subheading 4.2**. For Weel and Cdc25 we would like the total concentration to be scaled to 1, even after they have been diluted, and this can be specified in the "Rules" spreadsheet of the Model Builder. In the "Species" spreadsheet we create two new species and assign

Table 1
Parameter sets

Parameter	$\beta_{Marlovits}$	β_{local}	β_{global}
v_w	2	1.7	3.0
v_w^t	0.01	2.4e-4	3.5e-6
v_w^h	1	1.4	2.4
v_w^m	0.05	0.027	0.014
v_c	2	3.0	120
v_c^t	0.017	0.015	0.015
v_c^n	0.17	0.18	0.18
v_c^m	0.05	0.017	0.0027
K_{mw}	0.1	0.01	0.099
K_{mwr}	1	0.01	0.01
K_{mc}	0.1	0.14	20
K_{mcr}	1	0.14	3.4
E		0.018	0.059

The weighted sum of squares (the value of the objective function E) for each estimated set is shown in the last row. Parameter sets $\beta_{Marlovits}$ from Marlovits et al. *(1)*, β_{local} from the local parameter estimator, and β_{global} from the global parameter estimator.

Table 2
Initial conditions

Species	M-phase	Interphase
Ma	MT	0
Mi	0	MT
Ca	CT	0
Ci	0	CT
Wa	0	WT
Wi	WT	0

For example, in Interphase the initial value of inactive MPF (Mi) is set to the total amount of MPF (MT) while the initial value of active MPF (Ma) is set to 0. Initial conditions of the species to model extracts starting in M-phase or Interphase.

them values in the "Rules" spreadsheet with the rules CaScaled = Ca/Dilution and WaScaled = Wa/Dilution.

4.2. Defining Simulation Runs

In this section, we define simulation runs in the Run Manager. Each experiment has a line in the Run Manager and all the values set for the runs can be seen in **Fig. 12**. The Run Manager reads in the SBML file containing our model and the file containing basal parameter values and initial conditions. One way to specify these files is through the "File" menu.

Experiments from Kumagai and Dunphy Figures 3C and 4B *(28)*, Kumagai and Dunphy Figure 10A *(29)*, and Tang et al. Figure 2 *(29)* specify what state the extract was in when the experiment began, either interphase or M-phase. Initial conditions for the model are created to mimic these extract states, and the values of these initial conditions can be seen in **Table 2**.

For the initial conditions for M-phase and interphase, we create two runs in the Run Manager called "M-phase" and "Interphase," respectively (**Fig. 12**). All runs starting in M-phase will inherit from the M-phase basal run. Similarly, all runs starting in interphase will inherit from the Interphase run.

Experiments in Kumagai and Dunphy Figure 10A *(27)* and Tang et al. Figure 2 *(29)* add a buffer that dilutes the extracts by a factor of 0.83 and 0.67, respectively. We handle this by creating a simulation run for each case, called "Dilution = 0.83" and "Dilution = 0.67". These runs set the parameter Dilution to the correct value. Then we create a simulation run "Dilute" that applies the parameter Dilution to all species that are diluted (e.g., $C_T = C_T \cdot Dilution$, $W_T = W_T \cdot Dilution$, etc.). The initial conditions for the species are diluted by their assignments (**Table 2**). None of these runs are intended to be simulated. They exist just to be inherited by runs that use diluted species.

Table 3
Experimental data

Experiment	Species	Time	Concentration
Kumagai and Dunphy Figure 3C *(28)* Interphase	L2	2	1
		4	1
		16	1
Kumagai and Dunphy Figure 3C *(28)* M-phase	L2	4	0
		16	0

(continued)

Table 3 (continued)

Experiment	Species	Time	Concentration
Kumagai and Dunphy Figure 4B *(28)* Interphase	L	2 4 8	1 1 0.85
Kumagai and Dunphy Figure 4B *(28)* M-phase	L	2 4 8	0.75 0.51 0.21
Kumagai and Dunphy Figure 10A *(27)* Interphase	Ca	5 10 20 40	0.75 0.5 0.1 0
Kumagai and Dunphy Figure 10A *(27)* M-phase	Ca	1.25 2.5 5 10	0.8 0.9 1 1
Tang et al. Figure 2 *(29)* Interphase	Wa	7.5 15	0.5 1
Tang et al. Figure 2 *(29)* M-phase	Wa	2 5 7 10	0.5 0 0 0

4.3. Entering the Experimental Data

With this model we will attempt to reproduce the experimental data from Kumagai and Dunphy *(27, 28)* and Tang et al. *(29)*. The data from these papers (images of gels, the points on plots, etc.) are quantified in **Table 3**. These data are entered into PET via the "Edit Data" screen (**Fig. 16**). A basal set is defined in the "Edit Basals" screen from the basal file containing the Marlovits parameters. For each experiment, the "time series" transform is selected in the "Edit Data" screen, the measured species is selected, and the experimental data are entered so that the optimization code will be able to compare simulation output to the experimental data. Now a set of simulations can be run and we can see how well the Marlovits parameters fit the experimental data (**Fig. 21**).

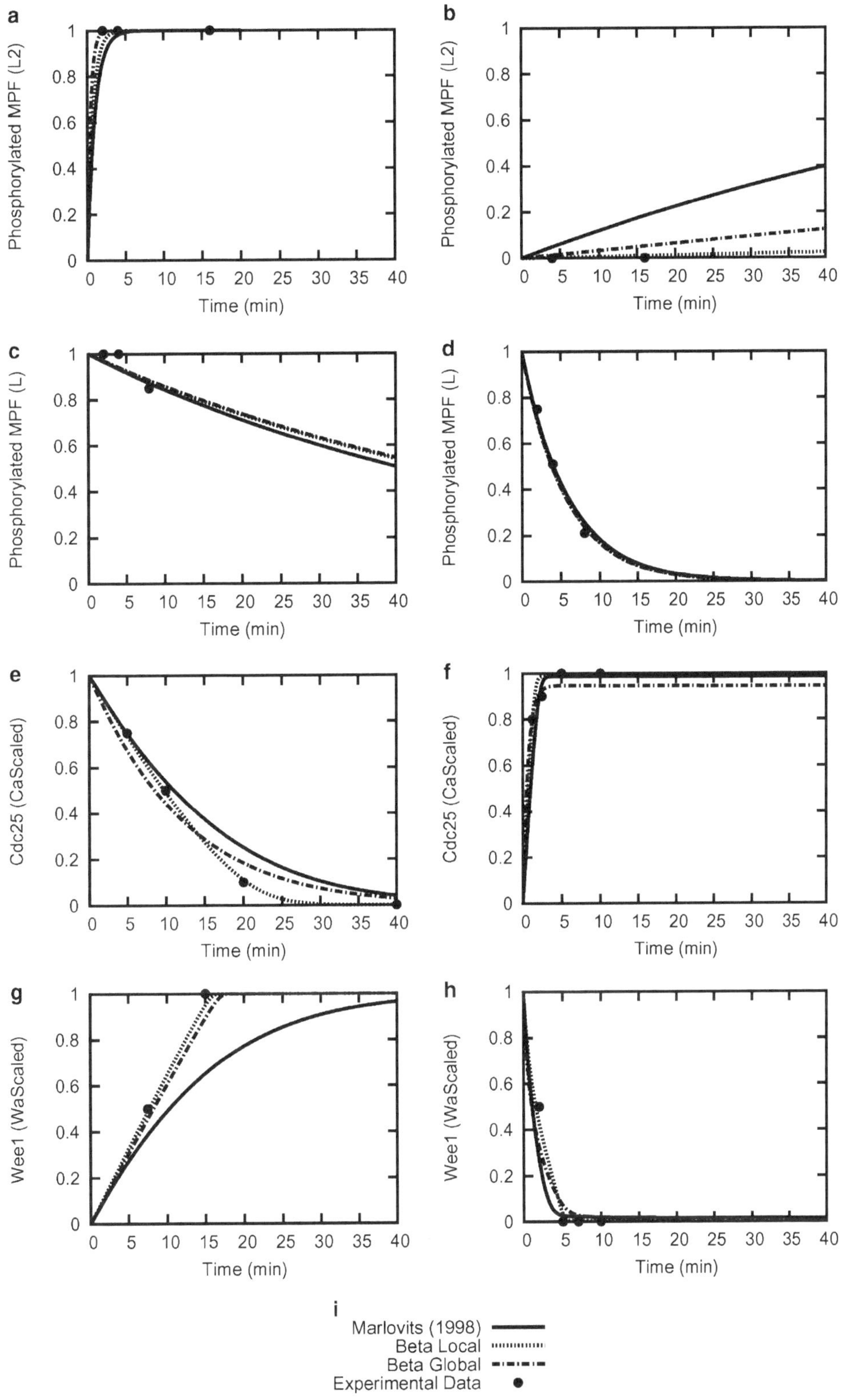

Fig. 21. The parameter set "Marlovits (1998)" ($\beta_{Marlovits}$), "Beta Local" (β_{Local}), and "Beta Global" (β_{Global}) are plotted along with the experimental data for comparison.

Table 4
Lower and upper bounds for the parameters

Parameter	Lower (VTDirect)	Lower (ODRPACK95)	Upper
V_w	1e-6	0	1e4
V_{wp}	1e-6	0	1e4
V_{wpp}	1e-6	0	1e4
V_{wppp}	1e-3	0	100
V_c	1e-6	0	1e4
V_{cp}	1e-6	0	1e4
V_{cpp}	1e-6	0	1e4
V_{cppp}	1e-3	0	100
K_{mw}	0.01	0.01	100
K_{mwr}	0.01	0.01	100
K_{mc}	0.01	0.01	100
K_{mcr}	0.01	0.01	100

VTDirect will only explore parameter space within these bounds. We use different lower bounds for VTDirect and ODRPACK95, as explained in the text

4.4. Performing Local Parameter Estimation

We choose the Marlovits parameters as an initial guess to be used by the local optimization algorithm ODRPACK95 and set some reasonable lower bounds on the parameters (**Table 4**). Only the simulation runs that we wish to fit to data are checked in the "Simulations" screen of PET, and only the parameters we wish to be estimated are checked in the "Estimator Settings" screen. We use the default settings for ODRPACK95, which, in practice, are usually adequate. As the initial guess we select the "Marlovits (1998)" basal set. The optimizer returns the parameters β_{local} in **Table 1**, and we can compare the results to the Marlovits set by running simulations on the basal set and on the fitted parameter values. (Running the simulation would actually show a window similar to **Fig. 19**, but here we show the plots more compactly in **Fig. 21**). We see from **Fig. 21** that the parameter estimator does return parameters that fit the data better. We can also see that the parameter values are close to the starting value of Marlovits (**Table 1**).

4.5. Global Parameter Estimation

In some cases, the user may not have a good starting point for the parameters, or the user might wish to explore parameter space in search of other good parameter sets. PET supports these cases by providing a global parameter estimation algorithm, VTDirect. VTDirect requires upper and lower bounds on the parameter values. In our example, we assume that we know little about the true values of the parameters. We give bounds that span several orders of magnitude, and we use a logarithmic scale to distribute the search evenly across these orders of magnitude. Since we use a logarithmic scale, we must set non-zero lower bounds. We set most lower bounds to 10^{-6}, which allows these parameters to get sufficiently close to zero to have a negligible quantitative effect on the model. The bounds are recorded in **Table 4**. VTDirect is run with the settings from Table 5, and the resulting parameter set is passed to ODRPACK95 for refinement. We reset the parameter bounds for the ODRPACK95 run to those of **Table 4**. ODRPACK95 does not use the logarithmic scale setting and therefore can have lower bounds of 0 for this run. The global refined parameter set is called β_{global} in **Table 1**.

4.6. Next Steps

Visually, the parameters generated by the global and local optimization runs both fit the experimental data (**Fig. 21**). The parameter sets (β_{local} and β_{global} in **Table 1**) are similar, except for the values of v_c, v_{cppp}, and K_{mc}. For K_{mc} = 20 and C_T = 1, the Michaelis-Menten rate law for reaction $C_i \rightarrow C_a$ in **Fig.** 1 should be replaced by a mass action rate law, $(v_c/K_{mc} \cdot M_a \cdot C_i)$. This change to the model is addressed in Zwolak et al. *(26)*, and we will not go through the analysis here.

Next, we can create another variation of the model by adding experimental data for timelags and thresholds, as discussed in Zwolak et al. *(25)*. Automated parameter estimation can be run to find parameter values that fit these new experiments, as well as the experiments discussed in this section. The model can continue to be refined and expanded in this way to test further hypotheses and achieve new goals.

The files for the modeling example and its variations are distributed with JigCell and PET and can be found at http://mpf.biol.vt.edu/MMRN_chapter/.

Table 5
Settings used by VTDirect

Settings	Value
EPS	1.0
Sum of Squares Tolerance	1.0e-10
Maximum Iterations	1.0e4
Maximum Evaluationsr	1.0e5

5. Summary

We have demonstrated how a modeler would enter all of the necessary information needed to define, simulate, and validate a model of a molecular regulatory network. Advanced support tools like the JigCell Model Builder make it easy to check the syntactic consistency and completeness of the model. This makes it possible to construct larger models than can be done "by hand" and thus opens the possibility of constructing more complex models than previously possible. The JigCell Run Manager provides a way to organize and manage the information needed to define the ensemble of simulation runs for validating the model against a specific set of experiments. PET provides a tool to help the user compare simulation output to experimental data. PET also provides automated tools for finding "best fitting" values of the rate constants in a model. Our example walks the reader through a complete cycle of entering the model, testing it for initial validity, and using parameter estimation to improve the model.

While tools such as JigCell and PET allow modelers to build and test larger models than were possible before, there is still a long way to go before it will be possible to build models that capture the complex regulatory systems within mammalian cells. Current models are defined as a single monolithic block of reaction equations, an approach that is reaching its limits. In the future, modelers will be able to express their models as a collection of interacting components, thus allowing them to build large models from smaller pieces. Improvements are also needed in simulators (including the ability to perform efficient stochastic simulations), in parameter estimation, and in computer performance.

References

1. G. Marlovits, C.J. Tyson, B. Novak, and J.J. Tyson (1998) Modeling M-phase control in *xenopus* oocyte extracts: the surveillance mechanism for unreplicated DNA. *Biophys. Chem.* 72, 169–184.
2. K.C. Chen, L. Calzone, A. Csikasz-Nagy, F.R. Cross, B. Novak, and J.J. Tyson (2004) Integrative analysis of cell cycle control in budding yeast. *Mol. Biol. Cell* 15, 3841–3862.
3. A.C. Hindmarsh (1983) ODEPACK: A systematized collection of ODE solvers, in *Scientific Computing*, ed. by R.S. Stepleman, North Holland Publishing Company, 55–64.
4. P. Mendes (1997) Biochemistry by numbers: Simulation of biochemical pathways with Gepasi 3. *Trends in Biochem. Sci.* 22, 361–363.
5. N.A. Allen, L. Calzone, K.C. Chen, A. Ciliberto, N. Ramakrishnan, C.A. Shaffer, J.C. Sible, J.J. Tyson, M.T. Vass, L.T. Watson, and J.W. Zwolak (2003) Modeling regulatory networks at Virginia Tech. *OMICS* 7, 285–299.
6. H. Sauro, M. Hucka, A. Finney, C. Wellock, H. Bolouri, J. Doyle, and H. Kitano (2003) Next generation simulation tools: The Systems Biology Workbench and BioSPICE integration. *OMICS* 7, 355–372.
7. J. Schaff, B. Slepchenko, Y. Choi, J. Wagner, D. Resasco, and L. Loew (2001) Analysis of non-linear dynamics on arbitrary geometries with the Virtual Cell. *Chaos* 11, 115–131.
8. Y. Cao, H. Li, and L. Petzold (2004) Efficient formulation of the stochastic simulation

algorithm for chemically reacting systems. *J. Chem. Phys.* 121, 4059–67.

9. M. Gibson, and J. Bruck (2000) Efficient exact stochastic simulation of chemical systems with many species and many channels. *J. Phys. Chem. A* 104, 1876–1889.
10. D. Gillespie (2001) Approximate accelerated stochastic simulation of chemically reacting systems. *J. Chem. Phys.* 115, 1716–1733.
11. M. Vass, C. Shaffer, N. Ramakrishnan, L. Watson, and J. Tyson (2006) The JigCell Model Builder: a spreadsheet interface for creating biochemical reaction network models. *IEEE/ACM Trans. Computat. Biol. and Bioinform.* 3, 155–164.
12. N. Allen, R. Randhawa, M. Vass, J.W. Zwolak, J.J. Tyson, L.T. Watson, and C. Shaffer (2007) JigCell, http://jigcell.biol.vt.edu/.
13. M. Vass, N. Allen, C. Shaffer, N. Ramakrishnan, L. Watson, and J. Tyson (2004) The JigCell Model Builder and Run Manager. *Bioinformatics* 20, 3680–3681.
14. J.W. Zwolak, T. Panning, and R. Singhania (2007) PET: Parameter Estimation Toolkit, http://mpf.biol.vt.edu/pet.
15. M. Hucka, A. Finney, H. Sauro, and 40 additional authors (2003) The systems biology markup language (SBML): a medium for representation and exchange of biochemical network models. *Bioinformatics* 19, 524–531.
16. M. Hucka, A. Finney, B.J. Bornstein, S.M. Keating, B.E. Shapiro, J. Matthews, B.L. Kovitz, M.J. Schilstra, A. Funahashi, J.C. Doyle, and H. Kitano (2004) Evolving a lingua franca and associated software infrastructure for computational systems biology: The systems biology markup language (SBML) project. *Syst. Biol.* 1, 41–53.
17. H. Sauro, and B. Ingalls (2004) Conservation analysis in biochemical networks: computational issues for software writers. *Biophys. Chem.* 109, 1–15.
18. B. Ermentrout (2002) *Simulating, Analyzing, and Animating Dynamical Systems: A Guide to XPPAUT for Researchers and Students*, SIAM.
19. StochKit (2005) Project website, www.cs.ucsb.edu/cse/StochKit.
20. E. Conrad (2007) Oscill8, http://oscill8.sourceforge.net/.
21. J. Zwolak, P. Boggs, and L. Watson (to appear) Odrpack95: A weighted orthogonal distance regression code with bound constraints. *ACM Trans. Math. Softw.* .
22. P.T. Boggs, J.R. Donaldson, R.H. Byrd, and R.B. Schnabel (1989) Algorithm 676: Odrpack: software for weighted orthogonal distance regression. *ACM Trans. Math. Soft.* 15, 348–364.
23. D. Jones, C. Perttunen, and B. Stuckman (1993) Lipschitzian optimization without the Lipschitz constant. *J. Optim. Theory. Appl.* 79, 157–181.
24. P.T. Boggs, R.H. Byrd, and R.B. Schnabel (1987) A stable and efficient algorithm for nonlinear orthogonal distance regression. *SIAM J. Sci. Stat. Comput.* 8, 1052–1078.
25. J.W. Zwolak, J.J. Tyson, and L.T. Watson (2005) Parameter estimation for a mathematical model of the cell cycle in frog eggs. *J. Comp. Biol.* 12, 48–63.
26. J.W. Zwolak, J.J. Tyson, and L.T. Watson (2005) Globally optimized parameters for a model of mitotic control in frog egg extracts. *IEE Syst. Biol.* 152, 81–92.
27. A. Kumagai, and W.G. Dunphy (1992) Regulation of the cdc25 protein during the cell cycle in xenopus extracts. *Cell* 70, 139–151.
28. A. Kumagai, and W.G. Dunphy (1995) Control of the cdc2/cyclin B complex in *Xenopus* egg extracts arrested at a G2/M checkpoint with DNA synthesis inhibitors. *Mol. Biol. Cell* 6, 199–213.
29. Z. Tang, T.R. Coleman, and W.G. Dunphy (1993) Two distinct mechanisms for negative regulation of the wee1 protein kinase. *EMBO J.* 12, 3427–3436.

Chapter 5

Rule-Based Modeling of Biochemical Systems with BioNetGen

James R. Faeder, Michael L. Blinov, and William S. Hlavacek

Summary

Rule-based modeling involves the representation of molecules as structured objects and molecular interactions as rules for transforming the attributes of these objects. The approach is notable in that it allows one to systematically incorporate site-specific details about protein–protein interactions into a model for the dynamics of a signal-transduction system, but the method has other applications as well, such as following the fates of individual carbon atoms in metabolic reactions. The consequences of protein–protein interactions are difficult to specify and track with a conventional modeling approach because of the large number of protein phosphoforms and protein complexes that these interactions potentially generate. Here, we focus on how a rule-based model is specified in the BioNetGen language (BNGL) and how a model specification is analyzed using the BioNetGen software tool. We also discuss new developments in rule-based modeling that should enable the construction and analyses of comprehensive models for signal transduction pathways and similarly large-scale models for other biochemical systems.

Key words: Computational systems biology, Mathematical modeling, Combinatorial complexity, Software, Formal languages, Stochastic simulation, Ordinary differential equations, Protein–protein interactions, Signal transduction, Metabolic networks.

1. Introduction

BioNetGen is a set of software tools for rule-based modeling *(1)*. Basic concepts of rule-based modeling and the BioNetGen Language (BNGL) are illustrated in **Fig. 1** – these concepts and the conventions of BNGL will be thoroughly discussed later in the text. Here, in explaining how to use BioNetGen to model biochemical systems, we will be primarily concerned with signal-transduction systems, which govern cellular responses, such as growth and differentiation, to signals, such as hormones and cytokines. In other

Ivan V. Maly (ed.) *Methods in Molecular Biology, Systems Biology, vol. 500*

DOI: 10.1007/978-1-59745-525-1_5

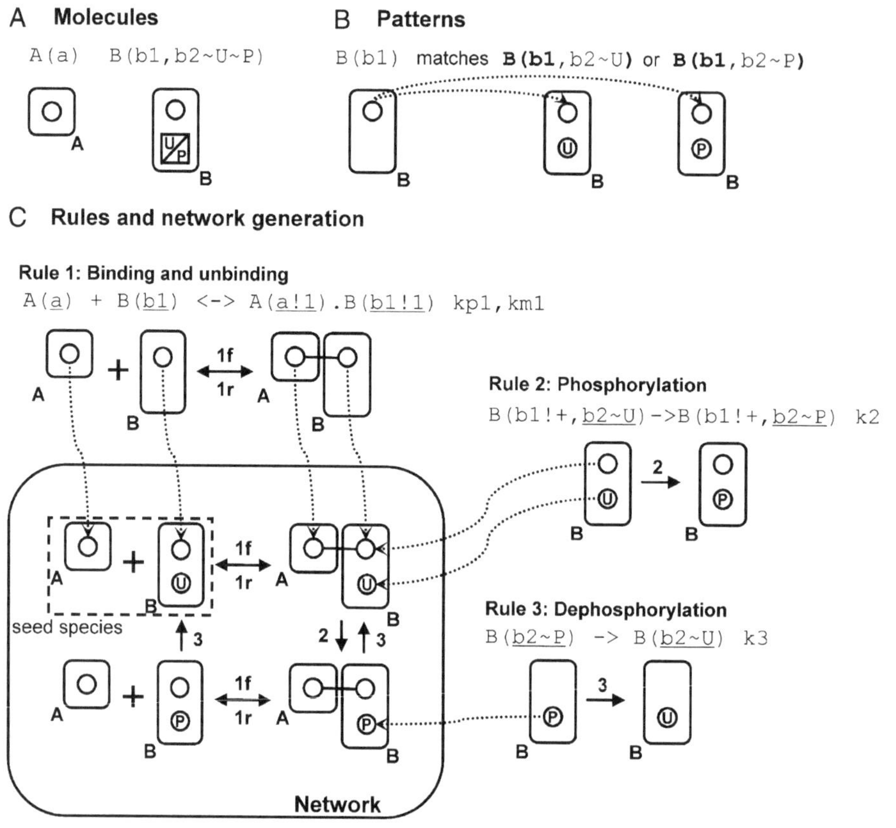

Fig. 1. Rule-based modeling concepts and their encoding in BioNetGen Language (BNGL). (**A**) The basic building blocks are molecules, which are structured objects, composed of components that represent functional elements of proteins and may have associated states that represent covalent modifications or conformations. Molecules may be assembled into complexes through bonds that link components of different molecules. (**B**) Patterns select particular attributes of molecules in species (shown in *bold*). The pattern shown here selects molecules of B with a free b1 binding site regardless of the phosphorylation or binding status of the b2 component. (**C**) Rules specify the biochemical transformations that can take place in the system and may be used to build up a network of species and reactions (*see* **Section 3.5** for a complete description of rule syntax). The reaction center (components undergoing direct modification) is underlined. (This is shown for clarity and is not part of BioNetGen syntax.) Starting with the seed species, rules are applied to generate new reactions and species by mapping reactant patterns onto species and applying the specified transformation(s). Species generated by new reactions may be acted on by other rules to generate new reactions and species, and the process continues until no new reactions are found or some other stopping criteria are satisfied.

words, signal-transduction systems are responsible for making decisions about the fates and activities of cells. Decision making in these systems is accomplished by dynamical systems of interacting molecules *(2)*. To develop predictive computational models of these complex systems, we must be able to abstract their relevant details in a form that enables reasoning about or simulation of the logical consequences of a set of interactions, which enables the testing of model predictions against experimental observations *(3)*. Analysis of predictive models can help to guide experimental investigations and may ultimately enable model-guided engineering and manipulation of cellular regulation *(4–6)*. Before beginning our discussion of BioNetGen, we will briefly recap features of signal-transduction systems that motivate a rule-based modeling approach and the general idea of rule-based modeling. For more thorough reviews of these topics see **refs.** *7, 8*.

A prominent feature of any signal-transduction system is an intricate network of protein–protein interactions *(9, 10)*. These interactions can have a number of consequences, including the posttranslational modification of proteins, the formation of heterogeneous protein complexes in which enzymes and substrates are colocalized, and the targeted degradation of proteins. For understanding and modeling the system dynamics of protein–protein interactions, the details that are most relevant are typically found at the level of protein sites, the parts of proteins that are responsible for protein–protein interactions. These interactions are mediated by evolutionarily conserved modular domains of proteins that have binding and catalytic activities, such as Src homology 2 (SH2) domains and protein tyrosine kinase domains, and by short linear motifs (e.g., immunoreceptor tyrosine-based activation motifs or ITAMs) *(11)* with binding activities that can often be switched on and off through posttranslational modifications, such as tyrosine phosphorylation *(12–14)*. A great deal of knowledge about the site-specific details of protein–protein interactions has accumulated in the scientific literature and is being actively organized in electronic databases *(15, 16)*, and new technologies, such as mass spectrometry (MS)-based proteomics *(17)*, can be applied to quantitatively monitor system responses to a signal at the level of protein sites on a large scale. For example, time-resolved measurements of the phosphorylation of individual tyrosine residues are possible *(18)*.

Despite the high relevance of the site-specific details of protein–protein interactions for understanding system behavior, models incorporating these details are uncommon. For example, the seminal model of Kholodenko et al. *(19)* and many of its extensions, such as the model of Schoeberl et al. *(20)*, for early events in signaling by the epidermal growth factor receptor (EGFR) do not track the phosphorylation kinetics of individual tyrosines in EGFR. Models that incorporate such details are generally difficult or impossible to specify and analyze using conventional methods, largely because of the combinatorial number of protein modifications and protein complexes that can be generated through protein–protein interactions *(7, 8)*. For example, a protein containing n peptide substrates of kinases can potentially be found in up to 2^n distinct phosphorylation states. This feature of protein–protein interactions, which arises because a typical protein involved in cellular regulation contains multiple sites of posttranslational modification and multiple binding sites, has been called combinatorial complexity and has been recognized as a significant challenge to our understanding of cellular regulation *(7, 21, 22)*. In a conventional model specification, which often takes the form of a list of the reactions that are possible in a signal-transduction system or the corresponding system of coupled ordinary differential equations (ODEs) for the chemical kinetics, each chemical species that can be populated and each reaction that can occur must be manually defined, which is infeasible for all but the simplest systems because of the vast numbers of chemical species and reactions that can usually be generated by protein–protein interactions.

Another limitation of conventional modeling is a lack of standards for explicitly representing the composition and connectivity of molecular complexes. The chemical species accounted for in a typical model are represented as structureless objects whose identities and properties are referenced only by name. Modelers attempt to name model parameters and variables such that their names suggest what is being represented, but conventions vary and are often inconsistent. A dimer of EGFR molecules may be represented as R-R or R:R – designations that abbreviate EGFR to R (for receptor) and that indicate the composition of the complex – or simply as D (for dimer). A dimer of EGFR molecules associated with the adapter protein Grb2 may then be represented as R-R-Grb2, D-Grb2, or even as R-Grb2 or simply by the index of a generic variable name (e.g., X_5). The latter examples obscure the fact that two receptor molecules are present in the complex. Model(er)-specific nomenclatures thus present a challenge to understanding a model, especially a large model, which becomes particularly problematic when one attempts to reuse or extend a model. In addition, information about how two molecules are connected is nearly always absent in a conventional model specification, even though in many cases there is detailed site-specific information available about the interaction. For example, interaction of EGFR and Grb2 occurs when the SH2 domain of Grb2 binds a phosphorylated tyrosine residue in EGFR, such as Y1068 *(23)*.

The limitations of conventional approaches to model specification noted above have prompted the development of formal languages specially designed for representing proteins and protein–protein interactions, the κ-calculus being an early and notable example *(24)*. One of these formal languages is the BioNetGen language (BNGL) *(8)*, which is based on the use of graphs to represent proteins and protein complexes and graph-rewriting rules to represent protein–protein interactions *(25, 26)*. BNGL allows site-specific details of protein–protein interactions to be captured in models for the dynamics of these interactions in a systematic fashion, alleviating both nomenclature and reusability issues *(8)*. BNGL also provides a means for specifying precise visualizations of protein–protein interactions *(25, 26)*. Below, we provide a thorough overview of the text-based syntax and semantics of BNGL, an understanding of which is essential for using the BioNetGen software (http://bionetgen.org). BioNetGen facilitates a rule-based approach to modeling biochemical reaction kinetics, an alternative to conventional modeling that largely overcomes the problem of combinatorial complexity *(8)*. We note that the current syntactical and semantic conventions of the κ-calculus are nearly identical to those of BNGL *(27)*.

In a rule-based approach to modeling, the molecular interactions in a system are abstracted as BNGL-encoded rules, which are precise formal statements about the conditions under which interactions occur and the consequences of these interactions. Rules also provide rate laws for transformations resulting from molecular interactions. At one extreme, a rule simply corresponds to an individual chemical reaction. However, a rule is far more useful when local context governs an interaction, and the rule can be specified such that it defines not a single reaction but a potentially large class of reactions, all involving a common transformation parameterized by the same rate law. The use of such rules to model protein–protein interactions can often be justified, at least to a first approximation, by the modularity of proteins *(12)*. Rules can be used to obtain predictions about a system's behavior in multiple ways. For example, they can serve as generators of a list of reactions. In other words, a set of rules, which can be viewed as a high-level compact definition of a chemical reaction network, can be used to obtain a conventional model specification *(1, 28, 29)*, which can then be analyzed using standard methods. Alternatively, rules can serve as generators of discrete reaction events in a kinetic Monte Carlo simulation of chemical kinetics *(21, 30, 31)*. A rule-based model is capable of comprehensively accounting for the consequences of protein–protein interactions, including all possible phosphoforms of a protein and the full spectrum of possible protein complexes implied by a given set of interactions. Such a model is specified using BNGL in a BioNetGen input file, which may also contain directions for processing the model specification. For example, actions may be defined for simulating a model and producing desired outputs. In the following, we will describe the elements of an example input file in detail.

Since our initial application of a rule-based modeling approach in 2001 to study signaling by the high-affinity IgE receptor *(32–34)*, the software that we have used in our work – initially a FORTRAN code called EQGEN – has evolved dramatically and has been applied to study a number of other biochemical systems *(35–39)*. The initial version of BioNetGen was released in 2004 *(1)*. The name "BioNetGen" is a mnemonic for "Biological Network Generator," but this name should not be interpreted to delimit the full range of the software's capabilities. The software not only generates reaction networks from rules, but also simulates such networks using a variety of methods. Iterative application of rules to a set of seed species (*see* **Fig. 1c**) may be used to generate a network in advance of a simulation, which may subsequently be carried out either by numerically solving ODEs or by implementing a stochastic simulation algorithm (SSA) *(40–42)*. Alternatively, rules may be applied during a simulation as the set of populated species grows, a procedure that has been called "on-the-fly" network generation and simulation *(28, 29)*. Finally, network generation may be avoided altogether by instantiating individual instances of chemical species and carrying out a discrete-event particle-based simulation, in which rules serve as event generators *(21, 30, 31)* (*see* **Subheading 3.7.2**). Simulation engines implementing such methods will soon be available within the BioNetGen framework and will be called through interfaces similar to those of the existing engines (*see* **Subheading 3.6**).

Later, we summarize essentially everything a modeler needs to know to start developing and analyzing rule-based models with BioNetGen. After an overview of the BioNetGen software distribution, we present a step-by-step guide to writing a BioNetGen input file, in which we carefully explain the elements of an example input file. Numerous tips and tricks can be found in the **Notes** section. Building on the basics, we then present several examples that illustrate more advanced BioNetGen capabilities. Finally, we briefly discuss new developments in rule-based modeling that should enable the construction and analyses of large-scale comprehensive models for signal-transduction systems.

2. Software

BioNetGen is a set of integrated open-source tools for rule-based modeling. A schematic of the software architecture is shown in **Fig. 2**. The software and documentation are available at http://bionetgen.org, a wiki site. Downloading the software or modifying the wiki pages requires user registration with a valid email address. The software is easy to install and runs with no compilation on Linux, Mac OS X, and Windows operating systems (*see* **Note 1**). BioNetGen can be also used online (without installation) from within the Virtual Cell modeling environment (http://vcell.org/bionetgen).

The components of BioNetGen include the network generation engine BNG2, which is written in Perl, the simulation program Network, which is written in C, a plotting program called PhiBPlot, which is written in Java, and a graphical front-end called RuleBuilder, which is also written in Java. The core

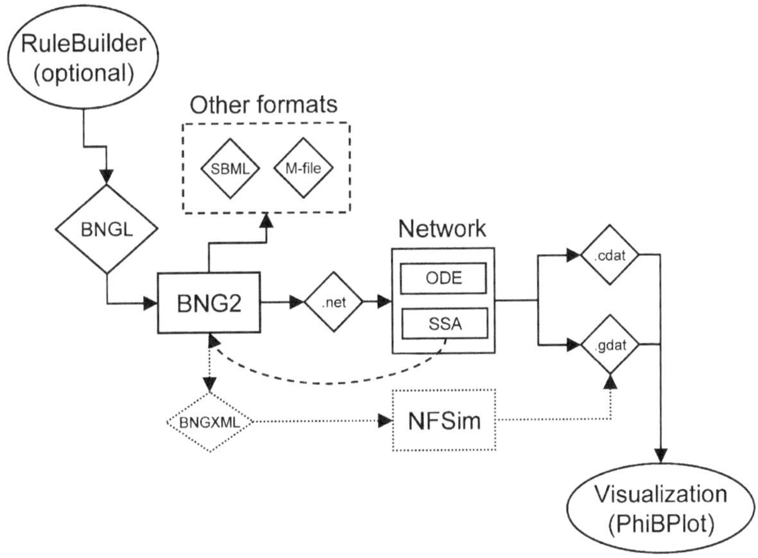

Fig. 2. Software architecture of BioNetGen. The BioNetGen language (BNGL) file specifies a rule-based model that can be processed by the BioNetGen core version 2 (BNG2) in a variety of ways. Iterative application of rules to an initial set of species can generate a reaction network that is passed to one of the simulation modules through the .net format or exported to formats (SBML, MATLAB) that can be read by other programs. In the near future, an XML-based encoding will be used to pass model specifications among additional software components, including a particle-based simulator called NFsim ("network-free" simulator) (Sneddon, M., Faeder, J. and Emonet, T., private communication). Simulation modules produce .cdat and .gdat files, which record the time courses of species concentrations and observables, respectively. The *dashed arrow* connecting the SSA module and BNG2 represents the on-the-fly network generation capability available for stochastic simulations.

component, BNG2, which has a command-line interface, processes BioNetGen input files to generate two kinds of outputs: a chemical reaction network derived by processing rules and/or the results of simulating a model (*see* **Note 2**). Input files are discussed below at length. Reaction networks are exported in a native .net format, in M-file format for processing by MATLAB (The MathWorks, Natick, MA), and in Systems Biology Markup Language (SBML), which is a community-developed standard for the encoding of biological models *(43)*. A network encoded in SBML can be processed by a variety of SBML-compliant software tools (for a list of these tools, see http://sbml.org). An example of an SBML-compliant tool that complements BioNetGen is COPASI *(44)*, which provides model analysis capabilities, such as parameter estimation methods, unavailable in the native BioNetGen environment. Simulation results are exported as tabular data in plain-text files that have the extension .cdat or .gdat. A .cdat file contains time series for concentrations of chemical species. A

.gdat file contains time series for observables defined in a BioNetGen input file or .net file (*see* **Subheading 3.3**). Simulations specified in an input file are preprocessed by BNG2 and then passed to Network, which is a simulation engine driver. Network interfaces with the CVODE package *(45, 46)*, a set of routines for solving stiff and nonstiff initial value problems for systems of ODEs. Network also provides an implementation of the direct method of Gillespie *(40)* for stochastic simulations. The command-line interface of Network allows a .net file to be processed directly without preprocessing by BNG2, but this option is unavailable for simulation in on-the-fly mode *(28, 29)*, which necessarily requires communication between BNG2 and Network. On-the-fly simulation is discussed further in **Subheading 3.7.2**. PhiBPlot is a utility for producing *x–y* plots from .cdat and .gdat files. The .cdat and .gdat files can also be processed by other plotting tools, such as Grace (http://plasma-gate.weizmann.ac.il/Grace). RuleBuilder provides a graphical user interface to BioNetGen. It also provides a drawing tool for creating and editing models that may be particularly helpful to new users.

BioNetGen has been integrated into the Virtual Cell (VCell) modeling environment (http://vcell.org) as a stand-alone application called BioNetGen@VCell. A BioNetGen service is callable from a VCell user interface and runs on a client computer. The VCell user interface can be used to visualize and export simulation results. Alternatively, a VCell BioModel can be automatically created from an SBML file generated by BioNetGen@VCell.

3. Methods

We will illustrate the method of constructing a rule-based model by stepping through the BioNetGen input file shown in **Listing 1**, which specifies a simplified version of a model for early events in EGFR signaling *(35)*. Additional examples can be found in the `Models2` directory of the BioNetGen distribution available from http://bionetgen.org, or on the Web at http://vcell.org/bionetgen/samples.html. A BioNetGen input file contains the information required to specify a model, including definitions of molecules, rules for molecular interactions, and model outputs, which we call "observables." An input file may also contain commands called "actions" that act on the model specification, such as generating the network of species and reactions implied by rules, performing simulations, and translating the model into other formats. The syntax of actions is borrowed from the Perl programming language. Model elements are specified in blocks delimited by "`begin`" and "`end`" tags as indicated in **Listing 1**.

Listing 1. Elements of the BioNetGen input file `egfr_simple.bngl`. Block names are shown in **bold**, and reaction centers are underlined for clarity in the `reaction rules` block.

```
begin parameters
  NA 6.02e23 # Avogadro's number (molecues/mol)
  f  1       # Fraction of the cell to simulate
  Vo f*1.0e-10 # Extracellular volume=1/cell_density (L)
  V  f*3.0e-12 # Cytoplasmic volume (L)
  # Inital amount of ligand (20 nM)
  EGF_init 20*1e-9*NA*Vo # convert to copies per cell
  # Initial amounts of cellular components (copies per cell)
  EGFR_init     f*1.8e5
  Grb2_init     f*1.5e5
  Sos1_init     f*6.2e4
  # Rate constants
  # Divide by NA*V to convert bimolecular rate constants
  # from /M/sec to /(molecule/cell)/sec
  kp1 9.0e7/(NA*Vo) # ligand-monomer binding
  km1 0.06          # ligand-monomer dissociation
  kp2 1.0e7/(NA*V)  # aggregation of bound monomers
  km2 0.1           # dissociation of bound monomers
  kp3 0.5           # dimer transphosphorylation
  km3 4.505         # dimer dephosphorylation
  kp4 1.5e6/(NA*V)  # binding of Grb2 to receptor
  km4 0.05          # dissociation of Grb2 from receptor
  kp5 1.0e7/(NA*V)  # binding of Grb2 to Sos1
  km5 0.06          # dissociation of Grb2 from Sos1
  deg 0.01          # degradation of receptor dimers
end parameters

begin molecule types
  EGF(R)
  EGFR(L,CR1,Y1068~U~P)
  Grb2(SH2,SH3)
  Sos1(PxxP)
  Trash()
end  molecule types

begin seed species
  EGF(R)                0
  EGFR(L,CR1,Y1068~U) EGFR_init
  Grb2(SH2,SH3)       Grb2_init
  Sos1(PxxP)          Sos1_init
end seed species

begin observables
  1 Molecules  EGFR_tot  EGFR()
  2 Molecules  Lig_free  EGF(R)
  3 Species    Dim       EGFR(CR1!+)
```

```
  4 Molecules  RP EGFR(Y1068~P!?)
  # Cytosolic Grb2-Sos1
  5 Molecules  Grb2Sos1  Grb2(SH2,SH3!1).Sos1(PxxP!1)
  6 Molecules  Sos1_act
EGFR(Y1068!1).Grb2(SH2!1,SH3!2).Sos1(PxxP!2)
end observables

begin reaction rules
  # Ligand-receptor binding
  1 EGFR(L,CR1) + EGF(R) <-> EGFR(L!1,CR1).EGF(R!1) kp1, km1
  # Receptor-aggregation
  2 EGFR(L!+,CR1) + EGFR(L!+,CR1) <-> EGFR(L!+,CR1!1).EGFR(L!+,CR1!1) kp2,km2
  # Transphosphorylation of EGFR by RTK
  3 EGFR(CR1!+,Y1068~U) -> EGFR(CR1!+,Y1068~P)  kp3
  # Dephosphorylation
  4 EGFR(Y1068~P) -> EGFR(Y1068~U) km3
  # Grb2 binding to pY1068
  5 EGFR(Y1068~P) + Grb2(SH2) <-> EGFR(Y1068~P!1).Grb2(SH2!1) kp4,km4
  # Grb2 binding to Sos1
  6 Grb2(SH3) + Sos1(PxxP) <-> Grb2(SH3!1).Sos1(PxxP!1) kp5,km5
  # Receptor dimer internalization/degradation
  7 EGF(R!1).EGF(R!2).EGFR(L!1,CR1!3).EGFR(L!2,CR1!3) -> Trash() deg\
  DeleteMolecules
end reaction rules

#actions
generate_network({overwrite=>1});
# Equilibration
simulate_ode({suffix=>equil,t_end=>100000,n_steps=>10,sparse=>1,\
    steady_state=>1});
setConcentration("EGF(R)","EGF_init");
saveConcentrations(); # Saves concentrations for future reset
# Kinetics
writeSBML({});
simulate_ode({t_end=>120,n_steps=>120});
resetConcentrations(); # reverts to saved Concentrations
simulate_ssa({suffix=>ssa,t_end=>120,n_steps=>120});
```

The five block types are "`parameters`," "`molecule types`," "`seed species`," "`reaction rules`," and "`observables`." The blocks may appear in any order. Actions to be performed on the model are controlled using commands that follow the model specification. All text following a "#" character on a line is treated as a comment, and comments may appear anywhere in an input file. Parsing of the input is line-based, and a continuation character, "\", is required to extend a statement over multiple lines. There is no limit on line length. Any BioNetGen input line may begin with an integer index followed by space, which is ignored during input processing but may be useful for reference purposes. For example, .net files produced by BioNetGen automatically index elements of each input block.

The following is a list of the general steps involved in constructing a BioNetGen model with the relevant section of the BNGL input file shown in parenthesis:

1. (`parameters`) Define the parameters that govern the dynamics of the system (rate constants, the values for initial concentrations of species in the biological system) (*see* **Subheading 3.1**).
2. (`molecule types`) Define molecules, including components and allowed component states (*see* **Subheading 3.2**).
3. (`seed species`) Define the initial state of system (initial species and their concentrations) (*see* **Subheading 3.3**).
4. (`observables`) Define model outputs, which are functions of concentrations of species having particular attributes (*see* **Subheading 3.4**).
5. (`reaction rules`) Define rules that describe how molecules interact (*see* **Subheading 3.5**).
6. (`actions`) Pick method(s) for generating and simulating the network (*see* **Subheading 3.6**).

Steps 1–5 may be done in any order and the entire protocol is likely to undergo multiple iterations during the process of model development and refinement. **Subheadings 3.1–3.6** describe the sections of the BNGL input file with specific reference to the model presented in **Listing 1**. **Subheading 3.7** then presents two additional models that illustrate the use of more advanced language features.

3.1. Parameters

Model parameters, such as rate constants, values for initial concentrations of chemical species, compartment volumes, and physical constants used in unit conversions can be defined in the `parameters` block (*see* **Note 3**). Both numerical and formula-based parameter assignments are illustrated in the `parameters` block of **Listing 1**, which illustrates how formulas may be used to clarify unit conversions and to define a global parameter that controls the system size (*see* **Note 4**). Parameters have no explicitly defined units, but must be specified in consistent units, as assumed by BioNetGen. We recommend that concentrations be expressed in units of copy number per cell and bimolecular rate constants be expressed on a per molecule per cell basis, as in **Listing 1**. This choice, which assumes that the reaction compartment is a single cell and its surrounding volume, allows one to direct BioNetGen to switch from a deterministic simulation to a stochastic simulation without changing parameter units.

3.2. Molecule Types

Molecules in a BioNetGen model are structured objects composed of components that can bind to each other, both within a molecule and between molecules. Components typically represent physical parts of proteins, such as the SH2 and SH3 domains of the adapter protein Grb2, or the PxxP motif of the guanine nucleotide exchange factor Sos1 that serves as a binding site for SH3 domains. Components may also be associated with a list of state labels, which are intended to represent states or properties of the component. Examples of component states that can be modeled using state labels are conformation (e.g., open or closed), phosphorylation status, and location. There is no limit on the number of components that a molecule may have or on the number of possible state labels that may be associated with a component (*see* **Note 5**).

BioNetGen allows users to *explicitly* enforce typing of molecules using the `molecule types` block, which is optional but recommended. The `molecule types` block defines the allowed molecule names, the components of each molecule type (if any), and the allowed states of each of these components (if any). Each molecule type declaration begins with the name of a molecule (*see* **Note 6**) followed by an optional list of components in parentheses (*see* **Note 7**). The tilde character ("~") precedes each allowed state value. In the input file of **Listing 1**, five molecule types are declared. These molecules have 1, 3, 2, 1, and 0 components, respectively. The component named Y1068 represents a tyrosine residue in EGFR that can be in either an unphosphorylated (U) or phosphorylated (P) state. For a molecule to be able to bind another molecule, at least one component must be defined. A molecule without components cannot bind or change states, but can be created or destroyed. Such a molecule essentially corresponds to a named chemical species in a conventional model (*see* **Subheading 3.5.6**). A component that appears in a molecule type declaration without a state label may be used only for binding and may not take on a state label in subsequent occurrences of the same molecule. In contrast, the potential binding partners of a component are not delimited in a molecule type declaration.

The namespaces for components of different molecules are separated, so it is permissible for components of different molecules to have the same name. If two components of the same molecule have the same name, however, they are treated as separate instances of an identical type of object. For example, the two Fab arms of an IgG antibody have identical antigen-binding sites, which could be modeled as `IgG(Fab,Fab)`.

3.3. Seed Species

The `seed species` block defines the initial chemical species to which rules are applied. This block may also be used to define the initial levels of populated species and identify species with fixed concentrations. Before discussing the details of the `seed species` block, we need to briefly explain how chemical species are represented in BNGL.

Chemical species are individual molecules or sets of molecules connected by bonds between components, in which each component that has allowed state values has a defined state. For example, a cytosolic complex of Grb2 and Sos1 in the model of **Listing 1** would be represented as `Grb2(SH2,SH3!1).Sos1(PxxP!1)`, where the "." character is used to separate molecules that are members of the same chemical species and the "!" character is a prefix for a bond name (any valid name is allowed, but we recommend using an integer, which makes BNGL expressions more readable). A shared name between two components indicates that the components are bonded. A complex of Grb2 and Sos1 that is associated with EGFR would be represented as `EGFR(L,CR1,Y1068~P!2).Grb2(SH2!2,SH3!1).Sos1(PxxP!1)`, where the bond with the name "2" in this expression indicates that the SH2 domain of Grb2 is connected to the phosphorylated residue Y1068 in EGFR (i.e., connected to component `Y1068` of the molecule `EGFR`, which is in the `P` state). Note that in a BNGL expression for a chemical species all components of each molecule are listed and each component that is allowed to have a state has one defined state chosen from among the set of possible states for that component. Wild card characters, which represent nonunique states and bonds, are not allowed in BNGL chemical species expressions. These wild card characters are discussed below in **Subheading 3.4**.

Finally, we note that the presence or absence of the `molecule types` block affects the way that molecules appearing in the `seed species` block are type checked (*see* **Note 8**).

Specifying the initial population level of a seed species is accomplished in the same way that a parameter value is assigned using either a numerical value or a formula, as can be seen in **Listing 1** (*see* **Note 9**). A species listed in the `seed species` block may also be designated as having fixed concentration (*see* **Note 10**).

Representation of molecular complexes in BNGL has been presented in this section to introduce the syntax of bonds, but, generally speaking, it is not necessary to define seed species that are complexes of molecules because they can be generated through a process of equilibration (*see* **Subheading 3.6**), provided that there are rules that generate these complexes. If a complex species is defined and no reaction rule is specified that causes dissociation of the complex, the complex will be indivisible. A multimeric protein composed of several polypeptide chains could be specified in this way.

3.4. Observables

The `observables` block is used to specify model outputs, which are functions of the population levels of multiple chemical species that share a set of properties. For example, if one could measure the tyrosine phosphorylation level of a particular protein, then one might be interested in determining the total amount of all chemical species containing the phosphorylated form of this protein. We call a function for calculating such a quantity an "observable." Observables are computed over a set of chemical species that match a search pattern or set of search patterns specified in BNGL (*see* **Fig. 1b**). Each observable is defined by a line in the `observables` block consisting of an (optional) index, one of two keywords that defines the type of observable (Molecules or Species), a name for the observable, and a comma-separated set of search patterns (*see* **Listing 1** and **Note 11**). Before we discuss the two types of observables and how they are computed, we will describe the basic syntax and semantics of patterns in BNGL, which are common to observables and reaction rules.

Patterns are used to identify a set of species that share a set of features, and their behavior is illustrated in **Fig. 1b**. Pattern specification includes one or more molecules with optional specification of connectivity among these molecules, optional specification of states of their components, and optional specification of how these molecules are connected to the rest of the species they belong to. Patterns are analogous to the regular expressions used in computer programming. A match between a chemical species and a pattern means that there exists a mapping (injection) from the elements of the pattern to a subset of the elements of the species. Roughly speaking, a species matched by a pattern includes this pattern as a part. Note that there may be multiple mappings of a pattern into a single species and that BioNetGen considers each mapping to be a separate match. The formal definition of a match in the graph formalism upon which BNGL is based was given by Blinov et al. *(26)*. Patterns are similar to species in that they are composed of one or more molecules and may contain components, component state labels, and edges. Unlike in species, however, the molecules in patterns do not have to be fully specified and the molecules do not have to be connected to each other by bonds specified in the pattern. The absence of components or states in a pattern excludes consideration of the missing elements from the matching process, as illustrated in **Fig. 1b**. In the model of **Listing 1** observable 1 is specified using the pattern `EGFR()`, which matches any species containing a molecule of EGFR, regardless of the state or binding status of any of its components.

When a component is specified in a pattern, both the absence and presence of a bond name affects matching. The specification of a component without an associated bond requires that the component is unbound in the corresponding match. For example, observable 2 in **Listing 1** uses the pattern, `EGF(R)`, which selects only species in which the R component of EGF is unbound. The specification of a component with an associated bond is used to select bound components. If a complete bond is specified, as in observable 5, which selects complexes of Grb2 and Sos1, then the component must be bound in the manner indicated by the pattern (*see* **Note 12**). An incomplete bond may also be specified using "!+", where the wild card "+" indicates that the identity of the binding partner of a component is irrelevant for purposes of matching. For example, observable 3 in **Listing 1** uses the pattern, `EGFR(L!+)`, which selects species in which the L component of EGFR is bound, regardless of the binding partner. A second wild card, "?", may be used to indicate that a match may occur regardless of whether a bond is present or absent (*see* **Note 13**), and is sometimes required for the correct specification of observables. For example, the two patterns `EGFR(Y1068~P)` and `EGFR(Y1068~P!?)` are not equivalent. The first pattern selects only EGFR molecules in which the Y1068 component is phosphorylated and unbound, whereas the second pattern selects all EGFR molecules in which the Y1068 component is phosphorylated. (The second pattern is more relevant for comparing model predictions against the results of Western blotting with anti-pY antibodies.) Examples of patterns from the `observables` block of **Listing 1** and their corresponding matches in the implied model are listed in **Note 14**.

We are now ready to discuss the two types of observables. An observable of the Molecules type is a weighted sum of the population levels of the chemical species matching the pattern(s) in the observable. Each population level is multiplied by the number of times that the species is matched by the pattern(s). An observable of the Species type is simply an unweighted sum of the population levels of the matching chemical species (*see* **Notes 15** and **16**). A Molecules type of observable is useful for counting the number of copies of a particular set of patterns in a system, e.g., the number of copies of receptors in receptor dimers. A Species type of observable is useful for counting the populations of chemical species in a system containing a particular pattern (or set of patterns), e.g., the number of receptor dimers, as specified by observable 3 in **Listing 1**. Changing the type from Species to Molecules for this observable would specify a function that gives the number of copies of receptors in receptor dimers. The values of observables computed by one of the simulation commands described below are written to a .gdat file (*see* **Note 17**).

3.5. Reaction Rules

The `reaction rules` block of a BioNetGen input file is used to specify rules, which describe the allowed ways in which species can be transformed and typically represent molecular interactions and the consequences of these interactions. Each rule is similar to standard chemical reaction notation in that it has four basic elements: reactant patterns, an arrow, product patterns, and a rate law specification (*see* **Note 18**). Patterns in rules have the same syntax and semantics as introduced above in our discussion of the `observables` block. Reactant patterns are used to select sets of reactant species to which the transformation implied by the rule will be applied. The arrow indicates whether the rule is applicable in forward direction only ("->") or in both the forward and reverse directions ("<->"). The product patterns define how the selected species are transformed by the rule and act as the reactant patterns when the rule is applied in reverse. Rules may transform a selected set of reactant species by adding or deleting molecules or bonds and by changing component state labels. Rules may not add or delete components of molecules (*see* **Note 19**). The default rate law for reactions produced by rules is an elementary rate law, in which the rate is given by the product of a multiplicity factor (usually an integer or ½) generated automatically by BioNetGen (*see* **Subheading 3.5.3**), the specified rate constant (which may be a numerical value or a formula), and the population levels of the reactants. This type of rate law is specified simply by appending a comma-separated numerical value or formula at the end of the line defining a rule, as illustrated in **Listing 1**. Nonelementary rate laws, such as Michaelis–Menten rate laws, may also be specified (*see* **Note 20**). For a rule that defines reverse reactions, a second numerical value or formula follows the first after a comma. Rules 1, 2, 5, and 6 in **Listing 1** provide examples of how the parameters of two elementary rate laws are defined on the same input line. It should be noted that the parameter of a default rate law is taken to be a single-site rate constant (*see* **Note 21**). Additional commands that modify the behavior of rules may appear after the rate law specification (*see* **Subheading 3.5.7**).

Consider the `egfr_simple.bngl` file illustrated in **Listing 1**. Each reaction rule is defined on one line of the input file. (Recall that long input lines can be continued using the "\" character.) The first six rules represent classes of reactions mediated by particular molecular interactions (e.g., rule 1 specifies a class of ligand-receptor binding reactions in which the R domain of the ligand associates with the L domain of the receptor), and the last rule represents a class of irreversible degradation reactions, which removes receptor dimers from the system while retaining cytosolic molecules bound to the receptor complex. Rules 3 and 4 also define classes of irreversible reactions, whereas the remaining rules define classes of reversible reactions. The molecularity of a reaction, *M*, is the number of species participating in the reaction. The molecularity of all reactions generated by a given rule is fixed and is equal to the number of reactant patterns, which are separated by "+" characters. The value of *M* for rules 1–7 in **Listing 1** is 2, 2, 1, 1, 2, 2, and 1, respectively. The "+" character is used on the right side of a rule to define the number of products produced by a reaction and the molecularity of reverse reactions (if the rule is reversible). In **Listing 1**, rules 1–6 each have one product, and the reverse reactions have 2, 2, 1, 1, 2, and 2 product(s), respectively. Reactions defined by rule 7 have a variable number of products because of the `DeleteMolecules` keyword, which is discussed later in this section.

We will now discuss the five basic transformations that can be carried out by a BioNetGen rule. These transformations are (1) add a bond, (2) delete a bond, (3) change a component state label, (4) delete a molecule, or (5) add a molecule. In each case, there is a direct correspondence between a transformation of a set of graphs and a biochemical transformation of the molecules represented by the graphs *(26)*. For example, adding a bond between the interacting components of two binding partners corresponds to connecting two vertices in the graphs representing these binding partners. In the following subsections, we will discuss each of these types of transformations and present examples. A transformation is specified implicitly by the difference between the product and reactant patterns in a rule. BioNetGen automatically determines a mapping from reactant molecules and their components to product molecules and their components, and from this mapping determines the set of transformations implied by a rule. Although we will note exceptions, we recommend in general that each rule apply only a single transformation. A user may manually override automatic mapping through the use of molecule and component labels, as discussed in **Subheading 3.7.1** (*see* **Note 22**) *(28, 38)*. Such labels have been used to create a database of carbon atom fates in metabolic reactions *(38)*.

3.5.1. Add a Bond

A rule may add bond labels (e.g., "`!1`") to specific components of reactant species selected by the reactant pattern(s) in the rule, which results in the formation of a new bond. Including a bond in a product pattern that is absent in the reactant pattern(s) specifies this action. The simplest example of such a transformation is provided by the rule "`A(a)+B(b)->A(a!1).B(b!1) k_bi`," which specifies the association of molecules A and B through the formation of a bond between components a in molecule A and b in molecule B. Note that the "+" character constrains the molecularity to 2, which means that a and b must belong to separate species, precluding binding of A to B when these molecules are part of the same complex. To specify intracomplex binding of a and b, we could specify the rule as "`A(a).B(b)<->A(a!1).B(b!1)` k_uni", where the "." character in the reactant pattern indicates that the molecules A and B are part of the same complex. Note that these two rules have bimolecular and unimolecular rate laws, respectively, because they have different molecularities, and thus the units of k_bi and k_uni necessarily differ. As noted earlier, it is the modeler's responsibility to specify values of model parameters using consistent units.

Let us consider rule 1 in **Listing 1**, which provides an example of a reaction rule for the reversible binding of a ligand to a receptor. We first consider application of the rule in the forward direction (application of the rule in reverse will be considered in **Subheading 3.5.2**). The reactant pattern EGF(R) selects ligand (EGF) molecules that have an unbound R component. Since EGF molecules in this model have only one component, the only species that is selected by this pattern is **EGF(R)** (Here, we adopt the convention that the image of a pattern in a matching species is shown in **bold**). The pattern EGFR(L,CR1) selects EGFR molecules with unbound L and CR1 components, regardless of the binding or phosphorylation status of the Y1068 component of EGFR. For example, the pattern would select all of the following possible species: **EGFR(L,CR1,**Y1068~U), **EGFR(L,CR1,**Y1068~P), and **EGFR(L,CR1,**Y1068~P!1). Grb2(SH2!1,SH3). By specifying the component CR1 in the pattern and indicating that this component is free (by the absence of a bond specification), we are requiring that the CR1 component be unbound. Because receptors must associate via the CR1 domain to form dimers, as specified by rule 2, this means that ligand can bind receptor monomers but not dimers through rule 1. Rule 1 can be made independent of the state of CR1 by simply omitting it from the pattern for EGFR. In other words, by specifying EGFR(L) instead of EGFR(L,CR1), ligand is allowed to associate with (and dissociate from) both monomeric and dimeric receptors. The general principle is that a reaction rule should only include molecules, components, state labels, and bond specifications that are either modified by a transformation or that affect the transformation. We call the component(s) directly modified by a transformation a *reaction center* and the rest of the information included in a rule the *reaction context*. For clarity, we will underline the reaction centers in the rules (*see* **Listing 1**). The process of rule application is illustrated in **Fig. 1** and further examples are listed in **Note 23**.

Let us now consider rule 2 of **Listing 1**, which specifies the reversible dimerization of ligand-bound EGFR and illustrates the use of bond wild cards in the reactant specification. The "!+" string following the L component of each EGFR means that the L component must be bound (albeit in an unspecified way) for the pattern to match and thus for the reaction to take place.

Another important feature of this rule is that it is symmetric with respect to interchange of the two reactant patterns, which is detected automatically by BioNetGen, which then ensures that generated reactions are assigned rate laws with correct multiplicity. Reaction multiplicity, which is a multiplicative factor in a rate law, is discussed in more detail below in **Subheading 3.5.3**. For many users, it is sufficient to note that BioNetGen automatically detects symmetries in rules and generates reactions with correct multiplicities.

3.5.2. Delete a Bond

Rules specify bond deletion when a bond that appears in the reactant patterns has no corresponding bond on the product side (*see* **Note 24**). Frequently, bond deletion rules are specified simply by making a bond addition rule reversible, as in the extension of the elementary bond addition rule above to "`A(a)+B(b) <-> A(a!1).B(b!1) k_a,k_d`". Bond dissociation step can also be specified using a unidirectional rule, as in "`A(a!1).B(b!1) -> A(a)+B(b) k_d`". The reversible rule syntax is provided solely as a matter of convenience; the functional behavior of the rules is identical whether an association/dissociation pair is specified as a single reversible rule or as two irreversible rules with the reactant and product patterns interchanged (*see* **Note 25**). Note that the molecularity of the products in the dissociation rule (2 in this case) has a restrictive effect analogous to that of the specification of molecularity in the association rule. When the rule is applied to a species selected by the reactant pattern, a reaction is generated only if removal of the specified bond eliminates all possible paths along bonds between A and B, i.e., if bond removal produces two separate fragments. Specifying bond dissociation that does not result in breakup of the complex requires a rule of the form "`A(a!1).B(b!1) -> A(a).B(b) k_d`". An example illustrating the different action of these two rules is provided in **Note 26**.

As an example of a bond deletion rule that has additional reaction context, let us consider the reverse of the dimerization rule discussed in **Subheading 3.5.1**,

```
EGFR(L!+,CR1!1).EGFR(L!+,CR1!1)→EGFR(L!+,CR1)+\
                                 EGFR(L!+,CR1) km2,
```

which breaks the bond between the CR1 components of two receptors in a complex. The contextual requirement that an L component of each EGFR also be bound is specified using the bond wild card "`L!+`". The molecularity of the products in the rule means that the rule will only be applied if breaking the bond results in dissociation of an aggregate. It is important to note here that the bond wild card "`!+`"can only be used to specify context; it is not permitted to break a bond that is only partially specified because such a rule would leave the molecularity unspecified.

3.5.3. Change a Component State Label

Rules specify a change in the state label of a component whenever the state label of a component changes in going from its appearance in the reactants to its corresponding occurrence in the products. State label changes may be used to represent covalent modification, a change in conformation, translocation between two compartments, or any other property of a molecule that might influence its subsequent reactivity. The simplest possible example of a rule specifying a state label change is rule 4 of **Listing 1**,

```
EGFR(Y1068~P)->EGFR(Y1068~U) km3
```

which encodes the dephosphorylation of a receptor tyrosine, through a change in the state label for Y1068 from "P", representing the phosphorylated state, to "U", representing the unphosphorylated state (*see* **Note 27**). It should be noted that just as for bond addition and deletion reactions, the rate constant should be specified as if only one instance of the reaction implied in the rule is possible for any given set of reactant species (*see* **Note 21**). BioNetGen will generate a distinct reaction for each distinct occurrence of the reactant pattern in a species. For example, consider the application of rule 4 to the following species in the EGFR network:

```
EGF(R!1).EGF(R!2).EGFR(CR1!3,L!1,Y1068~P)\
.EGFR(CR1!3,L!2,Y1068~P)
```

The two occurrences of the reactant pattern are shown in **bold**. During the process of network generation, this species is automatically assigned the index 11, which is used to reference species in the `reactions` and `groups` blocks of the resulting .net file. Because this species is symmetric, application of the rule generates two instances of the dephosphorylation reaction 11 → 8, and species 8 is

```
EGF(R!1).EGF(R!2).EGFR(CR1!3,L!1,Y1068~U)\
.EGFR(CR1!3,L!2,Y1068~P)
```

In this case, application of rule 4 to the first Y1068 appearing in species 11 generates the same species as application of the rule to the second instance (*see* **Note 28**). Upon generation of a reaction, BioNetGen checks to determine whether the reaction is identical to one that has already been generated. If so, the multiplicity of the reaction is incremented by one (*see* **Note 29**). So application of rule 4 to species 11 produces the reaction `11→ 8 2*km3`, where `2*km3` following the reaction refers to the constant portion of the elementary rate law that is used to compute the rate of the reaction. The multiplicity of the reaction is 2, and the rate is given by `2*km3*X11`, where X11 is the population level of species 11.

As in other reaction rules, additional contextual information can be supplied to restrict application of a rule. An example of a rule that uses contextual information in this way is rule 3 of **Listing 1**, which specifies phosphorylation of Y1068 within a receptor aggregate:

```
EGFR(CR1!+,Y1068~U)→ EGFR(CR1!+,Y1068~P) kp3
```

In this rule, the wild card operator "+" is used to specify that the phosphorylation reaction occurs only for a receptor that is part of a receptor dimer. Because in this model the CR1 domain can only bind to another CR1 domain, requiring CR1 to be bound, as specified here, is equivalent to requiring that another EGFR be present in the aggregate (*see* **Note 30**). Thus, the rule above models trans (auto)phosphorylation of Y1068 catalyzed by the protein tyrosine kinase domain in a neighboring copy of EGFR.

3.5.4. Add a Molecule

In addition to the operations described in the previous sections, rules may also specify the creation of new molecules as products, which could be used to model, for example, translational processes or transport across the cell membrane. As a simple example of how to introduce a source for a protein A, consider the rule

```
I()-> I()+A(a,Y~U)ksynth
```

where `I()` is a structureless molecule. The appearance of I on both the reactant and product sides of the rule means that its concentration will not change as a result of the reaction occurring. If the species "`I()`" is set to have a concentration of 1 in the `seed species` block and its concentration is not affected by any other rules, the rate constant `ksynth` will be have units of concentration/time and will define the synthesis rate of the species `A(a,Y~U)`. Note that BioNetGen does not allow the number of reactants or products in a reaction to be zero, which is why the molecule I must be included in this rule. Molecule addition is specified any time that a molecule appearing on the product side of a rule has no corresponding molecule on the reactant side. Appearance of a new molecule in the products generates an error unless the molecule is fully specified, i.e., all components of the molecule are listed and those components requiring a state label have a valid specified state label, and connected to the remainder of the pattern in which it appears. New molecules can also be combined with reactant molecules, as in the rule

```
B(b)-> B(b!1).A(a!1,Y~U)ksynth
```

which creates a new molecule of A bound to a B molecule.

3.5.5. Delete a Molecule

Rules may also specify degradation of specified molecules or of entire species matching a particular reactant pattern by omitting reactant molecules in the product patterns. Because degradation rules may specify deletion of individual molecules or entire species, the semantics of degradation rules are somewhat more complicated than those of other rules considered so far. Let us first consider the simplest form of a degradation rule

```
A()-> Trash() kdeg
```

which specifies degradation of any species in which the molecule A appears. Degradation of a species is specified whenever all of the reactant molecules used to select the species are omitted from the products. This rule also specifies the synthesis of a Trash molecule, which is necessary because BioNetGen require that at least one product molecule be specified. Note that the species `Trash()` acts as a counter for the number of A-containing species that have been degraded (*see* **Note 31**). If multiple molecules of A can appear within a single species, degradation reactions involving these species would have multiplicity equal to the number of occurrences of A in the degraded species. In other words, a species containing *n* copies of A will be degraded *n* times faster than a species containing only a single copy of A. If this behavior is not the desired, then the multiplicity can be held to one by specifying the `MatchOnce` attribute for the reactant pattern, as in

```
{MatchOnce}A()-> Trash() kdeg
```

As of this writing `MatchOnce` is the only recognized pattern attribute.

Rules can also specify the degradation of a set of molecules within a complex, which can be accomplished in one of two ways. First, one can specify the degradation of a molecule or molecules within a reactant complex by transferring to the products at least one of the molecules used to select the complex on the reactant side. The simplest example is the rule

```
A().B() -> B() kdeg
```

which specifies the deletion of the matching A molecule in the complex. When the rule is applied, the A molecule and all of its bonds will be deleted. If this action leaves behind only a single connected fragment containing the matched B molecule, a reaction will be generated. If, however, deletion of A leaves behind multiple fragments, no reaction will be generated. The keyword `DeleteMolecules` can be added to the rule following the rate law to bypass this constraint, as in

```
A().B()-> B()kdeg DeleteMolecules
```

which, when applied to the complex `C(c!1).A(a1!1,a2!2).B(b!2)`, would generate the reaction

```
C(c!1).A(a1!1,a2!2).B(b!2)->C(c)+B(b)kdeg
```

The deletion of the A molecule from the C-A-B chain produces a C fragment and a B fragment. The `DeleteMolecules` keyword can also be used when no molecules from the reactant pattern remain in the products. Thus, the species-deleting rule from the previous paragraph can be transformed into a molecule-deleting rule

```
A()->Trash() kdeg DeleteMolecules
```

which has the same action on the C-A-B complex as the rule above.

Rule 7 of **Listing 1** provides an example of how such a rule might be used to model endosomal degradation of signaling complexes in which some components of the complex are recycled. The rule specifies that the EGFR dimer and both associated EGF molecules are degraded, but the `DeleteMolecules` keyword means that additional molecules associated with the complex will be retained as products in any generated reactions. Thus, any Grb2 molecules that associate with such a dimer and any Sos1 molecules that bind to dimer-associated Grb2 molecules are (effectively) returned to the cytoplasm when the receptor complex is degraded.

3.5.6. Encoding Conventional Reactions

Addition and deletion actions may be combined within single rules to construct rules that describe conventional mass action kinetics involving structureless species. A typical rule of this type would be

```
A+B -> C kAB
```

which encodes the deletion of A and B and the addition of C. This rule will be valid only if the molecule C is defined to have no components, and it will have the intended meaning only if A and B are also structureless. Any standard reaction scheme can thus be trivially encoded in BioNetGen, although the power of the rule-based approach is lost. Structureless species may be useful as sources and sinks, and may also be used to represent small molecules or atoms. Note that `A` and `A()` are equivalent representations for a molecule or species A, in that neither representation specifies the substructure of A.

3.5.7. Commands for Modifying Rule Application

As described in **Subheading 3.5.5**, BioNetGen includes several commands that modify the application of rules. These commands have been introduced to address the need for specific behaviors that are difficult or impossible to specify using the semantics of patterns and transformation rules alone. In this section, we cover the `include/exclude` commands that provide a basic logic for extending the selection capabilities provided by patterns. In the future, we anticipate the development of a "pattern logic" that will provide these capabilities in a more general way.

The basic functionality of the `include_reactants` and `include_products` commands is to add criteria for the selection of reactant species to be transformed by a rule or the acceptance of products species generated by a rule. In other words, these commands provide an AND operator for pattern matching. The basic syntax of the include commands is illustrated by the rule

```
A(a)+B(b)->A(a!1).B(b!1) kab include_reactants\
(2,R1,R2)
```

which specifies that a bond will be created between a reactant species containing a free component a of a molecule A and a second reactant species containing a free component b of a molecule B only if the second reactant species also includes a molecule of *either* R1 or R2. The first argument of an `include` command is always a number corresponding to the index of a reactant or product pattern in the rule (1 for the first reactant/product, 2 for the second, etc.), and the remaining arguments are BNGL patterns, at least one of which must generate a match for the species to be selected. In logical terms, the effective pattern for the second reactant in this rule becomes "`B(b) AND (R1 OR R2)`". Any valid BioNetGen pattern may be used as an argument to an `include_reactants` or `include_products` command. Multiple include commands applying to the same reactant or product pattern can be specified to create additional selection criteria for a species, and thus function as additional AND operators. To generate similar behavior without the include command, two rules would have to be specified:

```
A(a)+B(b).R1A->(a!1).B(b!1) kab
A(a)+B(b).R2A->(a!1).B(b!1) kab
```

where the "." operator is used to test for the presence of an additional molecule in the second reactant complex. It is worth noting that the rule using the `include_reactants` command behaves slightly differently in this case than the pair of rules, because the latter may each generate multiple matches to the same reactant species if multiple molecules of either R1 or R2 are present. For instance, the pattern "`B(b).R1`" generates two matches to the species "`B(b.r!1).R1(r!1,d!2).R1(r,d!2)`" because R1 in the pattern can be mapped onto either of the two R1's in the complex. It is easy to specify two rules that have the same behavior as the one rule by extending the pattern "`B(b).R1`" to "`B(b.r!1).R1(r!1)`". Unfortunately, as illustrated in this example, subtle differences in the way that rules are specified can have dramatically different effects, which are sometimes difficult to anticipate. This problem will be alleviated in the future by extending BioNetGen to allow a user to differentiate between the reaction center (the part of a pattern affected by a transformation) and reaction context (the part of a pattern necessary for a transformation to occur) in rules.

The "`exclude_reactants(index,pattern1,pattern2,...)`" and "`exclude_products`" commands have the same syntax as the `include` commands but apply the logic "pattern_index AND ((NOT pattern1) OR (NOT pattern2) ...)", where pattern_index is the pattern used to specify the reactant or product with the specified index. Equivalent functionality can be obtained by the use of patterns alone, but in complex cases several patterns may be required to accomplish the same effect. It should be noted that when they appear in reversible reactions, include_reactants and exclude_reactants are automatically transformed into `include_products` and `exclude_products`, respectively, when the rule is applied in the reverse direction. Appearances of `include_products` and `exclude_products` commands are also similarly transformed.

3.6. Actions

BioNetGen is capable of performing two basic types of actions with a model specification in an input file: generate a chemical reaction network implied by the model specification and simulate the network (e.g., solve an initial value problem for the system of coupled ODEs that provides a deterministic description of the reaction kinetics in the well-mixed limit). These actions are controlled using commands that follow the model specification blocks we have discussed in the previous section (*see* **Listing 1**). Other commands export BioNetGen-generated networks in various formats. All of the available commands and the parameters that control them are summarized in **Table 1**, which also summarizes the general syntax.

3.6.1. Generating a Network

The commands shown in **Listing 1** illustrate the range of actions that can be performed on a BioNetGen model. The `generate_network` command directs BioNetGen to generate a network of species and reactions through iterative application of the rules starting from the set of seed species. At each step in this iterative process, rules are applied to the existing set of chemical species to generate new reactions. Following rule application, the species appearing as products in the new reactions are checked to determine whether they correspond to existing species in the network *(26)* (*see* **Note 32**). If no new species are found, network generation terminates.

Restrictions on rule application may be useful when rules sets would otherwise produce very large or unbounded networks (*see*

Table 1
Syntax and parameters for BioNetGen actions

Action/parameter[a]	Type[b]	Description	Default
generate_network		**Generate species and reactions through iterative application of rules to seed species**	
max_agg	int	Maximum number of molecules in one species	1e99
max_iter	int	Maximum number of iterations of rule application	100
max_stoich	hash	Maximum number of molecules of specified type in one species	-
overwrite	0/1	Overwrite existing .net file	0 (off)
print_iter	0/1	Print .net file after each iteration	0
prefix[c]	string	Set basename[e] of .net file to *string*	basename of .bngl file
suffix[c]	string	Append _*string* to basename of .net file	-
simulate_ode/simulate_ssa		**Simulate current model/network**	
t_end	float	End time for simulation	required
t_start	float	Start time for simulation	0
n_steps	int	Number of times after t=0 at which to report concentrations/observables	1
sample_times	array	Times at which to report concentrations/observables (supercedes t_end, n_steps)	-
netfile	string	Name of .net file used for simulation	-
atol[d]	float	Absolute error tolerance for ODE's	1e-8
rtol[d]	float	Relative error tolerance for ODE's	1e-8
steady_state[d]	0/1	Perform steady-state check on species concentrations	0
sparse[d]	0/1	Use sparse Jacobian/iterative solver (GMRES) in CVODE	0
readFile		**Read a .bngl or a .net file**	
file	string	Name of file to read	required
writeNET/writeSBML/writeMfile		**Write current model/network in specified format**	
setConcentration(species,value)		**Set concentration of species to value**	
setParameter(parameter,value)		**Set parameter to value**	

(continued)

Table 1
(Continued)

Action/parameter[a]	Type[b]	Description	Default
saveConcentrations()		Store current species concentrations	
resetConcentratons()		Restore species concentrations to value at point of last saveConcentrations command	

[a]General syntax is *action*({*scal val*,*array* [*x1*,*x2*,...],*hash* ⇒{*key1*⇒*val1*,*key2*⇒*val2*,...},...}).
[b]Scalar types are int, 0/1 (a boolean), string, and float. Multivalued parameters may be either arrays or hashes.
[c]The prefix and suffix parameters can be used with any command that writes output to a file.
[d]These parameters only apply to simulate_ode.
[e]*See* **Note 35**.

Note 33). These restrictions can be imposed using optional arguments to the `generate_network` command, which are shown in **Table 1**. The three basic restrictions that can be specified are an upper limit on the number of iterations of rule application (`max_iter`), an upper limit on the number molecules in an aggregate (`max_agg`), and an upper limit on the number of molecules of a particular type in an aggregate (`max_stoich`). An example of a command specifying all three restrictions in the order given above is

```
generate_network ({max_iter => 15,max_agg => 10,
                    max_stoich=>{L => 5,R > 5});
```

This command limits the number of iterations to 15, the maximum size of an aggregate to 10 molecules, and the maximum number of L or R molecules in an aggregate to be 5. An example illustrating the use of such restrictions is given in **Subheading 3.7.2**.

When network generation terminates, whether through convergence or when a stopping criterion is satisfied, the resulting network is written to a file with the .net extension (*see* **Note 34**). By default the basename of this file is determined from the basename of the input .bngl file. For example, the `generate_network` command in the file `egfr_simple.bngl` creates the file `egfr_simple.net` by appending the .net extension to the basename `egfr_simple`. The options `prefix` and `suffix`, which are taken by all commands that write output to a file, can be used to modify the basename of all files generated by the command (*see* **Note 35**). By default, `generate_network` will terminate with an error if the .net file it would produce exists prior to network generation. This behavior can be overridden by setting option `overwrite =>1`, as shown in **Listing 1**. This option can be useful during the debugging phase of model development.

3.6.2. Simulating a Network

Once a network has been generated, a simulation can be specified using the `simulate_ode` or `simulate_ssa` commands. The simulation specified in the example in **Listing 1** consists of three phases, which we now summarize and will be described in detail below. The first phase is equilibration, in which reactions that can occur prior to the introduction of the EGF ligand are allowed to reach steady state. Time courses produced by the first `simulate_ode` command, which terminates when the species concentrations pass a numerical check for convergence, are written to the files `egfr_simple_equil.gdat` and `egfr_simple_equil.cdat` (assuming the input file is named `egfr_simple.bngl`). Before the second phase of simulation, ligand is introduced (using `setConcentration`), the concentrations at the end of equilibration are saved (using `saveConcentrations`), and the network is written to an SBML file (using `writeSBML`). The second `simulate_ode` command then initiates a simulation of the dynamics following introduction of EGF ligand into the system. The results are written to the files `egfr_simple.gdat` and `egfr_simple.cdat`. The third phase is then preceded by a `resetConcentrations` command, which restores the concentrations to the initial values used in the second phase, i.e., following equilibration and introduction of EGF. The `simulate_ssa` command then initiates the third and final phase of simulation, a kinetic Monte Carlo simulation using the Gillespie algorithm, and results are written to the files `egfr_simple_ssa.gdat` and `egfr_simple_ssa.cdat`.

In the equilibration phase the population level of the ligand (`EGF(R)`) is zero, as specified in the `seed species` block of **Listing 1**. Network generation is unaffected by the population levels of the seed species, but in the absence of ligand the only reactions with nonzero flux are the binding and unbinding reactions of Grb2 and Sos1 in the cytosol, which are defined by rule 6. The purpose of the equilibration phase is then to allow the concentrations of free Grb2, free Sos1, and the cytosolic Grb2-Sos1 complex to reach steady-state levels, which we would expect to find in the resting state of the cell.

The first `simulate_ode` command propagates the simulation forward in time (in large time steps) and checks for convergence to a steady state. By going over each of the options used in this command, we will provide an overview of the operation and capabilities of the `simulate_ode` command. The "`suffix ⇒ equil`" appends "`_equil`" to the basename for output files of the simulation, which becomes here "`egfr_simple_equil`". This prevents output files from the equilibration phase from being overwritten by subsequent simulation commands. The end time (`t_end`) for the simulation is given a sufficiently large value to ensure that steady state is reached prior to the end of the simulation (*see* **Note 36**). The number of steps at which results are written to the output files is specified by the `n_steps` parameters, which is set to a relatively small value here because we are only interested in reaching steady state and not in tracking the time course. The interval between reporting of results is given by (`t_end/n_steps`), which is 10,000 s in this case. (Note that the `n_steps` parameter controls only the reporting interval and not the step size used by the CVODE solver, which uses adaptive time stepping). Results can also be reported at unevenly spaced intervals (*see* **Note 37**). The `sparse` option invokes fast iterative methods in the CVODE solver that can greatly accelerate the simulations (*see* **Note 38**). The `steady_state` flag causes a check for the convergence of the species population levels to be performed following each report interval, with the propagation terminating if the root mean square of the relative change in the population levels falls below a threshold, which is taken to be 10×`atol`, the absolute integration tolerance. Note that the basic operation of the `simulate_ssa` command is the same as that of the `simulate_ode` command. A summary of options available for the simulation commands is given in **Table 1**. Of the options discussed above, only `steady_state` and `sparse` are not available for use with `simulate_ssa`.

After completion of a simulation, the final population levels of all species in a network are saved and used by default as the initial population levels for subsequent simulation commands. In the example, we have modified or overridden this behavior by using the `setConcentration` (*see* **Note 39**) or `resetConcentrations` commands (*see* **Note 40**). Additional options are discussed in **Subheading 3.6.4**.

3.6.3. Viewing the Simulation Results

We now consider visualization of the output produced by the two simulation commands that follow equilibration. Each simulation is run from the same initial conditions, but the second is run using the `simulate_ssa` command, which produces a stochastic (discrete-event) trajectory using the direct method of Gillespie *(40)*. Trajectory data are written into two multicolumn output files for each simulation: a .gdat file that reports the value of each defined observable at each sample time and a .cdat file that reports the population level of every species in the network at each sample time. To avoid overwriting the data produced by `simulate_ode`, the `simulate_ssa` command sets the suffix parameter to "`ssa`", so that the basename of the file becomes "`egfr_simple_ssa`". Both data file types are in ASCII format, so they can be viewed in a text editor or imported into any number of different plotting and data analysis programs. The BioNetGen distribution includes the PhiBPlot plotting utility, which is a Java program that can be run by double-clicking on the file `PhiBPlot.jar` in the `PhiBPlot` subdirectory of the distribution or by typing "`java -jar path/PhiBPlot.jar [datafile]`"

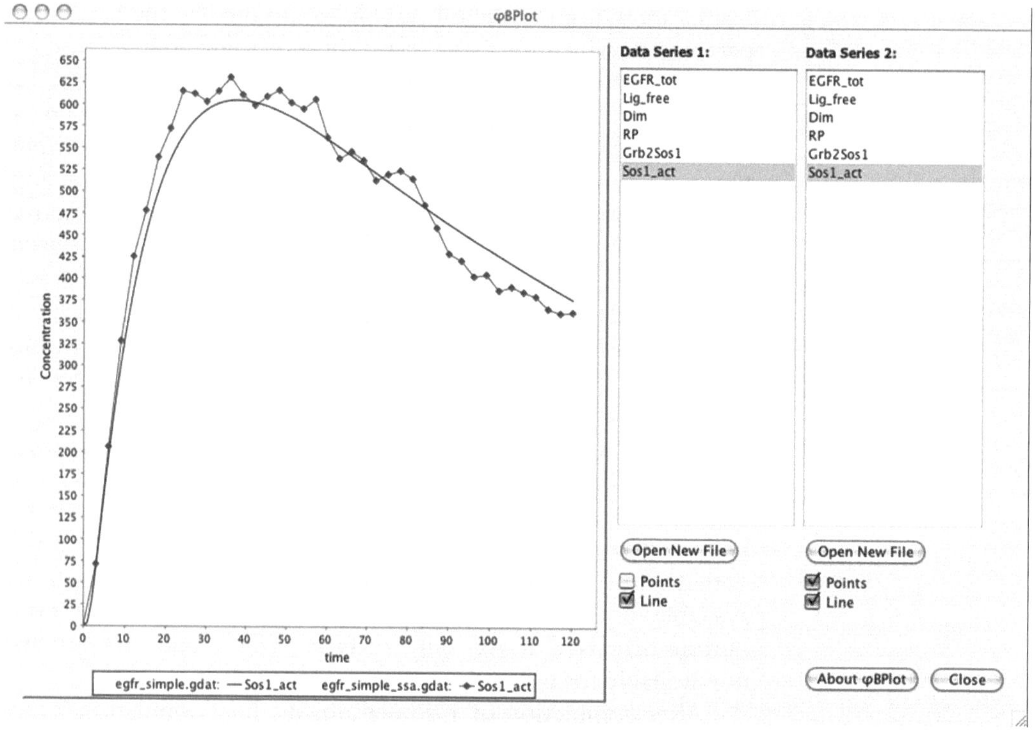

Fig. 3. Plotting BioNetGen simulation data in PhiBPlot. Data from up to two different files may be plotted simultaneously. Here, data for the `Sos1_act` observable from the ODE and SSA simulations is overlaid, showing the effects of fluctuations in the stochastic simulation.

on the command line. PhiBPlot can display data from up to two BioNetGen data files at a time and is useful for quickly visualizing the results of a BioNetGen simulation and for comparing the results of two (*see* **Fig. 3**).

3.6.4. Simulating a Previously-Generated Network

Network generation can be the most time-consuming part of processing a BioNetGen input file, and during repeated simulations of the same network (e.g., with varying parameters) one may wish to avoid regenerating the network. There are several ways to achieve this outcome. The first way, presented in the example above, is to run multiple simulations within the same input file using the `saveConcentrations`, and `resetConcentrations` commands in combination with the `setConcentration` and `setParameter` commands to vary initial conditions and parameters (*see* **Note 41**).

In some cases, however, it may be desirable to reload a network that was generated during a previous invocation of `BNG2.pl`. The `readFile` command provides a way to fully restore a previously generated network so that parameters and species concentrations can be modified using the `set` commands. The basic syntax is illustrated by the command

```
readFile({prefix=>"testread",file=>
"egfr_simple.net"}),
```

which restores the network generated in the example of **Listing 1** with population levels set to their postequilibration values (*see* **Note 42**). The `readFile` command, unlike other BioNetGen commands, resets the global basename to be the basename of the `file` argument, which is "`egfr_simple`" in the example given above. The `prefix` parameter is set here to override this behavior and to set the basename for subsequent simulations commands to "`testread`" rather than `egfr_simple`.

Reading a previously generated network from a file is always much faster than regenerating the network, but can still be time-consuming for very large networks. It may, therefore, be advantageous to pass the previously generated .net file directly to the simulation program by using the `netfile` argument to the `simulate_x command`, as in

```
simulate_ode({netfile=>"egfr_simple.net
t_end=>120,n_steps=>12});
```

The disadvantage of this method is that it does not permit the model parameters to be changed without directly editing the .net file (*see* **Note 43**).

3.7. Additional Examples

In this section, we discuss two example applications of BioNetGen. In the first example, we illustrate how BioNetGen can be used to extend a conventional model so that it can be used to interpret fluorescent labeling experiments. In the second example, we illustrate how BioNetGen can be used to produce a model for a system in which polymerization-like reactions are possible (e.g., a model for multivalent ligand-receptor interactions). The graphical formalism upon which BioNetGen is based was designed with these types of systems in mind (*25*). The structured objects (graphs) of BNGL allow the topological connectivity of (protein) complexes to be explicitly represented and tracked in a model.

3.7.1. Fluorescent Labeling

Here we illustrate how BioNetGen can be used to extend an existing (nonrule-based) reaction network. In some cases, one needs to add a property that is passed from one species to another in a reaction network. For example, many experiments involve fluorescent labeling, in which the system is injected with fluorescently-labeled proteins that can be monitored. Fluorescent species carry all the properties of nonfluorescent species, but can also be photobleached, losing fluorescence. Given a reaction network of nonfluorescent species, the network that includes both fluorescent and nonfluorescent species nearly doubles in size. For larger networks, this expansion will be error-prone if done manually.

Thus, it is desirable to be able to extend a model to enable tracking of fluorescent labels, and BioNetGen provides such a capability by allowing the definition of a mapping of component state labels from reactants to products. In addition to the application shown here, these mappings have been used to define carbon fate maps for many of the currently known reactions in metabolism *(38)*.

We consider a simple reaction network consisting of five species and described by four basic reactions (considering each direction as a separate reaction)

$$A + B \xleftrightarrow[k_{-AB}]{k_{+AB}} C$$

$$C + D \xleftrightarrow[k_{-CD}]{k_{+CD}} E$$

The label chemistry we want to describe works as follows: fluorescence is passed from A to C in reaction 1 and from C to E in reaction 2. This can be described by adding a component, which we will call "`f`", to the molecules A, C, and E. The f component in each molecule may be in either the "`off`" or the "`on`" state, as shown in the `molecule types` definitions of **Listing 2**. We then define rules for mapping the state of the f component between

Listing 2. BioNetGen input file for the fluorescent labeling example **(see Subheading 3.7.1)**.

```
begin parameters
NA 6.02e23 # Avogadro's number (molecues/mol)
f  0.1       # Fraction of the cell to simulate
Vo f*1.0e-10 # Extracellular volume=1/cell_density (L)
V  f*3.0e-12 # Cytoplasmic volume (L)
# Initial concentrations (copies per cell)
A_tot 10000
B_tot  8000
D_tot 50000
# Rate constants
# Divide by NA*V to convert bimolecular rate constants
# from /M/sec to /(molecule/cell)/sec
kpAB 3.0e6/(NA*V)
kmAB 0.06
kpCD 1.0e6/(NA*V)
kmCD 0.06
kpI  1.0e7/(NA*V)
kmI   0.1
end parameters

begin molecule types
A(f~off~on)
B()
C(f~off~on)
D()
E(f~off~on)
I()
```

```
end molecule types

begin seed species
A(f~off) A_tot
B()      B_tot
C(f~off) 0
D()      D_tot
E(f~off) 0
I()      0
end seed species

begin reaction rules
1 A(f%1) + B() <-> C(f%1) kpAB, kmAB
2 C(f%1) + D() <-> E(f%1) kpCD, kmCD
3 A(f~off) + I <-> A(f~on) kpI, kmI
end reaction rules

begin observables
Molecules A_f A(f~on)
Molecules C_f C(f~on)
Molecules E_f E(f~on)
Molecules Tot_f A(f~on) ,C(f~on),E(f~on)
end observables

generate_network({overwrite=>1});
# Equilibrate
simulate_ode({suffix=>equil,t_end=>10000,n_steps=>10,\
  steady_state=>1});
# Add indicator
setConcentration("I","A_tot/10");
simulate_ode({t_end=>200,n_steps=>50});
```

A and C (*see* rule 1 in `reaction rules` block of **Listing 2**) and between C and D (*see* rule 2 in `reaction rules` block of **Listing 2**) using the "`%`" character followed by a string to tag components (*see* **Note 22**). By not specifying the component state of f in the rules, we cause the component state to be mapped from the selected reactant molecule to the created product molecule. This trick allows us to avoid writing separate rules for the labeled and unlabeled species. (When mapping components in this way the user should be careful that the allowed state label values of the components are the same or an error will be generated.) The defined observables track the amount of label associated with each of the molecules that can be labeled (`A_f`, `C_f`, and `E_f`) and the total amount of label present in the system (`Tot_f`). The resulting network has 9 species and 10 reactions.

There are different ways in which labeled components may be introduced into the system. The simplest way would be to define an initial pool of labeled A molecules, i.e., define the species "`A(f~on)`" to have nonzero initial concentration. Here, we have chosen a somewhat more complex scenario in which the system is initially equilibrated without the label, followed by the introduction of an indicator molecule that adds label to A through a chemical reaction, the third rule in the input file. Following equilibration with no indicator present, the indicator concentration is set to be a fraction of the total number of A molecules using the `setConcentration` command. Results of simulation of the network following equilibration and introduction of the indicator molecule are shown in **Fig. 4**. The labeling reaction (rule 3) is fast compared with the other reactions, so that labeled A initially accumulates followed by a slower rise in the levels of labeled C and D molecules.

3.7.2. Polymerization

BNGL can be used to model the kinetics of molecular aggregates having different topological structures, such as chains, rings, and trees. Here, we present a simple model for the binding of a soluble multivalent ligand to a bivalent cell-surface receptor, such as a membrane-bound antibody. In this model, we consider a

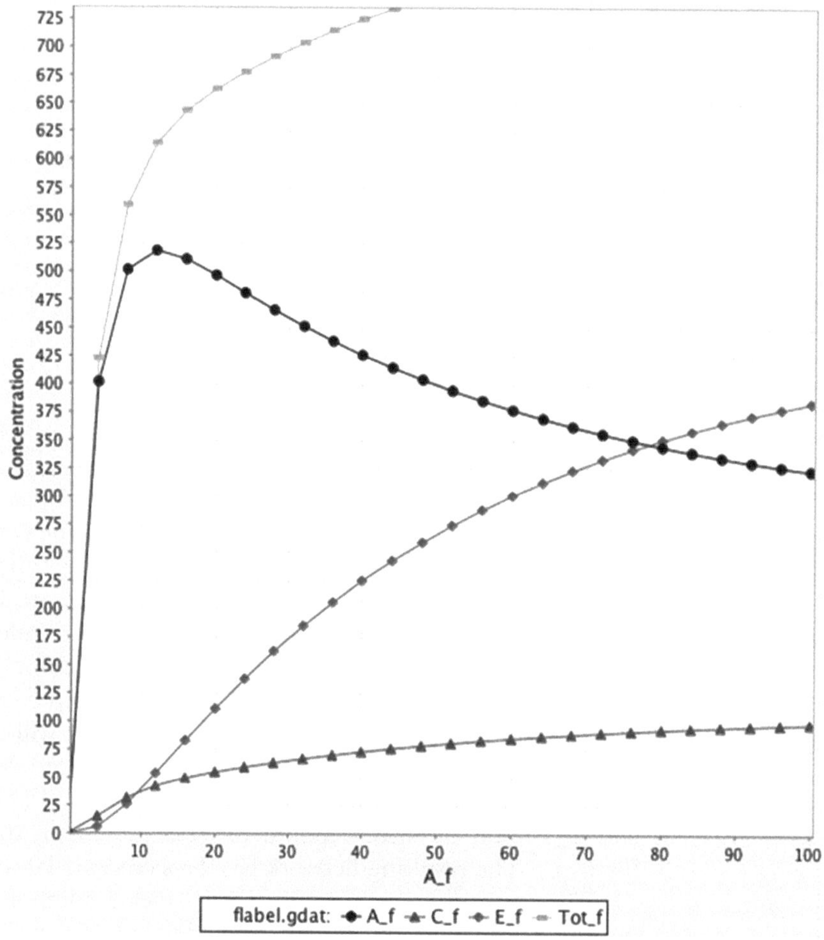

Fig. 4. Plot of simulation results obtained from BioNetGen input file for the fluorescent labeling example shown in **Listing 2** made using PhiBPlot (black and white rendition of color output). The plot shows time courses of the observables from the second `simulate_ode` command in the `actions` block of **Listing 2**.

bivalent ligand with two identical binding sites (L(l,l)) and a bivalent receptor with two identical binding sites (R(r,r)). The ligand may cross-link two receptors to form a dimeric receptor aggregate (R(r,r!1).L(l!1,l!2).R(r!2,r)), which can then interact with additional ligand via free receptor sites. Ligand-receptor interaction can form a distribution of linear chains of alternating ligands and receptors (R(r,r!1).L(l!1,l!2).R(r!2,r!3).L(l!3,l!4)....). Two simple rules, shown in **Listing 3**, provide an elementary model of bivalent ligand–bivalent receptor interaction under the assumptions that the length of a chain does not affect its reactivity and that rings do not form (*see* **Note 44**). A third rule that allows the formation of rings of any size is shown in **Listing 3**, but this rule is commented out (*see* **Note 45**). For a different example of polymerization in a biological context, *see* **Note 46**.

Listing 3. BioNetGen input file for binding of bivalent ligand to bivalent receptor (**see Section 3.7.2**).

```
setOption(SpeciesLabel,HNauty);
begin parameters
NA 6.02e23 # Avogadro's number (molecues/mol)
f  0.001       # Fraction of the cell to simulate
Vo f*1.0e-9 # Extracellular volume=1/cell_density (L)
V  f*3.0e-12 # Cytoplasmic volume (L)
L0  1e-9*NA*Vo # Conc. in Molar -> copies per cell
R0  f*3e5
kp1 3.3e/(NA*Vo)
km1 0.1
kp2 1e6/(NA*V)
km2 0.1
kp3 30
km3 0.1
end parameters

begin molecule types
 R(r,r)
 L(l,l)
end molecule types

begin reaction rules
# Ligand addition
1 R(r) + L(l,l) <-> R(r!1).L(l!1,l) kp1,km1
# Chain elongation
2 R(r) + L(l,l!+) <-> R(r!1).L(l!1,l!+) kp2,km2
# Ring closure
#3 R(r).L(l) <-> R(r!1).L(l!1) kp3,km3
end reaction rules
```

```
begin seed species
 R(r,r) R0
 L(l,l) L0
end seed species

begin observables
Species FreeL L(l,l)
Dimers  R==2
Trimers  R==3
4mers   R==4
5mers   R==5
6mers   R==6
7mers   R==7
8mers   R==8
9mers   R==9
10mers  R==10
gt10mers R>10
end observables

# Simulation of a truncated network
generate_network({overwrite=>1,max_stoich=>{R=>10,L=>10}});
simulate_ode({t_end=>50, n_steps=>20});

# Simulation on-the-fly
generate_network({overwrite=>1,max_iter=>1});
simulate_ssa({t_end=>50,n_steps=>20});
```

The observables block in **Listing 3** introduces a new syntax for using stoichiometry in the definition of observables, which is needed to track the aggregate size distribution in models that exhibit polymerization (*see* **Note 47**).

Because chains can grow to any length, unless stopping criteria are specified, the process of iterative rule application initiated by a generate_network command will not terminate until the user runs out of patience or the computer runs out of memory. We discuss here two methods of simulating a network that cannot be enumerated completely.

The first method is to specify any of the restrictions described in **Subheading 3.6.1** on the generate_network command, which will cause termination before all possible species and reactions have been generated. The first pair of actions in **Listing 3** shows how the max_stoich parameter can be used to limit the stoichiometry of complexes, producing in this case a network of 30 species and 340 reactions, which can be rapidly simulated using either the ODE or SSA methods. The accuracy of simulations on artificially truncated networks is, however, not guaranteed and may depend strongly on the parameter values. For the parameters shown in **Listing 3**, the population of clusters with more than about 5 receptors is small, and little error results from network truncation. However, if the value of the cross-linking parameter, `kp2`, is increased by a factor of 10, the cluster size distribution generated by the truncated network becomes inaccurate. The user must therefore be careful to check results for convergence, particularly when changing the parameter values over substantial ranges.

The second method, which is specified by the second pair of actions in **Listing 3**, is to do a minimal initial round of network generation and then allow the network to be generated as new species become populated during a stochastic simulation. The call to `generate_network` is required here to generate the reactions that can take place among the seed species; otherwise, an error will occur when a simulation command is invoked and there are no reactions in the network. With `max_iter` set to 1 only reactions involving seed species are initially generated. During simulation initiated with the `simulate_ssa` command, BioNetGen detects when a reaction event occurs that populates one or more species to which rules have not been previously applied and automatically expands the network through rule application. This behavior is built into the `simulate_ssa` command and no additional parameters need to be specified. The performance of on-the-fly simulation is highly dependent on the system parameters and on the number of molecules being simulated. Increasing the number of molecules while holding the concentrations fixed (accomplished by changing the parameter `f`) increases the size of the network that is generated by on-the-fly sampling. Because the network generation involves the computationally expensive step of generating and comparing canonical labels (*see* **Note 33**), the simulation performance can become poor if one attempts to simulate on-the-fly under conditions that lead to the possible formation of more than about 10^3–10^4 species. Simulation of the dynamics of 300 receptors up to steady state takes about 30 CPU seconds on a MacBook Pro with the 2.4 GHz Intel Core Duo processor and generates a network of about 50 species and 350 reactions.

In the near future, a third and more powerful option will be available for simulating large-scale networks, such as those that arise when polymerization is possible or when some of the signaling molecules have high valence (*see* **Note 48**). Work is currently underway to implement the discrete-event particle-based simulation method that has been recently developed, which extends Gillespie's method to consider rules rather than individual reactions as event generators *(30, 31)*. The main idea behind this method is that by tracking individual particles in a simulation rather than populations the need to explicitly enumerate the possible species and reactions is eliminated. The computational scaling of a stochastic, event-driven simulation using the particle-based approach becomes effectively independent of network size and has moderate (logarithmic) scaling with the size of the rule set. This rule-based kinetic Monte Carlo method offers significantly better performance than the earlier particle-based event-driven algorithm used in the StochSim software, which uses a less efficient event sampling algorithm that produces a high fraction of nonreactive events *(21)*. The planned incorporation of the rule-based kinetic Monte Carlo method will enable the efficient simulation of comprehensive models of signal transduction networks on the basis of molecular interactions, and, we hope, greatly increase the power of predictive modeling of such systems.

The plot in **Fig. 5** shows simulation results for the number of receptors in trimers as a function of time (in seconds) from the ODE simulation of the truncated network (smooth line) and the SSA simulation with on-the-fly network generation (jagged line). Following the initial equilibration period about 10–20% of the receptors are in trimers at any given time. The total time required for network generation and simulation is comparable in the two cases, with network generation consuming the vast majority of the CPU time.

3.8. Concluding Remarks

The information provided here serves as both an introductory guide and reference resource for the modeler interested in using BioNetGen to develop and analyze rule-based models of bio-

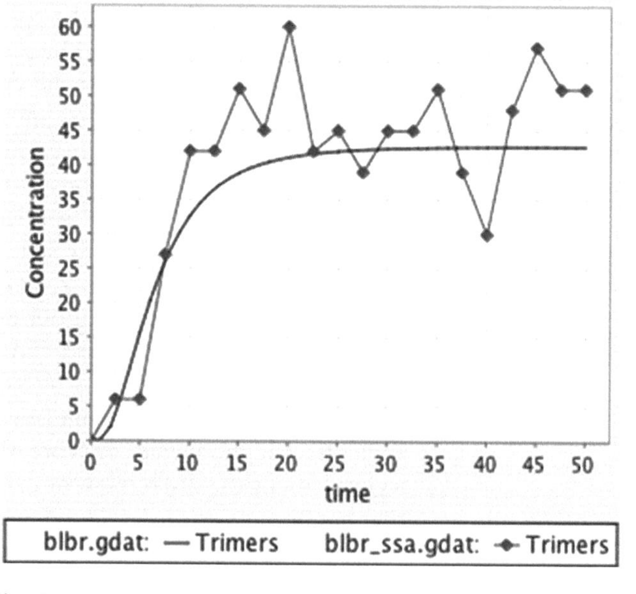

Fig. 5. Plot of simulation results obtained from BioNetGen input file for the bivalent ligand bivalent receptor binding model shown in **Listing 3** made using PhiBPlot. *Smooth solid line* is the curve obtained from the `simulate_ode` command; *jagged line* with *circles* shows results from the `simulate_ssa` command.

chemical systems. Several applications of BioNetGen have been presented and discussed, but the rule-based modeling approach enabled by BioNetGen can be used for a much broader range of purposes. We strive to be responsive to the needs of the BioNetGen user community and encourage users to contact us to share their experiences, to request new capabilities and features, and to report bugs. The BioNetGen web site (http://bionetgen.org) has a wiki format to allow users to contribute information and models. Updates of the information presented here will be announced at the wiki site. Rule-based modeling of biochemical networks is a rapidly evolving area of research and BioNetGen is therefore very much a work in progress, with new capabilities being added continually.

BioNetGen is an open-source project. Although contributions of code are welcome, the main reason the source code is made available is so that users can see how the code works and can confirm that model specifications are being processed as expected. Because of the difficulties of checking the correctness of a chemical reaction network or a simulation result generated automatically from rules, key elements of BioNetGen have been coded independently multiple times and crosschecked. After extensive testing, we are confident that the software is reliable. By following the guidance provided here, a modeler should be able to precisely use BNGL to obtain intended model specifications.

In the future, we hope to see the BioNetGen framework evolve to enable community-driven development of comprehensive models for cellular regulatory systems. The material components and interactions of a cellular regulatory system are typically too numerous and complicated for a single researcher to thoroughly document and capture faithfully in a model of comprehensive scope. For example, nearly 200 proteins are documented to be involved in EGFR signaling in the NetPath database (http://netpath.org). The ability to extend models through the composition of rules is a key factor that makes incremental construction of large-scale models a real possibility *(8, 27)*. To take advantage of collective intelligence for the construction of large-scale models, we are actively pursuing the following extensions of the BioNetGen framework: (1) implementation of methods capable of simulating models composed of a large number of rules *(30, 31)*, (2) manipulation and encoding of BNGL using an XML-based format proposed as an extension of SBML (http://sbml.org) to better facilitate electronic exchange and storage of models, and (3) development of conventions and database-related tools for annotating models and model elements (e.g., linking of molecule names in a model specification to amino acid sequences and other information in standard databases). However, for a long time to come, we foresee that a sound understanding of the material presented here will be useful for rule-based modeling with BioNetGen.

4. Notes

1. Users familiar with a command line interface on any of these systems should have no trouble following the instructions for using the software after reading this chapter. Other users may find the RuleBuilder application, which provides a graphical user interface to BioNetGen, more accessible. This application may be started by double-clicking on the `RuleBuilder-beta-1.51.jar` file in the RuleBuilder subdirectory of the BioNetGen distribution. The RuleBuilder Getting Started Guide in the same directory explains use of the software. Although this chapter focuses on the text-based interface, the basic concepts of BioNetGen modeling discussed here are essential for proper use of RuleBuilder.
2. BioNetGen is invoked in a command shell using `prompt> path/Perl2/BNG2.pl file.bngl`
3. The syntax of a line in the `parameters` block is `[`*`index`*`]` *`parameter`* `[=]` *`value`* where square brackets indicate optional elements, *`parameter`* is a string consisting of only alphanumeric characters plus the underscore character ("_") and containing at least one nonnumeric character. *`value`* may be either a number in integer, decimal, or exponential notation or a formula involving numbers and other parameters in C-style math syntax. *See* **Listing 1** for examples.
4. The size of the system being simulated can be scaled by changing the value of the parameter `f` in **Listing 1**. By scaling all of the initial populations and the volumes by this factor, the system size is scaled without changing the *concentrations* of any of the constituents. For a deterministic simulation, the simulation time and the behavior of the system (e.g., the value of any observable divided by `f`) is independent of `f`. For a stochastic simulation, however, the time required to carry out a simulation will be proportional to `f`, whereas the noise will be proportional to `1/sqrt(f)`.
5. In current BNGL each component may have at most one associated state label, which may take on an arbitrary number of discrete values, specified as strings. The state is thus a scalar variable that can be considered as an enum data type. Future planned extensions of BNGL include nesting of components to allow a single component to have multiple associated states and binding sites.
6. Names for all BioNetGen objects other than parameters, which includes molecules, components, state labels, bonds, labels, and observables may consist of alphanumeric characters and the underscore character ("_"), but may not include the dash character ("-"), which is sometimes used in the biologi-

cal literature as part of protein or domain names. It is not an allowed character here because in some contexts it may be confused with the arithmetic minus operator.

7. The syntax of a line in the `molecule types` block is

   [*`index`*] *`moleculeType`*

   where *`moleculeType`* has the syntax described in the text and illustrated in **Listing 1**.

8. If the `molecule types` block is present, all molecules in the `seed species` block must match the type declarations in the `molecule types` block. A molecule matches its type declaration if each of its declared components is present and each component state is a member of the declared set of possible states. If the `molecule types` block is not present, then the `seed species` block serves a typing purpose. The first instance of a molecule in the `seed species` block is taken to define the complete set of components in that molecule in the model, and only components that are assigned a state in the first occurrence may subsequently have defined states. For example, the Grb2 molecule implicitly defined by the species `Grb2(SH2,SH3)` may not have any states assigned to SH2 or SH3 components. However, the species `EGFR(L,CR1,Y1068~U)` defines the Y1068 component of EGFR as one that has an associated state label, which has at least one allowed value, "`U`", and potentially others to be defined later. Occurrences of additional allowed state labels may occur in the `seed species` block or in the `reaction rules` block, and in either case BioNetGen generates a warning message that additional allowed state values are being associated with the component.

9. The syntax of a line in the seed species block is

   [*`index`*] *`species`* [*`initialPopulation`*]

   where *`species`* has the syntax for a BioNetGen species as described in the text and illustrated in the `seed species` block of **Listing 1** and *`initialPopulation`* is a number or formula that specifies the amount of the species present at the start of the first simulation (default is zero).

10. The amount of a chemical species may be specified to have a constant value by prefixing the chemical species name in the `seed species` block with a "$" character, as follows: the expression "`$EGF(R) 1`" would set the amount of free EGF in the system to 1. This feature is useful for considering certain scenarios.

11. The syntax of a line in the `observables` block is

    [*index*] [*observableType*] *observableName* *pattern1*[, *pattern2*]...

    where *observableType* is either `Molecules` or `Species` (defaults to `Molecule` if omitted) *observableName* is a valid name for a BioNetGen observable, and each *pattern* is a valid BioNetGen pattern.

12. Recall that bond names are arbitrary and are used only to identify the bond endpoints. Thus, the bond names used in a pattern do not affect the resulting matches.

13. The "?" wildcard can also be used in state matching, but leaving component state out of a match is more commonly achieved by omitting the state label altogether. For example the patterns "`EGFR(Y1068)`" and "`EGFR(Y1068~?)`" are equivalent, i.e., generate the same matches.

14. For each pattern, selected matches to species in the model of **Listing 1** are listed with the image of the pattern elements shown in **bold**. (These are not meant to be exhaustive, just illustrative.) Note that some chemical species are matched multiple times by a given pattern.

    a. **EGFR()** matches **EGFR**(CR1,L,Y1068~U),EGF(R!1). **EGFR**(CR1,L!1,Y1068~U), EGF(R!1). EGF(R!2). **EGFR**(CR1!3,L!1,Y1068~U).EGFR(CR1!3, L!2,Y1068~U), and EGF(R!1).EGF(R!2).EGFR(CR 1!3,L!1,Y1068~U).**EGFR**(CR1!3,L!2,Y1068~U)

    b. **EGF(R)** matches **EGF(R)**

    c. **EGFR(CR1!+)** matches EGF(R!1).EGF(R!2). **EGFR(CR1!3**,L!1,Y1068~U). EGFR(CR1!3,L!2,Y1068~U), and EGF(R!1). EGF(R!2).EGFR(CR1!3,L!1,Y1068~P). **EGFR(CR1!3**,L!2,Y1068~U**)**

    d. **EGFR(Y1068~P!?)** matches EGF(R!1). EGF(R!2).**EGFR**(CR1!3,L!1,**Y1068~P**). EGFR(CR1!3,L!2,Y1068~U), and EGF(R!1). EGF(R!2).**EGFR**(CR1!3,L!1,**Y1068~P!4**). EGFR(CR1!3,L!2,Y1068~U).Grb2(SH2!4,SH3)

    e. **Grb2(SH2,SH3!1).Sos1(PxxP!1)** matches **Grb2(SH2,SH3!1).Sos1(PxxP!1)**

    f. **EGFR(Y1068!1).Grb2(SH2!1,SH3!2). Sos1(PxxP!2)** matches EGF(R!1).EGF(R!2). **EGFR**(CR1!3,L!1,**Y1068~P!4**).EGFR(CR1!3,L!2, Y1068~U).**Grb2(SH2!4,SH3!5).Sos1(PxxP!5)**

15. The sum corresponding to an observable is defined explicitly in the .net file that is generated by BioNetGen when an input file is processed. These sums are contained in the `groups` block of the .net file (*see* **Note 28**).

16. When an observable is defined by two or more patterns, the associated functions are computed as follows. For an observable of the Molecules type, the observable is a sum of the observables defined by each individual pattern in the set. For an observable of the Species type, the observable is an unweighted sum of the populations of chemical species matched by any of the patterns in the set. Multiple patterns can be useful for specifying observables that are functions of multiple sites on a molecule, e.g., the total phosphorylation level of a protein that can be phosphorylated at multiple sites.
17. The .gdat and .cdat files produced by BioNetGen simulation commands are ASCII text files that list the time courses of observables and concentrations, respectively, in a tabular format. The first line of each file is a header beginning with a "#" character, followed by a whitespace-separated list of strings identifying the contents of each column. The first column is "time" in both .gdat and .cdat formats. In a .gdat file the remaining columns list the observable names corresponding to each column. In the .cdat file, the remaining columns list the index of the species concentration corresponding to each column.
18. The syntax of a line in the `reaction rules` block is

    [*index*] *rPattern1* [*+rpattern2*] … *arrow pPattern1* [*+pPattern2*] … *rateLaw1*[*,rateLaw2*] [*command1*]…

    where each *Pattern* is a valid BioNetGen pattern, arrow is one of "–>" or "<–>," each *rateLaw* is a parameter or a rate law function (*see* **Note 20**), and commands have the syntax described in **Subheading 3.5.7**.
19. If a component of a molecule appears in a reactant pattern, the corresponding molecule in the product pattern, if it is not deleted, must include that component. Failure to include the full set of components referenced by the reactant pattern will produce an error. Thus, the rule "`A(a)-> A(b) kab`" produces an error, even if the A molecule has both components a and b.
20. Other rate laws are invoked by using one of the keywords for the allowed rate law types followed by a comma-separated list of numerical values or formulas in parentheses. As of this writing, the three recognized rate law types are "`Ele`", "`Sat`", and "`MM`". The formula for the Ele rate law is

    $$\mathrm{Ele}(k_1) = k_1 \prod_{i=1}^{M} x_i,$$

    where M is the molecularity of the reaction (i.e., the number of reactants) and xi is the population level of the ith reactant.

This rate law type is specified by default when only a numerical value or a formula is given following the product patterns in a rule, as described in Subheading 3.5. The current version of BioNetGen supports a few nonelementary rate law formulas, primarily to allow simulation of models from the literature that incorporate these rate laws. The formula for the Sat rate law is

$$\mathrm{Sat}(k_{\mathrm{cat}}, K_{\mathrm{m}}) = k_{\mathrm{cat}} \prod_{i=1}^{M} x_i \,/\, (K_{\mathrm{m}} + x_1)\,,$$

where x_1 is the population level of the reactant matching the first reactant pattern in the rule. Note that $V_{\max} = k_{\mathrm{cat}} x_2$ and K_{m} are the usual Michaelis–Menten parameters if M= 2 *(47)* and that these parameters should be specified in consistent units. An example of a rule that uses this rate law is

```
Prot(Y ~ U) + Kinase(aloopY ~ P)®Prot(Y ~ P) +
                    e(aloopY ~ P) Sat(kcat, Km)
```

The formula for the MM rate law is

$$MM(k_{\mathrm{cat}}, K_{\mathrm{m}}) = k_{\mathrm{cat}} x'_1 x_2 \,/\, (K_{\mathrm{m}} + x'_1)\,,$$

where $x'_1 = \left((x_1 + x_2 - K_{\mathrm{m}}) + \sqrt{(x_1 + x_2 - K_{\mathrm{m}})^2 + 4K_{\mathrm{m}} x_1}\right) / 2\,.$

Note that this rate law type is applicable only if $M = 2$. The MM rate law type is the same as the Sat rate law type when $M=2$ except that x_1 is replaced by x'_1 to account for the amount of "substrate" bound to "enzyme." In the near future it will be possible to define rate laws using arbitrary mathematical formulas.

21. A single-site rate law characterizes the rate of a reaction that involves the formation or dissolution of a single bond. In some cases, a reaction can occur in multiple ways that are indistinguishable. In these cases, the single-site rate law needs to be multiplied by a statistical factor to obtain the appropriate observable rate of the reaction. For example, if an antibody with two identical binding sites associates with a monovalent hapten, then there are two indistinguishable ways that this reaction could occur. If the single-site rate constant is *k*, then the observable rate at which the reaction occurs is 2 *k* [IgG] [hapten], where [IgG] is the concentration of bivalent antibody, [hapten] is the concentration of monovalent hapten, and the statistical factor of 2 accounts for the fact that hapten can add to either of the two sites on the antibody. BioNetGen in generating or simulating a reaction network automatically accounts for such statisti-

cal factors under the assumption that the rate law associated with a rule applies to a single-site reaction. A modeler should therefore be careful to always specify a single-site rate constant when writing a rule. Likewise, BioNetGen automatically adds a symmetry factor of 1/2 to account for reactions such as A + A → product(s), a factor of 1/6 to account for reactions such as A + A + A → product(s), etc. In general, when assigning a rate constant to the elementary rate law of a rule, one should assign the constant appropriate for a reaction of the form A+B → product(s) where in this reaction there is a unique path from the reactants to product(s). BioNetGen will automatically correct rates of reactions for statistical and symmetry factors. This feature is important because these factors often vary from reaction to reaction within a class of reactions defined by a single rule *(28)*.

22. Any component in a reaction rule may be tagged by adding the "`%`" character followed by the tag name. The scope of a tag is local to the rule in which it appears.

23. The application of rule 1 of **Listing 1** to the species {**`EGF(R)`**, **`EGFR(L,CR1`**`,Y1068~U)`, **`EGFR(L,CR1`**`,Y1068~P)`, **`EGFR(L,CR1`**`,Y1068~P!1).Grb2(SH2!1,SH3)`}, produces the following reactions:

```
EGF(R)+EGFR(L,CR1,Y1068~U)->
EGF(R!1).EGFR(L!1,CR1,Y1068~U) kp1

EGF(R)+EGFR(L,CR1,Y1068~P)->
EGF(R!1).EGFR(L!1,CR1,Y1068~P) kp1

EGF(R)+EGFR(L,CR1,Y1068~P!1).Grb2
(SH2!1,SH3)->\
EGF(R!2).EGFR(L!2,CR1,Y1068~P!1).Grb2(SH2!1,
SH3) kp1
```

where the images of the reactant patterns are shown in bold and the reaction centers are underlined. The rate law for an individual reaction has the same format as a rate law in a reaction rule (*see* **Note 18**).

24. The scope of a bond name is restricted to the pattern in which it appears. Bond names are not used in establishing the correspondence between reactant and product patterns. Thus, the rule "`A(a!1).B(b!1~U) -> A(a!2).B(b!2~P)`" has no effect on the bond between A and B even though in the specification the name of the bond changes between the reactant and product sides. Similarly, in the expression "`A(a!1).B(b!1) + C(c!1).D(d!1)`" the fact that both bonds have the same name has no consequence.

25. Internally, BioNetGen represents all reactions generated by rules as unidirectional and maintains this representation when generating a .net file or exporting networks to SBML and MATLAB M-file formats.

26. Consider the action of the following two rules on the initial species "`A(a1!1,a2!2).B(b1!1,b2!2)`", which describes a complex between A and B molecules connected by two bonds. Both rules break the bond between the a1 component of A and the b1 component of B. The first rule, "`A(a1!1).B(b1!1) -> A(a1)+B(b1)`," has a molecularity of two in the products, and thus does not apply to this complex because breaking the bond still leaves the complex held together by the bond between a2 and b2. The second rule, "`A(a1!1).B(b1!1) -> A(a1).B(b1)`", does not require dissociation of the resulting complex and generates the reaction "`A(a1!1,a2!2).B(b1!1,b2!2) -> A(a1,a2!1).B(b1,b2!1)`".
27. Describing dephosphorylation as a first order reaction involving only the substrate assumes that the responsible enzyme, a phosphatase is constitutively active and is present at an excess and unchanging level. Dephosphoryation reactions have been handled this way (*see*, e.g., **ref.** *33)* because the identities of the phosphatases acting on a particular substrate are often unknown.
28. The two products are

    `EGF(R!1). EGF(R!2).` **`EGFR`**`(CR1!3,L!1,`**`Y1068~U`**`).`

    `EGFR(CR1!3,L!2,Y1068~P)`

    `EGF(R!1).EGF(R!2).EGFR(CR1!3,L!1,Y1068~P).`

    `EGFR(``CR1!3,L!2,`**`Y1068~U)`**, which are isomorphic, as can be verified by switching the order of the two EGF and two EGFR molecules and renumbering bonds 1 and 2.
29. A correction is required for *rules* that are symmetric *(26)*. BioNetGen automatically detects rule symmetry generates reactions with the correct multiplicity. Consider the symmetric rule "`A(a) + A(a) → A(a!1).A(a!1) k`" applied to the set of species `{A(a,Y~U), A(a,Y~P)}`. The following reactions will be generated

    ```
    A(a,Y~U) + A(a,Y~U) → A(a!1,Y~U).A(a!1,Y~U)
    0.5*k
    A(a,Y~U) + A(a,Y~P) → A(a!1,Y~U).A(a!1,Y~P) k
    A(a,Y~P) + A(a,Y~P) → A(a!1,Y~P).A(a!1,Y~P)
    0.5*k
    ```

    where the first and third reactions are symmetric and thus have a multiplicity of ½, for the reason discussed above in **Note 21**. The second reaction has a multiplicity of 1 because there is only one way that a bond may be added to join the two A molecules.
30. This would not be the case if EGFR had another binding partner that could bind through the CR1 domain, such as another member of the ErbB family of receptors to which

EGFR belongs. Consider a simplified example of the protein "`A(Y~U,b)`", where the b component is a binding site that can bind either to the b site of a kinase B or to the b site of a kinase-dead mutant of B called Bi. The rule "`A(Y~U,b!+)-> A(Y~P,b!+)`" would not capture the described mechanism because both **`A(Y~U,b!1)`**`.B(b!1)` and **`A(Y~U,b!1)`**`.Bi(b!1)` would generate phosphorylation reactions under action of this rule. The most obvious way to address this problem is to explicitly specify that B must be present for the rule to apply, as in "`A(Y~U,b!1).B(b!1)-> A(Y~P,b!1).B(b!1)`".

31. Such counters can prevent the `steady_state` flag of `simulate_ode` from reaching steady state because the counter species will increase linearly in time if the steady-state concentration of the A-containing species is nonzero. If one wishes to preserve possible steady-state behavior, the concentration of the Trash species should be fixed by pre-pending a "$" character to its declaration in the `seed species` block (*see* **Note 10**). In other words, Trash should be declared as a seed species with fixed value using the line "`$Trash 0`" in the `seed species` block.

32. Species are compared during network generation by generating a string label for the species from a canonical ordering of the molecules, components, and edges *(26)*. A canonical ordering is one that guarantees that two graphs will generate the same label if and only if they are isomorphic *(48)*. In this way, the problem of determining graph isomorphism is reduced to string comparison and testing a species found in a new reaction for isomorphism with existing species is reduced to looking up its label in a hash table. For labeled graphs, such as those used in BioNetGen to represent species, the problem of canonical ordering is trivial if all labels in the graph are unique (lexical sorting will suffice). More powerful methods are needed if there are multiple occurrences of nodes with identical labels *(49)*. There are three different methods that BioNetGen can use to generate canonical labels and test species for isomorphism. To specify a canonical labeling method the command "`setOption("SpeciesLabel",`*`method`*`);`" is placed anywhere in the BNGL file outside of the input blocks and before the first action command. In this command *`method`* is `Auto`, `HNauty`, or `Quasi`. Use of this command is optional unless overriding the default method, which is `Auto`. The default method used by BioNetGen for generating canonical labels is called "Auto," which works by generating a quasi-canonical label that includes all information about the Species except the bonds, for which only the bond order of each Component is listed. These quasi-canonical labels are quick to generate, but they cannot distinguish all nonisomorphic species. Thus,

any two species that share a quasi-canonical label must be further checked for isomorphism directly, for which BioNetGen uses a variant of the Ullmann algorithm *(50)*. This method is always correct, but may be very slow if the number of identical Molecules or Components in a complex is greater than a handful, because it requires checking of a large number of permutations. A second exact method called "HNauty" is available that gives more robust performance when species are formed that involve substantial numbers of repeated elements. HNauty is a generalization of the Nauty algorithm of McKay *(49)* developed specifically to handle graphs representing species in BioNetGen *(51)*. HNauty is slower than Auto when most of the species in a network have low stoichiometry, but is sometimes required to simulate networks in which substantial oligomerization occurs. In some cases, such as when oligomers are restricted to linear chains, the quasi-canonical strings used as a filter by the Auto method turn out to be canonical. If that is the case, the "Quasi" method can be used to turn off the additional isomorphism check for species that match an existing quasi-canonical label, which can significantly accelerate network generation. This method should only be used when the user can confirm that the quasi-canonical labels are in fact canonical; otherwise, failure to resolve nonidentical species will result in unpredictable behavior.

33. If a rule set implies a large or unbounded network and a user attempts to generate the network, BioNetGen may not complete execution in a reasonable amount of time. In such cases a user has several options: (1) Restrict network generation using arguments to `generate_network`, as discussed in **Subheading 3.6**; (2) Use the "`print_iter 1`" option of the `generate network` command to cause BioNetGen to dump an intermediate .net file for each iteration of rule application and inspect the resulting .net file for indications of runaway polymerization that may be unintended; (3) Run a stochastic simulation with on-the-fly network generation (*see* **Subheading 3.7.2**); (4) Wait for the network-free simulation engine(s) to become available (*see* **Fig. 2**). (5) Use the macro model reduction module for BioNetGen, which uses the algorithms described in *(52–54)* to reduce the size of the network that needs to be generated to calculate the specified observables. The module is included in BioNetGen distributions 2.0.47 and later, and is invoked using the command "`MacroBNG2.pl --macro` *`file`*`.bngl`".

34. The .net file produced by BioNetGen is a BNGL file with the three additional blocks `species`, `reactions`, and `groups`, which contain the species, reactions, and observable function definitions that result from network generation. The syntax for the `species` block is identical to that

of the `seed species` block in the BNGL file. It contains a complete list of the species in the network and their concentrations at the current time. The syntax for each line in the `reactions` block is

```
index reactantListproductList [multiplicity*]
rateLaw
```

where the *`reactantList`* and *`productList`* are comma-separated lists referring to species by index, *`multiplicity`* is an optional factor multiplying the rate law, and *`rateLaw`* is either a single parameter (for an elementary type) or one of the additional types (*see* **Note 20**). An example of a reaction entry is

```
1 1,7 8 2*kp1
```

which specifies that species 1 and 7 undergo a bimolecular association to produce species 8 with an elementary rate law governed by the rate constant 2*kp1. The syntax for each line in the observables groups block is

```
index group Name speciesList
```

where the *`speciesList`* is a comma-separated list of species indices, each element of which has the form *`[weight*]speciesIndex.`* An example of a sum definition for observable 6 of the example model in **Listing 1** is

```
6 Sos1_act 13,16,18,20,22,2*23
```

In the sum, the population level of Species 23 has a weight of two, whereas the population levels of all other species have the default weight of one.

35. The `prefix` command sets the basename to be the value of its argument, whereas the `suffix` command appends its argument to the basename. For example, the command "`prefix⇒test`" would set the basename to `test`, and the command "`suffix⇒test`" would append "`_test`" to the basename. The scope of changes to the basename is local to the command in which the `prefix` or `suffix` commands appear. The basename for subsequent commands reverts to the basename of the file unless overriden by additional commands. The sole exception to this is the `readFile` action, which sets the global basename to either the value of the `prefix` command, if present, or to the basename of the `file` command.

36. In practice, a modeler should be careful to check by trial and error that `t_end` is sufficiently large to reach steady state. Recall that BioNetGen expects model parameters and variables to have consistent units, so times specified in simulation commands (e.g., by assigning a value to the `t_end` parameter) should be given in units consistent with those

of rate constants, which have units of inverse seconds in all examples presented here.

37. In addition to reporting simulation output at evenly spaced intervals, as specified using the `t_end` and `n_steps` parameters, BioNetGen can also report results at any set of times specified in the `sample_times` array. When this option is used, values of the `t_end` and `n_steps` parameters should not be specified. An example of nonuniform time sampling specified in this way is the command

```
simulate_ode({sample_times [1,10,100]});
```

which will result in observables and species concentrations being reported at $t = 0$ (the start time), 1, 10, and 100.

38. Rule-based networks tend to be sparse, that is, the vast majority of elements of the Jacobian matrix are zero (the elements of this matrix are $J_{ij} = {}^{\partial f_i(\mathbf{x})}/_{\partial x_j}$, where $\dot{x}_i = fi(x)$ is the ODE describing the kinetics of species i). This may not be the case for networks involving extensive oligomerization. Empirically, we have found that networks with more than a few hundred species tend to be accelerated by the use of sparse methods, with major gains occurring for networks of thousands to tens of thousands of species. The largest network that has been simulated with BioNetGen has about 50,000 species and 100,000 reactions. Above that point, the 2 gigabytes of memory addressable on 32 bit architectures is exceeded.

39. The `setConcentration` command has the syntax

```
setConcentration(species,value)
```

where *`species`* is a valid BioNetGen species specification (*see* **Subheading 3.3**) and *`value`* is a number or formula.

40. Note that if the initial concentration of a species is set to a parameter or a formula, changing the value of the parameter or of parameters in the formula using the `setParameter` after the first simulation is run will not affect the species concentrations, which are overwritten following the completion of the simulation.

41. It is straightforward to write scripts that utilize these commands to automate such tasks as parameter scans or averaging multiple stochastic simulations. The Perl script `scan_var.pl`, which is provided in the Perl2 directory of the BioNetGen distrbution, provides a simple example that can be used for scanning the value of a single parameter and could be easily extended to perform more complex actions.

42. A .net file with the name "*`basename`*`.net`" is automatically generated prior to execution of any simulation command and is read by the `run_network` program, which is executed as

a separate process. If a .net file with the same name already exists, it is overwritten. If multiple simulation commands are given in the same input file, it may be useful to use a different basename for each (using either the `prefix` or `suffix` commands), so that the input network to each simulation can be inspected later for information and debugging purposes. In the example shown in **Listing 1**, the first `simulate_ode` command produces the file "`egfr_simple_equil.net`", the second `simulate_ode` command produces the file "`egfr_simple.net`", and the `simulate_ssa` produces the file "`egfr_simple_ssa.net`".

43. BioNetGen's simulation engine, Network, has a command line interface that can be used directly, bypassing `BNG2.pl` altogether. Details of this interface are provided in the source code of `run_network.c`, which is located in the Network2 subdirectory of the distribution. A summary is provided by running the appropriate `run_network` executable in the `bin` directory of the distribution. In addition, `BNG2.pl` outputs the exact command used to execute `run_network` following the tag "`full_command:`"

44. Note that a nearly identical network of species and reactions can be generated by the single rule "`R(r)+L(r)<-> R(r!1).L(r!1) kp1,km1,`" where the difference is that in the single-rule network all reactions will have the same rate constant. This difference is important physically, because ligand that is bound to receptor is restricted to diffuse on the surface of the cell, whereas free ligand diffuses freely in three dimensions. Although restriction to the cell surface decrease the diffusion constant of the ligand, the effective concentration of receptor binding sites greatly increases resulting in a strong enhancement of the ligand-receptor binding rate *(55)*.

45. Rule 3 of **Listing 3** is the simplest ring closure rule that can be specified for this system, and permits the formation of all possible rings in this system, including a monomeric ring in which a single receptor binds the same ligand twice. To exclude this possibility, which may be sterically unfavorable, one could extend the rule to read "`L(l!1).R(r!1,r).L(l) <-> L(l!1).R(r!1,r!2).L(l!2) kp3,km3,`" which forces the ring closure to involve a ligand molecule other than the one to which the R molecule is bound. It is also possible to restrict the range of chain sizes that can undergo ring closure by explicitly including all of the molecules that form the ring, or by using a combination of `include_reactants` and `exclude_reactants` commands. For example, the adding the commands "`include_reactants(1,R.R)`" and "`exclude_reactants(1,R.R.R.R)`" to either ring closure rule would only allow the formation of rings containing two or three R molecules. This is desirable from a biophysical

perspective because the rate of ring closure may depend on the distance between the two endpoints of the chain that are being connected *(56)*. In the future, it will be possible to specify such relationships in a single rule using rate laws that depend on the specific properties of a species matched by a pattern in a rule.

46. Actin, which is one of the major components of the cytoskeleton, forms branched structures that play a critical role in many cellular processes including motility *(57)*. A simple model for the formation of branched actin structures is given by the definition of an actin molecule as “`A(b,p,br)`,” where the components b, p, and br represent the barbed end, the pointed end and the branching sites of actin respectively, a rule for chain elongation “`A(b)+A(p)<-> A(b!1).A(p!1) kp1,km1`”, and a rule for chain branching “`A(br)+A(p)<-> A(br!1).A(p!1) kp2,km2`.” The first rule generates linear filaments of actin, which become branched through the action of the second rule. Filaments may be extended either through the addition of monomers or by combination with another filament.
47. The basic syntax is “`molecule op number`,” where `molecule` is a molecule name, `op` is one of the comparison operators “`==`,” “`<`,” “`>`,” “⇐,” or “`>=`,” and `number` is a non-negative integer. This allows the stoichiometry of a single molecule type within a complex to be selected. If the observable is of type `Molecules` (the default), the observable will reflect the total number of molecules in species matching the selected stoichiometry. If the observable is of type `Species`, the observable will reflect the total population of species matching the selected stoichiometry. The current syntax allows stoichiometry of only a single molecule type to be considered at a time.
48. In our experience, the combinatorial explosion becomes a major bottleneck to generating and simulating networks in any realistic model that considers more than a handful of components. A recent model of EGFR signaling by Danos et al. *(27)* provides an example. The model considers 13 proteins, a small subset of the proteins that are active in EGFR signaling, and is composed of 70 rules that generate about 10^{23} species. Other rule-based models of growth factor signaling have produced similar eye-popping numbers *(58, 59)*. Even models that consider a few components may exhibit polymerization. For example, a simple model of Shp2 regulation constructed in BioNetGen involves only two molecule types, and yet must be truncated because the combination of binding and enzyme–substrate interactions generates infinite chains *(36)*. For the trivalent ligand bivalent receptor problem described in Yang et al. *(31)* there is a phase transition in which nearly all of the receptors coalesce into a single giant aggregate, which makes accurate truncation of the network effectively impossible.

Acknowledgments

Work on BioNetGen has been supported by NIH grants GM035556, RR18754, and GM76570 and DOE contract DE-AC52-06NA25396. J.R.F. also acknowledges support from the Department of Computational Biology at the University of Pittsburgh School of Medicine. Integration of BioNetGen into the Virtual Cell was supported by U54 RR022232 NIH-Roadmap grant for Technology Centers for Networks and Pathways. Special thanks to Byron Goldstein for the initial impetus that led to the development of BioNetGen and for his active and ongoing support. We thank the many people who have contributed to the development of BioNetGen and BioNetGen-compatible tools, including Jordan Atlas, Nikolay Borisov, Alexander Chistopolsky, Joshua Colvin, Thierry Emonet, Sarah Faeder, Leigh Fanning, Matthew Fricke, Bin Hu, Jeremy Kozdon, Mikhail Kravchenko, Nathan Lemons, Michael Monine, Fangping Mu, Ambarish Nag, Richard Posner, Amitabh Trehan, Robert Seletsky, Michael Sneddon, and Jin Yang. We also thank Gary An, Dipak Barua, Marc Birtwistle, James Cavenaugh, Ed Clarke, Vincent Danos, Jerome Feret, Andrew Finney, Walter Fontana, Leonard Harris, Jason Haugh, Michael Hucka, Sumit Jha, Jean Krivine, Chris Langmead, Paul Loriaux, Boris Kholodenko, Michael Saelim, Ed Stites, Ty Thomson, and Aileen Vandenberg for their helpful discussions and input. People contributing to the integration of BioNetGen with the Virtual Cell include James Schaff, Ion Moraru, Anuradha Lakshminarayana, Fei Gao, and Leslie Loew.

References

1. Blinov, M. L., Faeder, J. R., Goldstein, B., and Hlavacek, W. S. (2004) BioNetGen: software for rule-based modeling of signal transduction based on the interactions of molecular domains. *Bioinformatics* 20, 3289–3291.
2. Kholodenko, B. N. (2006) Cell-signalling dynamics in time and space. *Nat. Rev. Mol. Cell Biol.* 7, 165–176.
3. Aldridge, B. B., Burke, J. M., Lauffenburger, D. A., and Sorger, P. K. (2006) Physicochemical modelling of cell signalling pathways. *Nat. Cell Biol.* 8, 1195–1203.
4. Dueber, J. E., Yeh, B. J., Bhattacharyya, R. P., and Lim, W. A. (2004) Rewiring cell signaling: the logic and plasticity of eukaryotic protein circuitry. *Curr. Opin. Struct. Biol.* 14, 690–699.
5. Pawson, T. and Linding, R. (2005) Synthetic modular systems – Reverse engineering of signal transduction. *FEBS Lett.* 579, 1808–1814.
6. Bashor, C. J., Helman, N. C., Yan, S., and Lim, W. A. (2008) Using engineered scaffold interactions to reshape MAP kinase pathway signaling dynamics. *Science* 319, 1539–1543.
7. Hlavacek, W. S., Faeder, J. R., Blinov, M. L., Perelson, A. S., and Goldstein, B. (2003) The complexity of complexes in signal transduction. *Biotechnol. Bioeng.* 84, 783–794.
8. Hlavacek, W. S., Faeder, J. R., Blinov, M. L., Posner, R. G., Hucka, M., and Fontana, W. (2006) Rules for modeling signal-transduction systems. *Sci. STKE* 2006, re6.
9. Gomperts, B. D., Kramer, I. M., and Tatham, P. E. R. (2003) *Signal Transduction*. Elsevier Academic Press, San Diego, CA.
10. Hunter, T. (2000) Signaling: 2000 and beyond. *Cell* 100, 113–127.
11. Cambier, J. C. (1995) Antigen and Fc receptor signaling: The awesome power of the

immunoreceptor tyrosine-based activation motif (ITAM). *J. Immunol.* 155, 3281–3285.

12. Pawson, T. and Nash, P. (2003) Assembly of cell regulatory systems through protein interaction domains. *Science* 300, 445–452.
13. Pawson, T. (2004) Specificity in signal transduction: From phosphotyrosine-SH2 domain interactions to complex cellular systems. *Cell* 116, 191–203.
14. Seet, B. T., Dikic, I., Zhou, M. M., and Pawson, T. (2006) Reading protein modifications with interaction domains. *Nat. Rev. Mol. Cell Biol.* 7, 473–483.
15. Mathivanan, S., Periaswamy, B., Gandhi, T. K. B., Kandasamy, K., Suresh, S., Mohmood, R., Ramachandra, Y. L., and Pandey, A. (2006) An evaluation of human protein-protein interaction data in the public domain. *BMC Bioinformatics* 7, S19.
16. Mathivanan, S., Ahmed, M., Ahn, N. G., Alexandre, H., Amanchy, R., Andrews, P. C., Bader, J. S., Balgley, B. M., Bantscheff, M., Bennett, K. L., et al. (2008) Human Proteinpedia enables sharing of human protein data. *Nat. Biotechnol.* 26, 164–167.
17. Ong, S. E. and Mann, M. (2005) Mass spectrometry-based proteomics turns quantitative. *Nat. Chem. Biol.* 1, 252–262.
18. Olsen, J. V., Blagoev, B., Gnad, F., Macek, B., Kumar, C., Mortensen, P., and Mann, M. (2006) Global, in vivo, and site-specific phosphorylation dynamics in signaling networks. *Cell* 127, 635–648.
19. Kholodenko, B. N., Demin, O. V., Moehren, G., and Hoek, J. B. (1999) Quantification of short term signaling by the epidermal growth factor receptor. *J. Biol. Chem.* 274, 30169–30181.
20. Schoeberl, B., Eichler-Jonsson, C., Gilles, E. D., and Muller, G. (2002) Computational modeling of the dynamics of the MAP kinase cascade activated by surface and internalized EGF receptors. *Nat. Biotechnol.* 20, 370–375.
21. Morton-Firth, C. J. and Bray, D. (1998) Predicting temporal fluctuations in an intracellular signalling pathway. *J. Theor. Biol.* 192, 117–128.
22. Endy, D. and Brent, R. (2001) Modelling cellular behaviour. *Nature* 409, 391–395.
23. Jorissen, R. N., Walker, F., Pouliot, N., Garrett, T. P. J., Ward, C. W., and Burgess, A. W. (2003) Epidermal growth factor receptor: Mechanisms of activation and signalling. *Exp. Cell Res.* 284, 31–53.
24. Danos, V. and Laneve, C. (2004) Formal molecular biology. *Theor. Comput. Sci.* 325, 69–110.
25. Faeder, J. R., Blinov, M. L., and Hlavacek, W. S. (2005) Graphical rule-based representation of signal-transduction networks, in. *SAC '05: Proc. ACM Symp. Appl. Computing*, ACM, New York, NY, pp. 133–140.
26. Blinov, M. L., Yang, J., Faeder, J. R., and Hlavacek, W. S. (2006) Graph theory for rule-based modeling of biochemical networks. *Lect. Notes Comput. Sci.* 4230, 89–106.
27. Danos, V., Feret, J., Fontana, W., Harmer, R., and Krivine, J. (2007) Rule-based modelling of cellular signalling. *Lect. Notes Comput. Sci.* 4703, 17–41.
28. Faeder, J. R., Blinov, M. L., Goldstein, B., and Hlavacek, W. S. (2005) Rule-based modeling of biochemical networks. *Complexity* 10, 22–41.
29. Lok, L. and Brent, R. (2005) Automatic generation of cellular networks with Moleculizer 1.0. *Nat. Biotechnol.* 23, 131–36.
30. Danos, V., Feret, J., Fontana, W., and Krivine, J. (2007) Scalable simulation of cellular signalling networks. *Lect. Notes Comput. Sci.* 4807, 139–157.
31. Yang, J., Monine, M. I., Faeder, J. R., and Hlavacek, W. S. (2007) Kinetic Monte Carlo method for rule-based modeling of biochemical networks. *arXiv:0712.3773.*
32. Goldstein, B., Faeder, J. R., Hlavacek, W. S., Blinov, M. L., Redondo, A., and Wofsy, C. (2002) Modeling the early signaling events mediated by FceRI. *Mol. Immunol.* 38, 1213–1219.
33. Faeder, J. R., Hlavacek, W. S., Reischl, I., Blinov, M. L., Metzger, H., Redondo, A., Wofsy, C., and Goldstein, B. (2003) Investigation of early events in FceRI-mediated signaling using a detailed mathematical model. *J. Immunol.* 170, 3769–3781.
34. Faeder, J. R., Blinov, M. L., Goldstein, B., and Hlavacek, W. S. (2005) Combinatorial complexity and dynamical restriction of network flows in signal transduction. *Syst. Biol.* 2, 5–15.
35. Blinov, M. L., Faeder, J. R., Goldstein, B., and Hlavacek, W. S. (2006) A network model of early events in epidermal growth factor receptor signaling that accounts for combinatorial complexity. *BioSystems* 83, 136–151.
36. Barua, D., Faeder, J. R., and Haugh, J. M. (2007) Structure-based kinetic models of modular signaling protein function: focus on Shp2. *Biophys. J.* 92, 2290–2300.
37. Barua, D., Faeder, J. R., and Haugh, J. M. (2008) Computational models of tandem Src homology 2 domain interactions and application to phosphoinositide 3-kinase. *J. Biol. Chem.* 283, 7338–7345.
38. Mu, F. P., Williams, R. F., Unkefer, C. J., Unkefer, P. J., Faeder, J. R., and Hlavacek, W. S. (2007) Carbon-fate maps for metabolic reactions. *Bioinformatics* 23, 3193–3199.

39. Rubenstein, R., Gray, P. C., Cleland, T. J., Piltch, M. S., Hlavacek, W. S., Roberts, R. M., Ambrosiano, J., and Kim, J. I. (2007) Dynamics of the nucleated polymerization model of prion replication. *Biophys. Chem.* 125, 360–367.
40. Gillespie, D. T. (1976) A general method for numerically simulating the stochastic time evolution of coupled chemical reactions. *J. Comp. Phys.* 22, 403–434.
41. Gillespie, D. T. (1977) Exact stochastic simulation of coupled chemical reactions. *J. Phys. Chem.* 81, 2340–2361.
42. Gillespie, D. T. (2007) Stochastic simulation of chemical kinetics. *Annu. Rev. Phys. Chem.* 58, 35–55.
43. Hucka, M., Finney, A., Sauro, H. M., Bolouri, H., Doyle, J. C., Kitano, H., Arkin, A. P., Bornstein, B. J., Bray, D., Cornish-Bowden, A., et al. (2003) The systems biology markup language (SBML): A medium for representation and exchange of biochemical network models. *Bioinformatics* 19, 524–531.
44. Hoops, S., Sahle, S., Gauges, R., Lee, C., Pahle, J., Simus, N., Singhal, M., Xu, L., Mendes, P., and Kummer, U. (2006) COPASI--a COmplex PAthway SImulator. *Bioinformatics* 22, 3067–3074.
45. Cohen, S. D., and Hindmarsh, A. C. (1996) CVODE, A Stiff/Nonstiff ODE Solver in C. *Comp. Phys.* 10, 138–143.
46. Hindmarsh, A. C., Brown, P. N., Grant, K. E., Lee, S. L., Serban, R., Shumaker, D. E., and Woodward, C. S. (2005) SUNDIALS: Suite of nonlinear and differential/algebraic equation solvers. *ACM Trans. Math. Softw.* 31, 363–96.
47. Berg, J. M., Tymoczko, J. L., and Stryer, L. (2006) *Biochemistry.* W. H. Freeman, New York.
48. Gross, J. L., and Yellen, J. (eds.) (2003) *Handbook of Graph Theory.*CRC Press, Boca Raton, FL.
49. McKay, B. D. (1981) Practical graph isomorphism. *Congressus Numerantium* 30, 45–87.
50. Ullmann, J. R. (1976) An algorithm for subgraph isomorphism. *J. ACM* 23, 31–42.
51. Lemons, N. and Hlavacek, W. S. private communication.
52. Borisov, N. M., Markevich, N. I., Hoek, J. B., and Kholodenko, B. N. (2005) Signaling through receptors and scaffolds: Independent interactions reduce combinatorial complexity. *Biophys. J.* 89, 951–966.
53. Borisov, N. M., Markevich, N. I., Hoek, J. B., and Kholodenko, B. N. (2006) Trading the micro-world of combinatorial complexity for the macro-world of protein interaction domains. *BioSystems* 83, 152–166.
54. Borisov, N. M., Chistopolsky, A. S., Kholodenko, B. N., and Faeder, J. R. (2008) Domain-oriented reduction of rule-based network models *IET Syst. Biol.* 2, 342–351.
55. Lauffenburger, D. A. and Linderman, J. J. (1993) *Receptors: Models for Binding, Trafficking, and Signalling.* Oxford, New York, NY.
56. Posner, R. G., Wofsy, C., and Goldstein, B. (1995) The kinetics of bivalent ligand-bivalent receptor aggregation: Ring formation and the breakdown of the equivalent site approximation. *Math. Biosci.* 126, 171–190.
57. Pollard, T. D. and Borisy, G. G. (2003) Cellular motility driven by assembly and disassembly of actin filaments. *Cell* 112, 453–465.
58. Koschorreck, M., Conzelmann, H., Ebert, S., Ederer, M., and Gilles, E. D. (2007) Reduced modeling of signal transduction – A modular approach. *BMC Bioinformatics* 8, 336.
59. Heath, J., Kwiatkowska, M., Norman, G., Parker, D., and Tymchyshyn, O. (2007) Probabilistic model checking of complex biological pathways. *Lect. Notes Comput. Sci.* 4210, 32–47.

Chapter 6

Ingeneue: A Software Tool to Simulate and Explore Genetic Regulatory Networks

Kerry J. Kim

Summary

Here I describe how to use Ingeneue, a software tool for constructing, simulating, and exploring models of gene regulatory networks. Ingeneue is an open source, extensible Java application that allows users to rapidly build ordinary differential equation models of a gene regulatory network without requiring extensive programming or mathematical skills. Models can be in a single cell or 2D sheet of cells, and Ingeneue is well suited for simulating both oscillatory and pattern forming networks. Ingeneue provides features to allow rapid model construction and debugging, sophisticated visualization and statistical tools for model exploration, and a powerful framework for searching parameter space for desired behavior. This chapter provides an overview of the mathematical theory and operation of Ingeneue, and detailed walkthroughs demonstrating how to use the main features and how to construct networks in Ingeneue.

Key words: Software, Computer simulation, Biological models, Gene expression regulation, Genetic models, Signal transduction, Regulator genes, Kinetics.

1. Introduction

Many essential cellular and developmental functions are accom plished by groups of genes working together. Both traditional and recent high-throughput systems biology approaches have greatly expanded our knowledge of the regulatory interactions between genes. From this wealth of information, we are beginning to understand higher-level function arises in gene networks. However, it is difficult to synthesize the available information to produce useful hypotheses and predictions.

Computer simulations are useful and powerful tools for understanding genetic networks. Reconstituting the network of

Ivan V. Maly (ed.) *Methods in Molecular Biology, Systems Biology, vol. 500*

DOI: 10.1007/978-1-59745-525-1_6

regulatory interactions between genes into a set of rules or equations allows computers to predict and explore the consequences of those interactions. Computer simulations thus provide a bridge between molecular mechanism and network function. A common way to build computer models is to translate the biological interactions into a set of differential equations (for examples, *see* **refs.** *1–6*, for other ways of building models, *see* **refs.** *7–12)*. In such models, the rate of change of each biomolecule concentration is determined by adding its production and degradation terms, and the computer model calculates how those concentrations change over time.

The opportunities offered by computer modeling have prompted the development of software tools for both building and analyzing genetic networks. Specifically, computer simulations of networks are useful for:

- *Determining sufficiency of understanding*. Human intuition fails when confronted with a sufficiently complicated system, such as a genetic network. Computer simulations provide a rigorous and objective way to explore the consequences of the known interactions and to determine whether such interactions can account for the observed behavior/task. A model that reconstitutes all known interactions but that fails to reproduce the behavior of the real network indicates that there are essential but undiscovered regulatory interactions required.
- *Prediction and hypothesis generation*. With a working model of a network, it is easy to simulate various changes or perturbations to the network, to explore "what-if" situations or to perform virtual experiments on the model. By doing this, the model will generate predictions for the network behavior that, if interesting, can be tested experimentally.
- *Providing explanatory power and mechanistic insight*. A mechanistic model allows scientists to work through the causal chain of events in the operation of the network, which is particularly useful when the network exhibits nonintuitive behavior. By exploring the behavior of the model – aided by proper visualization tools – users can develop intuition for how the network works and fails under different conditions. For example, models of the yeast cell cycle show how cell cycle progression is tightly coupled to cell growth *(2)*. Additionally, the explanatory power offered by modeling is useful in teaching dynamical systems theory and network thinking to students.
- *Insight into network emergent behavior*. Networks are designed by evolution rather than an engineer, and general properties of networks are beginning to be determined. For example, one emerging design principle is that biological networks are

robust to quantitative variation: It has been shown that the *Drosophila* segment polarity network *(5, 13)* can continue to function despite enormous variation (several orders of magnitude) in the kinetics. Such robustness is not a property of most human-designed systems, as they often require carefully matched components.

1.1. Overview of Ingeneue

Ingeneue is a free, open source computer program that provides an intuitive and rapid way to build differential-equation-based models of genetic networks to tackle the problems listed above. It has tools to visualize the model output, systematically explore how quantitative changes alter model behavior, and automate searches for desired model behavior. Ingeneue was designed to be used by biologists without requiring extensive experience in programming and mathematics, and can be run on any lab computer with Java installed (Mac, PC, or Unix). To aid new users, Ingeneue comes with a manual, several tutorials, in-program help, and several networks. Ingeneue can be easily extended and modified, and the Ingeneue download contains the documented source code. Finally, all input and output files used by Ingeneue are human-readable plain text files that can be imported into other software packages (such as Mathematica, Excel, Matlab, etc.), for specialized analysis.

1.1.1. Ingeneue Simplifies Network Construction

Many programs can solve systems of differential equations rapidly and dependably (Mathematica, Matlab, etc.). The most difficult and challenging step is translating the known molecular facts, and the spatial arrangement of cells, into a system of differential equations. Ingeneue was designed to meet this challenge. Ingeneue can simulate networks in a single cell or in a 2D sheet of interacting cells as depicted in **Fig. 1**. **Figure 1A** shows the geometry of Ingeneue models: Each cell is hexagonal with seven compartments: one cytoplasmic compartment and six membrane compartments representing each side of the cell. Ingeneue models track the time evolution of the concentrations of all biomolecules in all cells and compartments.

Building a computational model from scratch is a significant undertaking, requiring one to specify and debug the system of equations (which can be thousands of equations in a large network with many cells), and also program the computer to solve and display the results. Ingeneue greatly simplifies the practice of building gene network models. One uses a text editor to write a plain text file that lists the important biomolecules (mRNA, proteins, and complexes) in the network and the regulatory interactions between them. Rather than writing differential equations, one tells Ingeneue what qualitative biological interactions occur: phosphorylation, translation, transcriptional activation, dimerization, etc., and the mathematical forms of these processes are already defined in

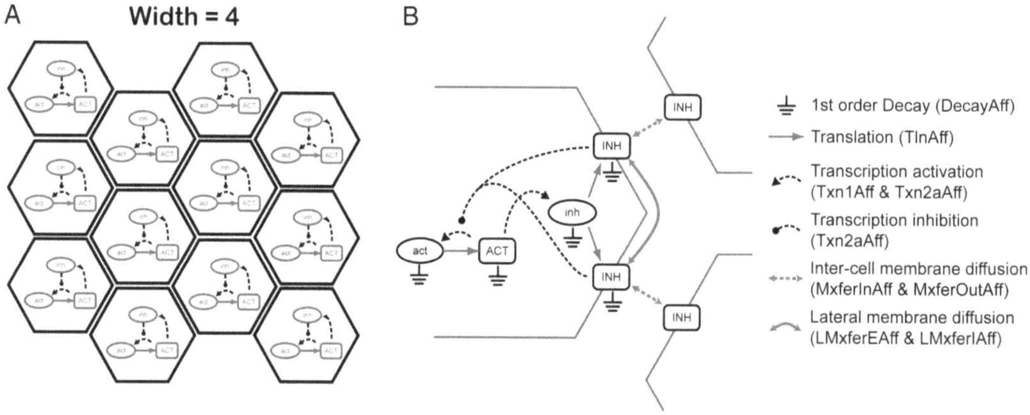

Fig. 1. Modeling a cellular layer with Ingeneue. (**A**) Geometry of a 3 × 4 cell Ingeneue model. Ingeneue simulates a two-dimensional sheet of hexagonal cells, with all cells containing the same genetic network. Cells have one cytoplasmic compartment, and six membrane-bound compartments (one for each side of the cell). Models have periodic (toroidal) boundary conditions, with the right edge wrapping around to the left, and the top connected to the bottom. (**B**) Schematic of a discrete cell Turing network. This is a two-gene network with a membrane-bound inhibitor and a cytoplasmic activator. The activator protein turns on transcription of both genes, while the inhibitor blocks the action of the activator. The nine nodes in the network are the activator mRNA (act), the inhibitor mRNA (inh), the cytoplasmic activator protein (ACT), and the six membrane-bound compartments of the inhibitor protein (INH). The INH can diffuse laterally to different sides of the same cell and diffuse across to neighboring cells. All nodes also are degraded at different rates. Legend shows the different biological processes corresponding to each *arrow*, and the name of the corresponding Ingeneue affector.

Ingeneue. Ingeneue allows users to specify networks in terms of predefined building blocks (called affectors), and Ingeneue assembles the mathematical system of differential equations. Ingeneue currently comes packaged with over 80 affectors (which are formulae for how molecules combine or interact), and additional affectors can be added easily by anyone familiar with Java programming.

This strategy building networks by selecting the right affector is faster and more intuitive than writing differential equations because it allows the model builder to focus on the biology rather than algebra. Should one want to modify the network, it is clear what interactions are already present and which to modify, bypassing the need to interpret complex lists of equations. Additionally, debugging an Ingeneue network is easier because mistakes are less likely to be made than typing the equations by hand.

1.1.2. Modeling Analysis Tools in Ingeneue

After constructing a model, one can explore the behavior of the model, and search for desired behavior. The behavior of the model depends on the parameters (rate constants, expression levels, etc.) and the initial conditions. Ingeneue offers a powerful tool for automated searches of parameter space for combinations that cause specific network behavior (such as making a particular spatial pattern). That is, one can specify the target network behavior and have Ingeneue systematically or randomly explore parameter values or initial conditions, and simulate the effects of various mutations or other perturbations. This is important, because in most networks, the parameters have not been experimentally measured, requiring exploration of parameter space *(13, 14)* to determine whether the model can reproduce the biology.

1.1.3. Development and Uses of Ingeneue

Ingeneue was developed initially to reconstitute the interactions between the core genes in the *Drosophila* segment polarity *(5, 13)* and neurogenic *(6)* networks. In these, the network exists in a sheet of cells that communicate through membrane-bound messenger molecules to stabilize spatial patterns of gene expression. The behavior of the model depended on the model parameters (reaction rates and other quantitative traits of the network), and Ingeneue searched for parameter sets that reproduce the observed pattern of gene expression. Each parameter was randomly varied over several orders of magnitude and, surprisingly, the model produced working behavior over the full range of nearly every parameter, and also showed a remarkable robustness to parameter changes.

Ingeneue is currently being used to investigate a cell cycle oscillator, pattern forming in *C. elegans* vulval development, the effects of ploidy (comparison of diploid vs. haploid networks), and the robustness of randomly wired networks. It has also been used in several courses, and has been a useful tool for teaching students how to build models of biological networks. Ingeneue has been freely available for download on the Web (with source code) since 2000.

2. Ingeneue Implementation

Ingeneue is written in the object oriented programming language Java, and can be modified easily by the scientific community. The source code is freely available and included with the Ingeneue download. This section describes the main design features underlying the mathematical implementation and software framework of Ingeneue models.

2.1. Network Construction

Ingeneue can simulate a network in a single cell or in a two-dimensional sheet of cells. Ingeneue uses periodic (toroidal) boundary geometry with the left edge touching its right edge and the top touching the bottom. Ingeneue models are specified in a plain-text file that contains all the information to simulate the network: the width and height of the sheet of cells, the relevant biomolecules (mRNA transcripts, proteins, complexes, etc.), the interactions between the biomolecules, initial conditions, and all parameters. In Ingeneue networks, Nodes refer to the concentration of a particular biomolecule in a specific compartment.

When Ingeneue reads a network file, it constructs the set of differential equations describing the model kinetics (*see* **Note 2**). The interactions between biomolecules are defined within a single cell, or between two neighboring cells, and Ingeneue then replicates the network in all cells in the sheet. To allow for communication between cells, Ingeneue membrane bound nodes can diffuse to neighboring cells, or interact with nodes on the opposing cell face (*see* **Fig. 1B** for an example).

2.1.1. Mathematical Framework

For most genetic networks, the differential equation describing how a node X (concentration within a particular compartment) changes over time is the sum of contributions from several biological processes:

$$\frac{dX}{dt} = \text{synthesis} - \text{degredation} \pm \text{conversion} \pm \text{fluxes}.$$

Figure 1B shows a detailed schematic for a hypothetical network. The inh mRNA in a cell depends on its rate of synthesis (transcription) and degradation. In terms of the biological processes, the inh mRNA concentration [inh] in **Fig. 1B** is given by:

$$\frac{d[\text{inh}]}{dt} = \text{Transcription}([\text{ACT}], K, v, T_{\text{M}}) - \text{Decay}([\text{inh}], H_{\text{inh}}),$$

where the rate of transcription is a function of concentration of the transcriptional activator [ACT] and the K, v, and T_{M} parameters described in **Subheading 2.2.1**. Likewise, degradation (first-order decay) depends on [inh] and its half-life H_{inh}. Ingeneue has the mathematical form for these processes predefined in objects called affectors that encapsulate how a single process changes a node. Most processes are independent of each other, thus the individual affectors can be added to describe how the node changes over time (Ingeneue has advanced features allowing for interactions using metaaffectors, see the Ingeneue documentation for details).

2.2. Hill Functions and Parameterization

The affectors that come with Ingeneue represent processes either with first- or second-order mass action kinetics (such as decay, translation, and binding/unbinding), or are built from Hill functions to quantify complex processes (such as transcriptional regulation or cooperative enzymatic reactions). The mathematical form to represent many biological processes is unknown, and Hill functions are an approximation that captures the qualitative behavior of many processes (saturation and monotonicity), and can be tuned with few parameters. The Hill functions Φ (representing activation) and Ψ (representing inhibition) are mathematically defined as:

$$\Phi([\text{Node}], K, v) = \frac{[\text{Node}]^v}{K^v + [\text{Node}]^v},$$

$$\Psi([\text{Node}], K, v) = 1 - \Phi([\text{Node}], K, v).$$

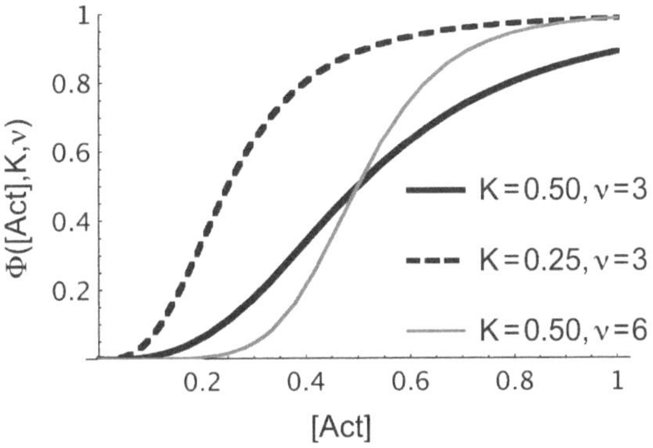

Fig. 2. Plots of the Hill function Φ for three different combinations of K and ν. The activator concentration is [Act] and Φ varies from 0 to 1. The apparent cooperativity, ν, determines the steepness of activation; high values are switch-like, lower values are more shallowly graded. K is the value of [Act] when Φ is 0.5; low levels of K correspond to strong activation (little activator needed for full effect).

Φ varies from 0 to 1. K is the value of [Node] for $\Phi = 0.5$, and ν determines the steepness or apparent cooperativity of the curve. Use of the Hill equation assumes that the components are in pseudo-steady-state equilibrium. **Figure 2** shows a plot of the Hill function Φ.

More complicated processes such as transcription under the control of multiple activators and repressors are constructed by combining Hill functions *(15)*. For example, consider transcription under the influence of one activator and one inhibitor with the inhibitor titrating away the activator according to a Hill function. In this, the remaining effective activator $[\text{EA}] = [\text{ACT}](1 - \Phi([\text{INH}], K_I, \nu_I))$ and the transcription rate is $\Phi([\text{EA}], K_A, \nu_A)$. Additional details for these complicated affectors can be found in the Ingeneue documentation for Affectors, a tutorial on writing Ingeneue affectors bundled with Ingeneue, and in *(15)*.

2.2.1. Nondimensionalization

Most affectors in Ingeneue are nondimensionalized and have a different form than one might expect from standard chemical reaction kinetics. Nondimensionalization is a standard technique to have all variables changed to dimensionless scalar quantities. This is not an approximation, but a rescaling that substantially reduces the number of free parameters in the model while allowing the model to maintain its full dynamical range of behaviors. The cost of this is that the meaning of many parameters is altered, as explained below. For in-depth discussion of nondimensionalization, see the supplements in *(13, 15)* and the dimensional analysis section in *(16)*.

Ingeneue uses a nondimensionalization strategy where most nodes are constrained to the range from 0 to 1. All nodes, time, and parameters become scalar quantities with no units/dimensions by using the following change of variables:

$$t = t_o \tau,$$
$$x(t) = x_o x(\tau),$$

where t and $x(t)$ are the dimensional time and node (concentration), t_o and x_o are scaling factors for nondimensionalization, and τ and $x(\tau)$ are the nondimensional time and node. The scaling factors t_o and x_o are arbitrary, and so we will set them to values so the maximum steady state due to synthesis and first-order decay for any node x is 1. To do this, one simply solves for the maximal steady-state value when synthesis is maximal and degradation is minimal. For example, the differential equation governing inh in **Fig. 1B** using the equations in **Table 1** is

Table 1
Mathematical form of affectors

Process and affector name	Formula	Nondimensional formula	Parameters
Transcription activated by 1 activator (ACT) (`Txn1AAff`)	$T_M \Phi([ACT], K, v)$	$\frac{\Phi(ACT(\tau), K', v)}{H'_{ACT}}$	T_M = maximum transcription rate K = half maximum [ACT] K' = half maximum ACT(τ) v = cooperativity of activation H'_{ACT} = rate of approach to steady state
Translation of mRNA into protein (`TlnAff`)	$[mRNA]R_T$	$\frac{mRNA(\tau)}{H'_{mRNA}}$	R_T = maximum translation rate H'_{mRNA} = rate of approach to steady state
First-order decay of a Node (`DecayAff`)	$-\frac{[Node]}{H_{Node}}$	$-\frac{Node(\tau)}{H'_{Node}}$	H_{Node} = mean lifetime of node H'_{Node} = rate of approach to steady state
Diffusion of membrane-bound Node to neighboring side (`LMXferEAff`) or cell (`MXferOutAff`)	$-D$ [Node]	$-D$ Node(τ)	D = fraction of Node that diffuses per unit time
Transcription regulated by 1 activator (ACT) and 1 inhibitor (INH) that can completely block the activator (Txn2aAff)	$T_M \Phi([EA], K_A, v_A)$, $[EA] = [ACT]\psi([INH], K_I, v_I)$	$\frac{\Phi(EA, K'_A, v_A)}{H'_{ACT}}$ $EA = ACT(\tau)\Psi(INH(\tau), K'_I, v_I)$	T_M = maximum transcription rate K_A = half maximum [ACT] K'_A = half maximum ACT(τ) v_A= cooperativity of activation K_I = half maximum [INH] K'_I = half maximum INH(τ) v_I = cooperativity of inhibition H'_{ACT} = rate of approach to steady state

$$\frac{d[\text{inh}]}{dt} = T_M \Phi([\text{ACT}], K, \nu) - \frac{[\text{inh}]}{H_{\text{inh}}},$$

where [inh] is the dimensional concentration, [ACT] is the dimensional concentration of activator protein, and H_{inh} is the mean lifetime of the inh mRNA. The maximal steady state of [inh], $[\text{inh}]_{ss}$, would occur when $d[\text{inh}]/dt = 0$:

$$\frac{d[\text{inh}]}{dt} = 0 = T_M \Phi([\text{ACT}], K, \nu) - \frac{[\text{inh}]_{ss}}{H_{\text{inh}}}.$$

Solving for $[\text{inh}]_{ss}$ yields:

$$[\text{inh}]_{ss} = T_M \Phi([\text{ACT}], K, \nu) H_{\text{inh}}.$$

$[\text{inh}]_{ss}$ is maximal when there is saturating [ACT] (when [ACT] >> K, $\Phi \to 1$):

$$\text{Max}([\text{inh}]_{ss}) = T_M H_{\text{inh}}.$$

Now, if we transform the original equation for [inh] into the nondimensional form with $\text{inh}_o = 1/\text{Max}([\text{inh}]_{ss})$:

$$\frac{d\,\text{inh}(\tau)}{d\tau} = \frac{\Phi(\text{ACT}(\tau), K', \nu) - \text{inh}(\tau)}{H'_{\text{inh}}}.$$

In the nondimensionalized equation, one parameter, T_M is gone, and the nondimensional parameters K' and H'_{inh} are now dimensionless, scalar quantities. K' is the fraction of the nondimensional ACT(τ) node for half-maximal activation of transcription, and H'_{inh} is the fractional rate at which inh(τ) approaches steady state. Unitless parameters such as cooperativity, ν, are unchanged by nondimensionalization. **Table 1** shows the nondimensional form of the affectors used in Ingeneue.

2.3. Searching Parameters for Desired Behavior

Ingeneue has a scripting language to search for parameter sets that produce some desired behavior. Using this, Ingeneue will assign a score to each model run with different parameter sets measuring how far that run was from the desired behavior according to selected criteria (correct pattern formation, steady state or oscillatory behavior, etc). Ingeneue can then save parameter sets with sufficiently good scores and subject them to additional analysis (i.e., simulating the effects of mutations or other perturbations).

3. Using Ingeneue

This section shows how to use most features and tools in Ingeneue. Not all features are discussed here, but most are self-explanatory or are discussed in the manual or tutorials that come bundled with Ingeneue.

3.1. Getting Ingeneue

1. Install Java version 1.4 or higher if it is not on your computer (*see* **Note 2**).
2. Download the latest version of Ingeneue from http://ingeneue.com. Follow the instructions on the Web site for installing Ingeneue.
3. Read the README.txt file in the Ingeneue root directory or from the Ingeneue Web site; this file contains last-minute information and details for using Ingeneue.

3.2. Loading and Running the Segment Polarity Network

Ingeneue comes with several networks located in the networks directory of Ingeneue. This section shows how to load the segment polarity network *(5, 13)* and how to use the Ingeneue interface to alter the model and visualize the results. The same procedure can be used for any network:

1. *Start Ingeneue* following the instructions on the Ingeneue Web site or the README.txt file. When Ingeneue starts, you will see a large window similar to **Fig. 3**. You can access the manual and several tutorials through the help frame on the

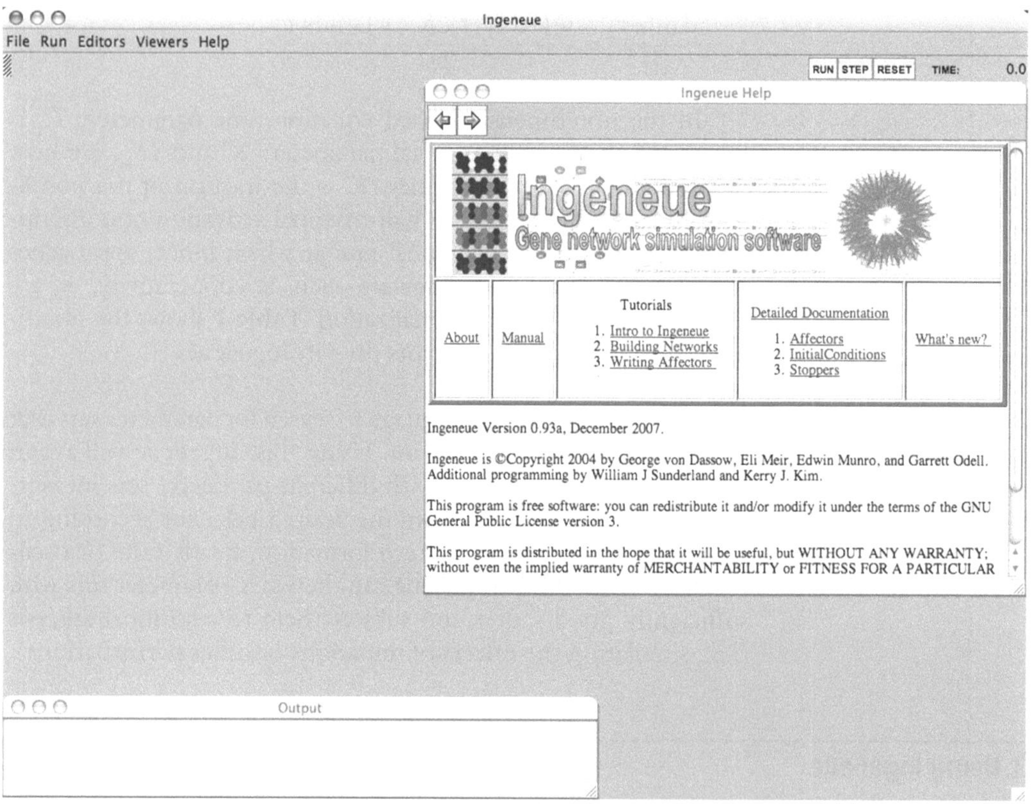

Fig. 3. The Ingeneue window. Your view may be slightly different because of your operating system, Java version, or Ingeneue version, but there should be two frames displayed the first time you run Ingeneue: (1) The *Ingeneue Help* frame provides access to the Ingeneue manual, tutorials, and detailed documentation. (2) The *output console* displays any messages or errors generated by Ingeneue. The *menu bar* is at the *top* of the window ("File," "Viewers," etc.). Model runs are controlled by the "Start," "Step," and "Reset" buttons in the *upper right* corner of the Ingeneue window.

right side (*see* **Note 3**). Information and error messages are displayed in the console frame at the bottom.

2. *Load a network file* by clicking on "File → Open" (i.e., click on "File" from the menu bar, then click on "Open"). A dialog box will appear. From the main Ingeneue directory, navigate to the "networks/segmentpolarity" subdirectory, and open the file "spg1_01.net." Two new frames will appear, one labeled "Network Viewer" (**Fig. 4A**) showing a schematic diagram of the segment polarity network, and one labeled "Cell Viewer" (**Fig. 4D**) showing several black hexagons.
3. *Run the network* by clicking on the "Run" button in the upper right corner of the screen. When you do this, the system of differential equations will be integrated forward in time for 1,000 min (*see* **Note 1**). The current simulation time (shown in the upper right corner in minutes) will rapidly increase, and the changing colors of the hexagons in the Cell Viewer show the progression of the pattern of gene/protein expression. In the Cell Viewer, each hexagon indicates one cell (similar to **Fig. 1A**), and the concentrations of selected Nodes are shown in each cell with brighter colors indicating higher concentration. You may press the Run button again to rerun the model. The "Reset" button sets the time to 0, and each time the Step button is pressed, the time will advance a small amount (the exact amount varies because of the adaptive-step size numerical integrator, *see* **Note 4**) and the Cell Viewer will be updated.
4. *Change parameter values* by clicking on "Viewers→Network Parameters." A new frame will appear (**Fig. 4C**) that lists all parameters in the model, their current values, and how they should be sampled in a random search: the range (minimum and maximum), and whether the sampling should be linear or logarithmic (low values have a higher probability than high). The "step" values are unused. Try changing one or more parameters (be sure you change the value, not the low, high, or step setting), then click on the "set all" button (parameter values you type in are not applied until you click a "set" button). Rerun the model and note how some changes alter the final gene expression pattern, while others do not. Press "reset all" before proceeding to the next step; this which will restore all parameters to default values (defined in the net file).
5. *Change initial conditions* by clicking on "Viewers→Node Viewer" or double clicking on any cell (hexagon) in the Cell Viewer. A frame will appear (**Fig. 4B**) showing the initial and current concentration for all nodes in the selected cell. The top of the frame shows a field of hexagons that represent the different cells in the network; node values can be viewed for any cell by clicking on the cell. Try changing the initial conditions for en and wg in several of the cells to see whether you can disrupt the pattern of gene expression.

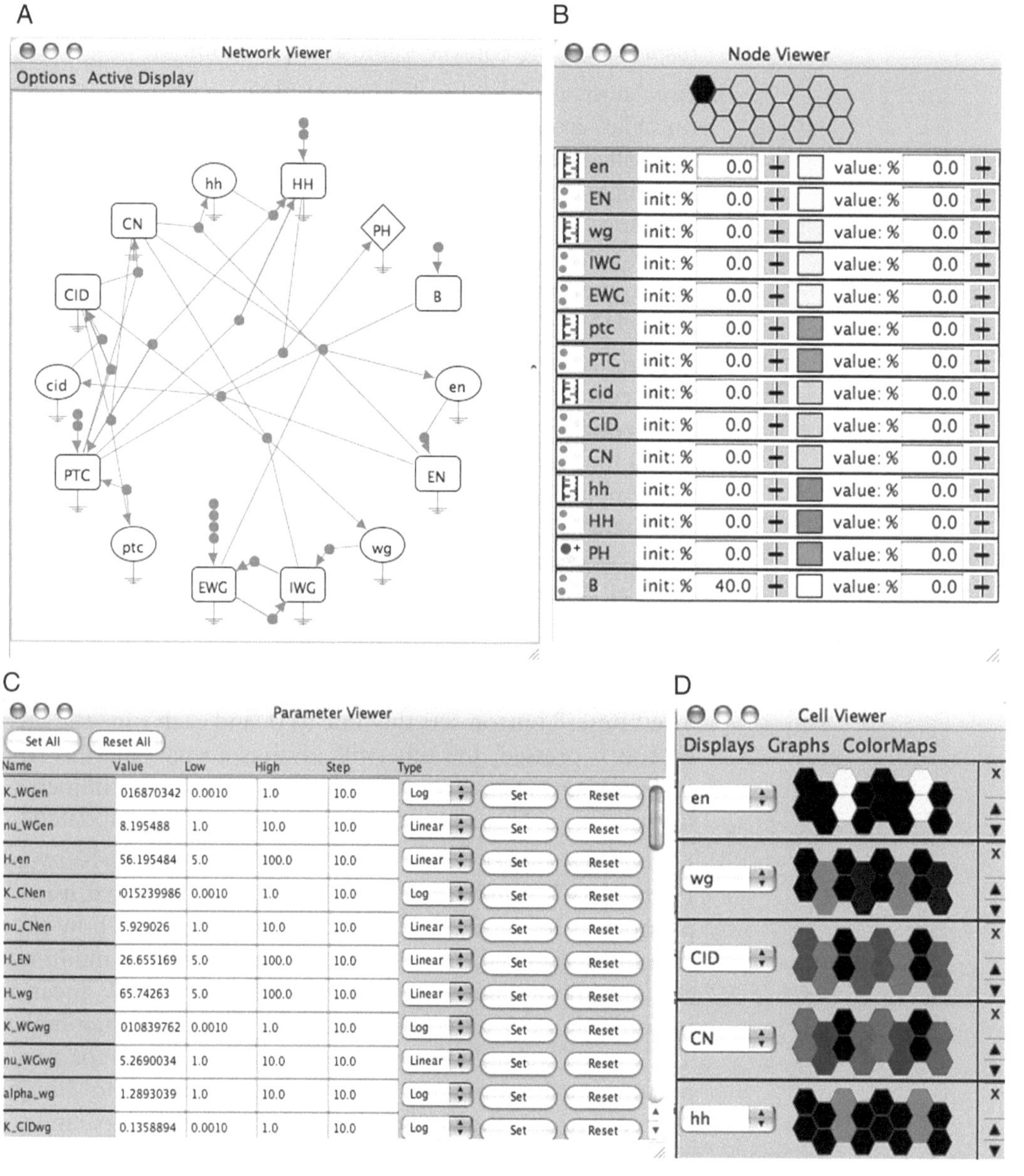

Fig. 4. Tools for visualizing network behavior and altering model parameters or initial conditions. These are accessible from the Viewers menu after a network is loaded. (**A**) *Network Viewer* shows a simplified graphical schematic of the concentrations (nodes) and interactions (affectors) in the network. Click on any node or affector for details. Selecting "turn on active display" under the active display menu will color-code each node according to its concentration, allowing you to visualize the model dynamics in a selected cell during the model run. (**B**) *Node Viewer* shows the current and initial value of each node in the selected cell. Cells can be selected by clicking on the hexagon at the top of the window. The initial and current levels of each node can be changed by entering a new value for init or value, respectively. (**C**) *Parameter Viewer* shows the current values of all model parameters, and allowed range. Parameters can be changed by entering a new value in the "value" column, followed by pressing the "set all" button at the top of the frame. The "reset all" button will restore all parameters to the value stored in the net file. (**D**) *Cell Viewer* indicates the concentration of selected nodes in all cells, with black being 0, and brighter colors meaning higher concentration. During model running, an animation for the time progression of the pattern of expression is shown.

3.3. Plots of Concentration Versus Time

1. *Plots of concentration vs. time* are shown by clicking on "Viewers → Time graph." A new frame will appear, as shown in **Fig. 5A**, giving you three options for plotting node values vs. time:
2. *Manual method*: To set up plots for one or a few specific nodes:
   (a) In the Time Graph frame, click on "New," and a new frame will appear (**Fig. 5A**). Click on the cell you wish to view in the field of hexagons, then click on the node you wish to plot, and select the color for plotting it. Click on "add" when you are done. Along the bottom of the graph, you will see the color legend for the graph. You can add as many nodes to the graph as you wish using this procedure.
   (b) To delete or alter the nodes in a graph, click on the node you wish to change in the legend at the bottom of the graph. This will summon the frame you used to choose the

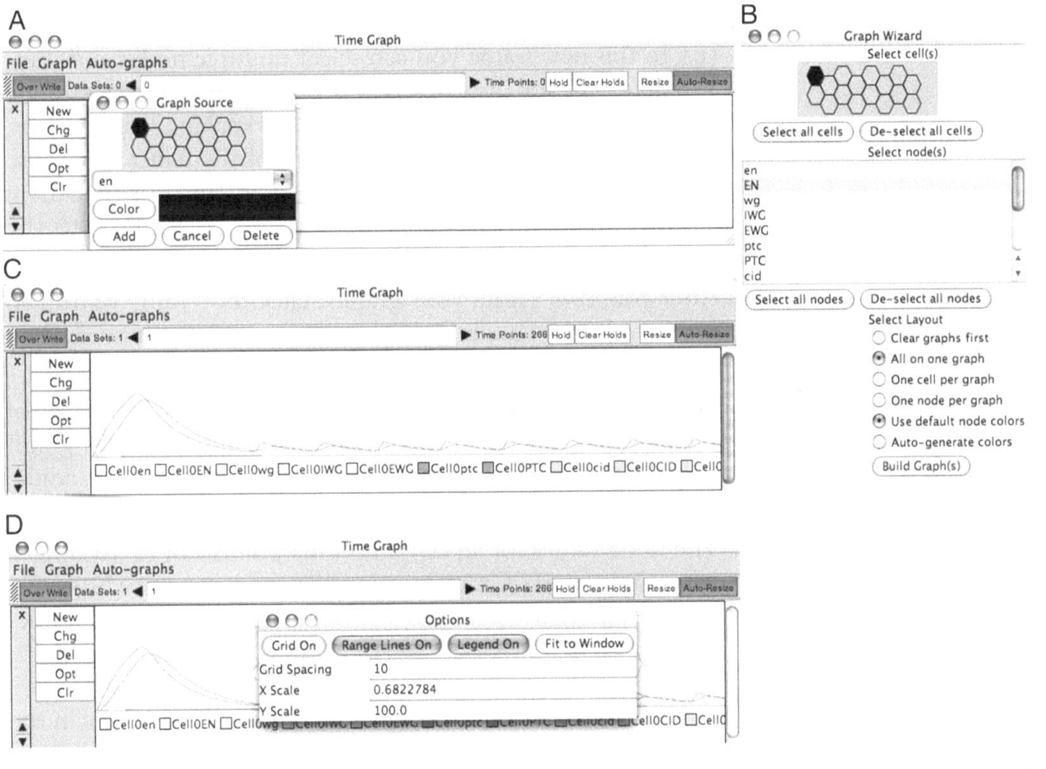

Fig. 5. Time graphs show a plot of node value vs. time for any node in any cell in the network. (**A**) Manual selection of node to plot; click on the "New" button on the *left*, then in the new frame (shown) click on a *hexagon* to select the cell, and choose a node and display color to add to the graph. (**B**) Graph wizard menu for adding multiple nodes to a time plot. Accessible by clicking on "Auto Graphs → Graph wizard." (**C**) After the nodes are selected, click on the "Run" button in the *upper right* corner, and the time plot will be displayed. (**D**) Time Graph options frame. Click on the "Opt" button on the *left* to the graphing options.

node, and you can now either delete the node from the plot by clicking on "delete," or alter the node by selecting a different cell, node, or color, followed by pressing "Add." To delete all the nodes from the graph, click on the "clr" button.

(c) To make additional graphs, select "Graphs→Add new graph."

3. *Automatic*: To plot all nodes in all cells, click on "Auto Graphs→Quick Graphs."
   (a) "All Nodes/All Cells option, one cell per graph" displays all nodes grouped into graphs by cell as shown in **Fig. 5C**.
   (b) "All Nodes/All Cells, one node per graph" displays all nodes grouped into graphs by node.
4. *Graph wizard*: Use this if you want to plot many nodes, but the Auto Graphs do not suit your needs:
   (a) Click on "Auto Graphs→Graph wizard." A new frame will appear as shown in **Fig. 5B**.
   (b) In this new frame you can select multiple nodes and multiple cells for display. Click on the node names to toggle selection of the desired nodes. After making your choices, use the buttons in the lower half of the panel to choose whether to plot all nodes in the same graph or split them into different graphs by node or by cell (similar to the automatic method above).
5. After you have set up your graphs, click the "Run" button to see the plotted results. Every time you run the model while the Time Graphs frame is open, the graphs will be updated.
6. Additional options are available to customize the display of the graphs by clicking on the "opt" button next to any graph (**Fig. 5D**) that allows you to display grid lines and adjust colors.
7. The node values for all cells and time points can be saved to a tab-delimited file for analysis in other programs (*see* **Note 5**). To save the data for all runs stored in the Time Plots frame, click on "File→Save data set" and enter a name for the file.

3.4. Finding Parameter Sets Using Automated Searches

This section demonstrates how to search for parameter sets in the segment polarity network that produce the biologically observed pattern of en, wg, and hh stripes. To do this, Ingeneue loads an iterator file that specifies how to vary parameters and the criteria that delineate "good" sets of parameter values. The real segment polarity network stabilizes the striped pattern of en, wg, and hh gene expression shown in **Fig. 4D**. Ingeneue comes with an iterator file that randomizes all parameters over their biological range and searches for this pattern of behavior (*see* **Note 6**).

1. There are two ways to use iterator files. Do either of the following and wait until Ingeneue finds at least three parameter sets (*see* **Note 7**).

   (a) *Command line*: Use this option for a fast search that does not use the graphical user interface. Exit Ingeneue and access the command prompt on your computer. From the Ingeneue root directory, type the following command all on one line (do not press enter between lines):

   ```
   java main.GeneNet networks/segmentpolarity/
   spg1_4cell.netnetworks/segmentpolarity/
   randomsampler.iter
   ```

   This tells Ingeneue to start by loading the network file "spg1_4cell.net" and then run the "randomsampler.iter" iterator file. The "spg1_4cell.net" network has a 1 × 4 cell array, which is topologically identical to the 2 × 8 model (because of the toroidal boundary conditions) and will run four times faster. "Good" parameter sets (*see* **Note 6**) are saved in the "output" subdirectory, and will have a filename with a prefix of "spg," followed by a timestamp. Ignore any warning messages that appear in the console.

   (b) *Using the graphical interface*: Choose "File→open" and open "spg1_4cell.net." Next, choose "File→open," and open "randomsampler.iter." A new frame will appear labeled "iterator control," shown in **Fig. 6A**. Click on "Run" in the iterator control frame. The model will be repeatedly run with randomized parameters, and you can see expression patterns changing as the search proceeds. Ingeneue prints a message every time a successful parameter set is found.

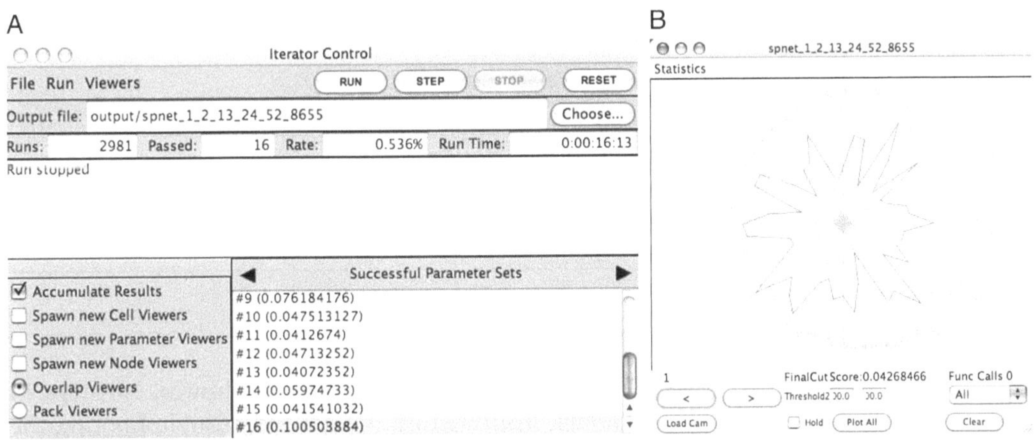

Fig. 6. Tools for searching and visualizing desired model behavior. These windows are opened when loading an iterator file or an iterator output file, and require that a network be loaded first. (**A**) *Iterator Control* is displayed when an iterator is loaded. The results of the search will be saved to the file name specified here. Click on "run" to begin the search. During the search, the successful parameter sets will be shown, as well as the number of attempts ("runs"), the number of successful sets ("passed"), and the percent of successful sets ("Rate"). Click on "stop" to halt the search. When halted, clicking on the successful parameter sets will load the parameter set into the current model. (**B**) *Cam viewer* shows a spoke and wheel plot for all successful parameter sets generated by an iterator. This window is displayed when you load an output file generated by an iterator.

3.5. Analysis and Visualization of Working Parameter Sets Found by Iterators

1. In Ingeneue, choose "File→Open," and open "spgl_4cell.net." Then, choose "File→Open," and navigate to the "output" subdirectory, and open the file that was generated in **Subheading 3.3** (the file will usually be called "spg_XX" where XX are numbers indicating the time of the run; alternatively, you can load the "spg.params" file from the "networks/segmentpolarity" directory). A new window will appear labeled with the name of your file as shown in **Fig. 6B** (*see* **Note 8**).
2. This window contains a wheel plot representing the values of all parameters in working parameter sets. Points near the center/edge of the wheel correspond to minimum/maximum of the range explored in the random parameter search. You can view different parameter sets by clicking on the < and > buttons. You can view a superposition of all parameter sets by clicking on the "Plot all" button.
   (a) By clicking "load cam," the displayed parameter values will be written to the model for more detailed visualization (i.e., you can then run the model to make time plots or view the pattern of expression). While the final pattern is the same for all parameter sets, the dynamic behavior of the model through time is different.
3. The "statistics" menu option has the following options:
   (a) *Summary*. Saves the mean, range, and variance for the parameter sets.
   (b) *Param tendencies*. Saves the range explored for each parameter in the parameter search.
   (c) *Cross corr*. Saves the cross correlation coefficients (*see* **Note 9**) for all pairs of parameters. Also prints out a list of the pairs of parameters that have the largest absolute value cross correlation coefficient.
   (d) *Dump*. Saves all parameter values into a tab-delimited plain text file.

4. Buliding Networks in Ingeneue

This section is a step-by-step walkthrough showing all steps in building a model, from writing the net file, troubleshooting the model, and searching for parameter sets using a new iterator. The steps outlined here are the same you would follow for building any network.

The model you will build in this section is the two-gene network shown in **Fig. 1B**, in a 2 × 2 field of cells. This is a modified Turing network that is capable of stabilizing spatial patterns of gene expression as well as oscillations and propagating waves of gene expression *(17, 18)*. You can either type in this example as you work through this section or find the completed files in the networks/turing subdirectory.

4.1. Writing Network Files

Ingeneue network files contain all the information necessary to build the network in the following order: Name of model, geometry, nodes, interactions, parameters, and initial conditions. Each line in the network file starts with an ampersand "&" followed by a keyword that tells Ingeneue what kind of information the line contains. Comments (which are ignored by Ingeneue) begin with "//" (*see* **Note 10**). The overall structure of a network file is as follows:

```
&Model
//Information about model geometry
   &Network networkname
   &Nodes
      //Information about nodes
   &endNodes
   &Interactions
      //Information about Interactions
   &endInteractions
   &Parameters
      //Information about parameters
   &endParameters
   &InitialConditions
      //Information about initialconditions
   &endInitialConditions
&endNetwork
```

Indentation is not necessary, but makes the file easier to read. The various sections of the network file are surrounded by `&section` and `&endsection`. The sections below describe how to write a net file.

4.1.1. Model Name and Geometry

1. Start your favorite text editor to begin writing the contents of the net file. You will specify the following:
   (a) `&Model` indicates the start of an Ingeneue net file.
   (b) `&width` and `&height` are both set to 2 to indicate a 2 × 2 array of cells.
   (c) `&numsides` is 6 for hexagonal cells.
   (d) `&Network turing` tells Ingeneue that the name of the network is "`turing`."
2. To specify a network of a 2 × 2 array of hexagonal cells, type the following:

```
&Model
&width   2
```

```
&height     2
&numsides   6
&Network turing
```

4.1.2. Nodes

Figure 1B shows the four nodes in the Turing network: the activator mRNA (act), the inhibitor mRNA(inh), the cytoplasmic activator protein (ACT), and the membrane-bound inhibitor protein (INH). By convention, names of mRNA are all lowercase, and names of proteins are all uppercase. The order in which you define the various nodes does not matter, but Ingeneue is case sensitive, meaning "`act`" is different from "`aCt`" and "`ACT`." There is no need to have the gene itself (the DNA) in Ingeneue models as the DNA concentration is usually unchanging (exceptions would be if one were modeling a cell cycle or a cell growth process).

1. Type the following line to indicate the beginning of the Nodes section: `&Nodes`
2. Give information for each of the nodes in the following format:
   (a) `&Nodename:` Name of the node starting with an ampersand (`&`).
   (b) `&Location:` Location in the cell: cytoplasmic (`cyto`) or membrane.
   (c) `&Color:` Color for display in the Cell Viewer. Choices are: `red, green, blue, cyan, magenta, pink, yellow, white,` and `orange`.
   (d) `&Show:` Either on (show by default in the Cell Viewer) or `off` (hidden).
   (e) `&Scale:` Magnification factor for plotting; if a node always has a low concentration, set this to a large value and the concentration will be multiplied by this factor for display. Normally, set this to `1`.
   (f) `&Type: rna` for a messenger RNA (mRNA); `protein`, or `complex` for dimers or other assemblies.
   (g) `&endNodename:` Tells Ingeneue that you have specified all the information about the Node.
3. The act node is a cytoplasmic mRNA, and will be displayed in the color cyan in the Cell Viewer. To do this, type in the following:

```
&act
  &Location    cyto
  &Color       cyan
  &Show        on
  &Scale       1
  &Type        rna
&endact
```

4. You will specify the other nodes in the same way. For simplicity, `&Scale` will be set to 1 for everything, and we'll show the mRNA the same color as the protein. Type the following:

```
&ACT
  &Location    cyto
  &Color       cyan
  &Show        on
  &Scale       1
  &Type        protein
&endACT
&inh
  &Location    cyto
  &Color       red
  &Show        on
  &Scale       1
  &Type              rna
&endinh
&INH
  &Location    membrane
  &Color       red
  &Show        on
  &Scale       1
  &Type        protein
&endINH
```

5. To indicate the end of the nodes section, type in the following line: `&endNodes`

4.1.3. Interactions and Affectors

The Interactions section specifies what processes affect each of the Nodes. **Table 2** shows a summary of the most commonly used affectors. Click on the "Help→Detailed Documentation→ Affectors" for instructions on how to use each affector. List the affectors that affect each node as follows:

```
&AffectorName <nodes> <params>
```

Where `<nodes>` are the nodes that modulate the affector and `<params>` are the parameters that affect the quantitative behavior of the affector. Each affector requires different nodes and parameters; see the documentation for the affectors to determine this. The order that you list affectors does not matter, but you must follow the order of parameters and nodes given to each affector according to the detailed documentation. Parameters can be named, however, you like; **Table 3** shows the naming conventions for commonly used parameters.

1. Type the following line to indicate the beginning of the interactions section: `&Interactions`
2. For each node, list all of the affectors that influence it. This process is simplified if you have drawn a wiring diagram similar

Table 2
Commonly used affectors

Affector name	Description	Affects
DecayAff	First-order decay of a node	All nodes
TlnAff	Translation of protein from mRNA	Proteins
Txn*Aff	Transcriptional affectors. Many affectors available due to wealth of transcriptional control logic	mRNA
LMXfer*Aff	Lateral diffusion of membrane-bound nodes to (LMXferEAff) or from (LMXferIAff) neighboring faces of same cell	Membrane Nodes
MXfer*Aff	Diffusion of membrane-bound nodes to (MXferOutAff) or from (MXferOutAff) facing side of adjacent cell	Membrane Nodes
Dimerize*Aff	Binding of two nodes to form a complex. Different affectors allow for binding within and between cells	Free monomer nodes
Dissociation*Aff	Unbinding of complex into component nodes. Opposite of Dimerize affectors above	Dimer node
Endo*Aff	Endocytosis of membrane-bound node into the cytoplasm. EndoEAff is for use on the membrane-bound node, EndoIAff for he cytoplasmic	Membrane and cytoplasmic nodes
Exo*Aff	Exocytosis of cytoplasmic node to membrane (placed on all membranes equally). Opposite of endocytosis affectors above.	Membrane and cytoplasmic nodes

to **Fig. 1B** that shows all the processes affecting each node. *See* **Table 2** and the detailed documentation for the affectors to determine what affectors to use for your network. The detailed documentation lists what nodes and parameters are required by each affector. The following affectors affect the act node:

(a) `Txn2bAff`: act mRNA transcription is activated by ACT, and inhibited by INH. To use this affector, first specify the inhibitor node (INH) and then the transcriptional activator (ACT). Next you will give the parameters for activation (half maximal activation and cooperativity), the half life of the act mRNA, and finally the inhibition parameters (half maximal inhibition and cooperativity).

(b) `DecayAff`: the act mRNA undergoes first-order decay. This affector requires you to specify the node under decay (act) and its half-life.

Table 3
Parameter meaning and naming convention

Prefix	Range	Meaning
`H`	1–200, Linear	Half life. Because of nondimensionalization, the time constant over which synthesis and decay act. Low values indicate concentration changes quickly, high is slowly
`K`	0.01–1, Log	Half maximal concentration of a regulatory node, parameter used in Hill functions. Low values indicate a strong activator (little activator needed for effect; *see* **Fig.** 2)
`nu`	1–10, Linear	Apparent cooperativity for activation. Low values are linear, high means switch-like behavior. Almost always appears in an affector when a *K* parameter (above; *see* **Fig.** 2) is used
`Mxfer`	0.0001–1, Log	Rate that node diffuses to facing side of opposite cell, somewhat like a diffusional permeability. Low values mean slow transfer/diffusion
`LMXfer`	0.0001–1, Log	Rate that node diffuses to adjacent sides of same cell. Low values mean slow transfer/diffusion
`Endo`	0.001–1, Log	Rate that membrane-bound node gets endocytosed into cytoplasm. Low values mean slow endocytosis
`Exo`	0.001–1, Log	Rate that cytoplasmic nodes get excocytosed into membrane. Low values mean slow exocytosis
`max`	1–1,000, Log	Relative stoichiometric amount of node. Needed because nondimensionalization normalizes all nodes to a max of 1, but stoichiometric consequences remain

3. Type the following for the interactions for the act mRNA, using the scheme from **Table 3** for naming parameters:

```
&act
  &DecayAff act H_act
  &Txn2aAff INH ACT K_ACTact nu_ACTact H_act
  K_INHact nu_INHact
&endact
```

4. The inh mRNA and the ACT protein interactions are pretty similar. Both undergo first-order decay (`DecayAff`). The ACT protein is simply translated from the act mRNA (`TlnAff`), and inh mRNA transcription is activated by ACT protein (`Txn1Aff`). Note that some of the parameters appear in more than one affector, and that translation is not an affector for the mRNA nodes (we assume translation alters the protein concentration, not the mRNA). Type the following:

```
&ACT
  &TlnAff act H_ACT
  &DecayAff ACT H_ACT
&endACT
&inh
  &DecayAff inh H_inh
  &Txn1Aff ACT K_ACTinh nu_ACTinh H_inh
&endinh
```

5. The INH protein has six affectors: first-order decay, translation, and two affectors each for lateral membrane diffusion and cell–cell diffusion.
   (a) MXferOutAff and MXferInAff are for diffusion to and from the facing side of the opposite cell. These are implemented as two separate affectors to allow for the possibility of unidirectional transport.
   (b) LMXferEAff and LMXferIAff are for diffusion to and from adjacent membrane sides of the same cell.
6. Type the following for the INH node:

```
&INH
  &TlnAff inh H_INH
  &DecayAff INH H_INH
  &MxferOutAff INH Mxfer_INH
  &MxferInAff INH Mxfer_INH
  &LMxferEAff INH LMxfer_INH
  &LMxferIAff INH LMxfer_INH
&endINH
```

7. End the interactions section by typing:

```
&endInteractions
```

4.1.4. Parameters

The parameters section lists all parameters (used by the affectors in the interactions section), their default value, their allowed range (maximum and minimum values), and whether random searches should explore this range linearly or logarithmically. For each parameter, specify these values in the following format:

```
&Paramname DefaultValue MinValue MaxValue Sampling
```

DefaultValue is the value for the parameter when the net file is first loaded, and the other fields determine the range (MinValue to MaxValue) and sampling (Linear or Logarithmic) for random parameter searches.

1. Type the following line to indicate the beginning of the parameters section:

```
&ParameterValues
```

2. Ingeneue requires that you define default values and ranges for all parameters you used in the interactions section. When building a network for the first time, just guess default values;

in the next section you will write an iterator to search for parameter sets producing oscillations. Typical ranges for parameters are shown in **Table 3**. Type the following (notice the use of comments, *see* **Note 11**):

```
  // Mean lifetimes:
&H_act    10.0 1.0 100.0    Linear
&H_ACT    10.0 1.0 100.0    Linear
&H_inh    10.0 1.0 100.0    Linear
&H_INH    10.0 1.0 100.0    Linear
  // Diffusion rates of INH:
&Mxfer_INH      0.1 0.0001 1.0    Logarithmic
&LMxfer_INH     0.1 0.0001 1.0    Logarithmic
  // Transcriptional activation:
&K_ACTact0.1 0.01 1.0      Logarithmic
&K_ACTinh0.1 0.01 1.0      Logarithmic
&K_INHact0.1 0.01 1.0      Logarithmic
&nu_ACTact      3.0 1.0 10.0      Logarithmic
&nu_ACTinh      3.0 1.0 10.0      Logarithmic
&nu_INHact      3.0 1.0 10.0      Logarithmic
```

3. End the parameters section by typing:

```
&endParameterValues
```

4.1.5. Initial Conditions

The last section in the net file specifies the initial values of all nodes in all cells. For the Turing network, we will set the initial values for the inh and INH nodes to 0 everywhere, but set act and ACT to 0.3 in two cells and 0.7 in two other cells. Ingeneue provides a set of objects called `InitialConditions` that allow you to construct spatial patterns of initial conditions. **Table 4** shows the most useful `InitialConditions;` details on how to use them can be viewed by clicking on "Help→Detailed Documentation→Initial conditions."

Table 4
Commonly used initial conditions

Name	Effect
`CellIC`	Allows setting of the concentration for a specific cell
`RowIC`	Sets the node value for a whole row (horizontal line) of cells
`ColumnIC`	Sets the node value for a whole column (vertical line) of cells
`IncrementingIC`	Produces a linear gradient from left to right
`CenterIC`	Sets the node value for a specific cell, and also sets the value in its adjacent neighbors

1. Type the following line to indicate the beginning of the parameters section:
   ```
   &InitLevels
   ```

2. For each of the nodes, enter the default expression level in all cells. We will modify the initial conditions in the next step. To set the initial level of act and ACT to 0.3 and the other nodes to 0, type:
   ```
   &BackgroundLevel act   0.3
   &BackgroundLevel ACT   0.3
   &BackgroundLevel inh   0.0
   &BackgroundLevel INH   0.0
   ```

3. To set act and ACT to 0.7 in the upper left and lower right cells (in the grid of 2 × 2 cells), we'll use the `CellIC InitialCondition`, (see the detailed documentation for `CellIC` for the meaning of its inputs). Type the following:
   ```
   &CellIC // High in upper left cell:
      &Node act
      &Value 0.7
      &XPos 0
      &YPos 0
   &endIC
   &CellIC
      &Node ACT
      &Value 0.7
      &XPos 0
      &YPos 0
   &endIC
   &CellIC // High in lower right cell:
      &Node act
      &Value 0.7
      &XPos 1
      &YPos 1
   &endIC
   &CellIC
      &Node ACT
      &Value 0.7
      &XPos 1
      &YPos 1
   &endIC
   ```

4. End the parameters section by typing:
   ```
   &endInitLevels
   ```

5. Type the following to indicate the end of the model:
   ```
   &endNetwork
   ```

6. Save the file to networks directory and call it "turing.net".

4.2. Debugging Net Files

Debugging is almost always a normal step in constructing any model, as your initial attempt to make a network will probably have some errors. Whenever Ingeneue loads a net file, it is checked for errors, and Ingeneue will print error messages in the output console if any errors are found. If you see any error messages or if the network fails to load (i.e., the Network Viewer frame does not appear), this is an indication there are mistakes in your net file.

Fix errors one at a time, starting from the first error message. After fixing an error, save the file, and reload it until no errors remain (you need not restart Ingeneue). Often, a single error in a net file may trigger several error messages, and fixing a single mistake may eliminate multiple messages. The most common errors are:

1. Misspelling (or miscapitalization) of parameter, node, or affector names. The error message from Ingeneue will usually indicate the misspelled variant of the name. Most commonly, these errors appear in the interactions section.
2. Incorrect format for an affector. Each affector requires a different set of nodes and different parameters. Ingeneue will report which affector is incorrectly used in the interactions section. Check the detailed documentation for affectors to verify that the affector has the correct nodes and parameters.
3. Forgetting or misspelling the `&endSection` tag. Each section of the net file (nodes, parameters, etc.) must be ended using the `&endSection`.

4.3. Writing Iterators to Automate Parameter Space Searches

Iterator files are scripts that instruct Ingeneue to find parameter sets that produce some desired behavior. This section describes how to write an iterator file that randomly picks values for all parameters searching for parameter sets that produce temporal oscillations in all four cells of the Turing network. The format of iterator files is similar to that of net files (commands begin with & and comments with //). It is easier to write iterator files from the inside-out – i.e., we will not be writing the iterator file from start to finish, but from most specific to most general.

Iterator files use stopper objects, which monitor, control, and score the results of the model according to different criteria. **Table 5** lists the most commonly used stoppers and their behavior. A detailed description for using these stoppers can be found by clicking on "Help→Detailed Documenation→Stoppers." By convention, lower scores (near 0) indicate a better fit to the behavior the stopper searches for, larger scores are worse fits. Here, we will use four stoppers (one for each cell) to search for temporal oscillations in the act node:

1. In a text editor, make a new empty plain-text file.

Table 5
Commonly used stoppers

Stopper name	Explanation
`MetaStop`	A container that checks for multiple conditions. Almost all models will use this. Allows scores to be combined in several ways (sum, maximum, etc.) from individual stoppers
`MetaStopTwo`	Similar to `MetaStop`, offers a few more features
`SimpleStop`	Halts integration after specified time. Does not return a score on the pattern
`OscillatorStop`	Detects periodic oscillations in a node. Lower scores indicate more regular oscillations. Useful for searching for oscillatory solutions
`ThresholdStop`	Scores a specific node in the specified cell relative to a threshold. Low scores indicate the threshold criteria were met and the node is far from the threshold. Useful for finding pattern forming solutions
`OscillatingStripeStop`	Looks for a column of cells with a high value of a Node, surrounded by columns of cells with low values of that Node. Gives low scores for nonoscillatory solutions. Useful for finding stable pattern forming solutions
`NoChangeStop`	Halts integration when a node's rate of change is very slow (i.e., when the absolute value of the 1st derivative becomes small) in all cells. Score is the derivative normalized to the node value

2. The `OscillatorStop` stopper allows Ingeneue to find temporal oscillations in a single node in a specific cell. Type the following to search for act oscillations in the upper left cell of the Turing network; see the detailed documentation for the `OscillatorStop` for the meanings of its inputs:
```
&Stopper OscillatorStop
    &StopTime          1000
    &Node              act
    &Position          1
    &MaxNumPeaks       3
    &TransientPeriod 250
    &MinAmplitude      0.1
    &Tolerance         0.2
    &NoDamping         True
&endStopper
```
3. The `OscillatorStop` searches for temporal oscillations in a single cell. To search for oscillations in all four cells, copy the above code three times, changing the `&Position` value to 2 (upper right cell), 3 (lower left), and 4 (lower right) so your iterator file contains the following:

```
&Stopper OscillatorStop
   << copied from above, &Position 1>>
&endStopper
&Stopper OscillatorStop
   << copied from above, &Position 2>>
&endStopper
&Stopper OscillatorStop
   << copied from above, &Position 3>>
&endStopper
&Stopper OscillatorStop
   << copied from above, &Position 4>>
&endStopper
```

4. Each `OscillatorStop` will return a score from 0 (large, regular temporal oscillations) to 1000 (no oscillations), based on the behavior of the act node. To have Ingeneue save parameter sets where all four cells are oscillating, modify your file to that shown below:

```
&MetaStop
   &StopTime 1000
   &Cutoff 0.2
   &StopMode And
   &ValueMode Max
   << code from above >>
&endStopper
```

5. The above code enclosed the four stoppers within a `MetaStop`, which allows Ingeneue to select for parameter sets meeting criteria from many stoppers. The `&Cutoff` line tells Ingeneue to save any parameter sets with an `OscillatorStop` score of less than 0.2 (lower scores mean better parameter sets). The `&ValueMode` line tells Ingeneue to apply the cutoff to the largest of the scores; this ensures that all four cells are oscillating. For details on the other input values to `MetaStop`, see the detailed documentation.
6. The iterator file also specifies the details of how the model should be run, and what information to save. Modify your file to that shown below:

```
&Evaluators
&Stopper FinalCut StartAtBeginning Savefi-
nalpars Integrator= CashKarp
   << code from above >>
&endEvaluator
```

7. The `&Stopper` line specifies how to run the model:
   (a) `FinalCut`: The name for the combination of stoppers we used. This can be set to anything.
   (b) `StartAtBeginning`: Resets the model time to 0 before each run.

(c) `SaveFinalPars`: Saves the parameter sets satisfying the stoppers.

(d) `Integrator= CashKarp`: Sets the numerical integrator for running the model (*see* **Note 4**).

8. You will need to specify how to randomize parameters and where to save the output files:

(a) `&Iterator`: Tells Ingeneue which iterator object to use; use UberIterator.

(b) `&OutputPathName`: The subdirectory within Ingeneue where the saved parameter sets should be stored. Enclose this in quotation marks.

(c) `&ParamsToVary`: A list of all parameters, and how to randomize them. The format and interpretation of this section is identical to the parameters section of the network file. You can copy the parameters section from the net file into this section.

9. The iterator file will have Ingeneue randomize parameters over the same range as in the net file, and save the good parameter sets to the output subdirectory. Type the following to set up the `UberIterator` for a random search of the Turing network:

```
&Iterator UberIterator
 &OutpathName "output"
 &ParamsToVary // Copied from Turing.net
   &H_act 10.0 1.0 100.0 Linear
   <<…rest of parameters from section
   4.1.4>>
   &nu_INHact 3.0 1.0 10.0 Logarithmic
  &endParamsToVary
  << code from above >>
&endIterator
```

10. Save the file to networks directory and call it "oscillator.iter".

Use this iterator to find parameter sets that produce temporal oscillations in the Turing network. The file "Turing.params" in the models/Turing directory has several parameter sets that have been found. Note differences in the details of the phasing of the oscillations between different parameter sets. From this point, you are encouraged to explore on your own (*see* **Note 12**).

5. Conclusions

Ingeneue is a powerful and useful tool for both building and exploring simple or arbitrarily complex genetic networks in order to provide valuable insight into how, and how robustly, networks accomplish pattern-forming, or temporal oscillations tasks. Ingeneue is extensible and is actively being developed to add additional features. Future versions of Ingeneue will provide more sophisticated graphical tools for constructing networks, a more flexible framework for specifying affectors without modifying source code, more mechanistically detailed simulation of transcriptional logic, and the ability to simulate evolution of genetic networks.

6. Notes

1. Ingeneue solves systems of ordinary differential equations in which an initial value is specified for every dependent variable (node). The existence and uniqueness of solutions to the system of equations hinge on mathematical details we will not cover here. The affectors supplied with Ingeneue have been constructed so that, given positive parameter values, every set of initial conditions determines a unique solution which extends indefinitely far into the future. Ingeneue does not solve boundary value problems in which target values for different dependent variables are assigned at different times.
2. See the installation instructions on the Ingeneue Web site for instructions on how to determine whether your computer has Java installed. Ingeneue will work if you have the Java Runtime environment (JRE), but you will be unable to make changes to the Ingeneue source (i.e., add new affectors); the Java Developers Kit (JDK) is required if you want to make changes (i.e., add affectors or stoppers) to Ingeneue. The JDK is included on computers running Mac OS X (be sure that you have all the latest updates installed).
3. All help files in Ingeneue are in html format, meaning they can be viewed (and printed) from any Web browser. Two versions of the Ingeneue manual are in the "manual" subdirectory: "manual.html" has the entire manual in a single file for easy printing, while "manual00toc.html" will allow you to navigate to the different sections of the manual. Tutorials are located in the "tutorials" subdirectory.
4. Integrators and the stop time for integration can be set from "Run→Options." Ingeneue has five numerical integrators, four of which are standard *(19)*. The default Cash-Karp integrator is best for most systems; experiment with other integrators if you desire faster integration. The integrators differ in accuracy and their speed; the semiexplicit APC usually is the fastest but has the lowest accuracy – for some models or parameter sets,

it may produce erroneous results, but it can be useful as a first cut to eliminate parameter sets that fail to produce the desired behavior. The relative performance and detailed implementation of the integrators is described in Supplement B of *(15)*. *See* **Note 7** for additional tips on increasing Ingeneue speed.

5. The format of the file is in five columns as follows (1) Data set; starts with 0. Multiple runs can be stored by deselecting the "overwrite" button. (2) Simulation time in minutes. (3) Cell number. Numbering starts with 0 and proceeds sequentially rightward and downward (i.e., in same order that you read words on a page). (4) The node number (order is same as specified in net file). (5) Value of node.
6. The iterator does the following: Randomize all parameters, then integrate to 200 min and check whether en, wg, and hh have the correct spatial pattern of expression. If the pattern is correct, integrate to 1,000 min, and Ingeneue considers a parameter set successful if the spatial pattern persists without oscillations. About 1 in 200 parameter sets is successful.
7. Searches are substantially faster when Ingeneue is run from the command line compared to when Ingeneue is run using the graphical interface; however, you will not be able to see the details of the search, so the graphical interface may be desirable for early searches. To speed up the graphical interface (1) Close any unneeded/unused frames, especially the Node Viewer, Cell Viewer, and Time Graphs. (2) Change the integration scheme (*see* **Note 4**) and reduce the integration time to the minimum necessary. (3) Use the newest Java and Ingeneue versions; newer versions usually are faster. (4) Close other programs while Ingeneue is running. (5) Reduce the dimensions of the model (width and height) to the lowest value possible (i.e., for the segment polarity network, a 1 × 4 field of cells is faster than 2 × 8).
8. Some Ingeneue frames are separate from the main Ingeneue window and can get "lost" behind it. If frames are disappearing, try minimizing or moving the Ingeneue window to see what is behind it.
9. The cross-correlation coefficient $r_{a,b}$ is mathematically defined as:

$$r_{a,b} = \frac{E\left((a - E(a))(b - E(b))\right)}{\sqrt{\mathrm{Var}(a)\,\mathrm{Var}(b)}},$$

where a and b are two distributions, E is the mean, and *Var* is the variance. The cross-correlation coefficient is useful to determine whether there is compensation between parameters in the model – i.e., a large negative r indicates parameter a usually increases when parameter b decreases. Because of symmetry, $r_{a,b} = r_{b,a}$. All parameters fully correlate with themselves; $r_{a,a} = 1$.

10. A tutorial bundled with Ingeneue explains how to write your own affectors. This requires knowledge of the Java programming language and is beyond the scope of this chapter. Existing affectors are extensively documented, and from this documentation it should be clear to Java beginners how to code new affectors by modifying existing ones. See the file `AffectorTemplate` in the Affectors directory; it has detailed comments to show you how to write your own affectors with any mathematical function you desire.
11. I strongly recommend that you use comments liberally in your network and iterator files to aid in later understanding and debugging. Typical areas to use comments: (1) At the very start of the file, identifying your name, the date and version of the model, and a brief summary of what the file does. (2) Group similar parameters together labeled by category (i.e., half-lives). (3) At the start of the `initialconditions` section describing the qualitative pre-pattern. Also, you can temporarily delete interactions by adding a "//" to the start of the line, causing Ingeneue to ignore that affector.
12. Things to explore: After finding many parameter sets, do you notice trends or restrictions on parameters (when visualizing, or using the summary/cross-correlation statistics)? How does the model behavior change if you change the periodicity of the network from 2 × 2 to 3 × 3? What types of striped patterns can the network produce? Try making the activator membrane-bound and diffusible as well and investigate the network dynamics.

Acknowledgments

I thank Garrett Odell, Edwin Munro, and George von Dassow for helpful discussions, and Garrett Odell and Li Hua for comments on this chapter. This work and continued development of Ingeneue are supported by NIGMS: 5 P50 GM66050.

References

1. Tyson, J. J., Novak, B., Odell, G. M., Chen, K., and Thron, C. D. (1996) Chemical kinetic theory: Understanding cell-cycle regulation. *Trends Biochem. Sci.* 21, 89–96.
2. Chen, K. C., Calzone, L., Csikasz-Nagy, A., Cross, F. R., Novak, B., and Tyson, J. J. (2004) Integrative analysis of cell cycle control in budding yeast. *Mol. Biol. Cell* 15, 3841–3862.
3. de Jong, H. (2002) Modeling and simulation of genetic regulatory systems: A literature review. *J. Comput. Biol.* 9, 67–103.
4. Rust, M. J., Markson, J. S., Lane, W. S., Fisher, D. S., and O'Shea, E. K. (2007) Ordered phosphorylation governs oscillation of a three-protein circadian clock. *Science* 318, 809–812.

5. von Dassow, G. and Odell, G. M. (2002) Design and constraints of the Drosophila segment polarity module: Robust spatial patterning emerges from intertwined cell state switches. *J. Exp. Zool.* 294, 179–215.
6. Meir, E., von Dassow, G., Munro, E., and Odell, G. M. (2002) Robustness, flexibility, and the role of lateral inhibition in the neurogenic network. *Curr. Biol.* 12, 778–786.
7. Schlitt, T. and Brazma, A. (2007) Current approaches to gene regulatory network modelling. *BMC Bioinform.* 8(Suppl 6), S9.
8. Aguda, B. D. and Goryachev, A. B. (2007) From pathways databases to network models of switching behavior. *PLoS Comput. Biol.* 3, 1674–1678.
9. Alberts, J. B. and Odell, G. M. (2004) In silico reconstitution of Listeria propulsion exhibits nano-saltation. *PLoS Biol.* 2, e412.
10. de Silva, E. and Stumpf, M. P. (2005) Complex networks and simple models in biology. *J. R. Soc. Interface* 2, 419–430.
11. Longabaugh, W. J., Davidson, E. H., and Bolouri, H. (2005) Computational representation of developmental genetic regulatory networks. *Dev. Biol.* 283, 1–16.
12. van Riel, N. A. (2006) Dynamic modelling and analysis of biochemical networks: mechanism-based models and model-based experiments. *Brief Bioinform.* 7, 364–374.
13. von Dassow, G., Meir, E., Munro, E. M., and Odell, G. M. (2000) The segment polarity network is a robust developmental module. *Nature* 406, 188–192.
14. Zwolak, J. W., Tyson, J. J., and Watson, L. T. (2005) Parameter estimation for a mathematical model of the cell cycle in frog eggs. *J. Comput. Biol.* 12, 48–63.
15. Meir, E., Munro, E. M., Odell, G. M., and Von Dassow, G. (2002) Ingeneue: A versatile tool for reconstituting genetic networks, with examples from the segment polarity network. *J. Exp. Zool.* 294, 216–251.
16. Edelstein-Keshet, L. (2004) *Mathematical Models in Biology.* SIAM, Philadelphia, PA.
17. Murray, J. (2004) *Mathematical Biology. II: Spatial Models and Biomedical Applications.* Springer, Heidelberg.
18. Miura, T. and Maini, P. K. (2004) Periodic pattern formation in reaction-diffusion systems: An introduction for numerical simulation. *Anat. Sci. Int.* 79, 112–123.
19. Press, W. H., Flannery, B. P., Teukolsky, S. A., and Vetterling, W. T. (1992) *Numerical Recipes in C: The Art of Scientific Computing.* Cambridge University Press, Cambridge.

Part III

Spatial Analysis and Control of Cellular Processes

Chapter 7

Microfluidics Technology for Systems Biology Research

C. Joanne Wang and Andre Levchenko

Summary

Systems biology is a discipline seeking to understand the emergent behavior of a biological system by integrative modeling of the interactions of the molecular elements. The success of the approach relies on the quality of the biological data. In this chapter, we discuss how a systems biology laboratory can apply microfluidics technology to acquire comprehensive, systematic, and quantitative data for their modeling needs.

Key words: Microfluidics, Systems biology, High-throughput screening, Microfabrication, Microfluidic large-scale integration, Cell culture, Single cell analysis.

1. Introduction

1.1. Systems Biology Approach

The last decade has witnessed the rapid development of technologies enabling the complete sequencing of the human genome. Now the challenge is to understand how the identified genes and their expressed products dynamically interact with each other and respond to environmental cues. A well-accepted, although not always adopted, view is genes and the corresponding proteins can and should be considered components of a system. Characterizing the properties of a single component is necessary but not sufficient for true understanding of the behavior of the entire system. A set of principles and methodologies have recently been developed to link the actions of individual biological molecules to an integrated physiological response. Unified under the name of systems biology, the general framework of the approach is to (a) comprehensively perturb and then systematically measure the temporal responses of all the molecular elements at the distinct levels of the biological system, (b) integrate the quan-

Ivan V. Maly (ed.) *Methods in Molecular Biology, Systems Biology, vol. 500*

DOI: 10.1007/978-1-59745-525-1_7

titative information into a network model to recapitulate the systems behavior, (c) formulate new hypotheses and test them experimentally, and (d) refine the model to include new findings and then repeat the cycle *(1–3)*. The insights generated by this emerging field not only change the way biology research is conducted but they can also directly impact human health. Systems biology has already revolutionized how one examines diseases by exploring the hypothesis that one can distinguish between a normal and diseased state by comparing the dynamic expression and activity patterns of genes identified as key nodes within the network *(1, 4)*. Although systems biology is rapidly approaching maturity, there still remains several challenges accompanying its ambitious vision *(5)*. This review will concentrate on a subset of these challenges, namely how the needs for comprehensive, systematic, and quantitative measurements *(6, 7)* can arguably be met by new developments in microfluidics technology *(1, 7–9)*. We begin by giving an overview of the measurements suitable for system level modeling.

1.2. Measurement for Systems Biology

1.2.1. Single Cell as a Basic Response System

Biological information is hierarchical in nature: gene and gene regulatory networks → protein → signal transduction pathways → pathway networks → cell → multicellular organism (**Fig. 1A**). Each distinct level in this hierarchy can be viewed as a system. The functions of each of these systems reside in the temporal interactions of the

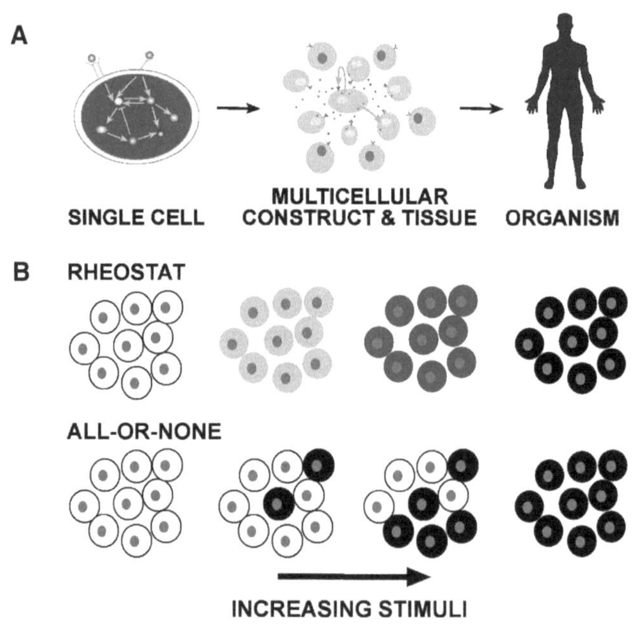

Fig. 1. Single cell is the established fundamental unit of measurement in the realm of systems biology. (**A**) Biological information is hierarchical in nature. (**B**) Two possible sources of graded response, which is not distinguishable if the measurement is made with biochemical assays that average the response.

internal components and adjacent systems in the hierarchy *(10)*. A single cell is the established fundamental unit of measurement in the realm of systems biology *(1)*. When a cell receives an input from the environment, the cell's internal networks of genes and proteins process the information and generate a physiological response. Heterogeneity of physiological responses among individual cells is often observed, even when the cell population is genetically homogeneous. If one extracts the average response from a population of cells instead of measuring it individually, one risks neglecting responses that can stochastically be significantly different from the mean, including all-or-none type behavior *(11, 12)*. The differences in how individual cells respond to the same perturbation can reveal the control mechanism and other system properties (**Fig. 1B**). In general, therefore, one cannot reliably generate information about the total wealth of regulatory network behavior based solely upon measurements of mixed population of cells; cell responses should be explored at the single-cell level, if possible and practical.

1.2.2. Cell Lines Versus Heterogeneous Cell Populations

For mammalian cells, single cells can be derived from multiple sources. The source of the single cell needs to be determined based on the goal of the model. The sources range from (a) biopsy samples, (b) monolayer cell culture, and (c) organ-type (or three-dimensional) culture. Compared with biopsies, cultured monolayer cells undoubtedly allow tighter controls, but their ability to reflect physiological functions is often debatable. Extensive cell–cell communication can occur in populations of cells, homotypic or heterotypic. Cells can communicate by diffusion of their actively secreted product, cell–cell contact, or exertion of forces. These parameters can influence cellular responses; therefore, it is desirable to preserve the spatial relationship of the cells and the surrounding fluidic environment. In addition to working with primary cells extracted from biopsy samples, a potential attractive practical solution is to construct an organ-type (or three-dimensional) cell culture *(13)*. This approach would aim at mimicking in vivo tissue geometry by recapitulating the arrangement of heterogeneous types of cells in three-dimensional spaces *(14, 15)*.

1.2.3. Context of Measurement Needs Consistency

To completely describe a system's response to a particular perturbation, a threshold number of components in the system must be measured over time. Eventually the measurements will need to be integrated into a coherent model for understanding and predicting the behavior of the system. To accomplish this goal, ideally multiple measurements should be made simultaneously from a single sample *(1)*. At the very least, the preparation of the samples and the context from which they are obtained need to be consistent across all the measurements. Any alterations invariably introduce new parameters, making integration of measurements more challenging than it already is.

1.2.4. Types of Measurements

Even though the interrogated biological molecules may be different in identity, in general the data falls into three main types: (1) concentration, (2) kinetic parameters (essentially time-series measurement of concentration), and (3) regulatory interactions. Specific modeling approaches will have specific data requirements. For example, ordinary differential equation (ODE)-based dynamics modeling requires highly accurate measurements of kinetic parameters and concentrations. Existing high-throughput technologies (i.e., DNA microarray, and mass spectrometry with isotope-coded affinity tags) have great coverage of nodes and can be quantitative, but they are limited to population measurements with low temporal resolution. Thus, these data are not suitable for ODE-based modeling. Technology for generating high-throughput measurements meeting the needs of systems modeling has yet to be developed. Finally, it is interesting to speculate on how many measurements are necessary to reverse engineer a system *(16, 17)*. The exact number is bound to be staggering, and our ability to develop automation technology to meet this quantity will be the limiting factor constraining the success achievable by the systems biology approach.

1.3. Why Microfluidics?

In addition to introducing truly novel measurement principles, ingenious adaptation of existing technologies can also generate powerful methods for acquiring the biological data necessary for modeling. Earlier, we have established the fundamental measurement scale to be defined by a single cell. The order of magnitude estimates of single cell length are 1–10 μm and a volume of several picoliters. However, manual and automated biochemical assays are usually performed in microliter to milliliter volumes of cellular matter, mainly due to limitations in conventional fluid handling techniques. The difficulties of handling small volumes of fluid include evaporation, loss during transfer, and high surface tension (resulting from the increased surface to volume ratio at small volume).

Microfluidics refers to an integrated system of miniaturized components handling small fluidic volumes on the nanoliter scale and below *(8, 18, 19)*. Standard components (channels, pumps, and mixers) utilized to transport and store reagents are micrometer in scale (**Fig. 2A**, hence the name of the technology. Because the feature size of the operating components is at the same length scale as a single cell, microfluidics is poised to meet the experimental needs of systems biology *(1, 7–9)*. Using recently developed microfluidic devices, some of which are described later, one can easily move single cells around within a network of tiny channels, change the fluidic environment to expose cells to complex spatiotemporal perturbation and detect the output, or lyse cells and transfer the picoliter volume of the intracellular contents to various "processing stations" on the chip (**Fig. 2B**) *(20)*. Computer programmable

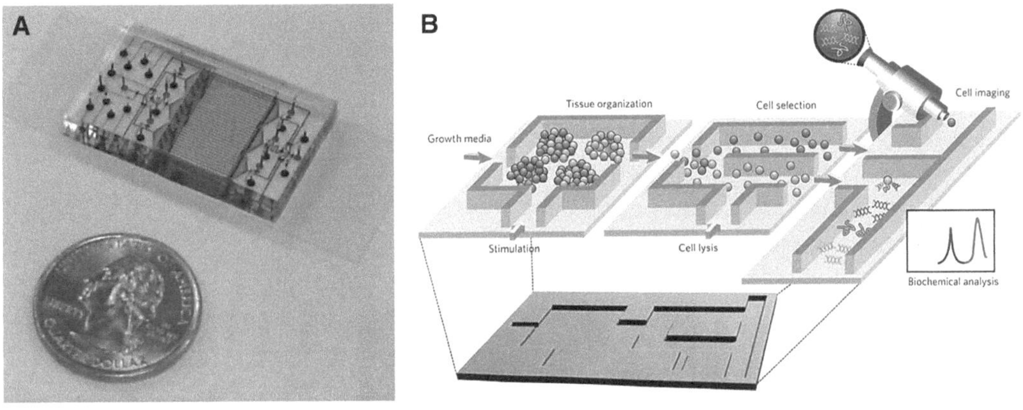

Fig. 2. Integrated microfluidics cell analysis system. (**A**) A high-throughput immunofluorescence staining device imaged next to a US quarter *(46)*. The fluidic layer is filled with red food dye (appears as gray in the image), and the control layer with blue food dye (appears as black in the image). (**B**) Conceptual depiction of how different microfluidic operations can be integrated with each other to carry out cell micropatterning, stimulation, sorting, lysis, and finally interrogation on a single chip (reproduced from **ref.** *20* with permission from Nature Publishing Group).

moving parts (i.e., micromechanical valves) can automate all fluidic operations *(21)*. The miniaturized nature of microfluidics allows multiple parallel runs of typically low-throughput biochemical assays, thereby converting them to powerful high-throughput methods *(8)*. Decreasing analyte volume down to the detection limit theoretically allows multiple measurements to be made simultaneously from the same single cell sample. Microfluidics offers high spatiotemporal resolution in fluidic control, and allows more accurate modeling of the in vivo environmental context prior to or during a measurement. Furthermore, it can be used to model the in vivo spatial relationship between cells within a functional multicellular ensemble *(20)*. Microfluidics is also viewed by many as the ideal technology to steer lysates from individual cells to measurement platforms based upon emerging nanotechnology-based methods *(1, 7, 22)*.

2. Material and Methods: How to Set Up Your Wet Bench to Use mLSI Microfluidics Chip

2.1. General Method

In this review we highlight a specific subset of microfluidic devices fabricated using soft lithography that are characterized by their ease of application in a typical laboratory setting and suitability for generating data for systems level modeling (**Fig. 3**). Soft lithography produces microscale features by molding soft polymer, using microfabricated silicon-based wafer templates (**Fig. 4A**) *(23)*. A single wafer allows multiple casting, thus significantly reducing the cost of mass production. Polydimeth-

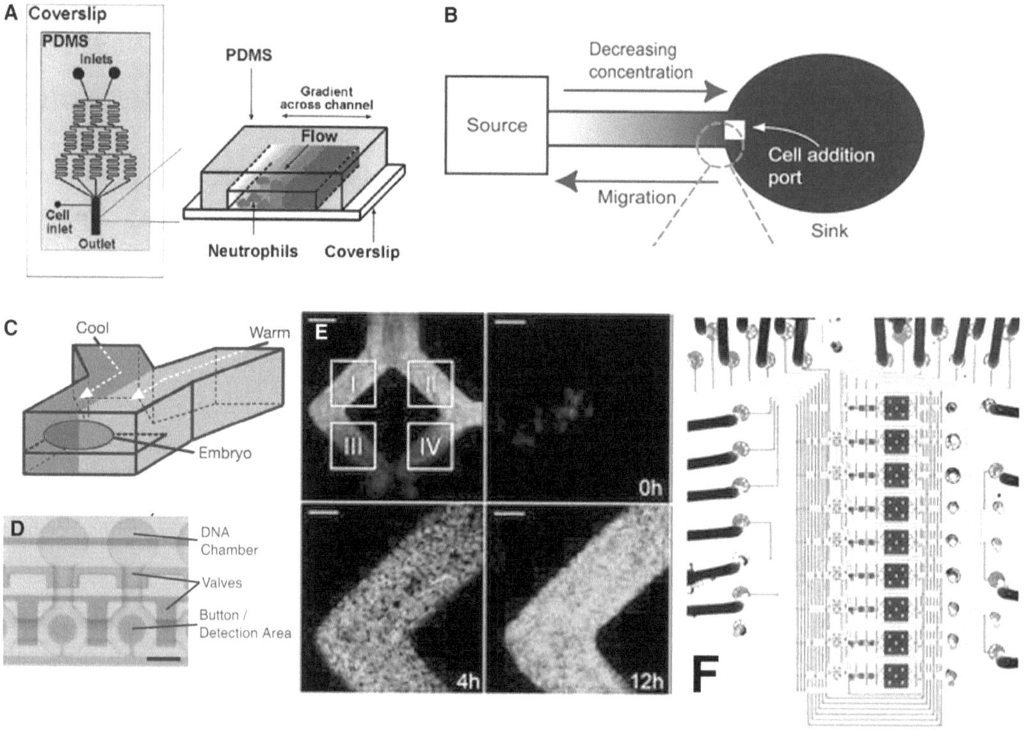

Fig. 3. Microfluidic devices. (**A**) Analysis of cell chemotaxis using a chip with a Christmas tree design for gradient generation (reproduced from **ref.** *36* with permission from Nature Publishing Group). (**B**) Diffusion-based and flow-free gradient generation (reproduced from **ref.** *39* with permission from Royal Society of Chemistry). (**C**) T-step stimulation device allowing generation of stepwise gradients (reproduced from **ref.** *41* with permission from Nature Publishing Group). (**D**) Unit chambers of a high-throughput microfluidic platform on the basis of mechanically trapping the TF-DNA binding pairs using micromechanical valves (reproduced from **ref.** *56* with permission from AAAS). (**E**) Progressive steps in visualization of self-organization of an *Escherichia coli* colony in a microfluidic chamber (reproduced from **ref.** *48* under the terms of Creative Commons License). (**F**) Single-cell isolation and genome-amplification chip (reproduced from **ref.** *60* under the terms of Creative Commons License).

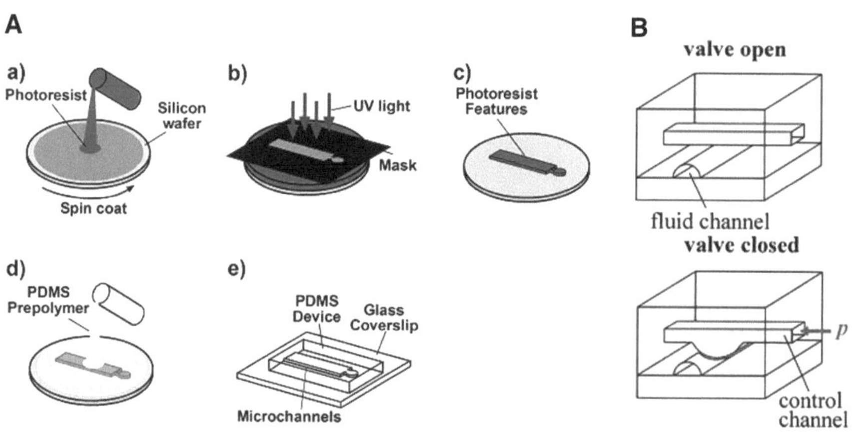

Fig. 4. Soft lithography. (**A**) Process flow: (**a**) Photoresist was spin-coated onto a silicon wafer and (**b**) the mask patterns were lithographically transferred onto the photoresist with an UV aligner and following development, (**c**) results in precisely fabricated microfeatures on the wafer. (**d**) This master is used as a mold for casting a ~5-mm-tall PDMS layer. (**e**) The elastomeric layer is released from the mold and laid flat on a clean glass coverslip, resulting in the final device. (**B**) The NanoFlex™ valve and its operation (reprinted from **ref.** *21* with permission from Royal Society of Chemistry).

ylsiloxane (PDMS) is a widely used polymer for chip construction. Gas-permeable and optically transparent, PDMS is highly appropriate for a variety of cellular studies. A PDMS fabricated surface can be chemically tailored to the needs of a specific application. The impressive versatility of PDMS microfluidic devices has motivated countless applications in various scientific disciplines *(9, 20)*. Such devices will likely make a significant impact on systems biology because of the potential to replace conventional biological automation paradigms *(8)*. Central to achieving this goal is the development of microfluidic large-scale integration (mLSI) technology – a simple fabrication process for constructing chips with hundreds to thousands of integrated micromechanical valves *(24)*. The NanoFlex™ valve *(25)* is the basic unit of fluidic handling in mLSI, essential for automation and parallel execution of multiple picoliter-scale operations (**Fig. 4B**). The NanoFlex™ valves stop and start flow on-chip by pneumatically deflecting a flexible membrane *(25)*, givin one the ability to manipulate fluid transport without accounting for the properties of the fluid. Functionality of a microfluidic device is encoded in its fluidic channel design, and in the near future, in-chip operation of all existing biochemical assays will be demonstrated. An almost infinite variety of layouts is possible for performing the same task, and should be tailored to the needs of the specific cellular system.

2.2. Peripheral Equipment

The microfluidic devices highlighted in this review generally do not possess stand-alone functions and need peripheral equipment to support their operation. Fundamentally, one needs to be able to actuate the devices through exercising fluidic pressure control and image the experimental outcomes generated by various cellular labels *(26, 27)*.

As PDMS-based microfluidic devices are optically transparent, they can be used in combination with virtually any type of optical microscopy, including epifluorescent, confocal, and multiphoton. Optical microscopes are currently a staple of almost any biological lab and are commonly available as shared equipment. For increased throughput and experimental versatility in using the devices, it is desirable to have a microscope equipped with a motorized XYZ stage driven by automation software. If on-chip live-cell imaging is to be performed, the microscope also needs to be enclosed by a temperature and CO_2 incubation chamber *(28)*.

Pressure-driven flow in a microfluidic device is generally in the range of nanoliter per minute. This flow rate can be achieved by placing the media reservoirs feeding the inlets and outlets at different hydraulic heights. Analogous to Ohm's law, the pressure difference is determined by multiplying the desired volumetric flow rate by the channel's resistance, with the latter frequently easily computable using standard hydraulic resistance formulas. Alternatively, a syringe pump can be used to maintain a constant flow rate if the internal resistance of the device varies over time.

Actuation pressure for turning on and off the NanoFlex™ valves requires a high external pressure source ~20–80 pounds per square inch (psi). This high pressure is fed into an array of miniaturized solenoid valves [e.g., available from The Lee Company *(29)*], which is interfaced to a computer via a National Instrumentation PCI card *(25, 30)*. One can write custom programs to automate valve switching. Alternatively, one can purchase a ready-made control module from Fluidigm *(31)*.

Micro-to-macro fluid adaptation is often a challenge in traditional glass- or silicon-based microfluidic devices. The flexible nature of PDMS allows easy punching of holes for flow channel access. Adaptors slightly larger than the cording tool generate tight seals sustaining up to ~40 psi of pressure with no leakage, even when no additional adhesive is applied. The hole punching procedure and the sources of the adaptors and corder are described in detail in *(32)*.

2.3. Chip Bonding and Bubble Degassing

A PDMS microfluidic chip is typically sealed with a glass coverslip bottom (**Fig. 4A**). This technique has the advantage of providing structural support to the otherwise flexible device. One has the option of reversibly bonding the two parts by oven baking at 80°C from a few hours to overnight. In our experience, this type of bond can withstand up to 10 psi of pressure, a value much greater than the pressure differential necessary for driving flow in the device. Prior to the operation of most devices, bubbles in the channels occasionally created during the initial filling of the device with cell medium or a buffer must be removed in a "dead-end priming step," as follows. The permeability of PDMS to nonpolar gases, including O_2, N_2, and CO_2, is not only beneficial for the rapid enrichment of the medium inside the chip with atmospheric components essential for cell survival, but can also be used to drive air bubbles out of the chip. This is achieved by uniformly increasing the hydrostatic pressure in the chip to higher than atmospheric levels and maintaining it at these levels throughout the experiment (note that we mean here the absolute hydrostatic pressure; pressure differential needed to drive flow can be superimposed on this absolute pressure level). After an experiment, the PDMS chip can be separated from the glass and cleaned with Alconox, water, and then with 70% ethanol and reused.

Irreversible bonding of a PDMS device and glass coverslip components can also be accomplished within a vacuum plasma chamber or using an inexpensive hand-held corona unit *(33)*. A vacuum plasma chamber is the established method for creating a permanent bond between PDMS and PDMS interfaces, or PDMS and glass interfaces. However, it is typically expensive, bulky, and requires high maintenance. The inexpensive hand-held corona unit is a validated and suitable alternative for the purpose of bonding the chips described in this chapter. Bonding is achieved by plasma (or corona) treating the cleaned bonding surfaces, and then the two surfaces are brought in contact with each other and left undisturbed for at least an hour for bonding to take effect. In addition to activating surfaces for bonding, plasma treatment has the advantage of temporarily increasing the hydrophilicity of PDMS for ~24 h, during which the channel surface is energetically unfavorable for bubble formation. Monomer extraction from the PDMS bulk can prolong the hydrophilic effect to weeks *(34)*. Irreversibly bonded devices are usually difficult to reuse after one operation.

3. Case Studies

3.1. Perturbing Single Cells

3.1.1. Spatial Concentration Gradient

Flow is predominantly laminar in microchannels *(19)*. Because chemicals in laminar streams flowing in contact with each other can only mix by diffusion, different concentrations of a chemical of interest present in different parts of the stream can gradually blend to create stable and reproducible concentration gradients with complex profiles *(35)*. On this principle, Jeon et al. demonstrated how one can use a prominent layout, the so-called *Christmas tree* structure (**Fig. 3A**), to investigate gradient sensing and chemotaxis in a neutrophil-like cell line *(36)*. This novel method filled an important void in the cell biology toolbox since previously there was no simple means to generate instantaneous and stable linear gradients. Armed with this technology, one can easily vary chemical concentration gradient parameters, such as shape, slope, and mean value to achieve a systems understanding of how single cells sense chemical gradients and respond to them. This strategy can also be adapted to generate gradients of substratum-bound extracellular matrix (ECM) components *(37)*. Dertinger et al. used this approach to investigate axon specification in rat hippocampal neurons, and discovered the quantitative threshold value of laminin gradient that biased the orientation of axon specification. Recently, we developed a high-throughput chip enabling the creation of composite gradients of both diffusible and surface-bound guidance cues to more realistically mimic the conditions growth cones encounter in vivo *(38)*. Applying this assay to *Xenopus* embryonic spinal neurons, we demonstrated how the presence of a surface-bound ECM gradient can finely tune the polarity of growth cone responses to soluble neurotropic factor gradients. The data generated from the chip experiments allowed us to develop a computational model to explain biochemical mechanisms responsible for converting the multiple gradient inputs into a binary turning decision (report in preparation).

Strategies also exist for exposing nonadherent or shear-sensitive cells to stable concentration gradients *(12, 39, 40)*. Paliwal et al. described a microfluidic device capable of testing yeast responses to pheromone gradients characterized by different mean concentration values and steepness *(12)*. Budding yeast, a nonadherent type of simple eukaryotic cell, has the ability to respond to graded pheromone levels by altering the expression of multiple genes and orienting its growth toward higher pheromone concentrations. The functional area of the device used in these experiments consists of an array of shallow parallel horizontal test chambers of various lengths, and two flow-through vertical channels adjacent to the opposite edges of the test chambers. The gradient inside the test chambers is created by diffusion between the two flow-through channels, each carrying a high or low concentration solution. As the design of the chip allowed for various test chamber lengths in a single experiment, the yeast cells were exposed to a range of linear pheromone gradients each with a different steepness but with the same mean concentration values. The rich datasets generated by this high-throughput device enabled ODE-based modeling of transcriptional regulation in the pheromone response. This integration of modeling and microfluidic experimentation revealed how bimodality in gene expression allows a cell population to adapt its transcriptional response in different pheromone gradients and mean concentrations.

A similar diffusion-based strategy can also create stable chemical gradients, but requires no active fluid flow (**Fig. 3B**). The device embodying this idea has two very large stationary reservoirs at the same hydrostatic pressure levels: the diffusion sink and source, at the opposite sides of a test channel of a much smaller height *(39, 40)*. Instability of the gradients in the test channel caused by stimuli addition can be eliminated by covering the sink and the source with a high fluidic resistance membrane, whose pores still allow diffusive transport of chemical species into the channel *(39)*. This design can be easily arrayed up for the screening of multiple soluble factors. In comparison to a conventional Boyden chamber assay, this platform has the advantage of better optical accessibility, which allows easier characterization of generated gradients as well as extraction of more quantitative cellular migration data. However, it is important to note that the transient time for gradient generation using a diffusion-based strategy is dependent on the molecular weight of the chemical. The characteristic diffusion time should be compared to the estimated chemotactic response time to determine whether the effect of transient gradient stimulation is negligible in comparison to stable gradient stimulation.

As described earlier, various microfluidic device designs take advantage of constructing a laminar flow with at least two adjacent streams carrying solutions with different concentrations of chemicals of interest. At the initial contact interface between the streams, due to the negligible time available for diffusion, the gradient profile is step-like (**Fig. 3C**). Very similarly, one can create a step-like temperature gradient using streams carrying solutions at different temperatures. This phenomenon has been exploited to uncover spatial control mechanisms in both intercellular and intracellular signaling networks. For instance, Lucchetta et al. used a simple T-shaped device to create a temperature step around a live *Drosophila* embryo *(41)*. This clever approach to investigating the robustness of early embryo patterning to temperature-dependent perturbations suggested that the mechanism of compensation during embryo development is not a simple reciprocal gradient system. This microfluidic device design can be scaled up, and combined with the molecular genetic toolbox of *Drosophila* to screen for essential genes involved in regulating the compensation mechanism during development. The T-shaped channel functional unit can also be integrated with downstream cell sorting and single-cell biochemical analysis operations to allow generation of cross-level measurements in the biological hierarchy (*see* **Subheading 1.2.1**). Within this type of integrated device, embryos perturbed in the temperature step can be transported to the downstream units by fluidic flow to undergo embryo dissociation and cell sorting, followed by lysing the sorted single cell individually to release the intracellular content; finally the nucleic acids and proteins are transported to the analysis modules for quantification. (The biochemical analysis modules are described in detail in **Subheading 3.2**) et al. used the same chip design to investigate spatial propagation of signals in single cells *(42)*. Single cancer cells were partially exposed to a stream carrying exogenous growth factor and the extent of spreading of an intracellular fluorescently tagged protein from the location of stimulation was monitored real-time. They discovered the key nodes controlling the local and global activation of their pathway of interest, receptor density, and endocytosis rate.

3.1.2. Temporally Varying Stimuli Input

Perturbing cells with transient or pulsed stimuli is especially useful for revealing the presence of feedback interactions *(43, 44)*. Various proof-of-concept devices have been developed with this modality *(45)*. Central to performing temporally varying perturbation are the aforementioned NanoFlex™ valves, which allow programmable on-chip switching of fluid access to different parts of devices. By using a multiplexed system of valves, stimulations of different durations can be delivered to individual subgroups of a population of cells (**Fig. 2A**). Subsequent to the resulting complex cell stimulation protocols, the intracellular content of single cells within each subgroup can be probed with immunocytochemistry to assess the relative changes in protein concentration in response to stimulation *(46, 47)*. This approach is akin to robotics-based high-content screening, but with the advantage of reduced sample consumption and better reproducibility due to higher precision in fluid delivery. This technology was used by Kaneda et al. to measure the kinetic parameters of a phosphorylated protein in primary cell lines harboring different expression levels of an epigenetically imprinted growth factor gene *(47)*. Signaling analysis revealed an unusual ligand hypersensitivity, which could be exploited to propose a novel in vivo chemopreventive strategy, primarily targeting colon cancer. Because of its low cell number and reagent requirements this chip has the potential to analyze primary cells derived from an individual patient, conferring the ability to distinguish between normal and diseased states by comparing the dynamic expression and activity patterns of the key nodes within the signaling network.

3.1.3. Mimicking the In Vivo Cell and Tissue Boundaries

The precise nature of microfabrication inherent in creating microfluidic devices can afford higher accuracy in approximating the natural boundaries surrounding cells and tissues. Thus, one can combine the advantages of controlling cellular chemical microenvironments discussed in the previous sections with the ability to specify the mechanical properties and geometries of the chambers enclosing groups of cells. For example, the initial stages of developing bacterial biofilms involve embedding cells in small naturally occurring cavities *(48)*. Cho et al. and Groisman et al. developed a series of devices (**Fig. 3E**) to discover and study the dynamical self-organization of bacterial cell colonies tightly packed in microchambers of different shapes and sizes *(48, 49)*. A simple mechanical model of cell–cell and cell–wall interactions explained the observed colony self-organization, leading to important insights on how young biofilms might be spatiotemporally organized, maximizing their survival chances. The chip permitted real-time microscopy at single-cell resolution, holding the promise for deciphering molecular sensing elements responsible for converting inputs from physical forces into physiological responses, such as mitosis or cell migration.

The close resemblance in length scale and shape of microchannels frequently used in various microfluidic devices to the body's own microfluidic transport system, the vasculature, implies considerable potential for studying a variety of blood flow-related problems, such as the behavior of medium-suspended blood cells *(50, 51)*. Microfluidic devices allow independent modulation of relevant parameters, such as flow rate, channel (vessel) diameter, hematocrit, and chemical and gas concentration. Higgins et al. used this approach to study sickle cell anemia, a prominent example of a single genetic mutation leading to pathology at the organism level. Although the molecular pathology of the disease is well characterized, sickle cell patients are heterogeneous in their clinical presentations *(52)*. The observed heterogeneity is usually attributed to differences harbored by the tissue microenvironment. Data acquired in this well-defined multiscale experimental model will likely facilitate further systems-based network analysis of the relationship between the cell microenvironment and genetic defects. Ultimately, this approach might provide a mechanistic basis for predicting the specific pathophysiological patterns in a patient-specific manner.

3.2. Intracellular Content Analysis

3.2.1. Microfluidic Digital PCR and RT-PCR

Microfluidic digital polymerase chain reaction (PCR) *(53, 54)* is a powerful method arising from the clever combination of digital PCR and microfluidics. Digital PCR *(55)* includes the partitioning of a complex pooled sample into single molecule templates for individual PCR amplifications. The practicality of such separate amplifications relies on the accuracy of parallel isolations of single molecular templates during the initial steps, which can be greatly improved and enhanced by microfluidics. Ottesen et al. applied microfluidic digital PCR to perform multigene profiling of the genomes of single bacterial cells harvested from the wild, and systematically determined the fraction of cells within complex ecosystems encoding the genes of interest *(53)*. An interesting extension of this technique is to perform microfluidic digital PCR using complimentary DNA templates generated from a single-cell-based reverse transcriptase (RT) reaction. Microfluidics is proven necessary in this instance because it overcomes the major limitation preventing RT-PCR from achieving its theoretical sensitivity required for single cell gene-expression analysis, namely the handling of the single cell content and the measurement of the output. Warren et al. used the microfluidic chip-based digital RT-PCR assay to systematically and quantitatively analyze transcription factor (TF) expression within a population of hematopoietic stem cells *(54)*. The ability to quantitatively characterize the developmental states of single cells at snapshots of time can thus bring us a step closer toward understanding the transcriptional regulatory networks governing the transition from stem cells to diverse terminally differentiated states.

3.2.2. Regulatory Interaction Measurement

Quantifying the affinities of molecular interactions in intracellular environments or in free solutions is critical to understanding the collective properties of regulatory networks. The technical challenges include systematically measuring the parameter space of the governing biochemical processes (e.g., the activities of interacting molecules and rate constants of individual reactions or transport events) and capturing transient, low-affinity binding events. In conventional assays, the weakly bound material is often rapidly lost during the rigorous washing steps. Maerkl and Quake developed a high-throughput microfluidic platform based on mechanically trapping TF-DNA binding pairs using micromechanical valves (**Fig. 3D**), thereby eliminating the off-rate problem *(56)*. They used a microarray to spot dilute series of DNA sequences to achieve a dense array of DNA templates coding for the TF and the target DNA. Then, the TF was synthesized in situ in the nanoliter volume microchamber followed by incubation with the target DNA. Finally, micromechanical valves were brought into contact with the surface, physically trapping surface-bound material (the TF-DNA pairs) while the solution-phase unbound molecules were washed away. Hopefully, their success in predicting biological functions by combining purely in vitro biophysical measurements with in silico modeling will soon become a standard practice as systems biology matures.

3.2.3. On-Chip Cell-Sorting Followed by Nucleic Acid Extraction and Purification

The microfluidic devices described in the last subheading partition the pooled intracellular content from externally lysed cells into nanoliter portions, where PCR reactions based on single molecule templates can take place. This powerful strategy is suitable for analyzing entities present at a single copy per cell, but many more abundant proteins or metabolites lose the information of their cellular origin upon release from membrane encapsulation. A microfluidic approach to solving this problem is to perform single cell sorting on-chip *(57–59)*, followed by lysing the single cells, and then analyzing the content with a high-affinity reporter system, which in principle does not require the size-separation step, e.g., through quantitative PCR for detecting DNA and RNA, antibodies for detecting protein (**Fig. 3F**, refs. *60–62*. The methods proposed by both Fu et al. and Takahashi et al. involve computational analysis of digitally acquired images, where positive identification of a target triggers a sorting on-chip valve to isolate the cell of interest *(57, 58)*. Wang et al. switched the streams optically to improve throughput *(59)*. Any of these cell sorters are powerful representative technologies for bridging the information acquired from microfluidic chip-based cellular assays described in the last section with other single-cell-based intracellular biochemical analyses described in this section.

4. Conclusion: Future Development Driven by the Needs of Systems Biology

In **Subheading 3**, we examined microfluidic devices that have immediate applications benefiting systems biology analysis. The discussed devices make measurements at a specific level of the biological hierarchy. Thus far, integration between modules and high-throughput cross-level measurements has not been achieved on a microfluidic platform. In the near future we envision the introduction of integrated devices capable of performing all operations simultaneously on one chip. In the first module, microfabricated multicellular constructs recapitulating physiological or pathophysiological function will be subjected to a panel of perturbations, such as small molecule inhibitors or growth factors. Their output responses can be detected by a variety of non-invasive imaging techniques. Multicellular constructs or single cells displaying the target behavior can then be transported to the second module where they can be sorted into individual cells and lysed independently to release their intracellular content. Finally, biochemical analysis of the states of the multiple nodes, the hypothetical regulators of the output response, will be performed at the last device module, thus linking molecular interactions to a physiological response at the single-cell level. Incorporating nanotechnologies *(22, 63)* capable of interrogating dynamics of biomolecules in label-free and higher sensitivity reactions will further enrich the functionalities of microfluidics. Akin to how technology has spearheaded the genome sequencing project, the multifaceted capabilities offered by the unifying platform of microfluidics will help realize the untapped potential of systems biology.

Acknowledgments

The authors would like to thank Benjamin Lin for reading the manuscript critically.

References

1. Hood, L., Heath, J. R., Phelps, M. E., and Lin, B. (2004) Systems biology and new technologies enable predictive and preventative medicine. *Science* 306, 640–643.
2. Kitano, H. (2002) Systems biology: a brief overview. *Science* 295, 1662–1664.
3. Ideker, T., Galitski, T., and Hood, L. (2001) A new approach to decoding life: systems biology. *Annu. Rev. Genomics Hum. Genet.* 2, 343–372.
4. Irish, J., Hovland, R., Krutzik, P. O., Perez, O. D., Bruserud, Ø., Gjertsen, B. T., and Nolan, G. P. (2004) Single cell profiling of potentiated phospho-protein networks in cancer cells. *Cell* 118, 217–228.
5. Levchenko, A. (2003) Dynamical and integrative cell signaling: challenges for the new biology. *Biotechnol. Bioeng.* 84, 773–782.
6. Szallasi, Z. (2006) Biological data acquisition for system level modeling – an exercises in the art of compromis, in *system Modeling in Cellular Biology: From Concepts to Nuts and Bolts* (Szallasi, Z., Stelling, J.R., and Periwal, V., eds.), MIT Press, Cambridge, MA, pp. 201–220.
7. Heath, J., Phelps, M. E., and Hood, L. (2003) Nanosystems biology. *Mol. Imaging Biol.* **5**, 312–325
8. Melin, J. and Quake, S. R. (2007) rofluidic large-scale integration: the evolution of design rules for biological automation. *Annu. Rev. Biophys. Biomol. Struct.* 36, 213–231.
9. Breslauer, D., Lee, P., and Lee, L. Mic Microfluidics-based systems biology. *Mol. Syst. Biol.* 2, 97–112.
10. Kholodenko, B. (2006) Cell-signalling dynamics in time and space. *Nat. Rev. Mol. Cell Biol.* 7, 165–176.
11. Ferrell, J. J. and Machleder, E. (1998) The biochemical basis of an all-or-none cell fate switch in *Xenopus* oocytes. *Science* 280, 895–898.

12. Paliwal, S., Iglesias, P., Campbell, K., Hilioti, Z., Groisman, A., and Levchenko, A. (2007) MAPK-mediated bimodal gene expression and adaptive gradient sensing in yeast. *Nature* 446, 46–51.
13. Nelson, C., Vanduijn, M. M., Inman, J. L., Fletcher, D. A., and Bissell, M. J. (2006) Tissue geometry determines sites of mammary branching morphogenesis in organotypic cultures. *Science* 314, 298–300.
14. Desai, T. (2000) Micro- and nanoscale structures for tissue engineering constructs. *Med. Eng. Phys.* 22, 595–606.
15. Bhatia, S. and Chen, C. (1999) Tissue engineering at the micro-scale. *Biomed. Microdevices* 2, 131–144.
16. Andrec, M., Kholodenko, B., Levy, R., and Sontag, E. (2004) Inference of signaling and gene regulatory networks by steady-state perturbation experiments: structure and accuracy. *J. Theor. Biol.* 232, 427–441.
17. Sontag, E., Kiyatkin, A., and Kholodenko, B. (2004) Inferring dynamic architecture of cellular networks using time series of gene expression, protein and metabolite data. *Bioinformatics* 20, 1877–1886.
18. Whitesides, G. M. (2006) The origins and the future of microfluidics. *Nature* 442, 368–373.
19. Beebe, D. J., Mensing, G. A., and Walker, G. M. (2002) Physics and applicaitons of microfluidics in biology. *Annu. Rev. Biomed. Eng.* 4, 261–286.
20. El-Ali, J., Sorger, P. K., and Jensen, K. F. (2006) Cells on chips. *Nature* 442, 403–411.
21. Haeberle, S. and Zengerle, R. (2007) Microfluidic platforms for lab-on-a-chip applications. *Lab Chip* 7, 1094–1110.
22. Helmke, B. P. and Minerick, A. R. (2006) Designing a nano-interface in a microfluidic chip to probe living cells: challenges and perspectives. *Proc. Natl. Acad. Sci. USA* 103, 6419–6424.
23. Xia, Y. and Whitesides, G. M. (1998) Soft lithography. *Annu. Rev. Mater. Sci.* 28, 153–184.
24. Thorsen, T., Maerkl, S. J., and Quake, S. R. (2002) Microfluidic large-scale integration. *Science* 298, 580–584.
25. Unger, M. A., Chou, H.-P., Thorsen, T., Scherer, A., and Quake, S. R. (2000) Monolithic microfabricated valves and pumps by multilayer soft lithography. *Science* 288, 113–136.
26. Meyer, T. and Teruel, M. N. (2003) Fluorescence imaging of signaling networks. *Trends Cell Biol.* 13, 101–106.
27. Xie, X. S., Yu, J., and Yang, W. Y. (2006) Living cells as test tubes. *Science* 312, 228–230.
28. Goldman, R.D. and Spector, D.L. (eds.) (2004) *Live Cell Imaging*. Cold Spring Harbor Laboratory Press, Cold Spring Harbor, NY.
29. The Lee Company (http://www.theleeco.com).
30. National Instrumentation (http://www.ni.com).
31. Fluidigm Corporation, USA (http://www.fluidigm.com).
32. Kartalov, E. P. and Quake, S. R. (2004) Microfluidic device reads up to four consecutive base pairs in DNA sequencing-by-synthesis. *Nucleic Acids Res.* 32, 2873–2879.
33. Haubert, K., Drier, T., and Beebe, D. (2006) PDMS bonding by means of a portable, low-cost corona system. *Lab Chip* 6, 1548–1549.
34. Vickers, J. A., Caulum, M. M., and Henry, C. S. (2006) Generation of hydrophilic poly (dimethylsiloxane) for high-performance microchip electrophoresis. *Anal. Chem.* 78, 7446–7452.
35. Jeon, N. L., Dertinger, S. K. W., Chiu, D. T., Choi, I. S., Stroock, A. D., and Whitesides, G. M. (2000) Generation of solution and surface gradients using microfluidic systems. *Langmuir* 16, 8311–8316.
36. Jeon, N. L., Baskaran, H., Dertinger, S. K., Whitesides, G. M., Van de Water, L., and Toner, M. (2002) Neutrophil chemotaxis in linear and complex gradients of interleukin-8 formed in a microfabricated device. *Nat. Biotechnol.* 20, 826–830.
37. Dertinger, S. K., Jiang, X., Li, Z., Murthy, V. N., and Whitesides, G. M. (2002) Gradients of substrate-bound laminin orient axonal specification of neurons. *Proc. Natl. Acad. Sci. USA* 99, 12542–12547.
38. Wang, C. J., Li, X., Lin, B., Shim, S., Ming, G.-L., and Levchenko, A. (2008) A microfluidics-based turning assay reveals complex growth cone responses to integrated gradients of substrate-bound ECM molecules and diffusible guidance cues. *Lab Chip* 8, 227–237.
39. Abhyankar, V. V., Lokuta, M. A., Huttenlocher, A., and Beebe, D. J. (2006) Characterization of a membrane-based gradient generator for use in cell-signaling studies. *Lab Chip* 6, 389–393.
40. Taylor, A. M., Blurton-Jones, M., Rhee, S. W., Cribbs, D. H., Cotman, C. W., and Jeon, N. L. (2005) A microfluidic culture platform for CNS axonal injury, regeneration and transport. *Nat. Methods* 2, 599–605.
41. Lucchetta, E. M., Lee, J. H., Fu, L. A., Patel, N. H., and Ismagilov, R. F. (2005) Dynamics of *Drosophila* embryonic patterning network perturbed in space and time using microfluidics. *Nature* 434, 1134–1138.
42. Sawano, A., Takayama, S., Matsuda, M., and Miyawaki, A. (2002) Lateral propagation of EGF signaling after local stimulation is dependent on receptor density. *Dev. Cell* 3, 245–257.

43. Bhalla, U. S., Ram, P. T., and Iyengar, R. (2002) MAP kinase phosphatase as a locus of flexibility in a mitogen-activated protein kinase signaling network. *Science* 297, 1018–1023.
44. Krishnan, J. and Iglesias, P. A. (2004)Uncovering directional sensing: where are we headed? *Syst. Biol.*1,54–61.
45. King, K. R., Wang, S., Jayaraman, A., Yarmush, M. L., and Toner, M. (2008) Microfluidic flow-encoded switching for parallel control of dynamic cellular microenvironments. *Lab Chip* 8, 107–116.
46. Cheong, R., Wang, C. J., and Levchenko, A. (2008) High-content cell screening in a microfluidic device. *Mol Cell Proteomics,* in press.
47. Kaneda, A., Wang, C. J., Cheong, R., Timp, W., Onyango, P., Wen, B., Iacobuzio-Donahue, C. A., Ohlsson, R., Andraos, R., Pearson, M. A., Sharov, A. A., Longo, D. L., Ko, M. S., Levchenko, A., and Feinberg, A. P. (2007) Enhanced sensitivity to IGF-II signaling links loss of imprinting of IGF2 to increased cell proliferation and tumor risk. *Proc. Natl. Acad. Sci. USA* 104, 20926–20931.
48. Cho, H., Jönsson.H., Campbell, K., Melke, P., Williams, J. W., Jedynak, B., Stevens, A. M., Groisman, A., and Levchenko, A. (2007) Self-organization in high-density bacterial colonies: efficient crowd control. *PLoS Biol.* 5, e302.
49. Groisman, A., Lobo, C., Cho, H., Campbell, J. K., Dufour, Y. S., Stevens, A. M., and Levchenko, A. (2005) A microfluidic chemostat for experiments with bacterial and yeast cells. *Nat. Methods* 2, 685–689.
50. Higgins, J. M., Eddington, D. T., Bhatia, S. N., and Mahadevan, L. (2007) Sickle cell vasoocclusion and rescue in a microfluidic device. *Proc. Natl. Acad. Sci. USA* 104, 20496–20500.
51. Runyon, M. K., Johnson-Kerner, B. L., and Ismagilov, R. F. (2004) Minimal functional model of hemostasis in a biomimetic microfluidic system. *Angew. Chem. Int. Ed. Engl.* 43, 1531–1536.
52. Loscalzo, J., Kohane, I., and Barabasi, A.-L. (2007) Human disease classification in the postgenomic era: a complex systems approach to human pathobiology. *Mol. Syst. Biol.* 3, 124.
53. Ottesen, E. A., Hong, J. W., Quake, S. R., and Leadbetter, J. R. (2006) Microfluidic digital PCR enables multigene analysis of individual environmental bacteria. *Science* 314, 1464–1467.
54. Warren, L., Bryder, D., Weissman, I. L., and Quake, S. R. (2006) Transcription factor profiling in individual hematopoietic progenitors by digital RT-PCR. *Proc. Natl. Acad. Sci. USA* 103, 17807–17812.
55. Vogelstein, B. and Kinzler, K. W. (1999) Digital PCR. *Proc. Natl. Acad. Sci. USA* 96, 9236–9241.
56. Maerkl, S. J. and Quake, S. R. (2007) A systems approach to measuring the binding energy landscapes of transcription factors. *Science* 315, 233–237.
57. Fu, A. Y., Chou, H.-P., Spence, C., Arnold, F. H., and Quake, S. R. (2002) An integrated microfabricated cell sorter. *Anal. Chem.* 74, 2451–2457.
58. Takahashi, K., Hattori, A., Suzuki, I., Ichiki, T., and Yasuda, K. (2004) Non-destructive on-chip cell sorting system with real-time microscopic image processing. *J. Nanobiotechnol.* 2, 5.
59. Wang, M. M., Tu, E., Raymond, D. E., Yang, J. M., Zhang, H., Hagen, N., Dees, B., Mercer, E. M., Forster, A. H., Kariv, I., Marchand, P. J., and Butler, W. F. (2005) Microfluidic sorting of mammalian cells by optical force switching. *Nat. Biotechnol.* 23, 83–87.
60. Marcy, Y., Ishoey, T., Lasken, R. S., Stockwell, T. B., Walenz, B. P., Halpern, A. L., Beeson, K. Y., Goldberg, S. M., and Quake, S. R. (2007) Nanoliter reactors improve multiple displacement amplification of genomes from single cells. *PLoS Genet.* 3, 1702–1708.
61. Hong, J. W., Studer, V., Hang, G., Anderson, W. F., and Quake, S. R. (2004) A nanoliter-scale nucleic acid processor with parallel architecture. *Nat. Biotechnol.* 22, 435–439.
62. Marcus, J. S., Anderson, W. F., and Quake, S. R. (2006) Microfluidic single-cell mRNA isolation and analysis. *Anal. Chem.* 78, 3084–3089.
63. Burg, T. P., Godin, M., Knudsen, S. M., Shen, W., Carlson, G., Foster, J. S., Babcock, K., and Manalis, S. R. (2007) Weighing of biomolecules, single cells and single nanoparticles in fluid. *Nature* 446, 1066–1069.

Chapter 8

Systems Approach to Therapeutics Design

Bert J. Lao and Daniel T. Kamei

Summary

A general methodology is described for improving the therapeutic properties of protein drugs by engineering novel intracellular trafficking pathways. Procedures for cellular trafficking experiments and mathematical modeling of trafficking pathways are presented. Previous work on the engineering of the transferrin trafficking pathway will be used to illustrate how each step of the methodology can be applied.

Key words: Drug design, Protein drugs, Trafficking, Mathematical model, Transferrin.

1. Introduction

Drug design has traditionally emphasized improving drug/receptor binding at the cell surface. Following receptor binding, however, many protein drugs are endocytosed into a cell and trafficked to various cellular destinations *(1)*. An internalized protein drug may be sorted to a degradative lysosome, for example, or recycled back to the cell surface where it can continue to function and exert its therapeutic effects. Such sorting decisions can have a significant impact on drug half-life and bioactivity *(2)*.

A focus on trafficking may be a promising approach for advancing drug design, since the trafficking pathways of protein drugs can limit their therapeutic function. This is because intracellular trafficking has evolved to suit the functions of physiological proteins naturally present in the body, not protein drugs. By modifying the molecular properties of a protein drug, however, the trafficking pathway of the drug can be manipulated in a rational manner so as to better suit its therapeutic function.

Ivan V. Maly (ed.) *Methods in Molecular Biology, Systems Biology, vol. 500*

DOI: 10.1007/978-1-59745-525-1_8

This chapter describes a broad methodology for modifying intracellular trafficking pathways to increase the potency of protein drugs. The methodology is organized into four steps: (1) perform a systems-level analysis of protein drug trafficking to establish a trafficking design goal, (2) apply a mathematical model of protein drug trafficking to identify molecular design criteria, (3) engineer the protein drug according to the most promising molecular design criteria identified from the mathematical model, and (4) experimentally validate the new trafficking pathway and improved protein drug properties.

In previous work, we have demonstrated that creating a new intracellular trafficking pathway for transferrin (Tf) increases its efficacy as a drug carrier *(3)*. This work will be used as a case study to demonstrate how each step of the methodology can be applied.

2. Materials

1. Transferrin (Sigma-Aldrich, St. Louis, MO).
2. HeLa cells (American Type Culture Collection, Manassas, VA).
3. $Na^{125}I$ (MP Biomedicals, Irvine, CA).
4. IODO-BEADS (Pierce Biotechnology, Rockford, IL).
5. Sephadex G-10 column (Sigma-Aldrich, St. Louis, MO).
6. 35-mm Dishes (Becton Dickinson and Company, Franklin Lakes, NJ).
7. MEM supplemented with 2.2 g/L sodium bicarbonate (Invitrogen, Carlsbad, CA).
8. Fetal bovine serum (Hyclone, Logan, UT).
9. Sodium pyruvate (Invitrogen, Carlsbad, CA).
10. Penicillin (Invitrogen, Carlsbad, CA).
11. Streptomycin (Invitrogen, Carlsbad, CA).
12. Cell incubator (VWR, West Chester, PA).
13. WHIPS: 20 mM HEPES, pH 7.4 containing 1mg/mL polyvinylpyrrolidone (PVP), 130 mM NaCl, 5 mM KCl, 0.5 mM $MgCl_2$, 1 mM $CaCl_2$ (all components from Sigma-Aldrich, St. Louis, MO).
14. Acid strip solution: 50 mM glycine–HCl, pH 3.0 containing 100 mM NaCl, 1 mg/mL PVP, 2 M urea (all components from Sigma-Aldrich, St. Louis, MO).
15. Mild acid strip solution: 50 mM glycine–HCl, pH 3.0 containing 100 mM NaCl, 1 mg/mL PVP (all components from Sigma-Aldrich, St. Louis, MO).

16. Z2 Coulter counter (Beckman Coulter, Fullerton, CA).
17. Packard Cobra Auto-Gamma counter (Packard Instrument Co., Downers Grove, IL).
18. NaOH (Sigma-Aldrich, St. Louis, MO).
19. Incubation medium: MEM supplemented with 20 mM HEPES, pH 7.4 containing 1% sodium pyruvate, 100 units/mL penicillin, 100 μg/mL streptomycin (all components from Sigma-Aldrich, St. Louis, MO).

3. Methods

3.1. Systems-Level Analysis of Protein Drug Trafficking

In the first step of the methodology, a systems-level analysis is performed to establish a goal for how the trafficking of a protein drug can be modified to improve its therapeutic properties. Formulation of this trafficking goal can be aided by identifying features of the trafficking pathway that influence its therapeutic function. Depending on the state of knowledge regarding the trafficking of the protein drug, it may also be useful to conduct cellular trafficking experiments to obtain further information.

3.1.1. Selection of Protein Drug for Trafficking Modification

Key considerations for the selection of a protein drug are the extent to which the drug undergoes intracellular trafficking and the degree to which this trafficking affects its therapeutic function. For example, some cytokines possess a significant trafficking component in that they can be internalized by a cell via receptor-mediated endocytosis following receptor binding, and then sorted to either a degradative lysosome or recycled to the cell surface *(4)*. Lysosomal degradation can have a substantial impact on protein half-life. Therefore, adjusting the balance of the endosomal sorting decision represents one method for modulating cytokine half-life. In fact, the trafficking of granulocyte colony-stimulating factor (GCSF) has previously been engineered to promote cellular recycling over lysosomal degradation, leading to an extension of GCSF half-life *(2)*.

We selected the serum iron transport protein Tf for trafficking modification, since the trafficking that Tf undergoes when it binds the Tf receptor (TfR) is integral to its therapeutic function. Tf has been studied extensively as a drug carrier because TfR is overexpressed in cancer cells, allowing the possibility of specific targeting of therapeutics to tumors and minimization of exposure of noncancerous cells to the therapeutic *(5)*. Tf undergoes receptor-mediated endocytosis upon binding to TfR, and is then trafficked to an endosomal compartment. The acidic pH of the endosome promotes iron release from Tf, and iron-free Tf is then recycled back to the cell surface *(6)*. Notably, iron-free Tf has little to no affinity for TfR at bloodstream pH, and must rebind iron in order to reenter the TfR trafficking pathway. This trafficking pathway can be exploited for drug delivery by conjugating therapeutics to Tf, which allows the therapeutics to access the interior of cells that overexpress TfR.

3.1.2. Cellular Trafficking Experiments

To obtain information about the trafficking pathway of the protein drug, cellular trafficking experiments can be performed. Radiolabeling the protein is an established approach, allowing one to obtain quantitative rate constants that characterize individual steps of the trafficking pathway. This information can be incorporated into a mathematical model of protein trafficking, which is discussed further in **Subheading 3.2**. Later, methods for performing cellular trafficking experiments to obtain rate constants for the Tf trafficking pathway (**Fig. 1**) are presented.

3.1.2.1. Measurements of Association ($k_{\text{FeTf,TfR}}$) and Dissociation ($k_{\text{FeTf,TfR,r}}$) Rate Constants

To determine the association rate constant ($k_{\text{FeTf,TfR}}$) and dissociation rate constant ($k_{\text{FeTf,TfR,r}}$) for iron-loaded Tf (FeTf) binding to TfR, two in vitro cell-surface binding studies can be performed. To isolate the binding events and minimize the trafficking processes, both of these experiments are conducted with HeLa cells on ice.

1. Iodinate iron-loaded Tf proteins with $Na^{125}I$ using IODO-BEADS. Purify radiolabeled Tf using a Sephadex G-10 column with bovine serum albumin present to block nonspecific binding (*see* **Note** 1).
2. Seed the HeLa cells on 35-mm dishes in MEM supplemented with 2.2 g/L sodium bicarbonate, 10% FBS, 1% sodium pyruvate,

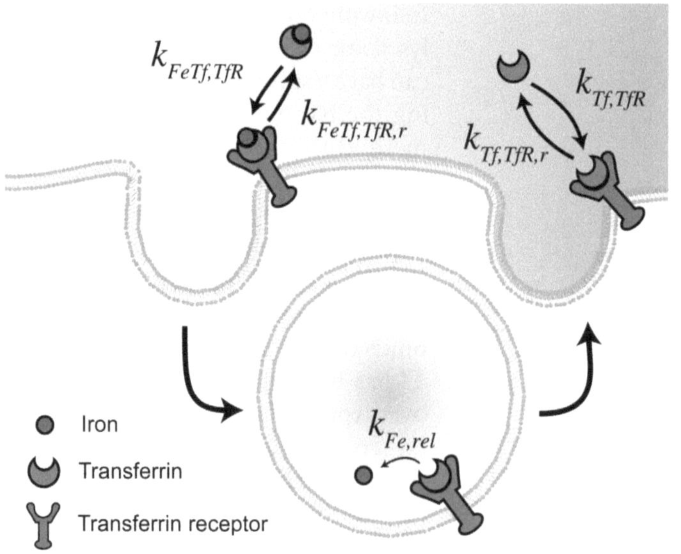

Fig. 1. Schematic of the Tf/TfR trafficking pathway.

100 units/mL penicillin, and 100 μg/mL streptomycin at a pH of 7.4.

3. Incubate cells overnight at 37°C in a humidified atmosphere with 5% CO_2 to a final density of 4×10^5 cells/cm^2.
4. Incubate cells with radiolabeled Tf in serum-free media at concentrations of 10, 30, 100, 300, and 1,000 ng/mL for a few hours to allow equilibrium to be obtained.
5. Take an aliquot of the medium to determine the concentration of Tf in the bulk media (L_{eq}) at equilibrium.
6. Aspirate the remainder of the media, and wash the cells with WHIPS to remove most of the nonspecifically bound Tf. Then, wash cells with acid strip solution to dissociate Tf from TfR. This collected sample will correspond to the number of cell-surface Tf/TfR complexes at equilibrium (C_{eq}) (*see* **Note 2**).
7. The equilibrium dissociation constant (K_D), as well as the total number of cell-surface TfR molecules (R_T), can be determined as follows. Since trafficking is minimized, the total number of cell-surface TfRs remains constant at its initial value, and we have the following relation:

$$R_T = R_{eq} + C_{eq}, \quad (1)$$

where R_{eq} is the number of free cell-surface TfRs at equilibrium. Solving the previous equation for R_{eq} and substituting into the following definition for K_D, we have the following equation:

$$K_D \equiv \frac{L_{eq}R_{eq}}{C_{eq}}. \quad (2)$$

Performing some algebraic manipulation yields the following equation:

$$\frac{C_{eq}}{L_{eq}} = -\frac{1}{K_D}C_{eq} + \frac{R_T}{K_D}. \quad (3)$$

A Scatchard analysis can then be performed, where a plot of C_{eq}/L_{eq} vs. C_{eq} will yield a straight line with the slope equal to $-1/K_D$ and the ordinate intercept equal to R_T/K_D.

Since the equilibrium dissociation constant is also equal to the ratio of the FeTf dissociation rate constant ($k_{FeTf,TfR,r}$) to that of the FeTf association rate constant ($k_{FeTf,TfR}$), we have the following relationship between these two rate constants:

$$k_{FeTf,TfR,r} = K_D k_{FeTf,TfR}. \quad (4)$$

The second binding experiment, which will now be described, will then be used to provide a second relationship between these two constants that can be solved for the two unknowns, $k_{FeTf,TfR}$ and $k_{FeTf,TfR,r}$.

1. Seed the HeLa cells onto 35-mm dishes and incubate them with radiolabeled Tf in serum-free media at the concentration equal to the K_D found earlier.
2. At 0, 3, 6, 9, 12, and 15 min, obtain an aliquot of the bulk media to determine the concentration of Tf in the bulk media (L) for a given time point.
3. Aspirate the media and perform WHIPS washes and acid strips to determine the number of cell-surface Tf/TfR complexes (C) for the same time point.
4. Since this experiment is not at equilibrium and there is only binding occurring, we have the following relationship that applies based on mass-action kinetics:

$$\frac{\mathrm{d}C}{\mathrm{d}t} = k_{\mathrm{FeTf,TfR}} LR - k_{\mathrm{FeTf,TfR,r}} C. \tag{5}$$

Combining **Eqs. 1**, **4**, and **5** yields the following equation:

$$\frac{\mathrm{d}C}{\mathrm{d}t} = k_{\mathrm{FeTf,TfR}} L\left(R_{\mathrm{T}} - C\right) - k_{\mathrm{FeTf,TfR}} K_{\mathrm{D}} C, \tag{6}$$

where R_T and K_D were determined from the first experiment. **Equation 6** can be used to fit $k_{\mathrm{FeTf,TfR}}$ to the experimental data, and the $k_{\mathrm{FeTf,TfR}}$ value can then be substituted into **Eq. 4** to yield the $k_{\mathrm{FeTf,TfR,r}}$ value. These association and dissociation rate constants correspond to those at 0°C, and not at 37°C, since the cells were placed on ice to minimize trafficking.

In both of these experiments, the number of cells on each dish is determined by using control dishes that do not contain radiolabeled Tf and counting the number of cells on these dishes with a Coulter counter (*see* **Note 3**).

3.1.2.2. Measurement of Internalization Rate Constant (k_{int})

Unlike the two binding experiments described in **Subheading 3.1.2.1**, this experiment is performed with the cells incubated at 37°C to allow internalization to occur.

1. Seed the HeLa cells onto 35-mm dishes and incubate them with radiolabeled Tf in serum-free media at the concentration equal to the K_D found from the experiment described in **Subheading 3.1.2.1**.
2. At short time periods (e.g., 0, 2, 4, 6, and 8 min), obtain the number of cell-surface Tf/TfR complexes (C) by performing the acid strip sampling described earlier (*see* **Note 4**).
3. At these same time points, obtain values for the amount of internalized Tf (C_i) by solubilizing the cells with 1M NaOH and placing the resulting solution in a tube to be placed in a gamma counter to quantify the total amount of Tf inside the cells.
4. The species balance on the number of internalized complexes per cell at any given time (C_i) is given by the following equation:

$$\frac{\mathrm{d}C_\mathrm{i}}{\mathrm{d}t} = k_\mathrm{int} C, \tag{7}$$

where k_int is the internalization rate constant. Integrating this ordinary differential equation (ODE) yields the following equation:

$$C_\mathrm{i} = k_\mathrm{int} \int_0^t C\,\mathrm{d}t. \tag{8}$$

Since C_i and C can be measured at each time point, C_i can be plotted vs. the integral of C to yield a line that has k_int as the slope.

3.1.2.3. Measurement of Recycling Rate Constant (k_rec)

1. Seed the HeLa cells onto 35-mm dishes at 37°C and incubate them with radiolabeled Tf in serum-free media at the concentration equal to the K_D found from the experiment described in **Subheading 3.1.2.1**.
2. Allow the trafficking processes to reach steady state by sampling C and C_i at 30-min intervals over a period of 3 h to determine when these quantities begin to plateau. Typically, trafficking processes associated with receptor-mediated endocytosis will attain steady state in a couple of hours.
3. Following achievement of steady state, wash the cells with ice-cold WHIPS and mild acid strip solution.
4. Add serum-free media containing an excess of Tf to prevent recycled Tf from rebinding to TfR. Incubate cells at 37°C, and obtain aliquots of the bulk solutions at different time points (0, 3, 6, 9, 12, and 15 min). Each aliquot will be passed through a filter, where the radioactivity associated with the retentate (i.e., stuck on the filter) will correspond to the radioactivity of the recycled Tf in the bulk solution (*see* **Note 5**).
5. For each time point, perform WHIPS and acid strip washes followed by 1M NaOH solubilization to determine C_i.
6. As in the case of the internalization rate constant, the recycling (k_rec) rate constant can be found by generating plots based on the following integral:

$$N_\mathrm{rec} = k_\mathrm{rec} N_\mathrm{c} \int_0^t C_\mathrm{i}\,\mathrm{d}t, \tag{9}$$

where N_c is the number of cells on a dish, and N_rec is the total number of recycled Tf molecules.

3.1.3. Trafficking Design Goal

When formulating the trafficking design goal, it is helpful to identify features of the protein drug trafficking pathway that may affect its therapeutic function. For example, degradative lysosomes are a common destination of intracellular protein trafficking pathways, and this may reduce the half-life of protein drugs *(1)*. Thus, routing protein drugs out of the lysosomal pathway to extend protein half-life has been a trafficking design goal for previous proteins where trafficking modification was pursued *(2, 7)*.

For the Tf system, cellular trafficking experiments indicated that Tf is trafficked rapidly through a cell *(8, 9)*. The rapid trafficking of Tf aids the physiological function of Tf because it allows iron to be delivered efficiently. However, the rapid trafficking also hinders the efficacy of Tf as a drug carrier, since it limits the time frame in which the drug can be delivered. For example, it has been estimated that for Tf conjugates of the gelonin cytotoxin that for every ten million conjugates that are recycled, only one molecule of gelonin is actually delivered into the cell *(9)*.

Therefore, our trafficking design goal for the Tf system was to extend the time frame in which drug delivery could be achieved by establishing a new Tf trafficking pathway, such that the time Tf spent associated with a cell was increased. This may raise the probability that Tf achieves its intended purpose by delivering the drug to a cell, increasing its efficacy as a drug carrier.

3.2. Mathematical Model of Protein Drug Trafficking

In the second step of the methodology, a mathematical model of protein drug trafficking is used to help determine how the trafficking design goal can be achieved. Mathematical models of cellular trafficking have previously been used to aid analysis and interpretation of experimental trafficking data *(8, 9, 10, 11)*. In establishing these models, the principles of mass action kinetics are applied to derive a system of ODEs that account for the binding, internalization, recycling, and degradation steps that make up the different elements of a trafficking pathway. Quantitative information obtained from cellular trafficking experiments, such as those described in **Subheading 3.1.2**, can be used to parameterize the individual steps of the trafficking pathway within the model.

A sensitivity analysis of the model may assist in formulating molecular design criteria for the protein drug. These design criteria specify how the drug can be engineered to alter its trafficking pathway so as to achieve the trafficking design goal.

3.2.1. Formulation of ODEs

Each ODE in the model is written as a species balance that describes the change in number over time of the protein drug in a given state within the trafficking pathway. For example, the protein drug in the extracellular medium that is bound to a receptor at the cell surface and internalized within the cell would constitute three different states of the protein drug, so three separate ODEs would be written accounting for those states. Each of the terms in an ODE represents a specific step within the trafficking pathway that affects the change in number of the species over time (*see* **Note 6**). Individual trafficking steps in the model are characterized by constants, which describe the rate at which a trafficking step occurs.

Table 1
Model parameters

Parameter	Description	Value	Ref.
$k_{FeTf,TfR}$	Association rate of FeTf for TfR	4×10^{7} $M^{-1}min^{-1}$	*(9)*
$k_{FeTf,TfR,r}$	Dissociation rate of FeTf from TfR	1.3 min^{-1}	*(9)*
$k_{Tf,TfR}$	Association rate of iron-free Tf for TfR	0 $M^{-1}min^{-1}$	*(12)*
$k_{Tf,TfR,r}$	Dissociation rate of iron-free Tf from TfR	2.6 min^{-1}	*(8)*
k_{int}	Internalization rate	0.38 min^{-1}	*(9)*
k_{rec}	Recycling rate	0.15 min^{-1}	*(10)*
$k_{Fe,rel}$	Tf iron release rate	100 min^{-1}	Est.[a]
n_{cell}	Cell number	4×10^{5} cells	
V_b	Bulk media volume	1×10^{-3} L	
N_A	Avogadro's number	6.02×10^{23} mol	

[a]The estimation of the iron release rate value was based on the observation that iron is completely released from internalized Tf prior to it being recycled to the cell surface.

For the Tf model, species balances were written for the following: (1) extracellular FeTf in the bulk media, (2) extracellular iron-free Tf in the bulk media, (3) cell-surface TfR, (4) cell-surface complexes of FeTf and TfR, (5) cell-surface complexes of Tf and TfR, (6) internalized TfR, (7) internalized complexes of FeTf and TfR, and (8) internalized complexes of Tf and TfR. Rate constants were obtained from cellular trafficking studies in which radiolabeled Tf ligand was used to obtain quantitative information on individual steps of the trafficking pathway (**Table 1**). The equations of the model are presented here:

Species balance for bulk extracellular FeTf

$$\frac{d\left(\mathrm{FeTf}_{\mathrm{bulk}}\right)}{dt}=\begin{pmatrix}-k_{\mathrm{FeTf,TfR}}\,\mathrm{FeTf}_{\mathrm{bulk}}\,\mathrm{Tf}R_{\mathrm{surf}}\\+k_{\mathrm{FeTf,TfR,r}}\,\mathrm{FeTf\_Tf}R_{\mathrm{surf}}\end{pmatrix}\frac{n_{\mathrm{cell}}}{V_{\mathrm{bulk}}N_{\mathrm{A}}}. \quad (10)$$

Species balance for bulk extracellular Tf

$$\frac{d\left(\mathrm{Tf}_{\mathrm{bulk}}\right)}{dt}=\begin{pmatrix}-k_{\mathrm{Tf,TfR}}\,\mathrm{Tf}_{\mathrm{bulk}}\,\mathrm{Tf}R_{\mathrm{surf}}\\+k_{\mathrm{Tf,TfR,r}}\,\mathrm{Tf\_Tf}R_{\mathrm{surf}}\end{pmatrix}\frac{n_{\mathrm{cell}}}{V_{\mathrm{bulk}}N_{\mathrm{A}}}. \quad (11)$$

Species balance for surface TfR

$$\frac{\mathrm{d}\left(\mathrm{Tf}R_{\mathrm{surf}}\right)}{\mathrm{d}t} = -k_{\mathrm{FeTf,TfR}}\mathrm{FeTf}_{\mathrm{bulk}}\mathrm{Tf}R_{\mathrm{surf}} - k_{\mathrm{Tf,TfR}}\mathrm{Tf}_{\mathrm{bulk}} \mathrm{Tf}R_{\mathrm{surf}} + k_{\mathrm{FeTf,TfR,r}}\mathrm{FeTf\_Tf}R_{\mathrm{surf}} + k_{\mathrm{Tf,TfR,r}}\mathrm{Tf\_Tf}R_{\mathrm{surf}} - k_{\mathrm{int}}\mathrm{Tf}R_{\mathrm{surf}} + k_{\mathrm{rec}}\mathrm{Tf}R_{\mathrm{int}}. \tag{12}$$

Species balance for surface FeTf/TfR complex

$$\frac{\mathrm{d}\left(\mathrm{FeTf\_Tf}R_{\mathrm{surf}}\right)}{\mathrm{d}t} = +k_{\mathrm{FeTf,TfR}}\mathrm{FeTf}_{\mathrm{bulk}}\mathrm{Tf}R_{\mathrm{surf}} - k_{\mathrm{FeTf,TfR,r}} \mathrm{FeTf\_Tf}R_{\mathrm{surf}} - k_{\mathrm{int}}\mathrm{FeTf\_Tf}R_{\mathrm{surf}} + k_{\mathrm{rec}}\mathrm{FeTf\_Tf}R_{\mathrm{int}}. \tag{13}$$

Species balance for surface Tf/TfR complex

$$\frac{\mathrm{d}\left(\mathrm{Tf\_Tf}R_{\mathrm{surf}}\right)}{\mathrm{d}t} = +k_{\mathrm{Tf,TfR}}\mathrm{Tf}_{\mathrm{bulk}}\mathrm{Tf}R_{\mathrm{surf}} - k_{\mathrm{Tf,TfR,r}}\mathrm{Tf\_Tf}R_{\mathrm{surf}} - k_{\mathrm{int}}\mathrm{Tf\_Tf}R_{\mathrm{surf}} + k_{\mathrm{rec}}\mathrm{Tf\_Tf}R_{\mathrm{int}}. \tag{14}$$

Species balance for internalized TfR

$$\frac{\mathrm{d}\left(\mathrm{Tf}R_{\mathrm{int}}\right)}{\mathrm{d}t} = +k_{\mathrm{int}}\mathrm{Tf}R_{\mathrm{surf}} - k_{\mathrm{rec}}\mathrm{Tf}R_{\mathrm{int}}. \tag{15}$$

Species balance for internalized FeTf/TfR complex

$$\frac{\mathrm{d}\left(\mathrm{FeTf\_Tf}R_{\mathrm{int}}\right)}{\mathrm{d}t} = +k_{\mathrm{int}}\mathrm{FeTf\_Tf}R_{\mathrm{surf}} - k_{\mathrm{rec}}\mathrm{FeTf\_Tf}R_{\mathrm{int}} - k_{\mathrm{Fe,rel}}\mathrm{FeTf\_Tf}R_{\mathrm{int}}. \tag{16}$$

Species balance for internalized Tf/TfR complex

$$\frac{\mathrm{d}\left(\mathrm{Tf\_Tf}R_{\mathrm{int}}\right)}{\mathrm{d}t} = +k_{\mathrm{int}}\mathrm{Tf\_Tf}R_{\mathrm{surf}} - k_{\mathrm{rec}}\mathrm{Tf\_Tf}R_{\mathrm{int}} + k_{\mathrm{Fe,rel}}\mathrm{FeTf\_Tf}R_{\mathrm{int}}. \tag{17}$$

3.2.2. Solving the System of ODEs

Systems of ODEs can be readily solved using several mathematical software packages, such as MatLab, Maple, and Berkeley Madonna. Numerical solution of the ODEs allows simulation of the protein drug trafficking pathway and predictions of how levels of the molecular species change over time.

Since we were interested in increasing the amount of Tf associated with a cell, it was helpful to formulate the ODE solutions in a way that allowed us to focus on $\mathrm{FeTf\_TfR}_{\mathrm{int}}$ and $\mathrm{Tf\_TfR}_{\mathrm{int}}$. Together, these two species comprise the amount of intracellular Tf in the model. Plotting the sum of $\mathrm{FeTf\_TfR}_{\mathrm{int}}$ and $\mathrm{Tf\_TfR}_{\mathrm{int}}$ vs. time allowed the evolution of internalized Tf to be visualized graphically (**Fig. 2**).

3.2.3. Sensitivity Analysis

A sensitivity analysis can be performed to identify molecular design criteria by varying the parameters of the model that correspond to the molecular properties of the protein drug. For example, to

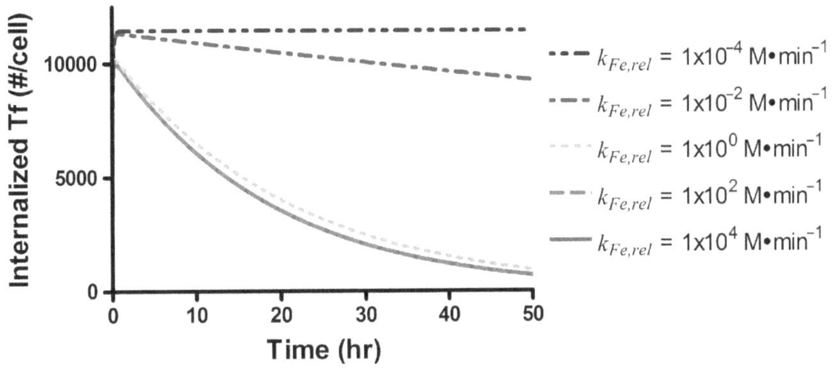

Fig. 2. Plot of internalized Tf vs. time, as predicted by solution of the mathematical model, for various values of the iron release rate, $k_{Fe,rel}$.

simulate the effects of increasing the binding affinity of a protein for its receptor, the association rate of the protein for its receptor within the model can be increased to observe how this changes the predicted levels of each molecular species. Molecular design criteria can be identified by observing which changes in protein molecular parameters lead to alterations of the trafficking pathway that are consistent with achieving the systems-level design goal.

Five different parameters were varied by several orders of magnitude within the Tf/TfR trafficking model to assess their impact on the level of cell-associated Tf. The five parameters varied were as follows: (1) $k_{FeTf,TfR}$, the association rate of FeTf for TfR, (2) $k_{FeTf,TfR,r}$, the dissociation rate of FeTf from TfR, (3) $k_{Tf,TfR}$, the association rate of iron-free Tf for TfR, (4) $k_{Tf,TfR,r}$, the dissociation rate of iron-free Tf from TfR, and (5) $k_{Fe,rel}$, the Tf iron release rate.

To compare levels of intracellular Tf for different parameter values, it was useful to quantify intracellular Tf with a single value by taking the area under the curve (AUC) of internalized Tf vs. time (**Fig. 3**). To address the trafficking design goal of increasing cellular association, we looked for changes in molecular parameters that increased the AUC value of internalized Tf vs. time. The results of the sensitivity analysis showed that cellular association of Tf was predicted to increase under the following three conditions: (1) the association rate of iron-free Tf for TfR is increased, (2) the dissociation rate of iron-free Tf from TfR is decreased, and (3) the iron release rate of Tf is decreased. These three conditions constitute our molecular design criteria.

3.3. Molecular Engineering of Protein Drug

In the third step of the methodology, the protein drug is engineered according to molecular design criteria ascertained from the mathematical model. It may be prudent to balance the selection of promising design criteria against considerations of cost and feasibility.

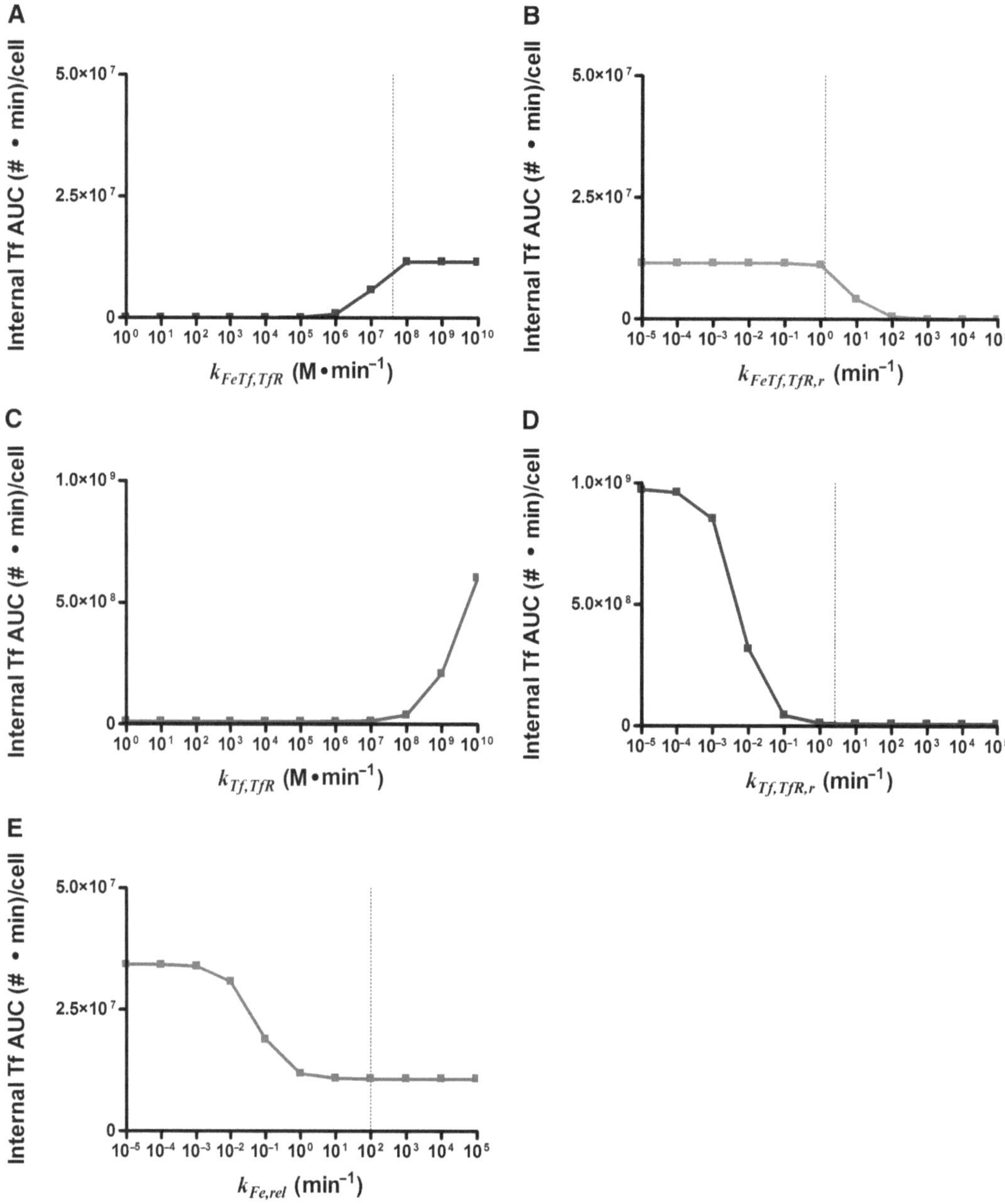

Fig. 3. Sensitivity analysis results for (**A**) $k_{FeTf,TfR}$, (**B**) $k_{FeTf,TfR,r}$, (**C**) $k_{Tf,TfR}$, (**D**) $k_{TfTfR,r}$, and (**E**) $k_{Fe,rel}$. Cellular association, as indicated by the AUC of internalized Tf vs. time, is plotted on the *y*-axis, and the value of the model parameter is plotted on the *x*-axis. The default value of the model parameter is shown by the vertical dashed line. Note the change in scale for (**C**) and (**D**).

For example, when examining the three molecular design criteria for Tf, increasing the affinity of Tf for TfR was predicted to result in substantially greater increases in cellular association than decreasing the iron release rate. However, increasing the affinity of a protein for its receptor is generally considered to be challenging. If such a criterion is pursued, one may be aided by the use of a high-resolution crystal structure of the protein/receptor complex. These complexes can be used to locate protein residues near the binding interface, which could potentially be mutated to increase affinity, without disrupting key residues essential for binding. This strategy has previously been used to identify three residues on the Fc region of IgG close to its neonatal Fc receptor (FcRn) binding region *(7)*. These residues were then randomly mutated to identify mutations, which increased the affinity of Fc for FcRn, leading to an extension of Fc half-life.

Application of modeling techniques to crystal structures has also been used in some instances to reduce the burden of experimental screening techniques by computationally identifying mutations that increase protein binding affinity for a receptor. This approach was used to successfully engineer a tenfold increase in the affinity of cetuximab, a therapeutic antibody used for treatment of colorectal cancer, for its epidermal growth factor target *(13)*. In general, however, the requirement of a high-resolution crystal structure of the protein/receptor complex makes the routine use of such modeling techniques prohibitive.

Since a high-resolution crystal structure of the Tf/TfR complex is not yet available, we decided instead to pursue lowering the Tf iron release rate as our molecular design criterion. An established methodology for lowering the Tf iron release rate by replacing its synergistic carbonate anion with oxalate was used *(14)*.

Unlike the more general trafficking and modeling procedures described earlier, the oxalate replacement procedure is specific to the Tf case study. Since this method is not applicable to other protein systems, it is not presented here.

3.4. Experimental Validation of Modified Trafficking Pathway and Improved Drug Properties

Finally, in the fourth step of the methodology, in vitro cellular trafficking experiments with radiolabeled protein ligands will be performed to validate the new trafficking pathway. In addition, assays are conducted to assess whether the new trafficking pathway translates into the desired improvement in therapeutic properties

3.4.1. Experimental Validation of Modified Trafficking Pathway

To see if inhibiting the iron release of Tf increased its cellular association, the amount of internalized Tf within HeLa cells was monitored over a 2-h period by performing cellular trafficking experiments with radiolabeled Tf.

1. After aspirating seeding medium from the HeLa cells, add incubation medium containing varying concentrations of radiolabeled iron-loaded Tf to each dish.
2. After 5, 15, 30, 60, 90, or 120 min, aspirate the incubation medium and wash the dishes five times with ice-cold WHIPS to remove nonspecifically bound Tf.
3. Add ice-cold acid strip solution to each dish. Place dishes on ice for 8 min and then wash again with an additional mL of the acid strip solution.

4. Following the removal of the specifically bound Tf on the cell surface by the acid strip washes, add NaOH (1 mL of 1 M) to the dishes for 30 min to solubilize the cells. After addition of another mL of NaOH, collect the two basic washes and measure the radioactivity with a gamma counter to determine the amount of internalized Tf.

   Oxalate Tf was found to associate with HeLa cells an average of 51% greater than native Tf at ligand concentrations of 0.1 and 1 nM *(3)*. This suggests that inhibiting Tf iron release alters the Tf trafficking pathway so as to increase its cellular association.

3.4.2. Assaying Improved Drug Properties

To address whether the increased cellular association of Tf improved its efficacy as a drug carrier, Tf was conjugated to diphtheria toxin (DT) and administered to HeLa cells in varying concentrations. Both native Tf and oxalate Tf conjugates were tested. Cell survivability was assessed using the MTT assay. Conjugates of oxalate Tf were found to be significantly more cytotoxic than conjugates of native Tf over a 48-h period. The IC_{50} value, the concentration of conjugate at which 50% inhibition of cellular growth was achieved, was found to be 0.06 nM for the oxalate Tf conjugate, compared with 0.22 nM for the native Tf conjugate *(3)*. This suggests that increasing the cellular association of Tf raises the likelihood of DT being delivered to HeLa cells in our in vitro cytotoxicity assay.

Like the oxalate replacement procedure, the cytotoxicity assay protocol is specific to the Tf case study, and is not presented here.

4. Notes

1. Since iodine-125 covalently binds tyrosines in this procedure, the presence of tyrosines at the receptor binding interface can have some effect on the binding affinity. If possible, visual inspection of a crystal structure can aid in identifying tyrosines at the receptor binding interface. If there are tyrosines, make sure that the mutants also preserve those residues to allow comparison.
2. Since the WHIPS washes may leave some nonspecifically bound ligand on the cell surface, another binding experiment generally needs to be performed in the presence of excess ligand to determine the number of remaining nonspecifically bound ligand molecules. Under these conditions, the unlabeled ligand will saturate the ligand receptors, and the amount of bound labeled ligand will correspond to the level of nonspecific ligand binding.

3. A hemacytometer may also be used. Once you choose the method (Coulter counter or hemacytometer), the key point is to be consistent for all of the experiments.
4. Tf recycles back to the cell surface in about 10 min. However, this experiment can be performed for a longer period of time as long as the ligand has not recycled back to the cell surface or has been degraded and exocytosed.
5. Other approaches for determining recycled vs. degraded ligand are the phosphotungstic acid assay and the trichloroacetic acid assay. In these assays, the radioactivity associated with the precipitate represents the recycled ligand.
6. Occasionally, it may be helpful to add ODE terms that represent trafficking steps that do not occur in the original trafficking pathway, but allow the protein drug to be simulated with modified properties. For example, to enable the simulation of iron-free Tf with an increased association rate for TfR, a term describing the association of Tf for TfR was added to the species balance of the Tf/TfR surface complex. This step does not occur in the original Tf trafficking pathway, since iron-free Tf does not naturally bind TfR.

Acknowledgments

This work was supported by the Sidney Kimmel Scholar Award.

References

1. Lauffenburger, D. A. and Linderman, J. J. (1993) *Receptors*. Oxford University Press, New York, NY.
2. Sarkar, C. A., Lowenhaupt, K., Horan, T., Boone, T. C., Tidor, B., and Lauffenburger, D. A. (2002) Rational cytokine design for increased lifetime and enhanced potency using pH-activated "histidine switching". *Nat. Biotechnol.* 20, 908–913.
3. Lao, B. J., Tsai, W. L., Mashayekhi, F., Pham, E. A., Mason, A. B., and Kamei, D. T. (2007) Inhibition of transferrin iron release increases in vitro drug carrier efficacy. *J. Control. Release* 117, 403–412.
4. Rao, B. M., Lauffenburger, D. A., and Wittrup, K. D. (2005) Integrating cell-level kinetic modeling into the design of engineered protein therapeutics. *Nat. Biotechnol.* 23, 191–194.
5. Weaver, M. and Laske, D. W. (2003) Transferrin receptor ligand-targeted toxin conjugate (Tf-CRM107) for therapy of malignant gliomas. *J. Neurooncol.* 65, 3–13.
6. Hentze, M. W., Muckenthaler, M. U., and Andrews, N. C. (2004) Balancing acts: molecular control of mammalian iron metabolism. *Cell* 117, 285–297.
7. Hinton, P. R., Johlfs, M. G., Xiong, J. M., Hanestad, K., Ong, K. C., Bullock, C., Keller, S., Tang, M. T., Tso, J. Y., Vasquez, M., and Tsurushita, N. (2004) Engineered human IgG antibodies with longer serum half-lives in primates. *J. Biol. Chem.* 279, 6213–6216.
8. Ciechanover, A., Schwartz, A. L., Dautry-Varsat, A., and Lodish, H. F. (1983) Kinetics of internalization and recycling of transferrin and the transferrin receptor in a human hepatoma cell line. Effect of lysosomotropic agents. *J. Biol. Chem.* 258, 9681–9689.
9. Yazdi, P. T. and Murphy, R. M. (1994) Quantitative analysis of protein synthesis

inhibition by transferrin-toxin conjugates. *Cancer Res.* 54, 6387–6394.

10. French, A. R. and Lauffenburger, D. A. (1997) Controlling receptor/ligand trafficking: effects of cellular and molecular properties on endosomal sorting. *Ann. Biomed. Eng.* 25, 690–707.
11. Kamei, D. T., Lao, B. J., Ricci, M. S., Deshpande, R., Xu, H., Tidor, B., and Lauffenburger, D. A. (2005) Quantitative methods for developing Fc mutants with extended half-lives. *Biotechnol. Bioeng.* 92, 748–760.
12. Lebron, J. A., Bennett, M. J., Vaughn, D. E., Chirino, A. J., Snow, P. M., Mintier, G. A., Feder, J. N., and Bjorkman, P. J. (1998) Crystal structure of the hemochromatosis protein HFE and characterization of its interaction with transferrin receptor. *Cell* 93, 111–123.
13. Lippow, S. M., Wittrup, K. D., and Tidor, B. (2007) Computational design of antibody-affinity improvement beyond in vivo maturation. *Nat. Biotechnol.* 25, 1171–1176.
14. Halbrooks, P. J., Mason, A. B., Adams, T. E., Briggs, S. K., and Everse, S. J. (2004) The oxalate effect on release of iron from human serum transferrin explained. *J. Mol. Biol.* 339, 217–226.

Chapter 9

Rapid Creation, Monte Carlo Simulation, and Visualization of Realistic 3D Cell Models

Jacob Czech, Markus Dittrich, and Joel R. Stiles

Summary

Spatially realistic diffusion-reaction simulations supplement traditional experiments and provide testable hypotheses for complex physiological systems. To date, however, the creation of realistic 3D cell models has been difficult and time-consuming, typically involving hand reconstruction from electron microscopic images. Here, we present a complementary approach that is much simpler and faster, because the cell architecture (geometry) is created directly in silico using 3D modeling software like that used for commercial film animations. We show how a freely available open source program (Blender) can be used to create the model geometry, which then can be read by our Monte Carlo simulation and visualization software (MCell and DReAMM, respectively). This new workflow allows rapid prototyping and development of realistic computational models, and thus should dramatically accelerate their use by a wide variety of computational and experimental investigators. Using two self-contained examples based on synaptic transmission, we illustrate the creation of 3D cellular geometry with Blender, addition of molecules, reactions, and other run-time conditions using MCell's Model Description Language (MDL), and subsequent MCell simulations and DReAMM visualizations. In the first example, we simulate calcium influx through voltage-gated channels localized on a presynaptic bouton, with subsequent intracellular calcium diffusion and binding to sites on synaptic vesicles. In the second example, we simulate neurotransmitter release from synaptic vesicles as they fuse with the presynaptic membrane, subsequent transmitter diffusion into the synaptic cleft, and binding to postsynaptic receptors on a dendritic spine.

Key words: Blender, MCell, DReAMM, MDL, Cell modeling, Cell architecture, Cell geometry, Stochastic, Diffusion-reaction.

1. Introduction

A quantitative understanding of cell and tissue function requires detailed models through which hypotheses may be generated and tested. As models become more complex, computer simulations provide tests of important questions that are beyond simple intuition

Ivan V. Maly (ed.) *Methods in Molecular Biology, Systems Biology, vol. 500*

DOI: 10.1007/978-1-59745-525-1_9

and are inaccessible to current experimental methods. In the best case, a tight coupling between models, simulations, and experiments can lead to breakthroughs in understanding.

Model development and simulation involve a number of distinct steps carried out with different software tools, similar to the different steps, methods, and equipment used in an experimental protocol. Realistic physiological models are particularly challenging because they encompass complex biochemistry taking place in small complex 3D spaces, and the different software tools required at each step can present steep learning curves. Nevertheless, it is increasingly necessary to develop and use such models to understand the physiology of disease, drug effects, and phenotypic changes produced by genetic manipulations.

Creation of a spatially realistic model begins with the definition of its cellular architecture, or geometry, followed by the addition of biochemical species and interactions within the geometry. The methods chosen for these initial steps depend in large part on the simulation approach to follow. For example, while some models may represent an entire cell as a single well-mixed compartment, we focus on stochastic diffusion and reactions within arbitrarily complex intra- and extracellular spaces. Triangulated surface meshes are used to represent cell and organelle membranes, and thus the meshes also define different volumes of intervening solution. To build a model, one must somehow create the surfaces and then define how many molecules of what types are present in different spatial regions. One must also provide the diffusion coefficient for each molecular species and define the network of biochemical interactions and associated rate constants. Having done so, one can investigate such problems as diffusion of neuromodulators and neurotransmitters through tortuous extracellular space in brain *(1–3)*, or neurotransmitter release (exocytosis) and synaptic transmission *(4–11)*. Simulations of exocytosis involve voltage- and/or calcium-triggered release of signaling molecules from synaptic or endocrine vesicles. The released molecules subsequently diffuse through some extracellular space and are detected by receptor protein molecules on the downstream cell or cells. Our program MCell (Monte Carlo Cell; **refs.** *4, 6, 7, 12)* was designed for such simulations, although it now is very general and can be used for a wide variety of diffusion-reaction models (e.g., **refs.***13, 14)*. The companion program DReAMM is used to visualize and animate the simulation results *(9, 12, 15)*.

In this chapter we present a protocol for development of an MCell model and simulations, providing step-by-step instructions for an example based on signal transduction at synaptic boutons. Creation of the initial 3D geometry has been a longstanding bottleneck for all such projects, especially when synaptic geometries are extracted (segmented) from electron microscopy data and subsequently converted into surface meshes for use in simulations *(6, 9, 10)*. Recently, however, we have developed a complementary and much faster approach based on the use of Blender *(16)*, open-source 3D modeling software (*see* **Note 1**), as well as plug-ins that we have developed to link Blender to MCell and DReAMM. As we illustrate here, arbitrary cell-like geometry can be designed directly *in silico* using Blender, and the resulting meshes can be exported to MCell and DReAMM for simulation and visualization. This time-saving approach makes detailed simulations and investigations available to virtually any laboratory in relatively short order.

Specifically, we show how to do the following:

1. Create the model geometry using Blender. In **Subheading 3.1**, a synapse is created on one of several dendritic spines (**Fig. 1**), similar to actual dendrite and spine ultrastructure (**Fig. 1**, inset). A synaptic cleft space separates the presynaptic bouton from the spine head, and the bouton contains two docked synaptic vesicles (*see* **Fig. 4E**). Particular regions of the vesicle membrane and pre- and postsynaptic membranes (*see* **Fig. 4F, G**) are defined for subsequent addition of calcium

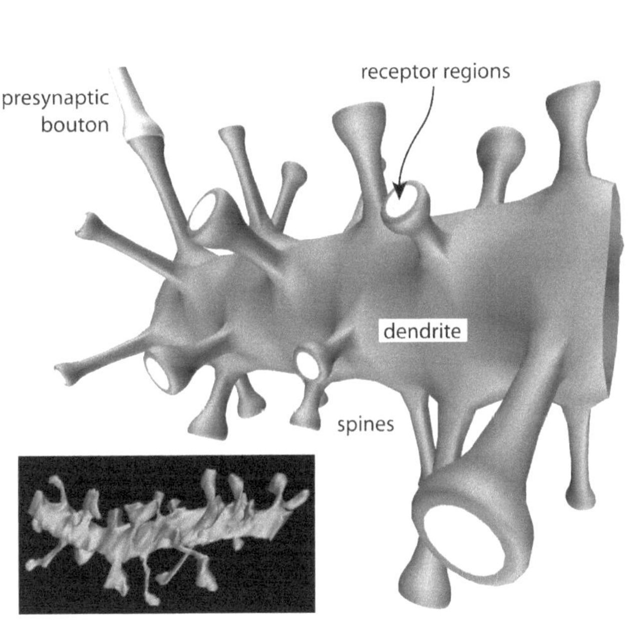

Fig. 1. Three-dimensional model of a spiny dendrite illustrates some of the structures to be created and used in simulations in this chapter. The model dendrite was created with Blender and visualized with DReAMM. Postsynaptic receptor regions (*light gray*) are located at the ends of the spine heads. One of the spine heads forms a synapse with a presynaptic bouton (*upper left*). For comparison, the inset shows a dendrite reconstructed from serial electron micrographs of rat brain. The data for the reconstruction were obtained as a publicly available VRML file from **ref.** *17*. The VRML data were subsequently imported into DReAMM and visualized as illustrated by the inset image.

binding sites, voltage-gated calcium channels (VGCCs), or ligand-gated neurotransmitter receptors, respectively, using MCell's MDL (**Subheadings 3.3** and **3.5**). In **Subheading 3.2**, the synaptic vesicles are modified with Blender to include an expanding fusion pore (*see* **Fig. 4H, I**), and the model is subsequently used for MCell simulations of neurotransmitter diffusion through the pore (**Subheading 3.5**).

2. Use Blender plug-ins to export the model geometry as triangulated surface meshes with region annotations for use with MCell and DReAMM.
3. Use MCell's MDL to specify (**Subheadings 3.3** and **3.5**) the following:
   (a) The types of molecules in the model (calcium ions, calcium binding sites on proteins, neurotransmitter molecules, postsynaptic receptor proteins), their diffusion constants, and their initial locations.
   (b) The stoichiometry, rates, and directionality of the stochastic reactions that can occur during simulations (conformational changes of calcium channels, calcium entry from open channels, calcium binding to protein sites on synaptic vesicles, neurotransmitter binding to postsynaptic receptors, channel opening and closing of bound receptors).
   (c) The reaction data to be saved when the simulations are run. This includes counts of molecules and reactions in specific compartments as a function of time, and the resulting text (ASCII) files can be analyzed or plotted using common graphing software and/or scripts.
4. Save MCell visualization data for DReAMM so that the model can be rendered (displayed) and checked. The importance of this step cannot be overemphasized, especially for complex models. For efficiency, the visualization files are written in a structured binary format that minimizes disk space and maximizes interactive speed and simplicity of use with DReAMM.
5. Run complete MCell simulations and visualize the results with DReAMM. In **Subheading 3.3**, we simulate a voltage-clamped presynaptic bouton in which calcium channels open stochastically, allowing calcium ions to enter. The ions then diffuse and bind to sites on the synaptic vesicles, simulating the presence of calcium-binding proteins (e.g., synaptotagmin) and some of the events leading to calcium-dependent neurotransmitter release (exocytosis). In **Subheading 3.5**, we simulate expansion of fusion pores between the vesicles and presynaptic membrane, with subsequent diffusion of neurotransmitter molecules into the synaptic cleft. Within the cleft,

neurotransmitter binds reversibly to the postsynaptic receptors. Receptors that reach a double-bound state can undergo a reversible conformational change to an open channel state (ligand-dependent channel gating).

6. Use DReAMM to animate simulation results.

The projects illustrated in this chapter are completely self-contained and can be used without reference to additional material. Many modeling features are illustrated, although spatial and biochemical details have been simplified for the sake of space and readability. Thus, the chapter provides an efficient introduction to development and use of spatially realistic stochastic cellular models and simulations, and, where necessary, provides links and citations to further discussion and examples.

2. Materials

2.1. Preliminary Issues

Building and simulating the models described in this chapter requires a computer with the programs Blender, MCell, and DReAMM installed. Like many open source programs, Blender, MCell, and DReAMM are available in precompiled binary executable form for a limited set of computer architectures and operating systems. This can dramatically simplify installation for users who have access to those particular systems and do not have experience in compiling ("building"), installing, and administering (large) programs obtained as source code. However, it is not possible to "prebuild" for all possible systems, and for this and other reasons it may be necessary or preferable to obtain the source code and build it "from scratch". This is generally not a problem for a UNIX user with experience in code development or system administration, but it can be challenging for a novice. Part of the difficulty is simply dealing with a command line interface (typing commands at a prompt in a window), understanding where system files are located (in which directories or "folders"), and understanding how directory and file access is granted or denied to different users on a multiuser system. This is a very real issue, especially as more and more experimental biologists are becoming more and more interested in using computational models and simulations to complement their wet lab work. Although the basics are straightforward, entire books are dedicated to the use and care of UNIX systems. In this chapter our focus is on building and using spatially realistic models. Hence, we assume a basic knowledge of UNIX and a command line interface, just as an experimental protocol must assume a basic knowledge of solutions, gels, radiochemicals, etc.

2.2. Computer Hardware and Operating System

The computer to be used can have either a 32- or 64-bit architecture (single or multiple cores), and should have some form of a UNIX operating system that includes developer features (C/C++ compiler, OpenMotif libraries, standard image tool libraries, etc.). A fully configured Linux or similar system should work without problem. Mac OS X is in fact a form of UNIX, so it also works well, but it must be installed with the developer libraries (*see* **Note 2**). The difference between a 32- or 64-bit system is unimportant unless very large MCell simulations are to be run, in which case a 64-bit system with a very large amount of physical memory (RAM) may be required. At the upper extreme for large models, the simplest present solution is to run MCell on a large shared memory architecture (hundreds of GBytes), and if no such local computers are available one can use a shared memory computer available at the Pittsburgh Supercomputing Center or other national facility. In practice, however, many models and simulations (including those for this chapter) can be run quite easily on present-day desktop computers with typical amounts of memory (e.g., 2–8 GB).

Efficient and high-quality visualization is critically important to spatially realistic modeling, and so the computer should have at least one high-resolution display (1,600 × 1,200 or higher). We routinely use two such displays configured as one 3,200 × 1,200 workspace, driven by a professional-level OpenGL graphics card (e.g., NVidia Quadro 3xxx or 4xxx series at present). Disk storage is generally not a problem in current desktop systems, where capacities approaching a terabyte or more are common. In principle, many projects can also be run on a laptop, although limited display space can be an inconvenience.

2.3. Download and Install Blender and Blender Plug-Ins

1. Download and install Blender from http://www.blender.org. This is the Blender development site, so the most recent version will be available. As an alternative, you may download all the software required for this chapter from https://www.mcell.psc.edu/download.html, in order to obtain the specific versions that match those illustrated here. This chapter is based on Blender version 2.45, and more recent versions may have changes to hot key (and other) functions. In addition, different versions of UNIX may assign different functions to particular keys (such as the *Alt* key), so there may be some inevitable differences from what is described in the following sections.
2. After installation, make sure that your path environment variable is set so that Blender can be started from any command line (type "blender" and hit *Enter* at a command line).
3. Download the two Blender plug-ins from https://www.mcell.psc.edu/download.html if you have not already done so. One

plug-in will be used to generate MCell MDL files (.mdl) from meshes created with Blender. The other can be used to generate DX files (.dx) for use with DReAMM (*see* **Note 3**). The DX format is also the default visualization file format output by MCell for use with DReAMM.

4. To make the Blender plug-ins functional, create a directory and copy or move both plug-ins into it. Start Blender and mouse over the lower edge of the top menu bar (**Fig. 2A**). Left click and pull down to uncover several button fields including one labeled "File Paths". Click on it and then locate the "Python" text box (**Fig. 2B**), which has two buttons next to it. Click the rightmost button, navigate to the plug-in directory you just created, and then select it as the default script location. Then click on the left button to make the scripts available to Blender. If desired, hide the pull down button fields.
5. Hit *Ctrl-u* to save these settings for future use.
6. Verify that on the Blender menu bar, "File -> Export" now shows entries for MCell (.mdl) and DReAMM (.dx).

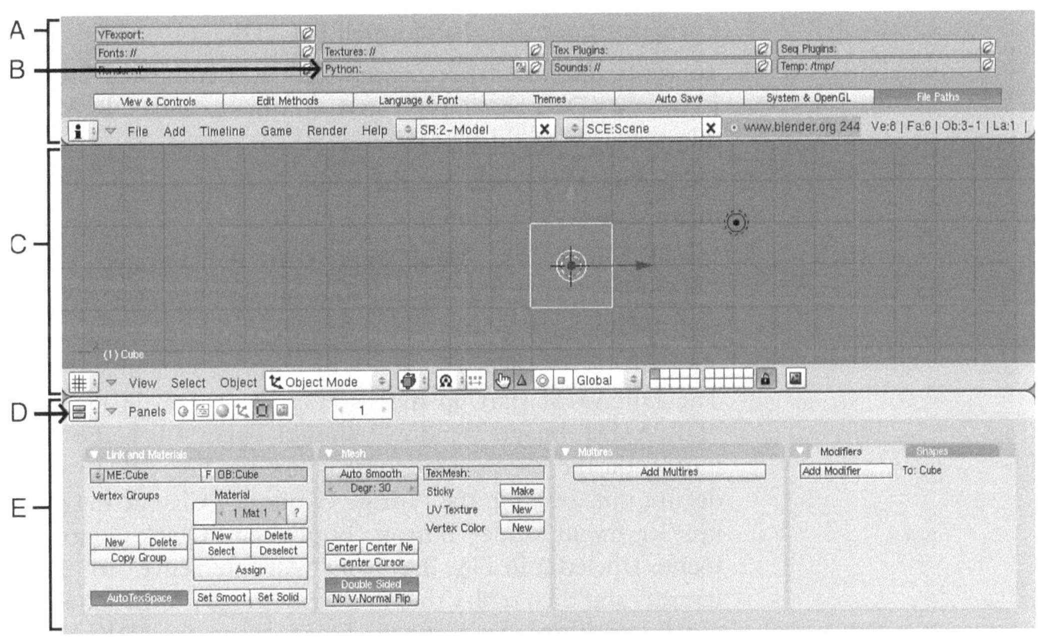

Fig. 2. Screen capture of Blender's main window. By default, the Blender interface consists of three areas (A, C, and E). At the top (A) is the "User Preferences" window, which is hidden initially. It allows the user to adjust Blender's appearance, performance, and file paths, including the location of the Python plug-ins directory (B). The "3D View" (C) is used for manipulation and visualization of mesh objects. The "Buttons Window" (E) offers a variety of operations that can be performed on the mesh objects. At the top left of the "Buttons Window" is a drop-down list (D) providing access to all of the available window areas.

2.4. Download and Install MCell

1. Download and install MCell from https://www.mcell.psc.edu/download.html. Throughout this chapter we assume the use of MCell version 3, or MCell3 *(12)*.
2. After installation, make sure that your path environment variable is set so that MCell can be started from any command line. To do so, enter the name of your executable for MCell at a command line (e.g., "mcell3"; the actual name may vary depending on your installation) and verify that MCell starts running and prints out a number of default start-up messages in the command window. In the absence of any command line arguments/options (as in this case), the start-up messages will include an error message stating that no MDL filename has been specified. This is normal and can be ignored.

2.5. Download and Install DReAMM

1. To use DReAMM you must first download and install PSC_DX, a visual programming, imaging, and data manipulation environment on which DReAMM is built. PSC_DX is a customized and improved version of OpenDX, or Open Data Explorer, originally developed by IBM. Note that DReAMM requires PSC_DX and will not run with OpenDX. Both PSC_DX and DReAMM can be downloaded from https://www.mcell.psc.edu/download.html. This chapter is based on PSC_DX and DReAMM version 4.1.
2. Compile (if necessary) and install PSC_DX.
3. To verify the installation, enter "dx" at a command line. A PSC_DX (Data Explorer) start-up menu and DReAMM splash image should appear.
4. Quit PSC_DX (click on the Quit button in the start-up menu).
5. Install DReAMM. Make sure that the start-up script ("dreamm") or a link to the start-up script is accessible in your path.
6. To verify the installation, enter "dreamm" at a command line. The DReAMM start-up menus (**Fig. 3**), splash image, and "DReAMM Image Window" should open. Click the Play button on the "Sequence Control" menu (**Fig. 3B**) to see a default time series in the "Image Window". Change the pop-up "Keyframe Mode" button (bottom of "Quick Controls" menu, labeled *8* in **Fig. 3A**) from "All Interactive" to "Keyframes" and then replay the time series to see it with animated camera positions. To quit DReAMM, go to the "DReAMM Image Window" menu bar and click on "File –> Quit". When prompted to save the project, click "No".

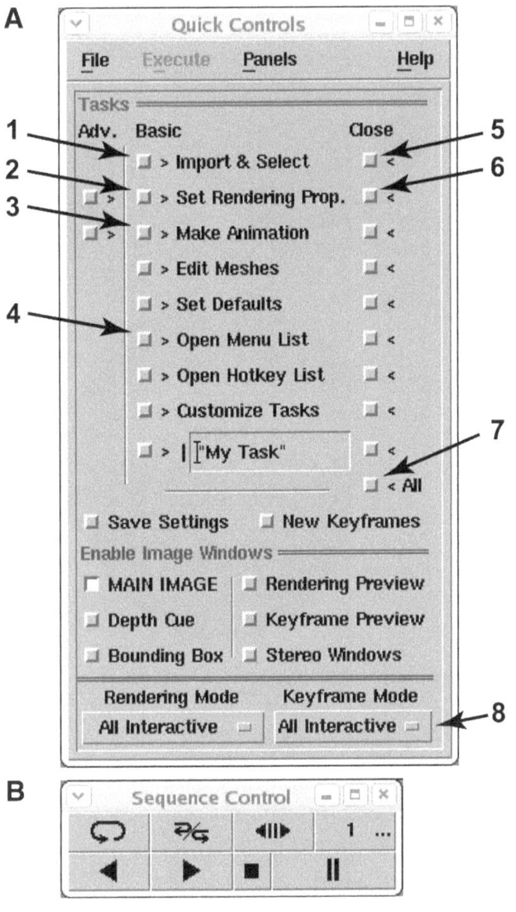

Fig. 3. Screen captures of start-up DReAMM controls. (A) The "Quick Controls" window organizes DReAMM's menus into task-based subsets that can be customized to suit user preferences. (1) Open menus to import data and select objects. (2) Open menus to assign rendering properties to objects. (3) Open menus to make an animation using keyframes. (4) Open a list of all menus that can be opened one-by-one. (5) Close the menus opened by (1). (6) Close the menus opened by (2). (7) Close all previously opened menus. (8) By default, the "Keyframe Mode" is set to "All Interactive", meaning that the camera can move about the 3D space freely. When set to "Keyframes", however, the camera will be locked into positions (keyframes) that were set using "Make Animation" menus. (B) "Sequence Control" for animations. Bottom row of buttons, left to right: Play in reverse, Play forward, Stop, Pause. Top row, left to right: Loop while either Play button is pressed, Palindrome (play to end or beginning and then reverse direction while either Play button is pressed), Single-step play mode, Frame number (pressing this button opens a "Frame Control" window that allows particular frames to be played). Note that Loop, Palindrome, and Single-step may all be used in combination.

3. Methods

3.1. Create Pre- and Postsynaptic Geometry with Blender

3.1.1. Create a Spine Head

In this section we will create a spine head by removing the upper portion of a sphere and then closing the opening (**Fig. 4A, B**). In later sections, the spine head will be copied and modified to create additional objects.

1. *Start Blender*: Enter "blender" at a command line. You will see two view ports, a large central "3D View" and a "Buttons Window"

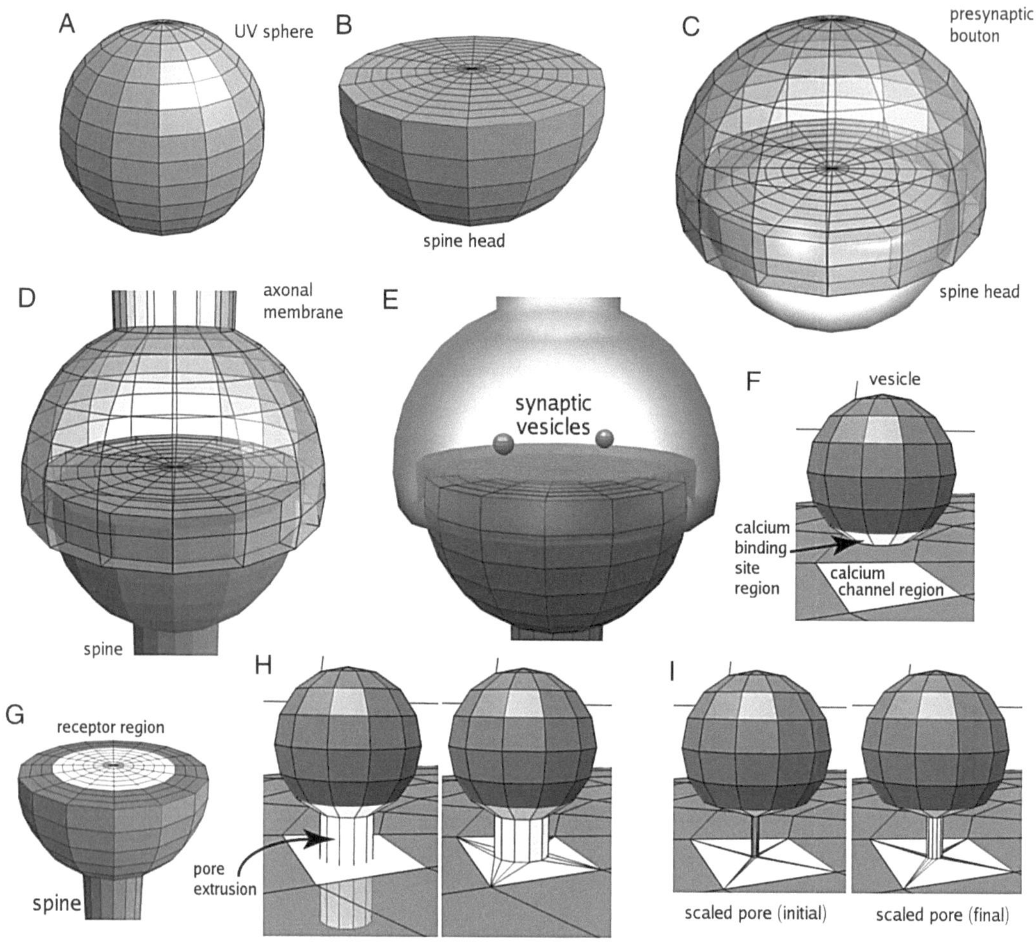

Fig. 4. Blender screen captures at key stages during mesh creation. A UV sphere (**A**, *see* **Note 6**) is cut nearly in half and closed off to create a dendritic spine head (**B**). The spine head then is duplicated, rotated, and manipulated to create a presynaptic bouton (**C**). Portions of the bouton and spine head are extruded to create axonal membrane and the spine neck (**D**). Two small spheres are added within the bouton and serve as synaptic vesicles (**E**). Regions on the presynaptic bouton and vesicle meshes are defined for voltage-gated calcium channels and calcium binding sites, respectively (**F**). A region on the spine head is defined for postsynaptic receptors (**G**). A fusion pore is extruded from each synaptic vesicle (**H**, *left*) and joined to the presynaptic bouton mesh (**H**, *right*). Finally, the fusion pore is scaled to the desired initial (**I**, *left*) and final (**I**, *right*) diameters.

widget at the bottom (**Fig. 2**). By default, the start-up 3D view shows the XY-plane parallel to the screen (*see* **Note 4**).

2. *Delete the default Blender start-up object*: At start-up, Blender creates a simple cube object. Delete this by hitting *x* (*see* **Note 5**), which will bring up a dialog box prompting to "Erase selected Object(s)". Click to confirm.
3. *Create a sphere*: Hit the *spacebar*. In the pop-up menu, mouse over "Add", then "Mesh", and click on "UVSphere" (**Fig. 4A, Note 6**). In the new pop-up menu titled "Add UV Sphere", make sure that "Segments" and "Rings" are set to "16", "Radius" is set to "0.25" (*see* **Note 7**), and then click "OK".

4. *Name the sphere*: Under the "Link and Materials" tab in the "ButtonsWindow", click in the text box that says "OB:Sphere" and change it so that it reads "OB:SpineHead".
5. *Change view*: Hit *1* (*on the number pad*) to switch to the XZ-view (*see* **Notes 8** and **9**).
6. *Deselect the sphere and make it semitransparent*: Hit *a* then *z*.
7. *Select the vertices to be removed*: Hit *b*, which will bring up a pair of cross-hairs used for selecting vertices. Clicking and dragging will create a rectangular area that follows the cross-hairs. Any vertices within this rectangle will be selected. Click and drag over all the vertices *above* (but not including) the equator of the sphere (the *XY*-plane in this view).
8. *Remove the faces that make up the top of the sphere*: Hit *x* and click on "Faces" in the "Erase" menu. The "3D View" should now show only the remaining lower portion of the sphere.
9. *Close the opening*: As shown in **Fig. 4B**, the desired result includes a set of new adjoining faces that will meet at a central vertex and close the opening (*see* **Note 10**). Begin by selecting the topmost vertices. Hit *b* and select the vertices by clicking and dragging the selector rectangle over the top edge. Next, hit *e* to extrude and click "Only Edges" in the "Extrude" pop-up menu. Hit *0* as the distance to extrude and hit *Enter* to confirm. Hit *s* to scale the extrusion, next hitting *0* to obtain the desired new radius, and then hit *Enter*. In the "Buttons Window" under "Mesh Tools", hit "Rem Double". This will remove all but one of the duplicated vertices and reconnect the triangles. Blender should inform you that 15 vertices have been removed, and the object should now be closed by a flat top.
10. *Subdivide the triangles that close the top*: At this point we need to create a set of concentric rings to be used in several subsequent operations (defining a region and creating an invagination). Hit *b* and once again select the topmost vertices, which now will also include the new central vertex. Next hit *7* (*on the number pad*) to change back to the *XY*-view (overhead view of the spine head). Hit *k* and, in the pop-up "Loop/Cut Menu", click "Knife (Multicut)". In the "Number of Cuts:" pop-up menu, click the right arrow until it reads "8" and then click "OK". In a clockwise motion, click and drag the knife-shaped cursor over all of the spoke-like edges radiating from the center point. All of the edges (spokes) need to be crossed; it does not matter where or in which order. After crossing all the edges, hit *Enter*. Blender will subdivide each edge into eight segments of equal length and create new coplanar faces arranged in concentric rings (**Fig. 4B**).

11. *Save your current mesh object*: Hit *F2* and save the mesh (*see* **Note 11**).

3.1.2. Create a Presynaptic Bouton

In this section we will duplicate the existing spine head and morph the copy into an invaginated presynaptic bouton.

1. *Reset the view*: Hit *1* (*on the number pad*) to switch again to the *XZ*-view. If you are continuing from the previous section and some vertices are currently selected, hit *a* once to deselect everything. Then hit *a* (again) to select all the vertices.
2. *Duplicate the spine head and rotate the copy*: Hit *Shift-d* to duplicate the existing spine head. After it is duplicated, Blender will automatically select only the new portion, which at this point is still considered part of the original object. Hit *r*, then type *180*, and hit *Enter* to rotate the new portion by 180° around the Y-axis (the current view axis). Hit *p* and click on "Selected" in the "Separate" menu to make the new rotated portion a separate object.
3. *Select and name the new object*: Hit *Tab* to switch to "Object Mode" and right click on the new object so that it is highlighted in pink (*see* **Note 12**). In the "Link and Materials" tab in the "Buttons Window", name the object by clicking in the text box containing the string "OB:SpineHead.001" and changing it to "OB:PresynapticBouton".
4. *Shift and scale the bouton*: Hit *g* to grab the object, *z* to constrain the shift (move) operation to the *Z*-axis, then type *0.15*, and hit *Enter* to confirm. This will move the bouton 0.15 units along the positive *Z*-axis. Hit *s* to scale the bouton, type *1.2*, and hit *Enter* to confirm. This will increase the size of the bouton by 20%.
5. *Create an invagination in the bouton*: Hit *Tab* again to switch to "Edit Mode". Hit *a* to deselect everything. Select the lowermost vertices of the bouton by hitting *b* and then clicking and dragging around them. Next, hit *Ctrl-Minus* (*on the number pad*) to perform a "Select Less" operation. In this case the outermost ring of vertices will be deselected. Hit *e* to extrude, click "Region" in the "Extrude" menu, *z* to constrain it to the *Z*-axis, type *0.075*, and hit *Enter*. This will move the remaining selected vertices 0.075 units in the positive *Z* direction, producing an invagination into which the postsynaptic spine head will fit (**Fig. 4C**).
6. *Save the current mesh objects*: Hit *F2* to save the current meshes (*see* **Note 11**).

3.1.3. Add Axonal and Dendritic Extensions

We now add cylindrical extensions to the rounded ends of both objects, to create a length of axonal membrane for the presynaptic bouton and a dendritic spine for the spine head.

1. *Confirm that you are in "Edit Mode"*: If you are continuing from the preceding step, you should already be in "Edit Mode". If necessary, hit *Tab* to switch from "Object Mode" to "Edit Mode".
2. *Confirm the view*: The operations to be performed in this step are most safely accomplished using a "side view", so if necessary hit *1* (*on the number pad*) for the *XZ*-view.
3. *Create a cylindrical axon segment on the presynaptic bouton*: If necessary, hit *a* to deselect all objects, then zoom in (*see* **Note 8**) on the upper pole of the presynaptic bouton until the top vertex is clearly visible. Select the vertex by right clicking on it. Now perform two "Select More" operations by hitting *Ctrl-Plus* (*on the number pad*) twice, until two concentric rings of vertices are highlighted. Hit *x* and click "Faces" on the "Erase" menu. This will remove the mesh faces selected at the pole of the presynaptic bouton, leaving an opening. Hit *b* and then select the vertices that line the opening by clicking and dragging around them. Next hit *e*, click "Only Edges" on the "Extrude" menu, then hit *z*, type *3.0*, and hit *Enter*. A cylindrical axon segment should now extend from the top of the bouton (**Fig. 4D**).
4. *Select the spine head*: Hit *Tab* to go into "Object Mode". Right click on the spine head (bottom) mesh to select it. Hit *Tab* to go back into "Edit Mode".
5. *Create a cylindrical spine on the spine head*: As in earlier **step 2**, zoom in and select the bottommost vertex (at the pole) by right clicking on it. Hit *Ctrl-Plus* (*on the number pad*) two times. Hit *x* and click "Faces" on the "Erase" menu. Hit *b* and then click and drag around the vertices that line the hole in the bottom of the mesh. Hit *e*, click "Only Edges" on the "Extrude" menu, hit *z*, type *-2.0*, and hit *Enter*. A cylindrical spine should now extend from the bottom of the spine head (**Fig. 4D**).
6. *Save the current mesh objects*: Hit *F2* to save the current meshes (*see* **Note 11**).

3.1.4. Add Synaptic Vesicles with Regions for Calcium Binding Sites

We will now create two small spheres to represent synaptic vesicles inside the presynaptic bouton. We will also define a region on each vesicle to contain calcium binding sites (**Fig. 4E, F**). In Blender, we define regions simply by assigning each one a uniquely named material. If desired, the regions can then be visualized by giving them separate colors. The regions assigned with Blender will be accessible automatically with MCell and DReAMM.

1. *Create the first synaptic vesicle*: New objects always appear at the position of the 3D cursor. Make sure that the cursor is now positioned at the origin by hitting *Shift-c*. If you are continuing from the previous section, hit *Tab* to enter "Object Mode". Hit the

spacebar, mouse over "Add", then "Mesh", and click on "UVSphere". In the "Add UV Sphere" menu, change "Segments" to "12", "Rings" to "8", "Radius" to "0.02", and then click "OK". This creates a small sphere centered at the origin.

2. *Rotate the vesicle into a more convenient orientation*: Hit *r*, then *x*, type *90*, and hit *Enter* to rotate the sphere 90° around the *X*-axis.
3. *Define the region for calcium binding sites*: Under "Link and Materials", click "New" (the rightmost option) to assign material properties to the whole vesicle. Default settings and a default name will be used automatically. Next hit *a* to deselect everything and then right click on the bottommost vertex of the vesicle to select only that vertex. Hit *Ctrl-Plus* (*on the number pad*) two times to select two additional concentric rings of vertices. Again click "New" under "Links and Materials" to create a second material. The words "2 Mat 2" appear in a box directly above. Click in the gray square to the left of these words, and a color selector will appear. The color selector includes a horizontal color bar and a larger saturation-value gradient box. Click in the upper right-hand corner of the gradient box so that a reddish shade is selected. Now click "Assign" to apply this red material to the selected faces.
4. *Name the new region*: Hit *F5* to change to the "Shading" section of the "Buttons Window", and, under "Links and Pipeline" in the text box titled "Link to Object", change the field that reads "MA:Material.002" (*see* **Note 13**) to "MA:CaBS_Reg" for the calcium binding sites region. Hit *F9* to return to the "Editing" section of the "Buttons Window".
5. *Move the vesicle*: Hit *Tab* to change into "Object Mode". Hit *g* to grab the entire vesicle, then *x*, type *–0.108*, and hit *Enter* to move the vesicle –0.108 units along the *X*-axis. Hit *g*, then *z*, type *0.105*, and hit *Enter* to move it 0.105 units along the *Z*-axis.
6. *Duplicate the vesicle and move the copy*: Hit *Shift-d* to duplicate the vesicle. Hit *g*, then *x*, type *0.216*, and hit *Enter* to move the copy 0.216 units along the *X*-axis away from the original.
7. *Rotate the vesicles into their final positions*: We want each vesicle situated above a single quadrangular face (quad) of the presynaptic mesh, so that the two quads can be defined as a region for VGCCs (*see* **Subheading 3.1.5** later). Thus, we now rotate the vesicles around the *Z*-axis. With the vesicle copy still selected, hold *Shift* and right click on the other vesicle to select it. Hit *r*, then *z*, type *11.25*, and hit *Enter* to confirm.
8. *Name the two vesicles*: Select the left vesicle by right clicking on it. To name it, look under "Link and Materials", click in the text box that reads "OB:Sphere", and change it to "OB:Vesicle1". To name the vesicle on the right, right click

on it and change the text box from "OB:Sphere.001" (*see* **Note 13**) to "OB:Vesicle2".

9. *Save the current mesh objects*: Hit *F2* to save the current meshes (*see* **Note 11**).

3.1.5. Define Region for VGCCs

We now define a presynaptic membrane region in which VGCCs will be located. This region will include two noncontiguous areas that underlie the synaptic vesicles. To expedite this operation, we will temporarily "clip" the existing meshes, i.e., make portions of them invisible. This will make it easier to see the particular remaining mesh faces that we want to include in the new regions.

1. *Clip the existing mesh objects*: Select the mesh for the presynaptic bouton by right clicking on it, and then hit *Tab* to change into "Edit Mode". Hit *Alt-b* to bring up cross-hairs for defining a clipping box. To define the box, click and drag a rectangle around both vesicles and the vertices on the presynaptic membrane directly beneath them. The rectangle will project through the plane of the screen to create a clipping box. Everything outside of the box will be invisible until *Alt-b* is hit again.
2. *Change the view*: Hit *7* (*on the number pad*) to switch to the *XY*-view (overhead). You should now be looking down on the synaptic vesicles above the presynaptic membrane. Hit *Ctrl-Tab* and select "Faces" in the "Select Mode" box that appears. Hit *z* to make the mesh opaque, so that subsequent color changes for regions will be easily visible.
3. *Define the region for VGCCs*: Click "New" under "Link and Materials" to assign a default material to the entire object. Then right click on the face directly below one of the vesicles (**Fig. 4f**; the face will extend beyond the vesicle's diameter). While holding *Shift*, right click on the corresponding face under the other vesicle (as noted earlier, the faces in a region do not need to be contiguous). Again, click "New" under "Link and Materials". Click in the gray box beside the words "2 Mat 2". Click in the green area of the horizontal color bar that appears, and then click near the upper right hand corner of the saturation-value gradient box above it. Now click "Assign", and the faces under the vesicles will change to a green color.
4. *Name the region*: Hit *Tab* to change into "Object Mode". Hit *F5* to change to the "Shading" section of the "Buttons Window" and, under "Links and Pipeline" in the text box titled "Link to Object", change the field that reads "MA:Material.003" (*see* **Note 13**) to "MA:VGCC_Reg". Hit *F9* to return to the "Editing" section of the "Buttons Window".
5. *Cancel the clipping box*: Hit *Alt-b*.

6. *Save the current mesh objects*: Hit *F2* to save the current meshes (*see* **Note 11**).

3.1.6. Define a Region for Postsynaptic Receptors

Similar to **Subheading 3.1.5**, we now define the region on the spine head (postsynaptic) membrane that will hold the ligand-gated neurotransmitter receptors (**Fig. 4G**).

1. *Reset view*: Hit *1* (*on the number pad*) to switch back to the *XZ*-view. If you are continuing from the previous section, the presynaptic bouton is still selected. Hit *h* to hide it.
2. *Define the postsynaptic receptor region*: Right click on the spine head mesh to select it and then hit *Tab* to switch to "Edit Mode". Then click "New" under "Link and Materials" to apply a default material to the entire mesh. Select the top row of faces by hitting *b* and then clicking and dragging a rectangle around them. Then hit *Ctrl-Minus* (*on the number pad*) three times to perform three "Select Less" operations, and thus deselect the three outermost rings of vertices. Click "New" under "Link and Materials". Click the gray color selector box and choose a blue shade. Click "Assign" to apply this material to the selected faces.
3. *Name the region*: Hit *Tab* to change into "Object Mode". Hit *F5* to change to the "Shading" section of the "Buttons Window" and, under "Links and Pipeline" in the text box titled "Link to Object", change the field that reads "MA:Material.004" (*see* **Note 13**) to "MA:Receptor_Reg". Hit *F9* to return to the "Editing" section of the "Buttons Window".
4. *Save the current mesh objects*: Hit *F2* to save the current meshes (*see* **Note 11**).

3.1.7. Add Multiple Spines to a Dendritic Shaft

In this section we create a dendritic shaft from a cylinder and then join it to the existing dendritic spine mesh. We then create replicated spines at different positions along the dendrite, and outline ways to change the dimensions of the spines to make a biologically realistic model similar to that shown in **Fig. 1**. Note that the presynaptic bouton mesh remains hidden throughout these operations since it will not be replicated. Thus, for the sake of subsequent illustrative simulations, we will end up with a single synapse onto one spine head, although the model dendrite will include multiple spines similar to those illustrated in **Fig. 1**.

1. *Create the dendritic shaft:* Hit *1* (*on the number pad*) for the *XZ*-view. Hit *Tab* to go into "Object Mode". Hit *Shift-c* to center the cursor at the origin. Hit the *spacebar*, mouse over "Add", then "Mesh", and click on "Cylinder". In the "Add Cylinder" menu that appears, set "Vertices" to "16", "Radius" to "1.00", "Depth" to "1.00", make sure that "Cap Ends" is deselected, and then click "OK". Hit *r*, type *11.25*, and hit *Enter*. This final rotation step around the axis of the cylinder

will align one of the cylinder faces with the base of the spine object.

2. *Assign material properties to the dendritic shaft*: In **step 8** later, the shaft will be joined to the dendritic spines, but we want the shaft to be a separate mesh region rather than an extension of the spine region. Thus, we need to apply new material properties to the shaft. Simply click "New" under "Link and Materials" to apply a new set of default material properties.
3. *Subdivide the faces along the length of the shaft*: For subsequent joining of spines to the shaft, it is preferable to subdivide the cylinder faces along their length. Hit *7* (*on the number pad*) for the *XY*-view. Click "Beauty" under "Mesh Tools". Hit *w* and, in the "Specials" menu, select "Subdivide Multi". In the "Number of Cuts" menu, select "2" and hit "OK". Each of the faces that made up the sides of the cylinder should now have been subdivided into three faces. Hit *1* (*on the number pad*) to change back to the *XZ*-view.
4. *Move the shaft to the end of the dendritic spine*: Hit *Tab* to go into "Object Mode". Hit *g*, then *z*, type *–3.0*, and hit *Enter* to confirm.
5. *Create multiple spines*: Hit *Shift-s* and click "Cursor –> Selection" in the "Snap" pop-up menu. Right click on the "Spine-Head" object to select it (from the current view the cylinder may seem to have disappeared, but it is still present). Hit *Tab* to go into "Edit Mode". Hit *a* and verify that the entire object is selected. Under "Mesh Tools" in the "Buttons Window", set "Degr" to "270.00", "Steps" to "3", and "Turns" to "1". Directly above these text boxes, click "Spin Dup". This will create three copies of the original spine head and place them at 90° increments around the cursor, which lies on the axis of the cylinder.
6. *Join the spines to the shaft*: Hit *Tab* to go into "Object Mode". All of the spines will be selected automatically following the preceding step. Hold *Shift* and right click on the shaft (cylinder) object to add it to the selected set of objects. Hit *w* and click "Union" in the "Boolean Tools" pop-up menu. This will perform a Boolean operation and merge the spine objects with the shaft object to create a new object (*see* **Note 14**). However, at this stage the original objects are still present and are still selected. Instead of deleting them, move them to another layer in case any changes are necessary later. This is done by hitting *m* (move), followed by *2* (*not on the number pad*), and clicking "OK". The original objects have now been moved to layer 2.
7. *Name the object created by the union operation*: Right click on the new merged object to select it. To name it, look under "Link and Materials", click in the text box that reads "OB:Tube.001", and change it to "OB:Dendrite".

8. *Replicate, extend, and merge the entire object*: Hit *Ctrl-a* and select "Apply scale and rotation" (*see* **Note 15**). Under the "Modifiers" tab in the "Buttons Window", click "Add Modifier" and select "Array". Change "Count" to "3". Change the "Relative Offset" of "*X*" to "0.000" and "*Y*" to "1.000". Click the "Merge" button. This operation will replicate the object twice (for a total count of 3), move the replicates along the *Y*-axis to the end of the previous piece, and then merge all of the pieces together. To see the result, change back to the *XY*-view [hit *7* (*on the number pad*)] and note how the shaft extends along the *Y*-axis.
9. *Generate a preliminary rotation of the spines*: Hit *1* (*on the number pad*) to change back to the *XZ*-view. Hit the *space-bar*, mouse over "Add", and select "Empty". This will create and select an empty object to which we will add a rotational transformation. First, hit *Alt-r* and click "Clear Rotation" to negate any preexisting rotations applied because the object was created using a certain view (*XZ*-view in this case). Next, hit *r*, type *45.0*, and hit *Enter* to rotate the empty object 45° around the current view axis. Now, right click on the dendritic shaft to select it. In the "Modifiers" tab, click "Object Offset" and then type "Empty" in the text box below. This will add the rotation of the "Empty" object to the dendritic shaft segments created in the preceding step. The amount of rotation is the product of the segment index and the specified 45°. The segment indices are (0, 1, 2), and so the first segment is not rotated, the second is rotated by 45°, and the third is rotated by 90° (and thus is realigned with the first segment).
10. *Finalize the rotation of the spines (array extensions)*: At this point we can view the preliminary rotation of the array extensions, but the rotation has not yet been finalized. Until it is finalized, we cannot make changes to the individual faces and vertices when in "Edit Mode". To finalize the rotation, click "Apply" in the "Array" modifier.
11. *Apply optional changes to the spine dimensions*: At this stage, each of the spine necks can be lengthened or shortened, and the spine head dimensions can be modified as well. For example, to lengthen a particular spine, hit *Tab* to go into "Edit Mode". Hit *b* and drag the select marquee around the vertices of the spine head. Then, along the bottom of the "3D View" pane, change the drop-down "Orientation" button from "Global" to "Normal". Hit *g*, then *z twice* (*see* **Note 16**), and then type in a positive value to lengthen the spine or a negative value to shorten it. Hit *Enter* to apply the change. In a similar fashion, individual spine heads can be scaled by selecting them, hitting *s*, and then entering a

value between 0 and 1 (shrink) or greater than 1 (expand). Finally, similar operations could be used to shrink or expand the diameter of the spine neck, or move the presynaptic bouton together with the postsynaptic spine head.

3.1.8. Smooth the Meshes

Sharp angles between mesh faces can cause inaccuracies and/or instabilities for many computational algorithms (e.g., computing a gradient on the mesh), and hence it is often desirable or necessary to smooth or otherwise "optimize" the mesh. In effect this amounts to low-pass filtering of the mesh to remove sudden (high-frequency) changes in shape (curvature). Smoothing is considerably less important for MCell simulations, because diffusing molecules in MCell move as discrete particles between meshes (volume molecules in solution) or on meshes (surface molecules in membranes), and the mesh *per se* is not used for gradient or other calculations. Nevertheless, here we illustrate smoothing in Blender to achieve a more "biological" appearance of the geometry (compare the objects in **Fig. 4** to the smoothed version in **Fig. 1**).

1. *Smooth the new mesh object that includes the dendrite and spines*: If you are continuing from the previous section, hit *Tab* and make sure you are in "Edit Mode". If necessary, hit *a* to select the entire new object that includes the dendritic shaft and all of the spines (the presynaptic bouton is still hidden). In the "Buttons Window", under "Mesh Tools", click "Smooth" five times. This will iteratively soften sharp edges between faces by moving edge vertices, but will not change the total number of vertices and faces.
2. *Unhide and select the presynaptic bouton mesh*: Hit *Tab* to go into "Object Mode". Hit *Alt-h* to make the "PresynapticBouton" reappear. Right click on the "PresynapticBouton" to select it.
3. *Smooth the presynaptic bouton mesh*: Hit *Tab* again to go into "Edit Mode". If necessary, hit *a* to select all of the presynaptic bouton mesh and, under "Mesh Tools", click "Smooth" five times. Hit *Tab* to go back into "Object Mode".
4. *Save the current mesh objects*: Hit *F2* to save the current meshes (*see* **Note 11**).

3.1.9. Export MDL Files for MCell Simulations

To complete this section we export the current mesh objects in MDL format for use in MCell simulations of presynaptic calcium influx and binding (**Subheading 3.3**). Export of MDL files utilizes the MDL plug-in for Blender that was installed in **Subheading 2.3**.

1. Click "File -> Export MCell (.mdl)". In the top text box that appears, navigate to the desired directory (folder) in which to

store the new files (you can create a new directory first if necessary). In the text box for the filename enter "Synapse.mdl".

2. Click "Export MDL" and click "OK". This will create five new MDL files in your specified directory. "Synapse.mdl" is the main file and will be read when the MCell simulation is started. It contains MDL statements that specify initial values for some important simulation parameters (*see* **Note 17**). It also includes statements for default visualization output for use with DReAMM (**Subheading 3.4**), and lists the remaining four files that also must be read (included) when the simulation is initialized. These remaining files are "Synapse_PresynapticBouton.mdl", "Synapse_Dendrite.mdl", "Synapse_Vesicle1.mdl", and "Synapse_Vesicle2.mdl", and they all contain mesh geometry and region information (*see* **Note 18**). Note that the names of the files correspond to the names of the mesh objects defined in the preceding sections.

3.2. Adding an Expanding Synaptic Vesicle Fusion Pore to the Model

In **Subheading 3.5** we will illustrate MCell simulations of neurotransmitter release and binding to postsynaptic receptors. The neurotransmitter molecules will diffuse through expanding fusion pores that connect the synaptic vesicles to the presynaptic membrane, so here we show how Blender can be used to create and scale the expanding pore structures. In brief, we create the pores with their initial and final dimensions (radius) and then morph between those limits to generate a set of intermediate pore structures. In a more realistic diffusion simulation project there might be several hundred intermediate structures, but in this simple example we will use only ten configurations including the initial and final. Each configuration will be written to a set of MDL files separate from those exported in the preceding section, and then in **Subheading 3.5** MCell will be used to read the succession of new mesh files using a feature called checkpointing (*see* **Note 19**).

1. *Join the meshes of both vesicles*: Hit *1* (*on the number pad*) for the *XZ*-view. Pan and zoom in on the presynaptic bouton so that the vesicles are clearly visible. Hit *z* to make the faces transparent. Now the vesicles should be visible inside the presynaptic bouton. Right click on the left vesicle to select it, and then, while holding *Shift*, right click on the other vesicle to select it as well. Next hit *Ctrl-j* and click "Join Selected Meshes" when prompted by the "OK?" dialog box. Hit *Tab* to go into "Edit Mode".
2. *Remove the bottom faces of each vesicle*: By removing the bottommost faces of each vesicle, we will create holes that can be extruded to create cylindrical pores, similar to the way that the axon and spine neck extensions were extruded in **Subheading 3.1.3**. Hit *Ctrl-Tab* and click "Vertices" in the "Select Mode" menu. Hit *a* to deselect everything. Right click on the

bottommost vertex of the left vesicle and, while holding *Shift*, right click the bottommost vertex of the right vesicle (use zoom and pan if necessary). Hit *Ctrl-Plus* (*on the number pad*) once to select one ring of vertices. Hit *x* and click "Faces" in the "Erase" menu.

3. *Extrude the pores*: Hit *b* and then click and drag the select marquee around the vertices remaining at the bottom of the left vesicle. Then hit *b* again and repeat for the right vesicle. Hit *e*, click "Only Edges", then hit *z*, type *-0.03*, and hit *Enter*. You should now see the extruded pores passing through the presynaptic bouton (**Fig. 4H**, left). The diameter of these temporary pores is approximately 0.015 units (*see* **Note 20**).
4. *Merge the fusion pores with the presynaptic membrane*: Hit *Tab* to go into "Object Mode". The vesicles should already be selected, so, while holding *Shift*, right click on the surrounding presynaptic bouton mesh. Now hit *w* and click "Difference" from the "Boolean Tools" menu. The vesicles, pores, and presynaptic bouton membrane should now form a continuous mesh (*see* **Note 14** and **Fig. 4H**, right).
5. *Save the original meshes*: The original objects are still present after performing the Boolean difference operation in the preceding step. Rather than delete them, move them to another layer for later use if any changes are required. Hit *m*, followed by *2* (*not the number pad*), and then click "OK".
6. *Name the object created by the difference operation*: Right click on the new object to select it. To name it, look under "Link and Materials", click in the text box that reads "OB:PresynapticBouton.001", and change it to "OB:PresynapticBouton".
7. *Scale the pores to their desired initial diameter*: The desired initial diameter for the pores is about 13.3% of the current diameter (compare **Fig. 4I**, left, with **Fig. 4H**). In the subsequent MCell simulation, this initial diameter will correspond to about 2 nm. Hit *Tab* to enter "Edit Mode". Hit *Ctrl-Tab* and click "Faces", so that we can now select faces rather than vertices as in preceding sections. Hit *b* and then click and drag the rectangular select marquee over the middle of a pore, but do not include the upper and lower vertices of the pore. This will select only the faces of the pore. Hit *7* (*on the number pad*) to change to the *XY*-view (overhead). Hit *s* and then *Shift-z* to simultaneously scale along the *X*- and *Y*-axes. Type *0.133* and hit *Enter*. Hit *1* (*on the number pad*) to change back to the *XZ*-view. Hit *a* to deselect. Repeat the selecting and scaling steps for the second pore.
8. *Take a snapshot of the initial pore configurations*: This snapshot subsequently will be used in **step 9** later when we morph the pores between their initial and final configurations. Hit

Tab to enter "Object Mode". Select the presynaptic mesh by right clicking on it. On the "Shapes" panel in the "Buttons Window", click "Add Shape Key". We have now defined a "Basis Key" that corresponds to the initial pore configuration. All subsequently defined "Shape Keys" will be relative to the "Basis Key".

9. *Take another snapshot to be modified for the final pore configuration*: Click "Add Shape Key" again. Now we will rescale the pore dimensions to their final diameter (**Fig. 4I**, right; about 10 nm in the subsequent MCell simulation). Hit *Tab* to enter "Edit Mode". Hit *b* and then click and drag the rectangular select marquee over the middle of a pore as in earlier **step 6**. Hit *7* (*on the number pad*) to change to the *XY*-view (overhead). Hit *s*, then *Shift-z*, type *5.0*, and hit *Enter*. Hit *1* (*on the number pad*) to change back to the *XZ*-view. Repeat for the second pore. At this point the first snapshot contains the initial pore configuration, and the second snapshot contains the final pore configuration.
10. *Interpolate between the snapshots*: Hit *Tab* to enter "Object Mode". Click on the icon in the upper left-hand corner of the "Buttons Window" (**Fig. 2D**) and select the "Action Editor" from the drop-down list. Click on the arrow beside the word "Sliders" near the bottom of the window, and sliders for the list of available "Shape Keys" will appear. In this simple case, only "Key 1" is present. Now we must map the final pore configuration (Key 1) to the endpoint of a timeline, and the "Basis Key" to the beginning of the timeline. Identify the vertical green line to the right of the "Shape Key" slider. Click on the line, drag it to the right, and then release it at the point marked 10 (requesting ten snapshots). Now move the slider next to "Key 1" from "0.00" to "1.00", and a diamond-shaped marker will appear at position 10 on the timeline. Next drag the green line back to 1 and then drag the slider from "1.00" back to "0.00". A diamond marker will now appear at position 1 on the timeline. Hit *Shift-Alt-a* to see an animation of the interpolated pore configurations. Hit *Esc* to stop playback.
11. *Save the current mesh objects*: Hit *F2* to save the current meshes (*see* **Note 11**).
12. *Save the interpolated mesh snapshots as MDL files*: We will now export MDL files for use with MCell in **Subheading 3.5**. As in earlier **Subheading 3.1.9**, click "File -> Export -> MCell (.mdl)". Navigate to the desired directory (folder) in which to store the new files (use a different directory from that used in **Subheading 3.1.9**). In the text box for the filename enter "VesicleFusion.mdl". Click "Export MDL", then "Enable Anim.", and "Iterate Script". Also change "Stop" to "10". Once these changes have been made, click "OK" (*see* **Note 21**).

13. *Save the current mesh objects and quit Blender*: Hit *F2* to save the current meshes (*see* **Note 11**) and quit Blender by clicking on the X in the upper right-hand corner of the window.

3.3. MCell Simulations of Presynaptic Calcium Influx and Binding

In **Subheading 3.1** we used Blender to create a set of pre- and postsynaptic meshes and then exported the meshes as MDL files for use with MCell. As outlined in **Subheading 3.1.9**, five files were created: the main file "Synapse.mdl" and the four geometry files "Synapse_PresynapticBouton.mdl", "Synapse_Dendrite.mdl", "Synapse_Vesicle1.mdl", and "Synapse_Vesicle2.mdl". In Blender the meshes were composed of quadrangular faces (**Fig. 4**), and the absolute spatial dimensions were arbitrary. In MCell, however, the spatial units will be interpreted as microns. In addition, MCell's collision detection algorithms require triangular faces, so each quadrilateral face in Blender was automatically split into two triangles when the meshes were exported. We will now supplement the exported MDL files in order to populate the meshes with molecules, define reactions between molecules, and provide commands that control how the MCell simulations will be run. For convenience we will use separate MDL files for many of these distinct operations. The final simulations then will be controlled from the main file that reads or "includes" all the subordinate files in the proper order when the simulation is initialized.

3.3.1. Define Molecules

We first create a new MDL file to describe the molecules included in the model. It will specify whether they exist in solution (a "volume" molecule) or on a surface (a "surface" molecule), and their diffusion coefficients.

1. At a command line, change into the directory where you exported the MDL files in **Subheading 3.1.9**.
2. Create a new file called "Molecules.mdl" (*see* **Note 22**).
3. Define the molecules: Enter the following block of text (*see* **Notes 23** and **24**):

```
DEFINE_MOLECULES {
Ca {DIFFUSION_CONSTANT_3D = 1E-6}
VGCC_C {DIFFUSION_CONSTANT_2D = 0}
VGCC_O {DIFFUSION_CONSTANT_2D = 0}
CaBS {DIFFUSION_CONSTANT_2D = 0}
CaBS_Ca {DIFFUSION_CONSTANT_2D = 0}
}
```

This simple model includes only diffusing calcium ions, VGCCs, and calcium binding sites that might, for example, be based on synaptotagmin molecules located on the synaptic vesicles. In MDL statements like those above, the names of the molecules are specified by the user and thus are usually chosen to be easily recognizable. The only (obvious) restriction is that a user-specified name may not be an exact match of an MDL keyword. In this example, the calcium ions are simply named "Ca", and since they are to be diffusing volume molecules they are given a nonzero 3D diffusion coefficient (cm^2/sec). We require both a closed and open state for the VGCCs, named "VGCC_C" and "VGCC_O", respectively. The channels will be stationary surface molecules and so are given a 2D diffusion coefficient with a value of 0. Finally, we require unbound and bound states for the calcium binding sites, named "CaBS" and "CaBS_Ca", respectively. Similar to the channels, the binding sites will be considered part of static surface molecules, and hence are given a 2D diffusion coefficient with a value of 0.

4. Save the file and quit.

3.3.2. Add Molecules to Mesh Regions

The only molecules that are to be present when the simulation begins are the closed VGCCs ("VGCC_C") and the unbound calcium binding sites ("CaBS"). The remaining molecules or states will be generated by reactions during the simulation. We add ten "VGCC_C" molecules to the presynaptic mesh region "VGCC_Reg" that was defined in **Subheading 3.1.5** (**Fig. 4F**). Similarly, we add ten "CaBS" molecules to the "CaBS_Reg" region defined on the synaptic vesicles in **Subheading 3.1.4**. The actual locations of the molecules within these regions will be randomized by MCell when the simulation is initialized (*see* **Note 25**).

1. Create a new file called "RegionModifications.mdl" (*see* **Note 22**).
2. Define the numbers and locations of molecules present at simulation start-up (*see* **Note 26**):

```
MODIFY_SURFACE_REGIONS {
PresynapticBouton[VGCC_Reg] {
MOLECULE_NUMBER { VGCC_C, = 10 }
}
Vesicle1[CaBS_Reg] {
MOLECULE_NUMBER { CaBS' = 10 }
}
Vesicle2[CaBS_Reg] {
MOLECULE_NUMBER { CaBS' = 10 }
}
}
```

These MDL statements modify (add molecules to) the preexisting mesh regions defined automatically when the "Synapse_PresynapticBouton.mdl", "Synapse_Vesicle1.mdl", and "Synapse_Vesicle2.mdl" files were exported from Blender. The comma or apostrophe that follows the molecule's name specifies how the molecule is oriented when it is added to the surface (a surface molecule may have a reactive domain, e.g., a binding site, on the front and/or back of the surface; *see* **Note 17**).

3. Save the file and quit.

3.3.3. Add Reactions

1. Create a new file called "Reactions.mdl" (*see* **Note 22**).
2. Define the reactions: Enter the following block of text:

```
DEFINE_REACTIONS {
VGCC_C' -> VGCC_O' [5E5]
VGCC_O' -> VGCC_C' [500]
VGCC_O' -> VGCC_O' + Ca' [1E3]
Ca' + CaBS' -> CaBS_Ca' [1E7]
CaBS_Ca' -> Ca' + CaBS' [500]
}
```

These MDL statements specify the stoichiometry, rates, and directionality for the reactions in the simulation. In the first line, closed VGCCs are able to undergo a conformational change to the open state with a first-order mass action rate constant of 5E5 s^{-1}. The reverse transition occurs in the second line, albeit at a much slower rate. A channel in the open state is also able to generate diffusing calcium ions in the presynaptic bouton (third line). Hence, in any given simulation time step, an open channel may close, generate one or more calcium ions, or simply remain open, all based on relative probabilities. This method for generating diffusing calcium ions from open channels is far more efficient than explicitly simulating separate pools of extracellular and intracellular calcium ions that pass through the open channel. In the fourth line, calcium ions bind to the calcium binding sites with a bimolecular mass action rate constant of 1E7$M^{-1}s^{-1}$. The last line specifies calcium unbinding with a first-order rate constant of 500 s^{-1}.

In all of these reactions, the apostrophes again specify the directionality of the reactions with respect to the orientation of the surface molecules. This is why calcium ions produced by an open channel enter the presynaptic bouton rather than the synaptic cleft space (*see* earlier **Subheading 3.3.2**, **step 2** and **Note 17**).

3. Save the file and quit.

3.3.4. Specify Reaction Data Output

MCell is able to count and save many different types of events during a simulation. In this simple example, we will just count the number of molecules present during each time step throughout the entire simulation space (world). The results for each molecule will be written to a separate ASCII file containing two columns. The first gives the simulation time in seconds, and the second gives the counted quantity.

1. Create a new file called "ReactionData.mdl" (*see* **Note 22**).
2. *Specify the desired reaction data output*: Enter the following block of text:

```
REACTION_DATA_OUTPUT {
{COUNT[VGCC_C, WORLD]} => "./reaction_data/VGCC_C.dat"
{COUNT[VGCC_O, WORLD]} => "./reaction_data/VGCC_O.dat"
{COUNT[CaBS, WORLD]} => "./reaction_data/CaBS.dat"
{COUNT[CaBS_Ca, WORLD]} => "./reaction_data/CaBS_Ca.dat"
{COUNT[Ca, WORLD]} => "./reaction_data/Ca.dat"
}
```

In this case, each file (.dat suffix) will be created automatically in a subdirectory called "reaction_data".

3. Save the file and quit.

3.3.5. Add Spatial Partitions to Speed Computation

During a simulation, diffusing molecules follow random walk steps that must be traced to detect possible collisions with surfaces and other molecules. This becomes very time-consuming unless each molecule looks only in its local environment first, and continues into adjoining space only if necessary. To define the local environments, the simulation world is partitioned into subvolumes. In effect, the partitions are transparent planes along the *X*-, *Y*-, and *Z*-axes, and the subvolumes are the cuboidal spaces created between the partitions.

1. Create a file called "Partitions.mdl" (*see* **Note 22**).
2. *Specify the locations of partitions along each axis*: Enter the following block of text:

```
PARTITION_X = [[-1.25 TO 1.25 STEP 0.1]]
PARTITION_Y = [[-1.25 TO 1.25 STEP 0.1]]
PARTITION_Z = [[0 TO 1 STEP 0.1]]
```

3. Save the file and quit.

3.3.6. Add the Include Files and Set the Number of Iterations

We now add all the pieces together by referencing ("including") the newly created MDL files in the main simulation file "Synapse.mdl". Thus, when the simulation is started using the main file, all of the subordinate MDL files will be read in the proper order.

1. Open the main file "Synapse.mdl" in a text editor (*see* **Note 22**).
2. *Reference the subordinate MDL files using INCLUDE statements*: Before the first preexisting INCLUDE statement add the following text:

```
INCLUDE_FILE = "Partitions.mdl"
INCLUDE_FILE = "Molecules.mdl"
INCLUDE_FILE = "Reactions.mdl"
```

Now, after the last preexisting INCLUDE statement add:

```
INCLUDE_FILE = "RegionModifications.mdl"
INCLUDE_FILE = "ReactionData.mdl"
```

3. *Change the iteration number*: By default, the simulation is set to run for only one iteration with a default time step of one microsecond. This will generate visualization output for the start-up conditions, and thus would allow verification of initial mesh and molecule placement using DReAMM (*see* **Subheading 3.4** later). This is an extremely useful step for large models that take a long time to simulate. In this case, however, we will increase the number of iterations so that we can see the appearance of diffusing calcium ions and occupation of calcium binding sites at later times. Change the first line of the file from:

   ```
   iterations = 1
   ```
   to
   ```
   iterations = 5000
   ```
4. Save the file and quit.

3.3.7. Run the Simulation

Assuming that you have installed MCell with the name "mcell3", run the simulation simply by entering:

mcell3 Synapse.mdl

at the command line in the directory where you created the MDL files. MCell will start and display initialization messages, display updates as iterations complete, and then display a variety of run-time summary statistics when the simulation is finished.

3.4. Visualize MCell Results with DReAMM (Part 1)

It is crucial to check any MCell model visually using DReAMM to verify that all components (location of molecules, reactions, geometry, etc.) have been set up properly (*see* **Note 27**). Here, we outline the essential steps for the model created in the preceding section.

3.4.1. Import the MCell Visualization Data and Select Mesh and Molecule Objects

1. *Start DReAMM*: Enter "dreamm" at a command line. The "DReAMM Image Window", "Quick Controls", and "Sequence Control" should all appear. In principle you can start DReAMM from within any directory and then navigate to the visualization files output by MCell. To simplify this example, however, start DReAMM from the same directory in which you ran the MCell simulation in **Subheading 3.3**.
2. *Import visualization data*: Click the "Import & Select" button near the top of the "Quick Controls" window (labeled

1 in **Fig. 3a**). Two menus (windows) should open. In the "Import & Select Objects" menu, click the ellipsis ("...") in the "Viz Data File" text box. Navigate to the "Synapse_viz_data" directory that was created by MCell and select the file named "Synapse.dx". Click "OK". DReAMM will automatically read the visualization files referenced by "Synapse.dx", display the names of available objects, and import the data for the first time step. By default, however, DReAMM will not display any data until it is selected.

3. *Select all meshes to be displayed*: The lower left-hand side of the "Imported Objects" menu will now display the names of the meshes we created previously in Blender and used in MCell: "World.Dendrite", "World.PresynapticBouton", etc. These objects thus are available to be rendered (displayed) and for other operations. In the lower center section locate the field named "Choose Operation", click on "Add All", and then click the "Apply Operation" button to select all of the meshes. All of the object names should now appear in the "Current Objects" list to the right, and all of the mesh objects now will be displayed in the "DReAMM Image Window" using DReAMM's default mesh colors and "Software Rendering Mode" (*see* **Note 28**).
4. *Select all volume molecules*: Click "Volume Molecules" in the central "List" field of the "Import and Select" menu, and the "Imported Objects" list will change to show the volume molecules in the model. In this case only "Ca" is present. Hit the "Apply Operation" button, and "Ca" will appear in the list of "Current Objects". However, no calcium ions are yet visible in the "Image Window" because no calcium ions are present at the beginning of the simulation, and we are currently viewing the first time step.
5. *Select all surface molecules*: Click "Surface Molecules" under "List" and the "Imported Objects" list will change again, this time showing all surface molecules in the model (calcium channels and binding sites). Click "Apply Operation", and all the surface molecule names will appear in the list of "Current Objects". At this point the surface molecules present at the beginning of the simulation are being rendered using DReAMM's default molecule rendering properties (white pixels). This may be difficult to see if there are few molecules or they are sparsely distributed but is the least expensive display option. In the following section we will customize the display properties so that the molecules can be seen easily.
6. *Center the view*: Select the "DReAMM Image Window" and hit *Ctrl-f* to center the view of the displayed objects. To see the synapse from a side view, hit *Ctrl-v* to bring up the "View Control" menu and then select "Bottom" under "Set View" (*see* **Note 29**).

7. *Close menus*: Click the "Close" button to the right of the "Import & Select" field on the "Quick Controls" menu (labeled *5* in **Fig. 3A**).

3.4.2. Visualize the Calcium Binding Site Regions

1. *Open the rendering properties menus*: Click "Set Rendering Prop". under "Quick Controls" (labeled *2* in **Fig. 3A**). Four separate menus should appear.
2. *Turn on the preview window*: In the "Rendering Properties" menu, click on the "Enable Preview" button in the center and the "Rendering Preview" window will appear. It displays the current rendering properties that can be applied to selected objects, including color (separate front and back colors for mesh objects), lighting, and shading (*see* **Note 30**).
3. *Make the presynaptic bouton semitransparent*: By default, the "Rendering Properties" menu will display the names of the imported mesh objects, and the first will be highlighted. Click on "World.PresynapticBouton" to select it, and then, if necessary, click on any others that remain highlighted to deselect them. In the lower left-hand corner, change "Opacity" from "1.0" to "0.3" (*see* **Note 31**) and then click on the "Once" button next to "Apply Operation" in the center of the menu. In the "DReAMM Image Window", the mesh for the "PresynapticBouton" will now be semitransparent, revealing the vesicles within it. Furthermore, the change in opacity is also reflected in the "Rendering Preview" window.
4. *Visualize the mesh regions using a colormap*: In order to distinguish different mesh regions in DReAMM, we use a colormap to render and display the object. When a region is defined in Blender, each mesh face within the region is automatically assigned a unique numerical metadata tag (value). The tag values begin with 0 for the first region, and are incremented thereafter. Similarly, in MCell's MDL, triangles that belong to a particular region can be assigned a numerical VIZ_VALUE. Meshes exported from Blender to MCell automatically inherit region VIZ_VALUEs from the metadata tags assigned by Blender. DReAMM subsequently uses the VIZ_VALUEs and colormaps to visualize regions. The default colormap visible in the "Colormap Editor" menu uses a stair-step pattern ranging from purple to red for tag values within the indicated numerical limits. We will now change the upper limit so that the default colormap will work for our mesh regions. To do so, click in the upper box that displays a numerical value, type "1.4", and then hit *Enter* (*see* **Note 32**).
5. *Apply the colormap to the synaptic vesicles*: In the "Rendering Properties" menu, select "World.Vesicle1", then "World.Vesicle2", and finally deselect "World.PresynapticBouton". Change "Use Color and Opacity Map" from "No" to "Yes"

and “Color Dependence” from “Vertices” to “Elements” (*see* **Note 33**). Then click “Apply Operation Once”. The bottom of the vesicles, i.e., the region “CaBS_Reg”, should now be yellow because it was assigned a VIZ_VALUE of 1. The remainder of each vesicle is purple because it has a VIZ_VALUE of 0. You may need to zoom in to see this clearly (*see* **Note 29**).

3.4.3. Highlight Different Molecules and Visualize the MCell Time Series

By default, the visualization data for each simulation time step (frame) has been saved, and DReAMM will automatically read and display the selected data for each frame in sequence. During playback calcium ions will appear and diffuse, and calcium binding sites will become occupied. For maximum speed they will all be rendered as white pixels under default conditions, so in the later step we will change the rendering properties for each molecule so that they may be distinguished clearly. We will also change the color of the postsynaptic mesh (dendrite and spines).

1. *Play the time series data*: Simply press the Play button on the “Sequence Control” menu (right arrowhead, **Fig. 3B**) to begin playback of the MCell simulation time series. No visible changes will be evident until several hundred frames have been displayed, so skip ahead if desired by clicking on the “Sequence Control” button that displays the frame number. Then use the pop-up “Frame Control” menu to select a subset of the available frames, and/or change the interval between the displayed frames.
2. *Reset the opacity*: In the “Rendering Properties” menu, change “Opacity” back to “1.0”. This change will be visible in the “Rendering Preview” window but will not affect any objects until we assign properties to them.
3. *Choose a color from the Color Library*: In the “Color Library” or “Rendering Properties” menu, toggle the “Source” selector from “Rendering Properties” to “Color Library”. This will change the source of the colors displayed in the “Rendering Preview” window from the color fields of the “Rendering Properties” menu to the color selected in the “Color Library” menu. In the “Display List Filter” text box of the “Color Library” menu, change the search string from “*” to “*yellow*” and hit *Enter* (*see* **Note 34**). Select “lightyellow3” from the list and hit the “Load” button. The RGB (Red, Green, Blue) values for “lightyellow3” are now listed in the “Front Color” and “Back Color” (half intensity) fields of the “Rendering Properties” menu, and the corresponding hue is also visible in the “Colormap Editor”.
4. *Assign the new color to the dendrite*: In the “Rendering Properties” menu select “World.Dendrite”, deselect “World.Vesicle1” and “World.Vesicle2”, and then click “Apply Operation Once”. The dendrite mesh, which includes the spines and spine heads, will now be rendered with “lightyellow3”.

5. *Assign a yellow spherical glyph to the calcium ions*: In the "Rendering Properties" menu, click on "Molecules" in the upper middle section by the word "List". The list of object names will switch to all of the molecules (volume and surface) in the model. In the lower right-hand corner change "Glyph" from "pixel" to "sphere (simple)" and both "Height" and "Radius" to "0.0025". At the lower left manually change "(Front) Color" to yellow by entering RGB values of 1, 1, and 0, respectively ("Back Color" is ignored for glyphs). Make sure that only "Ca" is highlighted in the list of object names. Assign the yellow glyph properties to calcium ions by clicking "Apply Operation (Once)".
6. *Assign a black arrow glyph to the unbound calcium binding sites*: Surface molecules have an XYZ location that lies on a surface mesh element, and they also have an orientation with respect to the plane of the mesh element. Thus, to visualize surface molecules we will use a directional (asymmetric) glyph (*see* **Note 35**). First, in the "Rendering Properties" menu, select "CaBS" and deselect any other highlighted molecules. Change the glyph to "arrow (simple)". Change the "Height" to "0.01" and the "Radius" to "0.0025". To make the arrow glyphs black, change all of the "(Front) Color" RGB values to 0. Click "Apply Operation Once". Outward pointing arrows should now be visible at the position of the CaBS molecules within the CaBS_Reg of each synaptic vesicle (**Fig. 5A**). The precise number will depend on the time step that you are viewing. At the beginning of the simulation (frame 1) there are ten unbound calcium binding sites. Later, the number will change as calcium ions bind and unbind.
7. *Assign a cyan arrow glyph to the bound calcium binding sites*: In the "Rendering Properties" menu, select "CaBS_Ca" and deselect "CaBS". Change the front RGB values to 0, 1, and 1, respectively. Click "Apply Operation Once". The bound calcium binding sites will now be visible as cyan, outward pointing arrows. As outlined in the preceding step, the number will depend on the time step that you are viewing (**Fig. 5B**).
8. *Assign a red arrow glyph to the closed VGCCs*: In the "Rendering Properties" menu, select "VGCC_C" and deselect "CaBS_Ca". Change the front RGB values to 1, 0, 0, respectively. Click "Apply Operation Once". The VGCCs (in the presynaptic membrane beneath the synaptic vesicles) that currently are closed will now appear as red arrows pointing toward the interior of the presynaptic space (**Fig. 5A**).
9. *Assign a green arrow glyph to the open VGCCs*: Similar to the preceding step, select "VGCC_O" and deselect "VGCC_C". Change the front RGB values to 0, 1, 0, respectively, and click "Apply Operation Once". The VGCCs that currently are open will now appear as green arrows (**Fig. 5B**).

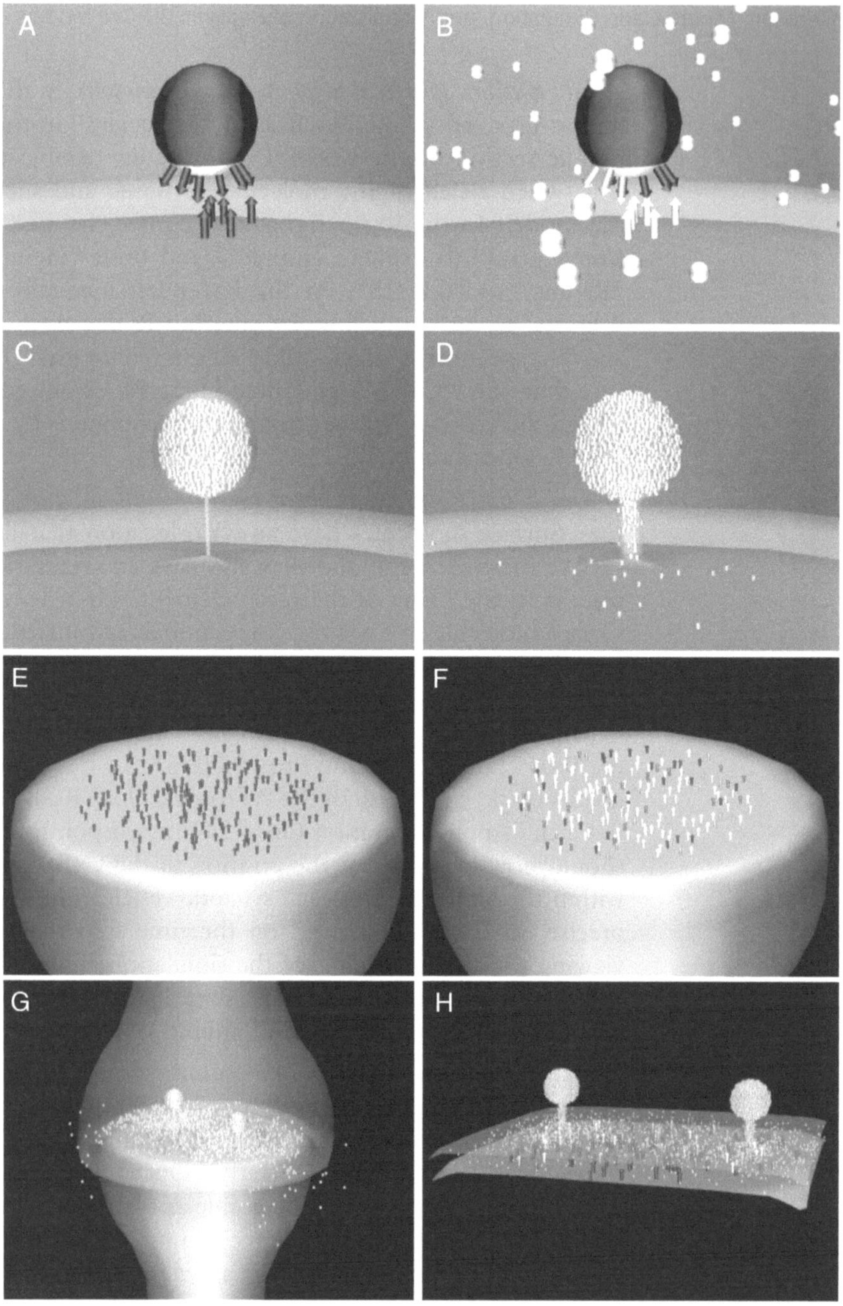

Fig. 5. Screen captures from simulations as visualized with DReAMM. (Note that addition of colors to objects is described in the text, whereas this image has been converted to grayscale and is described here accordingly). (**A**) At the beginning of the first simulation (**Subheading 3.3.7**), the bottom of the synaptic vesicle is populated with unbound calcium binding sites (*black downward pointing arrows*). Closed voltage-gated calcium channels (*black upward pointing arrows*) are located directly underneath the vesicle on the presynaptic membrane. (**B**) Later, calcium channels open (*white upward pointing arrows*) and release calcium ions (*white spheres*) into the presynaptic bouton. Some calcium ions then bind to available sites on the vesicle (*white downward pointing arrows*). (**C**) At the beginning of the second simulation (**Subheading 3.5.7**), neurotransmitter molecules (*white spheres*) fill the synaptic vesicles, and then diffuse out as the simulation proceeds and the fusion pore expands (**D**). For clarity, the postsynaptic receptors are not shown in **C** and **D** even though they are present in the simulation. (**E**) All of the postsynaptic receptors are in the unbound state (*black arrows*) at the start of the second simulation (presynaptic bouton not shown). (**F**) Later, a mixture of single-bound, double-bound, closed, and open receptors is present as indicated by different colors.

10. *Play the time series data*: Press the Stop button (square) on the "Sequence Control" menu (**Fig. 3B**) and then press the Play button. This will restart the time series from the first frame (of the selected interval, if you are using the "Frame Control" settings). Use the Pause (parallel lines), Single-step (double lines and arrowheads), Reverse (left arrowhead), Loop, and Palindrome (loop forward and backward) buttons to modify playback.

3.4.4. Save DReAMM Settings for the Simulation Objects

All DReAMM settings such as choice of rendered objects, rendering properties (including colormaps), etc., can be saved to a file for subsequent reuse.

1. *Save settings*: In the "Read/Write Settings" menu, click on the ellipsis ("...") button next to the "Write File" text box. In the pop-up menu, navigate to the directory where you would like to save the file, enter the name "SynapseCustom.dx", and then click "OK". Click the "Write Once" button in the "Read/Write Settings" menu to save the file (*see* **Note 36**).
2. *Quit DReAMM*: Quit DReAMM by clicking the X in the upper right-hand corner of the "DReAMM Image Window". When prompted, "Do you want to save the project file", click "No".

3.5. MCell Simulations of Fusion Pore Expansion and Neurotransmitter Release

We now adapt the MCell model of **Subheading 3.3** to include the expanding fusion pores (**Fig. 4I**) created with Blender in **Subheading 3.2**. We will add neurotransmitter molecules that originate within the vesicles and diffuse out through the expanding pores. We will also add postsynaptic receptors in the form of ligand-gated ion channels. As mentioned previously (**Subheading 3.2** and **Note 19**), MCell simulation of the expanding pores will use a feature called checkpointing. In brief, we will run a series of MCell simulations, saving the locations and states of all molecules after each run in the series. The first run will use the initial pore configuration and will proceed for a certain number of iterations, allowing neurotransmitter molecules to begin diffusing. The second run will use the next pore configuration but will use the molecule locations and states from the previous run as initial conditions. This pattern then will continue for all ten of the expanding pore configurations. In principle many other parameters can also change between checkpoint runs, and there are a variety of ways to automate setup of the files. Here, we use a simple example for the sake of illustration.

Fig. 5. (Continued) (**G**) The complete synapse is shown at a late time point, with the presynaptic bouton semitransparent and the spine head opaque. Neurotransmitter molecules can be seen diffusing within the synaptic cleft and escaping into the surrounding volume. (**H**) Image clipping is used to provide a better view of the vesicles, synaptic cleft, diffusing neurotransmitter molecules, and postsynaptic receptors.

3.5.1. Define Molecules

As in **Subheading 3.3**, we first create a new MDL file to describe the molecules included in the model. Before starting, make sure that you are in the directory created in **Subheading 3.2** when the MDL files for the expanding pore were exported from Blender.

1. Using a text editor, create a new file called "Molecules.mdl" (*see* **Note 22**).
2. *Define the molecules*: Enter the following block of text (*see* **Notes 17** and **24**):

```
DEFINE_MOLECULES {
nt {DIFFUSION_CONSTANT_3D = 1E-6}
nt_R_0B {DIFFUSION_CONSTANT_2D = 0}
nt_R_1Ba {DIFFUSION_CONSTANT_2D = 0}
nt_R_1Bb {DIFFUSION_CONSTANT_2D = 0}
nt_R_2B_C {DIFFUSION_CONSTANT_2D = 0}
nt_R_2B_O {DIFFUSION_CONSTANT_2D = 0}
}
```

In these MDL statements, we define a diffusing volume molecule ("nt", neurotransmitter) and five different states of a stationary neurotransmitter receptor (stationary surface molecules). The receptor represents a ligand-gated ion channel with two independent binding sites. "nt_R_0B" is the receptor in its unbound state; "nt_R_1Ba" and "nt_R_1Bb" are the two single-bound states; "nt_R_2B_C" is the double-bound, closed channel state, and "nt_R_2B_O" is the double-bound, open channel state.

3. Save the file and quit.

3.5.2. Add Molecules to Mesh Regions

We next add unbound receptors to the postsynaptic receptor region defined in **Subheading 3.1.6** (**Fig. 4G**). The actual locations of the molecules within the region will be randomized by MCell when the first simulation of the checkpoint sequence is initialized. Recall that this region extends to all of the spine heads and hence we add a total of 2,400 receptors distributed randomly across 12 spine heads.

1. Create a new file called "RegionModifications.mdl" (*see* **Note 22**).
2. Define the numbers and locations of molecules present at simulation start-up:

```
MODIFY_SURFACE_REGIONS {
Dendrite[Receptor_Reg] {
MOLECULE_NUMBER { nt_R_0B' = 2400}
}
}
```

3. Save the file and quit.

3.5.3. Add Reactions

1. Create a new file called "Reactions.mdl" (*see* **Note 22**).
2. *Define the reactions*: Enter the following block of text:

```
DEFINE_REACTIONS {
nt' + nt_R_0B' -> nt_R_1Ba' [1E7]
nt_R_1Ba' -> nt' + nt_R_0B' [1E4]
nt' + nt_R_0B' -> nt_R_1Bb' [1E7]
nt_R_1Bb' -> nt' + nt_R_0B' [1E4]
nt' + nt_R_1Ba' -> nt_R_2B_C' [1E7]
nt_R_2B_C' -> nt' + nt_R_1Ba' [1E4]
nt' + nt_R_1Bb' -> nt_R_2B_C' [1E7]
nt_R_2B_C' -> nt' + nt_R_1Bb' [1E4]
nt_R_2B_C' -> nt_R_2B_O' [1E4]
nt_R_2B_O' -> nt_R_2B_C' [1.5E3]
}
```

In the first two lines, a neurotransmitter molecule binds reversibly to the first binding site on the unbound receptor. In the next two lines, a neurotransmitter molecule binds to the second binding site on the unbound receptor. Next, a neurotransmitter molecule binds reversibly to the second site when the first site is already occupied, generating the double-bound closed channel state. Similarly, the double-bound closed channel state can also be generated when a neurotransmitter molecule binds to the first site when the second is already occupied. Finally (last two lines), the double-bound receptor can change conformations reversibly between the closed and open channel states. As outlined in **Subheading 3.3**, the apostrophes specify the directionality of the reactions with respect to the orientation of the surface molecules. In this case the neurotransmitter molecules are able to bind to receptor molecules from within the synaptic cleft space.

3. Save the file and quit.

3.5.4. Specify Reaction Data Output

1. Create a new file called "ReactionData.mdl" (*see* **Note 22**).
2. *Specify the desired reaction data output*: Enter the following block of text:

```
REACTION_DATA_OUTPUT {
  {COUNT[nt_R_0B, WORLD]} => "./reaction_data/nt_R_0B.dat"
  {COUNT[nt_R_1Ba, WORLD]} => "./reaction_data/nt_R_1Ba.dat"
  {COUNT[nt_R_1Bb, WORLD]} => "./reaction_data/nt_R_1Bb.dat"
  {COUNT[nt_R_2B_C, WORLD]} => "./reaction_data/nt_R_2B_C.dat"
  {COUNT[nt_R_2B_O, WORLD]} => "./reaction_data/nt_R_2B_O.dat"
  {COUNT[nt, WORLD]} => "./reaction_data/nt.dat"
}
```

Each file (.dat suffix) will be created automatically in a subdirectory called "reaction_data".

3. Save the file and quit.

3.5.5. Add Initial Distributions of Neurotransmitter Molecules

MCell's MDL supports several different ways to release molecules at different times and locations during simulations. In this case, the neurotransmitter molecules will originate inside each of the synaptic vesicles, and then will diffuse out through the expanding pore into the synaptic cleft. To create the initial distributions of molecules within the synaptic vesicles, we will use a simple method (once for each vesicle) that places a specified number of volume molecules at random locations within spherical bounds of specified diameter. By default, this will occur when the simulation is initialized, so the molecules will begin diffusing immediately. Each vesicle will initially contain 3,000 neurotransmitter molecules.

1. *Add the neurotransmitter release sites*: Use a text editor to modify the first of the main MDL files for the expanding fusion pore, "VesicleFusion_1.mdl" (*see* **Subheading 3.5.6** later). Add the following text at the end of the INSTANTIATE block (before its closing curly brace; *see* **Note 37**):

```
first_release_site SPHERICAL_RELEASE_SITE {
LOCATION = [-0.106, 0.021, 0.105]
MOLECULE = nt
NUMBER_TO_RELEASE = 3000
SITE_DIAMETER = 0.032
}
second_release_site SPHERICAL_RELEASE_SITE {
LOCATION = [0.106, -0.021, 0.105]
MOLECULE = nt
NUMBER_TO_RELEASE = 3000
SITE_DIAMETER = 0.032
}
```

2. Save the file and quit.

3.5.6. Final MDL File Setup

1. *Reuse the "Partitions.mdl" file*: Copy the "Partitions.mdl" file from **Subheading 3.4** to the current directory for the expanding pore MDL model.
2. *Add INCLUDE statements to the main MDL files for the checkpoint sequence*: When the pore expansion series was exported from Blender, a set of main and subordinate (included) MDL files was created. These files are numbered in sequence, e.g., the main files are named "VesicleFusion_1.mdl", "VesicleFusion_2.mdl", etc. As in **Subheading 3.3.6**, we now need to

add the remaining (newly created) INCLUDE file references to each of the ten main MDL files for the expanding pore. This can either be done by hand for each of the files, or all at once by entering the following two commands in succession at the command line:

sed -e "9aINCLUDE_FILE = \"Partitions.mdl\"\nINCLUDE_FILE = \"Molecules.mdl\"\nINCLUDE_FILE = \"Reactions.mdl\"\n" -i VesicleFusion_[1-9]*.mdl

sed -e "16aINCLUDE_FILE = \"RegionModifications.mdl\"\nINCLUDE_FILE = \"ReactionData.mdl\"\n" -i VesicleFusion_[1-9]*.mdl

3.5.7. Run the Expanding Pore Simulation

Highly accurate simulation of neurotransmitter diffusion through the pore would require many iterations and a very fine time step so that the average random walk step length would be much smaller than the pore radius at all times *(4, 5)*. In this simple example, however, each of the ten main MDL files is set to run for only one iteration with a relatively long time step (one microsecond). This will allow quick visualization of the pore expansion, but to see the transmitter escape and bind to receptors, we will need to run the final checkpoint segment for many more iterations.

1. *Increase the number of iterations for the final run of the checkpoint sequence*: Using a text editor, open the last of the main MDL files, "VesicleFusion_10.mdl". Change the first line from:

   ```
   iterations = 10
   ```

   to

   ```
   iterations = 500
   ```

   Then change the line:

   ```
   CHECKPOINT_ITERATIONS = 1
   ```

   to

   ```
   CHECKPOINT_ITERATIONS = 491
   ```

2. Save the file and quit.
3. *Run the simulation*: When the MDL files were exported from Blender, a separate "script" file was also created to run the ten MCell simulations in succession (rather than starting each by hand). The script file ("VesicleFusion.py") is written in a high-level command language called Python. Run the Python script by entering:

   ./VesicleFusion.py

   at the command line. You will see MCell's default run-time messages as the checkpoint runs execute in sequence, and the reaction and visualization files will be created.

3.6. Visualize MCell Results with DReAMM (Part 2)

We now use DReAMM to visualize the results of the expanding pore simulations. We will reuse the settings file created previously in **Subheading 3.4.4**, and will also apply some additional customizations and animation settings.

3.6.1. Import the MCell Visualization Data and Import the Settings File

1. *Start DReAMM*: Enter "dreamm" at the command line.
2. *Import visualization data*: Click "Import & Select" on the "Quick Controls" menu (labeled *1* in **Fig. 3A**). In the "Import & Select Objects" menu, click the ellipsis ("...") by the "Viz Data File" text box. Navigate to the "VesicleFusion_viz_data" directory and select the file "VesicleFusion.dx". Click "OK".
3. *Import the customization settings file*: In the "Read/Write Settings" menu, click the ellipsis ("...") by "Read File" and then navigate to and select the file "SynapseCustom.dx" that was created in **Subheading 3.4.4**. Click "OK". The "Dendrite" and "PresynapticBouton" meshes will appear with the rendering properties defined previously (*see* **Note 38**).
4. *Adjust the view*: Hit *Ctrl-f* to center the view. Now open the "View Control" menu (*Ctrl-v* from the "DReAMM Image Window") and then switch to a left view by setting "Set View" to "Left".
5. *Select all volume molecules*: In the "Import & Select Objects" menu, change "Add Selected" to "Add All". Next, click on "Volume Molecules" and click "Apply Operation Once".
6. *Select all surface molecules*: Click on "Surface Molecules" and click "Apply Operation Once".
7. *Close menus*: Click the "Close" button by "Import & Select" in the "Quick Controls" menu (labeled *5* in **Fig. 3A**).
8. *Apply previously defined volume molecule properties*: Click the "Set Rendering Prop". button in the "Quick Controls" menu (labeled *2* in **Fig. 3A**). In the "Rendering Properties" menu, click "Molecules" next to "List" (top center), then select "Ca", and click "Load from selected object". All of the rendering properties previously associated with "Ca" are now loaded in the "Properties" fields and in the "Colormap Editor" menu. To apply the same properties to the neurotransmitter molecules, select "nt" and deselect "Ca". Hit "Apply Operation Once", and all the "nt" pixels will change into yellow spheres (**Fig. 5C**).
9. *Apply previously defined surface molecule properties*: As in the preceding step, load the previously defined "VGCC_C" properties and apply them to "nt_R_0B". Then, load the properties for "VGCC_O" and apply them to "nt_R_2B_O". The unbound receptors will now appear as red arrows, and the double-bound open channel receptors will appear as green arrows.

10. *Assign rendering properties for the remaining receptor states*: Manually change "(Front) Color" to yellow by entering RGB values of 1, 1, and 0, respectively. Make sure that only "nt_R_1Ba" is highlighted in the list of object names. Click "Apply Operation (Once)". Now enter RGB values of 0, 0, 1 for blue, select only "nt_R_1Bb", and click "Apply Operation Once". Finally, enter RGB values of 0, 1, 1 for cyan, select only "nt_R_2B_C", and click "Apply Operation Once". The single-bound receptors now will be yellow or blue arrows, and the double-bound closed receptors will be cyan arrows. All surface molecules should now appear as colored arrows similar to their appearance in **Subheading 3.4.3** (**Fig. 5C–F**). If desired, try using the "receptor_1" or "receptor_2" glyphs for a more realistic appearance.
11. *Close menus*: Hit the "Close" button by "Set Rendering Prop". in the "Quick Controls" menu (labeled *6* in **Fig. 3A**).

3.6.2. Use Image Clipping

To see the expanding pore more clearly, we will use a clipping box (*see* **Note 39**) and visualize only the meshes and molecules in and around the synaptic cleft.

1. *Set the default view*: Hit *Ctrl-f*.
2. *Select the "Left" view*: Hit *Ctrl-v* to bring up the "View Control" menu and then select "Left" under "Set View".
3. *Align the clipping box*: Click on "Open Menu List" in the "Quick Controls" window (labeled *4* in **Fig. 3A**). This will open the "Menu List", which allows you to open individual DReAMM menus. Scroll down and click on "Image Clipping" to open the corresponding menu, and then click the button labeled "Align Clipping Box with Current View". This will orient the clipping box to the current viewing direction, so that the box's front and back faces are parallel to the screen and the left, right, top, and bottom limits are parallel to the corresponding screen edges.
4. *Set the clipping distance*: We now need to set the depth at which the clipping box begins; that is, the distance from the current camera location (look-from point) to the nearest point on the front face of the box. This is most easily done by "picking" a point on an object, and DReAMM will then calculate the distance to a plane that cuts through the picked point. In the "View Control" menu, select "Pick" under "Mode" and then choose "Clipping Distance" under "Pick(s)". Next, in the "DReAMM Image Window", click on the head of the presynaptic bouton. Note that the distance from the camera to the picked point now appears in the "Picked" field of "Near Distance" in the "Image Clipping" menu.

5. *Show the clipping box*: In the "Image Clipping" menu, click on the "Show Clipping Box" button. A yellow semitransparent box will appear and, due to its large default size, will fill the "DReAMM Image Window". With the image window active, hit *Ctrl-z* for "Zoom" mode (*see* **Note 29**) and then right click and drag to zoom out. If necessary, zoom in (left click and drag) or out again to see the entire clipping box in the image window.
6. *Adjust the clipping box size*: By default, the clipping box dimensions are 20 × 20 × 0.1 microns (width × height × thickness). Shrink the box's width and height by modifying the "Left", "Right", "Upper", and "Lower" limits in the "Image Clipping" menu. Continue until the box encloses only the synaptic vesicles and cleft, and then hide the box by clicking again on the "Show Clipping Box" button.
7. *Apply image clipping*: Click the "Apply Image Clipping" button. Only a small slice through the synaptic region will remain visible.
8. *Pan/zoom to reorient and magnify the view*: With the image window active, hit *Ctrl-g* for "Pan/Zoom" mode (*see* **Note 29**). Center the mouse pointer on the visible structures, and then left click and drag to enclose the objects within the marquee. When you release the mouse button, the view will be reoriented and magnified simultaneously.
9. *Rotate the view*: Hit *Ctrl-r* for "Rotate" mode. Point near the 3 o'clock position in the image window, and then right click and drag in a counterclockwise direction. The clipped objects will follow the pointer by rotating in the plane of the screen (a constrained rotation around the current view direction). Continue to the 12 o'clock position so that the synaptic cleft will be approximately horizontal. Next, point at the objects and left click and drag to perform a free rotation around the current look-to point. Drag primarily to the left or right to obtain an oblique view of the clipped objects.
10. *Adjust the clipping box thickness and fine-tune the dimensions*: In the "Image Clipping" menu, increase the thickness of the clipping box. A setting of about 0.35 microns should include both synaptic vesicles and most of the cleft space. With some minor adjustments to the box's dimensions, you can obtain a view similar to that shown in **Fig. 5H**. You may also find that you have to fine-tune the "Near Distance" for the box. This can be done by entering a value in the "Specify" field (start with a value very close to the indicated "Picked" value that was obtained in **step 4**) and then clicking on the "Specify" button.
11. *Play the time sequence*: Click the Play button on the "Sequence Control" to watch the pore open and the neurotransmitter

escape to bind to receptors. If necessary, use the Frame Controls to skip ahead in time (*see* **Subheading 3.4.3, step 10**).

3.6.3. Animate the Visualization

DReAMM includes many features to create sophisticated animations using keyframes (*see* **Note 40**). In this final section, we briefly introduce some of the possibilities by illustrating how to rotate the camera around the clipped objects.

1. *Temporarily turn off image clipping*: Again click the "Apply Image Clipping" button in the "Image Clipping" menu. By turning off image clipping temporarily, we will speed up and better illustrate the operations required to animate the camera.
2. *Open the "Make Animation" menus*: Click the "Make Animation" button on the "Quick Controls" menu (labeled *3* in **Fig. 3A**).
3. *Activate the "DReAMM Keyframe Editing Image Window"*: In the "Edit Keyframes" menu, click the "Show Camera Positions" and "Show Spline/Rotation Points" buttons. This will open a new image window that shows the model objects as well as glyphs representing the current camera and keyframe camera locations. Note that this new image window is fully interactive and operates with the same hot keys used in the main "DReAMM Image Window". With the new window active, hit *Ctrl-f* to center the view. Now hit *Ctrl-r* and rotate the view to see the model objects as well as the camera glyphs. The current camera location for the "DReAMM Image Window" view is shown by the common base of two orthogonal (perpendicular) arrows colored red and blue (look and up directions, respectively). By default, one keyframe camera location is also present, as indicated by orthogonal gold (look) and green (up) needle glyphs (line segments) with the numeral "1" alongside.
4. *Define the current view from the "DReAMM Image Window" as the starting point for the animation*: In the "Edit Keyframes" menu, click "Start New List" to begin a new list of keyframe data. The current camera location and view from the "DReAMM Image Window" will overwrite the existing default keyframe information. This will be indicated visually in the "Keyframe Editing Image Window" as the first keyframe camera glyph (labeled "1") becomes coincident with the current camera position glyph.
5. *Set up the rotation keyframes*: One of the most commonly desired animations is a simple rotation of the camera around some selected model objects. DReAMM allows you to do this operation in only a few simple steps. First, in the "Edit Keyframes" menu, click on the drop-down list that currently says "Add to End" and change it to "Compute Rotation Points".

Now click the "Apply Edit: Once" button. A set of magenta needle glyphs will appear in the "Keyframe Editing Image Window", showing potential camera locations and directions for the rotation. By default, a complete rotation is generated with 1° per step. Next, change the "Compute Rotation Points" drop-down setting to "Add Rotation to Range". Click the "Apply Edit: Once" button again. This finalizes the keyframe camera positions as indicated by the appearance of numbered keyframe camera glyphs.

6. *Close menus*: Click the "Close All" button in the "Quick Controls" menu (labeled *7* in **Fig. 3A**).
7. *Play the animation*: In the lower right-hand corner of the "Quick Controls" menu, change "Keyframe Mode" from "All Interactive" to "Keyframes" (labeled *8* in **Fig. 3A**). Click the Play button on "Sequence Control" (**Fig. 3A**) and watch the pore open and molecules diffuse while the camera rotates around the objects. If you leave the "Keyframe Editing Image Window" active, you will simultaneously see the current camera glyph moving from keyframe-to-keyframe around the model objects. To speed up animation playback, disable the "Keyframe Editing Image Window" by reopening the "Edit Keyframes" menu and clicking once again on the "Show Camera Positions" and "Show Spline/Rotation Points" buttons.
8. *Turn on image clipping*: Reopen the "Image Clipping" menu and click the "Apply Image Clipping" button again to see the animation of the clipped model objects. You can turn image clipping on and off during playback of the animation.
9. *Save your DReAMM settings*: As in **Subheading 3.4.4, step 1**, save your current DReAMM settings so that you can reload them and replay the animation at a later time.

4. Notes

1. There is presently a distinction between "modeling" software (e.g., Blender) and computer-aided design (CAD) software. Both allow the user to design 3D structures interactively, but "modeling" programs are specialized for animation and morphing of "smoother" or, in the present context, more "biological" shapes. CAD software, on the other hand, is oriented toward the design of objects for architectural or mechanical engineering, and may be integrated with computer-aided manufacturing (CAM) software and machinery.

2. Developer libraries typically are not installed by default with Mac OS X. The same problem can arise with some Linux distributions such as Ubuntu, Debian, or others.
3. The DX file format (.dx) was originally created by IBM for its commercial DataExplorer software, a large visual programming and visualization environment. Eventually DataExplorer (DX) was released as open source code (OpenDX), which we have now improved and expanded into PSC_DX for use with DReAMM and large-scale MCell models. We continue to use the native DX file format because it is very general and efficient for use with hierarchical assemblies of arbitrary mesh objects, each of which may be associated with multiple datasets of arbitrary type, and also annotated with a variety of metadata tags. Although we do not explicitly illustrate export of .dx files from Blender in this chapter, it is easily done using the supplied plug-in. Direct export from Blender to DReAMM can be very useful for rendering and creation of sophisticated animations, as well as specialized mesh editing operations. DReAMM can also export MDL files for use with MCell.
4. In Blender, the *X*-, *Y*-, and *Z*-axes are red, green, and blue, respectively, and the cardinal views (*XZ*-, *YZ*-, or *XY*-plane) can be accessed using number keys on the number pad (*1*, *3*, or *7*, respectively). While in a cardinal view, a visual key for the axes is present in the lower left corner of the "3D view" window. For example, if you hit *1* (*on the number pad*) for the *XZ*-view, you will see a horizontal red line labeled X and a vertical blue line labeled Z, indicating that the *XZ*-plane lies in the plane of the screen. The *Y*-axis is perpendicular to the screen in this view.
5. Most Blender hot keys work only when the cursor is located inside the "3D View" window.
6. A UV sphere is composed of quadrangular faces between lines of longitude and latitude, similar to a globe. Icosahedral spheres (composed of hexagons and pentagons; triangulated by Blender) are another option, but UV spheres are easier to use for the region definition and extrusion operations in this chapter.
7. Spatial units are arbitrary in Blender, but, once objects are exported to MDL files for use with MCell, dimensions will be interpreted as microns.
8. Although the cardinal views are sufficient for the operations in this chapter, it is often helpful to rotate around an object for a better view. To rotate around the currently selected object, click and drag using the middle mouse button. Certain operations may also require zooming and panning. Scroll up with the middle mouse button to zoom in, and scroll down to

zoom out. To pan, hold *Shift* and one of the following keys on the number pad: *8* (up), *2* (down), *4* (left), or *6* (right).

9. If you accidentally hit one of the number keys on the main keyboard instead of the number pad, you will move to a different layer and your current view will disappear. If this occurs, simply hit *1* on the main keyboard to return to layer 1.
10. To make the faces and central vertex, we first replicate the original topmost vertices using an extrude operation. Ordinarily this would be used to create a cylindrical extension, and in a later operation we extrude vertices in that fashion to create the spine shaft. In this case, however, we will extrude using a length of 0, so the new vertices will be exactly coincident with the original vertices. Then we will scale the radius of the extrusion to a value of 0, and this will move all of the new vertices to the desired central point in the plane of the opening. Finally, we will remove all of the unnecessary duplicate vertices, leaving all of the new triangles connected to a single central point.
11. It is useful to save periodic snapshots of all Blender files (.blend) using distinct names. This allows you to start over from a previous snapshot if something unexpected occurs.
12. "Object Mode" is used for operations on entire objects (e.g., selecting, moving, scaling), while "Edit Mode" is used for operations on selected faces or vertices.
13. The numerical suffix for the material name may differ (e.g., if a .blend file is reloaded). If this occurs, just ensure that you change the material name for the correct object.
14. There is an error (a bug) that sometimes appears when using Blender's Boolean operations. Under some conditions, the mesh vertices are reconnected incorrectly, leaving undesired faces and/or holes. This should not occur if the steps of the chapter are followed precisely. If the problem does occur, however, it can be fixed by deleting any extra faces and/or making the vertices around a hole into one or more faces as necessary.
15. Blender uses both "Global" and "Local" axes. There is only one set of "Global" axes, but each object has its own unique "Local" axes, which may or may not coincide with the "Global" axes. By using "Apply scale and rotation" in this step, the "Local" axes become coincident with the "Global" axes. This ensures that the objects will be aligned with the "Empty" object created in **Subheading 3.1.7**, **step 9**.
16. Each surface mesh element (face) has a front and back. The direction perpendicular (normal) to a face defines the face's normal vector. Vertices shared by more than one face have an associated normal vector defined as the average of the face normals. When objects are scaled (shrunken or enlarged),

each vertex is moved along its normal (inward or outward, respectively), and so the faces become smaller or larger. On the other hand, when a group of faces is moved rather than scaled, all of the vertices are moved along parallel vectors. In the particular case of this step, we want a spine head to move "out" or "in" along the axis of the spine neck. To do so, we can move the head along a vector defined by the average of its face normals. Since each spine head is radially symmetric around the spine axis, the average of the face normals will be the spine axis itself.

17. A full introduction to the MDL syntax is beyond the scope of these stand-alone examples. Please consult the MCell Reference Guide for additional information (http://www.mcell.psc.edu).

18. Every surface mesh object is composed of individual polygons. To describe any mesh in a general but compact way, all of the vertices (*x,y,z* coordinates) are listed first (see VERTEX_LIST in later example), and then the triangular faces are listed next (ELEMENT_CONNECTIONS). Each face is described by an array of index numbers that refer to the vertex list (a triple of vertex indices). For example, a face that connects the first three vertices could be listed as [0,1,2] (zero-based indices are used; the order in which the vertices are listed, together with the right hand rule, determines the front and back sides of the face). Thereafter, mesh regions can be named and defined using arrays of index numbers that refer to the list of faces. Thus, [0,1,2,3,4] would define a region composed of the first five faces. In an MDL file for one or more mesh objects, each object is defined by a POLYGON_LIST with a user-specified name, and that includes vertices, triangular connections, and regions (if any). For example:

```
my_mesh_name POLYGON_LIST {
VERTEX_LIST {
(list of all x,y,z coordinates)
}
ELEMENT_CONNECTIONS {
(list of all triangular faces, each defined by
   three vertex index numbers)
}
DEFINE_SURFACE_REGIONS {
my_first_region_name {
ELEMENT_LIST = [list of face index numbers]
}
my_second_region_name {
```

```
(...)
}
(...)
}
}
```

19. Checkpointing is a general term in large-scale computing and simply means that a running process (e.g., simulation) is stopped temporarily so that it can be restarted later. At the checkpoint, the current state of the process is written to a file that is subsequently read when computation resumes. With MCell, the checkpoint file includes the locations and states of all molecules, as well as other run-time parameters. When the simulation is restarted after a checkpoint, a complete set of input MDL files is read in addition to the checkpoint data. Thus, any new changes made in the MDL files (e.g., mesh geometry) are incorporated into the continuing simulation. In the example of this chapter, ten different sets of MDL files are used over a checkpointing sequence, and the radius of the fusion pore increases in each step. All of the diffusing neurotransmitter molecules are contained within the expanding pore or the synaptic cleft space later, and so diffusion from the vesicle through the pore can be simulated without concern that molecules escape from the changing pore geometry.
20. To check dimensions, click the "Edge Lengths" button in the "Buttons Window" under "Mesh Tools 1". This panel is on the far right of the "Buttons Window", and, depending on window size and resolution, you may have to scroll over to see it. If this is the case, scroll "up" using the middle mouse button while the cursor is over the "Buttons Window". Select "Edge Lengths", and values will appear for all highlighted edges (to three decimal places). Additional measurement capabilities are available through other online plug-ins for Blender.
21. This step may take some time to complete, depending on the speed of your computer. Also, you may find that Blender runs slowly after this step. This is due to a caching problem that can be eliminated simply by reopening the final saved .blend file and continuing.
22. MDL files are plain text files and can be edited using any available text editing program. Popular choices on Unix-like operating systems include vim, emacs, pico, or nano. In addition, the Writer program within OpenOffice (similar to MS Word within MS Office) can be used as long as the files are saved in plain text format.

23. A "block" denotes a set of MDL commands enclosed in curly braces, and some blocks may be nested (blocks within blocks).
24. As in other cell modeling languages, a "molecule name" in MCell's MDL is quite general and may actually correspond to a functional state, such as an open or closed conformation of a molecule that includes a transmembrane ion channel. The relationships between molecules and states are defined in subsequent reaction statements.
25. Molecules can be added to surface regions on meshes in several ways. In one approach the molecules are added directly by hand-editing a DEFINE_SURFACE_REGIONS block within a POLYGON_LIST (*see* **Note 18**). This can be quick and easy for small meshes, but hand-editing large mesh files is generally undesirable. Thus, in this chapter we illustrate use of the MODIFY_SURFACE_REGIONS command. In addition to other functions, MODIFY_SURFACE_REGIONS allows addition of molecules to mesh regions defined previously in other (included) MDL files. To add molecules, one specifies the name of the mesh region to be modified, the name of the molecules to be added, the number to be added, and the molecules' orientations with respect to the front and back of the surface.
26. In this example molecules are added to surface regions at the beginning of the simulation. It is also possible for molecules to appear and disappear during a simulation as reaction steps occur, and additional molecules may also be "released" in closed volumes and on surface regions at specified times during running simulations. When a molecule is produced by a reaction, its initial location depends on the location of the reactant(s), and may also reflect user-specified directionality with respect to a surface.
27. It is important to do visual checks often, especially before running long simulations with large models. Visual checks are typically done by running a short number of iterations for rapid visualization with DReAMM.
28. By default, DReAMM will use rendering properties that minimize the time and memory required to display images. It will also use "Software Rendering", which means that the calculations required to generate the images are performed on the computer's CPU, rather than the graphics processor on the video card. DReAMM can also use "Hardware Rendering", which does take advantage of graphics hardware and under many conditions will be much faster. However, as models increase in size and complexity, rendering in Hardware may become slower than in Software, or may not

be possible at all if insufficient memory is available on the video card. You can switch back and forth between Software and Hardware rendering by clicking "Options -> Rendering Options…" on the "DReAMM Image Window" menu bar, and then clicking on the "Software" or "Hardware" button under "Rendering Mode" in the pop-up "Rendering…" menu. Either Software or Hardware rendering may be used throughout the examples of this chapter, except when Software rendering must be used with "Image Clipping" (**Subheading 3.6.2** and **Note 39**).

29. With DReAMM, hot keys function only if both "Num Lock" and "Caps Lock" are off on the keyboard. Most DReAMM hot keys select different viewing modes and require that the "DReAMM Image Window" is selected. A sequence of view changes can be undone and redone using *Ctrl-u* and *Ctrl-d*, respectively, and a default view of centered objects can be obtained with *Ctrl-f*. Commonly used viewing modes include the following:
    - "Rotate" (*Ctrl-r*) – left click and drag for free rotation around the current look-to point, or right click and drag for constrained rotation around the current look direction.
    - "Navigate" (*Ctrl-n*) – left click, hold, and move the pointer toward an object to move the camera toward the object; middle click, hold, and move the pointer to pivot the camera (change the look-to point without changing the look-from point), or right click, hold, and move the pointer to move the camera away from objects. While in "Navigate" mode, the "View Control" menu includes sliders for forward or backward speed of motion and relative pivot speed, and also a selector button that allows the camera to look in different directions while traveling forward or backward (e.g., looking 45° to the right while traveling forward). For best results, use "Navigate" mode while using Hardware rendering.
    - "Zoom" (*Ctrl-z*) – left click and drag to zoom in on a region outlined by a centered marquee, or right click and drag to zoom out by fitting the current display into a centered marquee.
    - "Pan/Zoom" (*Ctrl-g*) – similar to "Zoom" mode except that you reorient the view and zoom simultaneously, by first pointing to the object of interest and then clicking and dragging a marquee around it.
30. Custom rendering properties are used to visualize particular molecules, region boundaries, and molecules obscured by meshes. Here, we illustrate the use of custom rendering properties to visualize the proper location and orientations

of the synaptic vesicles and their calcium binding sites, the VGCCs, and calcium ions.

31. Values can be changed by clicking the up/down arrows or by directly entering the desired value into the field and hitting *Enter*.
32. In a "real" model, use of a colormap ensures that all regions were designed properly. Any errors must be fixed and the modified model should be rerun and retested until all regions are verified visually.
33. To see a sharp transition between mesh elements with different colors, change the "Color Dependence" selector from "Vertices" to "Elements". If "Vertices" is used instead, color changes between mesh elements are blended to give the mesh a smooth appearance.
34. You must hit *Enter* after typing in a DReAMM text box before the changes will be recognized.
35. A directional glyph like an arrow is often a good choice when checking the orientation of surface molecules.
36. DReAMM settings files contain the list of currently selected objects, all keyframe data, and all assigned rendering properties, including colormaps. They can be reloaded later for the same or different visualization data, allowing reuse and exchange of settings. You can also choose to reload particular portions of a settings file. For example, you may wish to load rendering properties previously assigned to another model, but not the selected objects or keyframe information from the other model.
37. The locations (*X*,*Y*,*Z* coordinates) for each release site can be obtained with Blender or DReAMM, and are approximately centered within each vesicle. With Blender, you can display a median value for a selected set of vertices by hitting *n* and then clicking "Global" in the "Transform Properties" pop-up menu. With DReAMM, you can use the "Probes" menu to activate 3D cursors that can be moved around in space, with a display of their positions.
38. The settings file created previously includes rendering properties assigned for meshes and molecules. Since the meshes in the new model share the same names as the meshes in the previous model, the rendering properties are applied immediately. On the other hand, the molecules in the new model do not have the same names as the molecules in the previous model. Thus, at this point the new molecules are still rendered using default properties (pixels). In subsequent **Subheading 3.6.1**, **steps 8 and 9**, the previously defined rendering properties will be reassigned (copied) to the new molecules.

39. DReAMM includes two different types of clipping operations accessed through separate menus. "Data Clipping" allows you to specify separate *X*-, *Y*-, and/or *Z*-limits for different objects ("Meshes", "Wireframes", "Boundaries", "Volume Molecules", or "Surface Molecules"). Molecules with positions that lie within the limits are included for rendering, as are mesh elements (triangles) with included centers of mass. Thus, when using "Data Clipping" the amount of data to be rendered is reduced. If only a portion of a very large model is visualized, rendering speed can increase enormously. However, meshes clipped in this manner may have a jagged edge because complete triangles are removed. "Data Clipping" works when using either Software or Hardware rendering. In contrast, "Image Clipping" allows you to specify an arbitrarily oriented clipping box anywhere in space, and all objects contained within the box are shown with smooth cut edges. This increases rather than decreases the amount of computation, and with current versions of DReAMM this must be done with software rendering. For large datasets, "Data Clipping" and "Image Clipping" can be used in combination.
40. DReAMM animations are designed using interpolated keyframes. Each keyframe includes information about the camera position and view, lighting, depth cueing, and stereo visualization settings. In general, keyframes can be added and removed in a variety of ways, and all of the keyframe data parameters can be interpolated using a spline function over the entire keyframe sequence or a specified portion thereof. For additional information, see the DReAMM animation tutorials in *(15)*.

Acknowledgments

The authors thank Aji Janis and Gary Blumenthal for careful reading and testing of the examples in this chapter. This work was supported by NIH R01 GM068630 (JRS), P41 RR06009 (JRS), and F32 GM083473 (MD).

References

1. Tao, L. and Nicholson, C. (2004) Maximum geometrical hindrance to diffusion in brain extracellular space surrounding uniformly spaced convex cells. *J. Theor. Biol.* 229, 59–68.
2. Hrabe, J., Hrabetova, S., and Segeth, K. (2004) A model of effective diffusion and tortuosity in the extracellular space of the brain. *Biophys. J.* 87, 1606–1617.

3. Tao, A., Tao, L., and Nicholson, C. (2005) Cell cavities increase tortuosity in brain extracellular space. *J. Theor. Biol.* 234, 525–536.
4. Stiles, J. R., Van Helden, D., Bartol, T. M., Jr., Salpeter, E. E., and Salpeter, M. M. (1996) Miniature endplate current rise times <100 μs from improved dual recordings can be modeled with passive acetylcholine diffusion from a synaptic vesicle. *Proc. Natl. Acad. Sci. USA* 93, 5747–5752.
5. Stiles, J. R., Bartol, T. M., Jr., Salpeter, E. E., and Salpeter, M. M. (1998) Monte Carlo simulation of neurotransmitter release using MCell, a general simulator of cellular physiological processes, in *Computational Neuroscience* (Bower, J. M., ed.), Plenum, New York, NY, pp. 279–284.
6. Stiles, J. R., Bartol, T. M., Salpeter, M. M., Salpeter, E. E., and Sejnowski, T. J. (2001) Synaptic variability: new insights from reconstructions and Monte Carlo simulations with MCell, in *Synapses* (Cowan, W. M., Stevens, C. F., and Sudhof, T. C., eds.), Johns Hopkins University Press, Baltimore, MD, pp. 681–731.
7. Stiles, J. R. and Bartol, T. M. (2001) Monte Carlo methods for simulating realistic synaptic microphysiology using MCell, in *Computational Neuroscience: Realistic Modeling for Experimentalists* (De Schutter, E., ed.), CRC Press, Boca Raton, FL, pp. 87–127.
8. Pawlu, C., DiAntonio, A., and Heckmann, M. (2004) Postfusional control of quantal current shape. *Neuron* 43, 607–618.
9. Stiles, J. R., Ford, W. C., Pattillo, J. M., Deerinck, T. E., Ellisman, M. H., Bartol, T. M., and Sejnowski, T. J. (2004) Spatially realistic computational physiology: past, present, and future, in *Parallel Computing: Software Technology, Algorithms, Architectures & Applications* (Joubert, G. R., Nagel, W. E., Peters, F. J., and Walter, W. V., eds.), Elsevier, Amsterdam, pp. 685–694.
10. Coggan, J. S., Bartol, T. M., Esquenazi, E., Stiles, J. R., Lamont, S., Martone, M. E., Berg, D. K., Ellisman, M. H., and Sejnowski, T. J. (2005) Evidence for ectopic neurotransmission at a neuronal synapse. *Science* 309, 446–451.
11. He, L., Wu, X. S., Mohan, R., and Wu, L. G. (2006) Two modes of fusion pore opening revealed by cell-attached recordings at a synapse. *Nature* 444, 102–105.
12. Kerr, R.A., Bartol, T.M., Kaminsky, B., Dittrich, M., Chang, J.J.C, Baden, S.B., Sejnowski, T.J., and Stiles, J.R. Fast Monte Carlo simulation methods for biological reaction-diffusion systems in solution and on surfaces. *SIAM J. Sci. Comput.* 30, 3126–3149.
13. Kerr, R. A., Levine, H., Sejnowski, T. J., and Rappel, W. J. (2006) Division accuracy in a stochastic model of Min oscillations in Escherichia coli. *Proc. Natl. Acad. Sci. USA* 103, 347–352.
14. Koh, X., Srinivasan, B., Ching, H. S., and Levchenko, A. (2006) A 3D Monte Carlo analysis of the role of dyadic space geometry in spark generation. *Biophys. J.* 90, 1999–2014.
15. www.mcell.psc.edu and pages therein.
16. www.blender.org.
17. Synapse Web, Kristen M. Harris, PI, http://synapse-web.org/. The publicly available VRML file that contains the data for the reconstruction can be downloaded from www.synapse-web.org/anatomy/Ca1pyrmd/radiatum/index.stm.

Chapter 10

A Cell Architecture Modeling System Based on Quantitative Ultrastructural Characteristics

Július Parulek, Miloš Šrámek, Michal Červeňanský, Marta Novotová, and Ivan Zahradník

Summary

The architecture of living cells is difficult to describe and communicate; therefore, realistic computer models may help their understanding. 3D models should correspond both to qualitative and quantitative experimental data and therefore should include specific authoring tools such as appropriate visualization and stereological measures. For this purpose we have developed a *problem solving environment for stereology-based modeling* (PSE-SBM), which is an automated system for quantitative modeling of cell architecture. The PSE-SBM meets the requirement to produce models that correspond in stereological and morphologic terms to real cells and their organelles. Instead of using standard interactive graphing tools, our approach relies on functional modeling. We have built a system of implicit functions and set operations, organized in a hierarchical tree structure, which describes individual cell organelles and their 3D relations. Natural variability of size, shape, and position of organelles is achieved by random variation of the specific parameters within given limits. The resulting model is materialized by evaluation of these functions and is adjusted for a given set of specific parameters defined by the user. These principles are explained in detail, and modeling of segments of a muscle cell is used as an example to demonstrate the potential of the PSE-SBM for communication of architectural concepts and testing of structural hypotheses.

Key words: Implicit modeling, Cell architecture, Muscle cell, Stereology, 3D structure, Visualization, Automatic model generation, XML.

1. Introduction

The structure and the function of muscle cells are related in many intricate ways that are difficult to understand, describe, and communicate to others. Within this scope, computer modeling might be very instrumental for synthesis and verification of recent

Ivan V. Maly (ed.) *Methods in Molecular Biology, Systems Biology, vol. 500*

DOI: 10.1007/978-1-59745-525-1_10

knowledge, as well as for testing specific hypothesis. Here, we describe a novel modeling approach aimed to capture the cell architecture, that is, the manner how the muscle cells are constructed of their organelles *(1–3)*. We employed the methods of geometrical modeling implemented with the use of recent computer hardware and computer graphics tools to provide biologists and biophysicists with an environment for virtual cell modeling *(4–7)*. This approach has numerous potential applications. In this work we present an example of modeling striated muscle cells to demonstrate its principles and the potential of geometrical modeling for representation of 3D volume models.

Our effort is aimed at creation, verification, and visualization of complex models of muscle cells. A typical cell consists of hundreds or even thousands of various organelles. Thus, creation of such model organelle-by-organelle, using the traditional interactive techniques, would require unacceptably long time and, moreover, would not ensure the typical stochastic properties. Therefore, in our approach, the cell model is created in an automated process that allows simultaneous generation of numerous variants of models based on the same specification but differing in their random representations.

We call this automated geometry modeling system the *problem solving environment for stereology-based modeling* (PSE-SBM). The PSE-SBM meets the essential requirement, namely, to produce models that correspond in quantitative terms to volume and surface densities (VSD) of cell organelles known from stereological and morphological analysis of electron-microscopic images of real cells. It is obvious that such goals cannot be reached by conventional graphic modeling tools, as they require either interactive graphical input (e.g., Blender *(8)*, Truespace *(9)*, etc) or sophisticated techniques for capturing 3D volume data, like electron or confocal microscopy – all techniques being very demanding for manpower, time, and costs. In fact, our approach, dissimilar to interactive graphics tools, resembles functional modeling in a sense that it shares three features – usage of mathematical functions, adjustment of their parameters according to a set of input parameters, and visualization of the generated models in a user-defined way. We define a system of functions and set operations that describe each individual cell organelle and spatial relations among them. The resulting model, defined by the user through a set of specific parameters, is then just a combination of these functions. The functions employed here for modeling are known as implicit functions *(10, 11)* and their evaluation defines the outer surfaces of each organelle.

1.1. Implicit Modeling

Implicit surfaces (implicits) are a convenient geometric modeling tool for image synthesis and computer-aided geometric design. The set of techniques, known today as implicit modeling, was used for the first time by Blinn *(12)*. Currently, there are several types of implicit modeling systems that are oriented toward specific classes of objects. The first class stands for skeleton-based models. A simplest skeleton is a point or a set of isolated points. Implicit surfaces, built up on skeletal points, are known as blobs *(12)*, soft objects *(13)*, and metaballs *(14)*. Techniques presented by point skeletons were later extended by lines, curves, and polygons and generalized to convolution surfaces *(15, 16)*. The modeling systems implementing the approach based on convolution surfaces are often limited by the choice of skeletal elements according to technical difficulties in evaluating convolution integrals. Nevertheless, convolution surfaces are widely used in geometrical modeling of organic structures mainly due to their ability to represent smooth shapes *(17–19)*.

The second class is represented by implicit surfaces defined by analytical functions. These include algebraic (e.g., plane, quadrics) and nonalgebraic functions (e.g., superquadrics) *(20)*. Later, Pasko et al. generalized the representation of implicits by combination of all the aforementioned approaches into a single framework named functional representation (or F-rep) of the geometric object *(21)*.

To modify either the functional value or the coordinates of any given implicit surface function (implicit function), several unary modifiers *(10, 11, 21, 22)* can be applied. Furthermore, complex objects can be created via constructive solid geometry by Boolean set-theoretic operations. The basic set-theoretic operations can be defined using the min and max operators, which are not differentiable. Therefore, several analytical expressions that approximate these operators were proposed *(21, 23, 24)*.

Implicit surfaces are particularly well suited for construction of blends. A blend is a surface that forms a smooth transition between intersecting surfaces. The blends can be classified into global and local. In general, global blends include linear, hyperbolic, and superelliptic ones *(11)*. Local blends limit the domain in which the blending operation is performed by definition of an extra displacement function, which is added to the given set-theoretic operation *(25, 26)*.

Implicit surfaces are suitable for approximation of real world data. Muraki used the Blobby model to fit volumetric data *(27)*. Reconstruction of surface models using the methods based on thin plate splines *(28)* and radial basis interpolants *(29)* from unparallel slices was explored in several works *(30–32)*.

1.1.1. Modeling Environments for Biological Structures

Geometric modeling of 3D biological structures enables biologists to grasp and understand complex features of biological objects by coupling them with specific biophysical processes and by means of suitable visualization methods *(33)*.

Prusinkiewicz *(34)* presented characteristics useful for describing biological models from a computer scientist's point of view. Here, the models were classified into two groups: structure-oriented and space-oriented models. The structure-oriented models typically describe where each component of the structure is located, while the space-oriented models describe what is located at each point of space. Further, the structure and the space that embeds such models may be continuous or discrete. Models may have different topology (nonbranching filaments, a branching structure, a network, a 2D surface, a 3D solid object). In the case of time-evolving structures, the model may occupy constant space or may expand over time. The time-dependent neighborhood relationships between individual modules may be fixed or variable. Communication between the modules may have the form of lineage (information transfer from a parent to its offspring) or of interaction between coexisting modules.

Implicit surfaces support a free-form modeling methodology, which is an advantage in modeling of biological structures. Several modeling environments for implicits have been developed *(35–37)*. These represent the most promising branches in the field of implicits and their utilization in modeling of biological structures. Additionally, some support for implicits is available also in the well-known VTK toolkit *(38)*, including objects of the quadric subclass, CSG implicits, and several other primitive classes.

General progress in computing power and speed has accelerated the existing polygonization algorithms *(39, 40)*, which now allow users to visualize their results nearly in real time. High-quality rendering of implicit surfaces is provided by POV-Ray – the Persistence of Vision Raytracer *(41)*. It reads in a text file with a description of the scene (object specifications, lighting, and a camera) and generates an image by a rendering technique based on ray tracing.

1.2. Study of Cell Architecture

Cell architecture, i.e., the internal organization of cell organelles or structures that perform specific functions, is inferred mostly from electron microscopic images that provide resolution necessary to see all organelles and their relations *(2)*. Electron micrographs, obtained by an electron beam passing through a very thin (40–100 nm) slice of a tissue sample, depict only a minute volume of the tissue. A morphological concept of spatial cell organization emerges after inspection of numerous samples. As individual images are not representative for the whole tissue, quantification of cell structures is performed on many images by the methods of morphometry and stereology. Morphological description involves typical features, such as shapes, positions, and variability of cell components. Morphometric description reflects the average sizes and distances of cell structures as seen in their images. Stereological description provides VSD of cell components relative to a unit of cell volume. The design of the architecture of a specific cell is then expressed using this set of complex characteristics *(2)*.

1.3. Method Overview

In the concept described in this work, cell modeling starts with definition of a model, written down in a file using the model description language (MDL, **Subheading 3.1**). Such a file is then wrapped by job cell configuration (jcc, **Subheading 3.2**) syntax, in which specifications for processing of the input MDL file by the cell generator tool and specification of the contents of output files are provided. After issuing a cell generator command with an input jcc file, a set of cell models is produced; this takes approximately seconds. Cell models can be then verified (stereological quantification) to reveal dissimilarities between the expected and estimated stereological values. Finally, the user can adjust the initial configuration and rerun cell generation so that plausible stereological values are obtained (**Subheading 3.3.2**). Another important criterion of model quality is the appearance or fidelity of the model (**Subheading 3.3.1**). Visual inspection, an analog of morphological analysis, can reveal possible structural inconsistencies. Although the PSE-SBM allows direct visualization of created cell models, it is useful to convert models to their boundary or volumetric representations and take the advantage of interactive rendering techniques. Direct rendering of implicit surface models by a ray-tracing technique produces images for presentation purposes. The diagram in **Fig. 1** summarizes all components of the modeling environment.

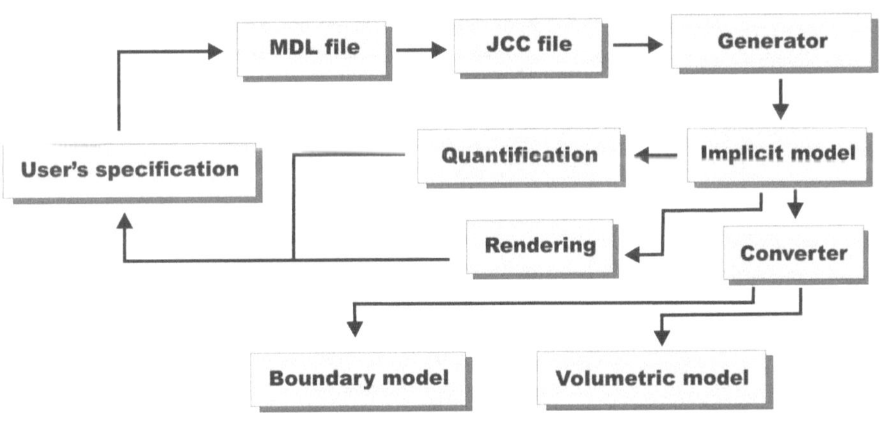

Fig. 1. The iterative modeling procedure: *User's specification* – an interactive specification of a PSE-SBM job, *MDL file* – a set of geometrical and statistic parameters in XML format that cover organelle descriptors, *JCC file* – wraps the MDL configuration describing the output to be produced, *Cell generator* – reads input JCC and MDL files and generates the cell model represented by means of XISL-based implicit functions, *Implicit model* – each organelle is represented by an individual implicit function, *Quantification* – estimation of VSDs, *Rendering* – images obtained by direct ray tracing of the implicit cell models, *Converter* – reads an input implicit cell model and converts it to either boundary or volumetric representation.

2. Materials

The methodology of implicit modeling is not the main topic of this paper. However, to gain a better insight, we introduce here the PSE-SBM system together with an introduction to cell and organelle modeling techniques. Further, we present a computing method for estimation of VSD of organelles.

2.1. Xisl

According to our goal aimed on stereology-based implicit modeling, we have developed XISL – a scripting language – and tools for representation of implicit surfaces *(42)*. The XISL suite is intended to assist developers in construction of implicit models of arbitrary cell or tissue type. Here, implicit models are specified in declarative text files by means of the extensible markup language (XML), where each implicit function class (a primitive, an operation, etc.) is defined by its appropriate tag(s). This ensures clear and self-explanatory notation of complex implicit objects (**Fig. 2**).

The XISL implicits are defined by means of the functional representation *(21)*. In functional representation an object is defined by the inequality $f(x_1, \ldots, x_n) \geq= 0$. In the three-dimensional case, an object defined by such inequality is usually called an implicit solid and an object defined by the equation $f(x_1, \ldots, x_n) = 0$

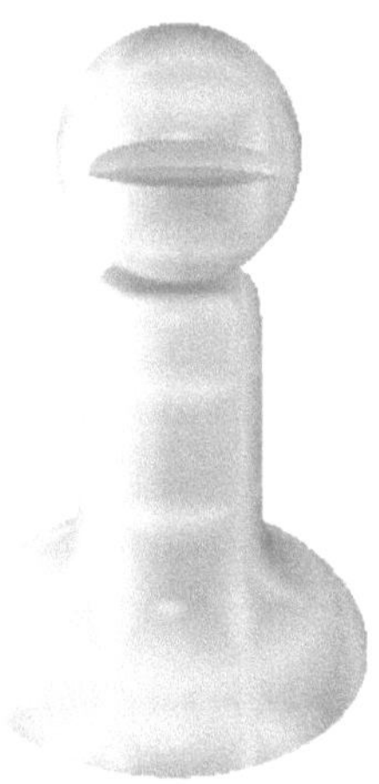

```
<defObject name="pawn">
    <intersectionRf>
        <unionRf>
            <blendedUnionRf a0="0.2" a1="0.3" a2="0.3">
                <aEllipsoid>
                    <vec3 x1="0" x2="0" x3="0"/>
                    <vec3 x1="0.5" x2="0.2" x3="0.5"/>
                </aEllipsoid>
                <blendedDifferenceRf a0="-0.1" a1="0.3" a2="0.2">
                    <aTube>
                        <vec4 x1="0" x2="0" x3="0" x4="0.25"/>
                        <vec4 x1="0" x2="1.3" x3="0" x4="0.20"/>
                    </aTube>
                    <aEllipsoid>
                        <vec3 x1="0" x2="1.4" x3="0"/>
                        <vec3 x1="0.3" x2="0.4" x3="0.3"/>
                    </aEllipsoid>
                </blendedDifferenceRf>
            </blendedUnionRf>
            <translation x="0" y="1.25" z="0">
                <getObject name="head"/>
            </translation>
        </unionRf>
        <aPlane>
            <vec3 x1="0" x2="0" x3="0"/>
            <vec3 x1="0" x2="1" x3="0"/>
        </aPlane>
    </intersectionRf>
</defObject>
```

Fig. 2. Demonstration of the XISL language (script) that defines the implicit "pawn" object and its rendition.

is called an implicit surface. The function *f* can be defined analytically, by means of a function evaluation algorithm, or by tabulated values and an appropriate interpolation procedure. The important property of implicit solids is unambiguous point-object classification. If $\boldsymbol{X} = (x_1, \ldots, x_n)$ is a point in $\boldsymbol{E}^n$, then it is classified as follows: for $f(\boldsymbol{X}) > 0$ it is inside the object, for $f(\boldsymbol{X}) = 0$ it lies on the boundary of the object, and for $f(\boldsymbol{X}) < 0$ it is outside of the object.

The general definition of XISL objects allows implementation of various forms of implicits. Each implicit function is represented via an *n*-ary hierarchical tree, the leafs of which stand for arbitrary implicit primitives and the inner nodes stand for unary, set-theoretic, blending, and interpolation operations.

Several modeling systems based on implicits were developed *(35, 37)*; nevertheless, XISL is a compact, extensible, and operating system-independent package.

2.2. Cell Modeling

The PSE-SBM modeling system allows users to define cell models by means of a high-level XML-based MDL. MDL specification defines organelles by their probability of occurrence and by the mean values of quantitative descriptors such as size and shape, including their variation.

An MDL specification, which defines the cell in a language understandable by the general user, has to be transformed to the low-level scene XISL description. This transformation has to be developed specifically for each type of a 3D scene, that is, for each cell type. In the following, we present application of this approach to striated muscle cells, which represent very complex and highly organized structures.

The initial step in cell modeling involves creation of the central skeleton of the cell, which in the case of muscle cells is represented by a system of parallel cross-sectional graphs (c-graphs) distributed along the longitudinal axis with perturbations specified in the MDL configuration. The longitudinal axis is defined by the orientation of myofibrils. The transversal cell direction is perpendicular to the longitudinal cell axis. An important modeling feature is that each organelle has a preferred orientation along one of these two directions. The c-graphs are used to create the myofibrillar system by means of the F-rep of polygons *(43)* and interpolation. Other organelle types (i.e., mitochondria, t-tubules, sarcoplasmic reticulum, and terminal cisterns) are defined by their own skeletons derived from the c-graphs. Specification of skeletons reflects the observed properties of real cells and their organelles and is specified by appropriate morphological rules in the MDL configuration. This concept is illustrated in **Fig. 3**.

The power of implicits resides in their ability to provide organic-like shapes easily. Cellular structures are mostly rounded objects without sharp edges and other details. Modeling of such

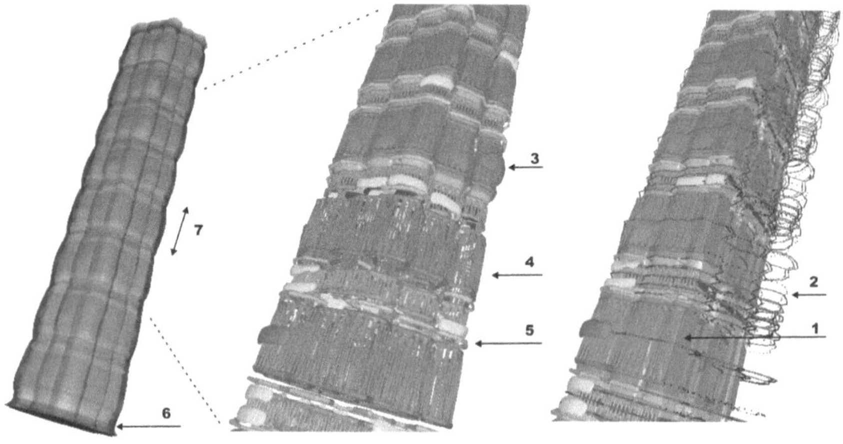

Fig. 3. An example demonstrating eight consecutive sarcomeres of a muscle cell. A sarcomere is indicated by number 7 on the left. For better clarity, the sarcolemma is hidden and, also, the bottom part of the myofibrillar system is clipped off by a transversal plane (*middle*). The complex system of underlying skeletons is made visible by clipping by a longitudinal plane (*right*). The myofibrillar system (1) is defined by means of the c-graphs (2). The remaining organelles include mitochondria (3), sarcoplasmic reticulum (4), t-tubules (5), and sarcolemma (6).

shapes can be achieved by combination of round primitives and proper operations applied on them. For instance, a special class of blend operations is suitable for creating smooth transitions between input round primitives such as algebraic primitives and skeleton-based primitives. In addition, implicit surfaces also directly support point to object classification, object-to-object collision detection, and object deformation, which are useful features in modeling of multiorganelle cells. An important computational requirement is rapid transmission of large models over the network. Therefore, it is crucial to have model representation with low demands on storage capacity. Thankfully, a noticeable feature of implicits is their compressibility. In contrary to other methods, such as boundary representation where an object is represented by a mesh, for the implicit representation it is sufficient to store a function as a set of symbolic terms that represent the function evaluation process (*see* **Fig. 2**). For example, the size of a compressed file containing about thousand of XISL-defined cell organelles did not exceed 1 MB, while a polygonal representation of such model at an appropriate resolution reached several tens of MBs.

2.3. Organelle Representation

Skeletal muscle cells of mammals are populated with various types of organelles, e.g., myofibrils, mitochondria, sarcoplasmic reticulum, terminal cisterns, sarcolemma, t-tubules, etc., which differ in size, shape, and topology. Therefore, in their modeling, different approaches should be employed.

Myofibrils, the contractile fibrils, are thin and long cylindrical objects, segmented into Z, I, and A bands that give rise to the striated pattern of muscle cells when viewed under a microscope.

Muscle cells contain many myofibrils that occupy more than 50% of their volume. Myofibrils are organized in parallel bundles spanning the whole length of the cell. In cross sections, myofibrils have rounded polygonal shape of about 1 μm in diameter. In the model, they are defined by means of cross-sectional graphs (c-graphs) in a system of parallel transversal modeling planes. The resultant formula of a myofibril is then obtained by interpolation between the neighboring c-graphs along the longitudinal axis. Myofibrils vary only slightly along their longitudinal axis and thus their contours in cross section remain nearly unchanged. At the I bands, the myofibril is slightly thinner than at the A bands. With respect to this, it is sufficient to define a single (initial) c-graph representing the basic myofibrillar topology across the cell. When required, this initial c-graph can be obtained from the c-graph database using a special syntax in the MDL configuration. The initial c-graph is then distributed along the longitudinal axis with slight perturbations as depicted in **Fig. 4**.

The thickening/thinning at the A/I band boundaries is specified by means of a user-defined factor. Longitudinal distribution of c-graphs is derived from the known sarcomere length (the distance between two neighboring Z bands), the number of modeled sarcomeres, and the relative length of each band within a sarcomere.

Mitochondria are closed, membrane-bound, elliptically shaped organelles of irregular smooth forms and variable sizes. To capture their varying elliptical shape, we developed a new method based on implicit sweep objects. The basic components of sweep objects are 2D sweep templates and 3D sweep trajectories. Several works addressed the problem of creation of generalized sweep objects *(43–45)*. In general, to create a sweep object, a transformation is created that maps a 2D template point c to a 3D position p. The

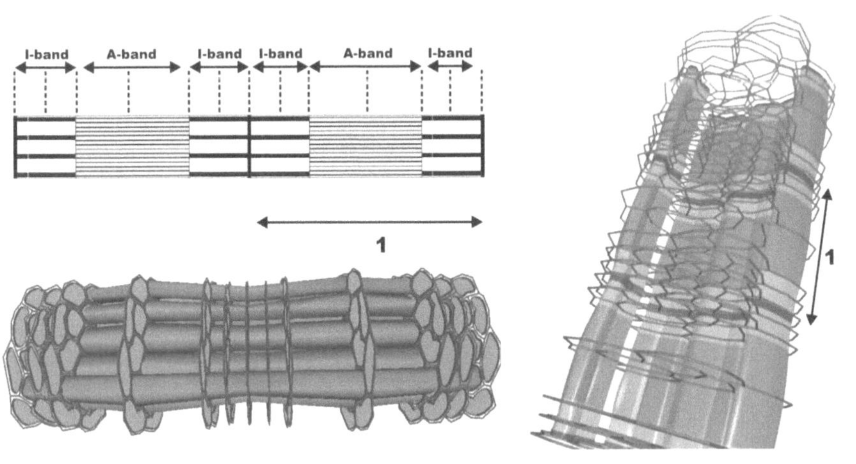

Fig. 4. Modeling of myofibrils. *Left*: an example of c-graph distribution on two sarcomeres (1 – a single sarcomere). Dashed vertical lines at the top represent the distribution of the c-graphs. *Right*: myofibrils and the corresponding c-graphs within three sarcomeres; for better clarity, the myofibrils located in the middle of the cell are hidden.

transformation maps the center of the 2D template to the point that lies on the trajectory. To preserve the cross-sectional elliptical shape of mitochondria, the 2D template is defined as an implicit ellipse with variable dimensions. To represent a 3D trajectory, we adopted the uniform quadratic B-splines.

Transversal tubules (t-tubules) form a planar network around and between myofibrils. In practice, t-tubules can be represented by a network of connected tubes with slightly varying diameter that pass randomly between the myofibrils. In fast skeletal muscles they run typically near the I/A band boundaries. T-tubules are positioned by means of line segments derived from a set of edge line segments of the c-graphs.

Sarcoplasmic reticulum (SR) is a rich but delicate membranous structure consisting of two compartments. The terminal cisterns of the SR are juxtaposed to the longitudinal sides of t-tubules. The network of longitudinal SR grows between the neighboring terminal cisterns along the myofibrils. It is represented by a system of tiny randomly interconnected and longitudinally oriented tubules, creating a typical mesh (*see* **Fig. 3**). Sarcoplasmic reticulum is modeled in two steps. The first step is building of the system of longitudinal tubules, which is achieved by extended interpolation *(46)* between sets of circular implicit shapes. The terminal cisterns, forming a smooth junction to the system of longitudinal tubules, are modeled in the second step. The resultant sarcoplasmic reticulum is obtained by union operation on these two components.

Sarcolemma is a membrane envelope tightly surrounding the muscle cell, which defines the cell volume. It is represented in a similar way as the myofibrils except that the polygonal skeleton is created automatically from the peripheral edges of all c-graphs.

Figure 5 demonstrates all aforementioned organelles, except for the sarcolemma, which is hidden in order to see the cell interior.

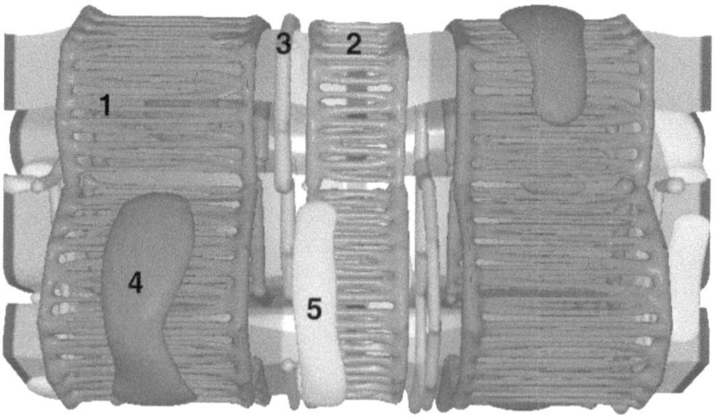

Fig. 5. An example of the model of two sarcomeres. 1 – A-band SR, 2 – I-band SR, 3 – t-tubules following the I/A band interface, 4 – A-band mitochondrion, 5 – I-band mitochondrion.

2.4. Computation of Stereological Densities

Our ambition was to build models that fulfill both geometric and stereological characteristics of a real cell. This is in contrast to traditional approaches, which care solely for the visual appearance and ignore the quantitative characteristics. Therefore, computation of volume and surface area of all objects in the model (the volume and surface densities) is required for evaluation of the model fidelity and for eventual estimation of corrections of input parameters in the MDL file. For this purpose we have developed and implemented a new Monte Carlo-based method for numeric evaluation of volumes and surfaces of all organelles over the whole volume of the model *(3)*. The same tools can be used for simulation of stereological experiments and testing its feasibility.

2.5. Software Tools

For purposes of platform heterogeneity and further PSE-SBM parallelization, our tools are based on command line invocations. Here, we demonstrate the Windows versions of these tools, which are deployed within PSE-SBM. The basic set of tools includes the following:

- `cell_model_generator` (cellJob), which reads an input MDL configuration wrapped by high-level XML syntax and generates cell (XISL) models, executable scripts (*.bat), and the POV-ray files aimed at model visualization.
- `cell_converter` (cell2diff), which provides for conversion of implicit (XISL) cell representation into boundary or volumetric representation; nevertheless, it can be also used in a window GUI mode as an interactive model preview tool using the boundary representation.
- `volume_surface_estimator` (xislVSI2xml), which is required for scripts generated by the `cell_model_generator`, estimating VSDs.

Besides these tools, the user can download and install external (third-party) software required for model visualization. The detailed instructions of software acquisition can be found in *(47)*.

3. Methods

3.1. MDL Specification

Let us assume that the user wants to build a model of a muscle cell with four sarcomeres, each of 1,770 nm in length. The length and number of sarcomeres determine the longitudinal model dimension. The transversal dimension is derived from the number and diameter of myofibrils that should be included in the model. We have prepared a c-graph database containing c-graphs of 4–22 myofibrils. Thus, a user can choose the initial topology of myofibrils from this database or create its own c-graph using an external graphical tool and store it in the proper format. Each sarcomere segment should start and end at the Z band. Next, the longitudinal size of each myofibrillar band has to be specified. Individual bands are represented by colored zones within each sarcomere. Let these dimensions be 700 nm for A bands, 2 × 500 nm for each I band at both sides of the A band, and 70 nm for Z bands; the sum of these lengths is 1,770 nm, as requested. As observed in real cell images, a myofibril cross-sectional size varies along the longitudinal axis; therefore, the user has to define the scaling parameters that are applied to the corresponding c-graphs of each myofibril. The aforementioned specification can be writen in the MDL form as follows:

```
<MDLconf name="modelA">
<data length="1770" sarcNum="4" iBandFract=
"0.56" aBandFract="0.40" zBandFract="0.04"/>
<cgraph file="../mp/mp22/mp_22_4.txt"
aBandMod="50" iBandMod="70" zBandMod="70"
alterMin="5" alterMax="10" iBandScale-
Min="2.95" iBandScaleMax="3.02" aBandScale-
Min="3.07" aBandScaleMax="3.14"/>
</MDLconf>
```

The **<data** .../> tag specifies the size of a single sarcomere (*length*), number of sarcomeres (*sarcNum*), and relative lengths of the I band, A band, and Z band in fractions of sarcomere length (*iBandFract, aBandFract, and zBandFract*).

The second tag **<cgraph** .../> specifies the input c-graph, where *file* represents the input c-graph file name with path (containing 22 myofibrils in this example); *aBandMod*, *iBandMod*, and *zBandMod* define appropriate distances between neighboring myofibrils for each myofibrillar zone. To use random c-graph distortion, one can specify the *alterMin* and *alterMax* attributes, which randomly shift c-graph points within limits of these values. The attributes *iBandScaleMin*, *iBandScaleMax*, *aBandScaleMin*, and *aBandScaleMax* represent the interval from which the scaling values are selected, which are then applied to each c-graph.

The use of only these two tags would result in generation of only the myofibrils. A more complex tag specification is needed to describe other cell organelles, mainly due to their high variability and stochastic character. For example, the user wishes to add organelles with the following properties:

- Mitochondria, which are located at the I and A bands with a diameter (short axis) of 30–50 nm at the I band and 40–60 nm at the A band. Further, it is required that the mitochondria occur more frequently at the I band than at the A band.
- Sarcoplasmic reticulum, positioned both at the I and A bands, with the following geometric parameters: longitudinal SR

tubules of 15–20 nm in radius and the neighboring SR tubules seeded at intervals of 40 nm. The occurrence probability of SR is 100% at myofibrils unless there is not enough space.

- T-tubules running at the border of the I and A bands, where the tubule radii are in the range of 20–40 nm, and they should cover 60% of the segments of the c-graphs.
- The distance between the sarcolemma and the c-graph outer edges is 150 nm.

These attributes are clearly comprehensible to general users; however, within the MDL configurations a user has to specify also the basic geometrical parameters such as skeleton sizes or the sizes of blending areas. Besides, the *cell_model_generator* tool involves a set of hidden parameters regarding implicit modeling, which cannot be adjusted by an unacquainted user. These inner parameters have been tuned by the developer specifically for modeling of striated muscle cells.

Now, by having the organelle parameters specified, the user appends the corresponding tags to the MDL configuration file "modelA". The resultant MDL configuration is as follows:

```
<MDLconf name="modelA">
<data length="1770" sarcNum="4" iBandFract="
0.59" aBandFract="0.40" zLineFract="0.01"/>
<cgraph file="../mp/mp22/mp_22_4.txt"
aBandMod ="50" iBandMod="70" zBandMod="70"
alterMin="5" alterMax="10" iBandScale-
Min="2.95" iBandScaleMax="3.02" aBandScale-
Min="3.07" aBandScaleMax= "3.14"/>
<sarc offset="150"/>
<mitch_IBandT0 prob="0.4" minLength="100"
maxLength="200" minSize="40" maxSize="80"
longVar= "30"/>
<mitch_ABandT0 prob="0.1" minLength="150"
maxLength ="250" minSize="60" maxSize="100"
longVar="100"/>
<t_tubule_hIA prob="0.6" minSize="16" max-
Size ="20"/>
<srA space="110" spaceEps="40" minSize="15"
maxSize="20" blendSize="1" blendImpact="0"/>
<srI space="110" spaceEps="40" minSize="15"
maxSize="20" blendSize="1" blendImpact="0"/>
</MDLconf>
```

Here, the <**sarc**../> tag specifies the sarcolemma with the offset attribute specifying the basic distance between the c-graph and the control polygon of sarcolemma.

The <**mitch_I(A)BandT0**.../> tag defines the occurrence of the I(A) band mitochondria with attributes *prob* specifying the probability of occurrence used in creation of the skeleton, `min-`

`Length` and `maxLength` specifying the allowed sizes of the longitudinal elliptic shape, `minSize` and `maxSize` defining the permitted transversal elliptic shape thicknesses, and `longVar` corresponding to maximal allowed longitudinal deviation within the skeleton location.

The **<t_tubule_hIA**.../> tag defines the occurrence of t-tubules that are created at the I/A band interfaces; the `prob` attribute represents the probability of occurrence of the t-tubule within the c-graph, and the `minSize` and `maxSize` specify the allowed interval of t-tubule radii.

The <**srA, srI**,.../> tags define the sarcoplasmic reticulum. Here, the attributes describe the basic SR elements – seeds, their spacing (`space`), spacing variation (*spaceEps*), permitted dimensions (`minSize, maxSize`), and the amount of blending material (`blendSize, blendImpact`) that smoothly joins the longitudinal SR tubes with terminal cisterns.

Importantly, all dimensions (size, length, etc.) are given in nm, and the relative ratios and probabilities are given in fractions. The aforementioned MDL configuration was stored in a text file mdl.mcc for further use.

3.2. Starting the Model Generation

Now, to be able to use the MDL configuration file by the cell generator tool, the configurations in the file have to be wrapped by the XML syntax. Here, the user can specify additional requirements starting with the `cellJob` tag, which, e.g., enables to specify the number of models to be generated per single configuration, the number of custom lookups per single model, and the precision of volume and surface area computations. The following example illustrates usage of the cellJob tag:

```
<cellJob name="cellX0" scriptType="1" npV
="1000" npS='100000'>
<povray zoomFactor="1.2" camRotX="60" cam-
RotY="150" camRotZ="20" suffix="y150a">
<clip id="SARC" type="0" v1="0.4" v2="0.4"
v3="0.5" v4="1" v5="0" v6="0"/>
<povray>
<povray zoomFactor="0.9" camRotX="60" cam-
RotY="150" camRotZ="0" suffix="y150b"/>
<conf file="mdl.mcc" prefix="X0"
conf="modelA" count="80"/>
</cellJob>
```

The attributes in the individual tags are as follows:

The <**cellJob**...> tag is the topmost element. The *name* attribute identifies the name (`cellX0`) of the job; the `scriptType` attribute defines the type (1 – for Grids, 2 – for Linux, and 3 – for Windows) of the output scripts that perform VSD estimation for each generated cell model. The `npV` and `npS` attributes represent the precision of VSD estimation, i.e., the number of counting points per each organelle, used in the numerical integration.

The **<povray...>** tag defines POV-ray visualization files *(41)* for each generated model; e.g., a single tag instance corresponds to a single output POV-ray file. The `zoomFactor`, `camRotX`, `camRotY`, and `camRotZ` represent the basic camera settings used for visualization of the model. The suffix attribute defines the string appended to the resultant POV-ray file name.

The **<clip...>** tag is an optional tag, which specifies the organelle classes that are intended to be clipped off in the output POV-ray files. The `type` attribute defines the applied clipping primitive (0 – cube or 1 – plane) specified in 3D space by a set of parameters *v1*, ..., *v6*.

The **<conf...>** tag is required by the MDL configuration. The `file` attribute specifies the file, from which the MDL configuration, labeled by the `conf` attribute, will be processed. The output cell (XISL) files will be prefixed by the `prefix` attribute. The `count` attribute defines the number of models requested to be generated by this MDL configuration.

By storing this syntax in the file `cellFile.jcc`, the procedure of the cell model generation (`cellJob`) can be executed by the command line statement "`cellJob cellFile.jcc jobX0`." For instance, by executing the model generator on the presented input script, the output files will include 80 cell (XISL) models, 160 POV-ray files (two per each model), and 80 script files that will compute VSDs. Typically, all models and associated files are created by a desktop PC in few seconds. The user can attach additional MDL configurations using the **<conf.../>** tag.

3.3. Model Inspection and Verification

3.3.1. Visual Inspection

The first aspect that corresponds to the model fidelity is derived from model visualization. The main purpose of visualization is to allow users to ascertain the fidelity of the model, i.e., the architecture of the cell, the topology, appearance, shape and size of organelles, the proportions among them, etc. In other words, the plausibility of the cell structure (i.e., correct understanding and MDL implementation) can be assessed.

Because of the stochastic nature of model generation and/or due to wrong implicit surface behavior, some errors in the model may arise. The SR or mitochondria may be generated with breaks or penetrating each other in some cases. We solved this by designing the underlying skeletons in a way that minimized collisions of organelles. In the case of SR, the underlying skeleton is pushed to the nearest myofibril and the mitochondrial skeleton is shifted in the opposite direction. Nevertheless, a problem may still arise when the input MDL parameters produce too narrow intermyofibrillar free spaces and thus preclude translation of underlying skeletons.

To provide for optimal inspection of the model, either to evaluate its fidelity or to reveal its failures, or for impressive model presentations, we equipped the PSE-SBM system with several model visualization tools, namely, the high-quality ray-tracing technique for representation of implicits, the interactive polygonal model preview for boundary inspection, and the interactive volumetric rendering technique for volumetric representation. The basic representation is the implicit one, in which the models are initially generated. To convert the implicit cell model to boundary or volumetric representation, one can use the `cell_converter` software tool, which takes as an input a special file specifying the XISL cell model and the procedure for model sampling.

In implicit representation, each object is represented by an implicit function that unequivocally separates the interior and the exterior of the object. It is still a challenge to render such objects directly utilizing only the enumeration of implicit functions. Direct rendering of implicit surfaces can be achieved by means of the ray-tracing technique, which is computationally demanding and unsuitable for interactive applications. Nevertheless, the XISL package enables users to render implicit surfaces employing the ray-tracing method by means of the generated POV-ray files, which can be edited according to custom demands in a POV-ray editor. Technically, the list of XISL implicits, written down in POV-ray file, is wrapped by a special syntax. The POV-ray tool is suitable for rendering of final high-quality images (*see* **Figs. 3** and **5**).

Objects defined by implicit functions are usually approximated by boundary models that can be interactively rendered (**Fig. 6**). In boundary representation, objects are defined by a set of triangles that approximate the implicit surface. The `cell_converter` tool includes adaptive reduction of the number of triangles to maximize performance while maintaining the surface

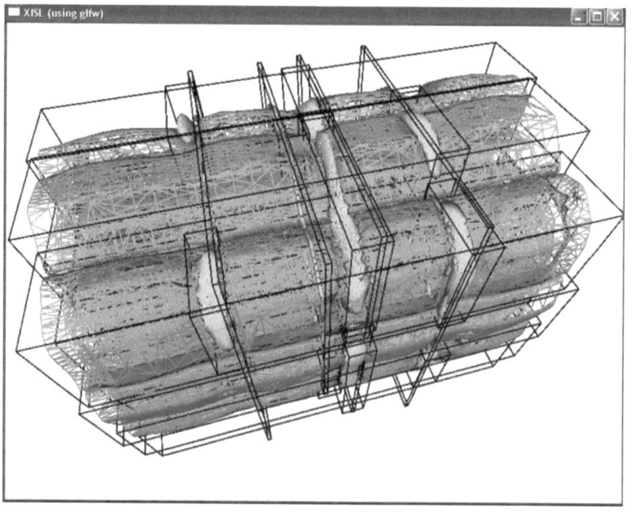

Fig. 6. An interactive preview of a cell model using model triangulation. The user can adjust the number of visible triangles in order to obtain faster rendering (*LOD* level of detail). In each LOD, the result can be stored in the *ply* format, which can be subsequently used as an input of other graphical applications.

fidelity. Such triangular meshes can be stored in the `ply` format *(48)*, which is well supported by standard 3D graphic editors.

The model can be also represented as a 3D sampled volume, where each 3D point represents a scalar value. This is a practical format for spatiotemporal simulations of various kinds. The `cell_converter` tool allows creation of such discretely sampled 3D volume data. The user should specify intensities to be assigned to each organelle class, resolution, and dimensions of the volume. Here, we employ the *f3d* format *(49)* capable of storing 3D Cartesian, regular and rectilinear data, and supporting different kinds of voxel types. Moreover, a set of tools is available for manipulation and rendering of such data. The volumetric cell models can be inspected by means of *f3dviewer* and *f3dvr* tools. The *f3dvr* tool is a sophisticated application for interactive rendering of 3D volume data. It provides for several rendering techniques covering calculation of lighting models or suppression of homogeneous regions in the data set. The data are displayed in the form of artificially created slices generated in parallel through the volume and subsequently blended together. Furthermore, the application allows distinguishing between objects defined by distinct scalar values (intensities), by means of the so-called transfer functions. The transfer functions allow defining color and transparency of objects, and emphasizing or suppressing selected objects. However, it is not always easy to classify the structures only by their intensities. To enhance the classification, multidimensional transfer functions have been developed that additionally utilize gradient magnitude, curvature, or other attributes of input intensities.

In the case of cell models we benefit from the fact that they are created in silico and, therefore, the individual organelle types can be assigned nonoverlapping intensities (**Fig. 7**). With respect to this fact, we have developed a rendering technique (*iso mode*), which combines nearest-neighbor filtering and trilinear filtering with utilization of one-dimensional transfer functions. A transfer function editor window enables users to draw curves representing transfer functions that specify the resultant red, green, blue, and opaque rendering ingredients (**Fig. 8**). The convenient transfer functions can be saved for later use. Some transfer functions have been pregenerated for fast rendering of individual organelle classes and their combinations.

3.3.2. Model Inspection from the Stereological Point of View

Models generated by the PSE-SBM system can include hundreds to thousands of organelles. The important criteria that assess the model credibility are the VSD characteristics. For computation of VSDs, evaluation scripts, pregenerated in the cell model generation step, have to be run. As an example, distributions of volume densities and surface densities computed from 80 models are presented in **Figs. 9** and **10**. This computation is an excessively time-consuming process depending on the amount of created organelles

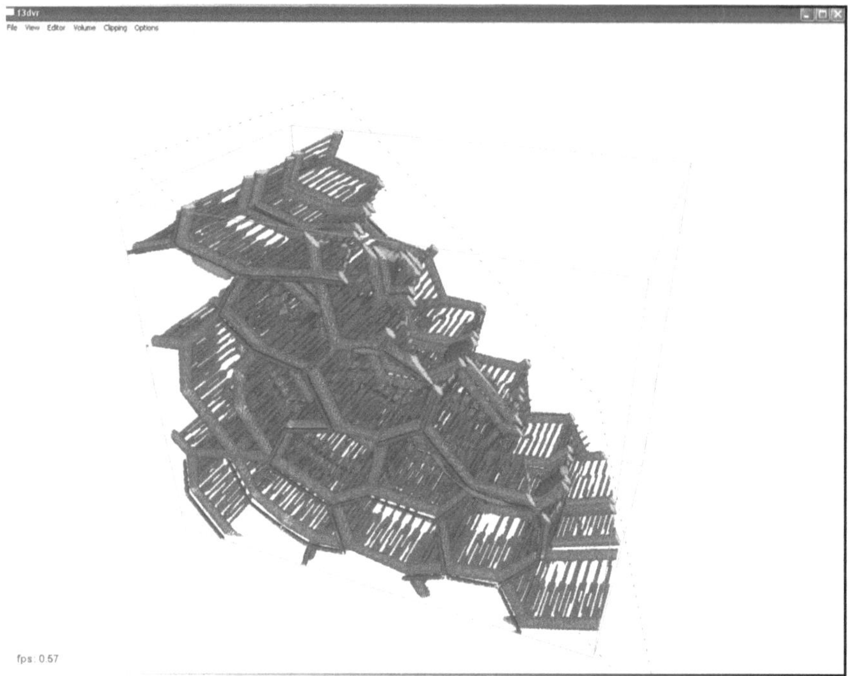

Fig. 7. The interactive presentation of the model by the *f3dvr* tool that enables to inspect the organelles of interests in a 3D volumetric representation.

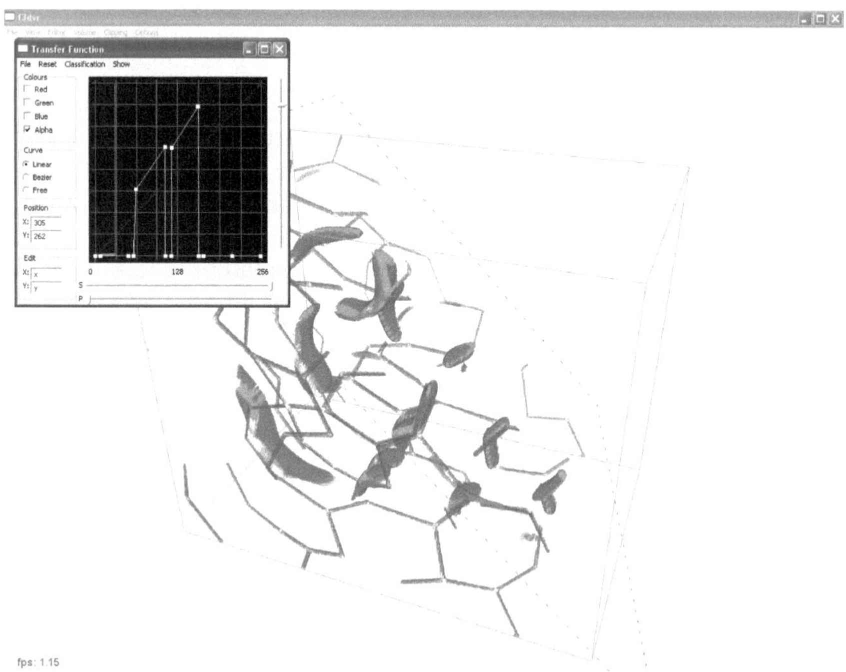

Fig. 8. The transfer function editor window. *Left* – on the *x* axis are intensities and on the *y* axis are color and opaque values. The color and opaqueness is defined for each intensity. The highest value (255) corresponds to total opaqueness and the lowest (0) value corresponds to total transparency. Here, a user wants to display t-tubules, defined in the intensity interval <60, 110>, and mitochondria, defined in the intensity interval <120, 160>, and sets the rest values (i.e., intensity ranges <0, 59>, <111, 119>, and <161, 255>) to 0 and the t-tubule and mitochondrion intensities to visually suitable values, e.g., 255. In practice, it is also convenient to adjust the curves defining the colors of colored models.

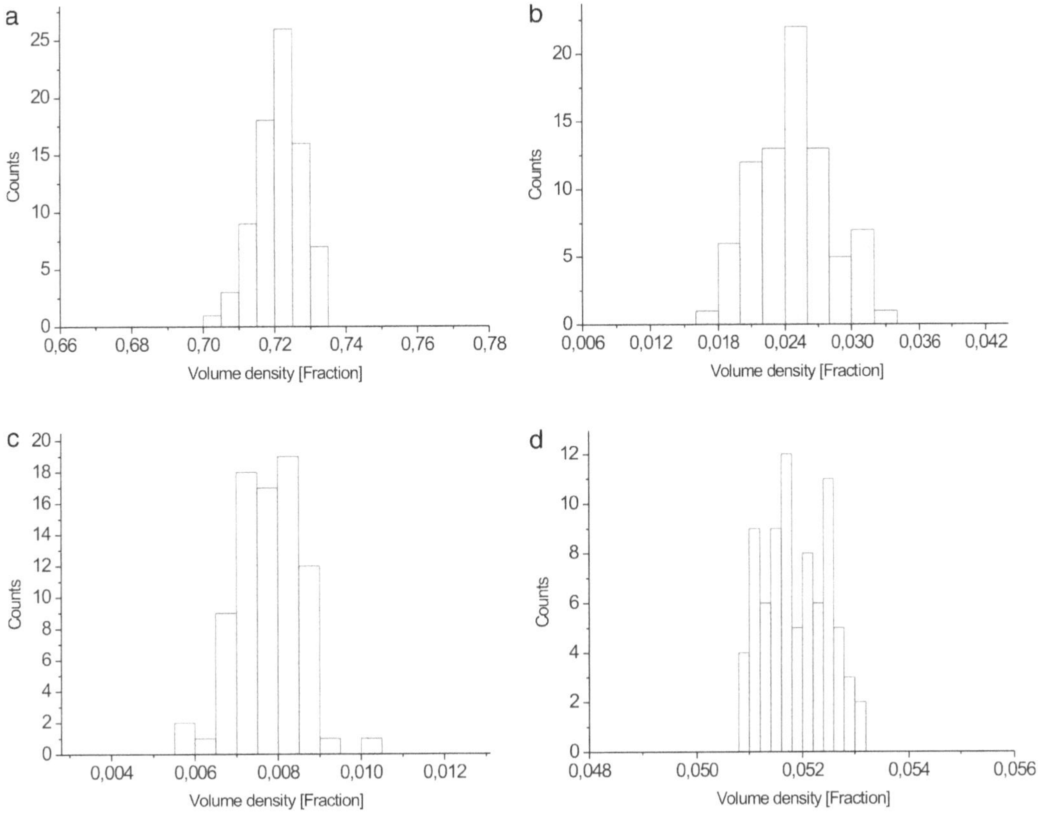

Fig. 9 The distribution of volume densities obtained from 80 models generated from the same MDL configuration. Each graph represents histogram of volume densities per organelle class: (a) myofibrils, (b) mitochondria, (c) t-tubules, (d) sarcoplasmic reticulum.

and the required precision. In this example, each VSD estimation took approximately 20 min per model for npV = 1,000 and npS = 100,000.

It may happen that the user is not satisfied with the resulting VSDs, that is, with the created models. There are several possibilities how to change the MDL configurations in order to obtain the required VSDs. For instance, let us assume that the volume density of mitochondria in the generated models is too low. To increase their volume density, i.e., their relative volume, the user can choose from the following possibilities:

1. Make the input c-graph thinner by setting the attributes `iBandScaleMin` and `iBandScaleMax` in the **<cgraph**…/> tag to lower values, which results in reduction of the global cell dimensions. However, this adjustment also raises VSDs of the other organelles.

2. Increase the probability (the `prob` attribute in the **<mitch_I(A)BandT0**…/> tag) of mitochondria occurrence. Note that too big a probability may generate mitochondria very frequently, and thus make their occurrence unrealistic.

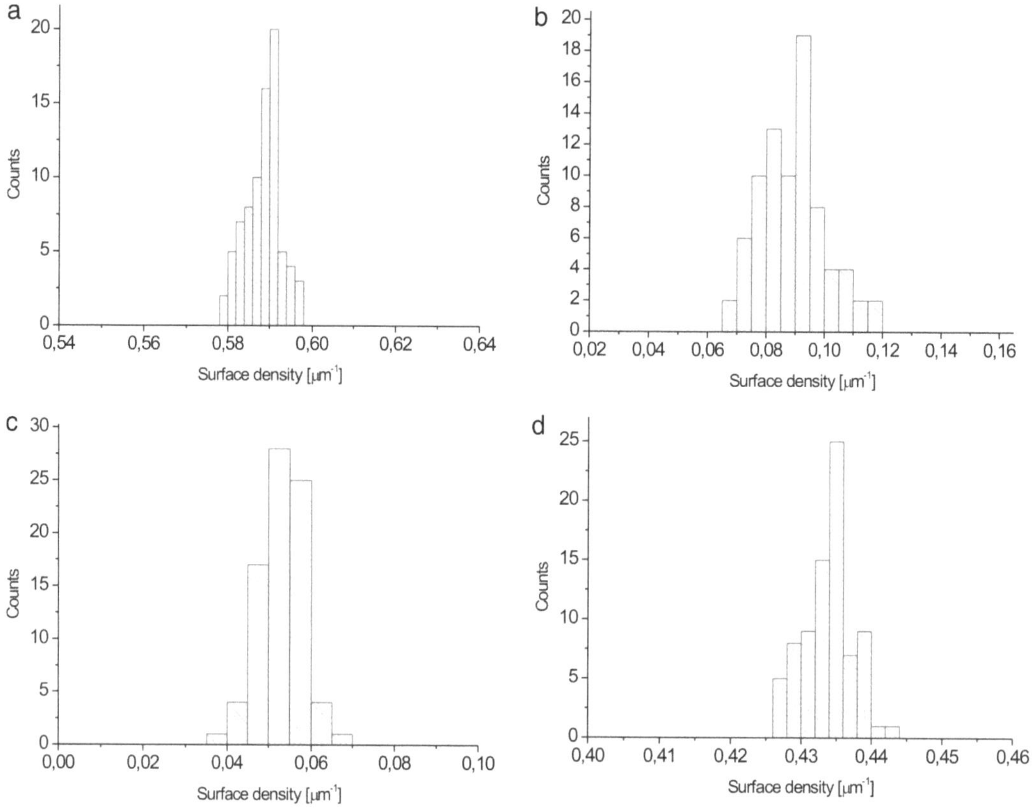

Fig. 10. The distribution of surface densities obtained from 80 models generated from the same MDL configuration. Each graph represents histogram of surface densities per organelle class: (**a**) myofibrils, (**b**) mitochondria, (**c**) t-tubules, (**d**) sarcoplasmic reticulum.

3. Increase one or both of the mitochondrion dimensions (`length, size`). Nevertheless, for too large values, the cell generator may not find suitable skeletons of the proper length, which will result in a small number of large mitochondria.

Therefore, the recommended way to solve this problem is to adjust a combination of all three parameters. Nevertheless, a user can produce several new MDL configurations as an input, where each of them corresponds to one of the possible mitochondrion adjustments. Such MDL configurations, wrapped by the jcc syntax, are then repetitively processed by the cell generator. It is recommended to label models created from multiple configurations differently for easy recognition by means of the prefix attribute in the <**conf**…> tag.

Stereology is traditionally used to quantify geometric properties of cell organelles on the basis of 2D images prepared by electron microscopy *(50)*. However, such stereological experiments might involve errors of different origin that are difficult to assess. Now, with credible synthetic models at hand, testing of stereological hypotheses in silico is possible. Randomly or specifically oriented sets of sections through the model can be produced by the PSE-SBM system and used for stereological analysis by exactly the same method as in the case of real cells. We applied this approach to assessment of the error of volume and surface density estimation in fast skeletal muscle cells performed by students of biology. It was found that the major source of errors resided in problematic assignment of specific loci to a single organelle, as in relatively thick sections it was often hard to distinguish between them.

3.4. Grid Version

Computation of the VSDs is a very time-consuming process requiring up to several hours on a single desktop computer. Therefore, we took the advantage of the grid environment *(7)* that, in addition to the large computational power, offers additional benefits. The task can be specified interactively via a web interface that is also used as a portal to the grid environment. The web interface and the GUI portal (**Fig. 11**) help to transfer/translate requests to the PSE-SBM. Moreover, through the web portal, the user can observe the state of running jobs and see the intermediate results (models) in the form of rendered images and evaluated stereological parameters. The crucial task is to retrieve the models of interest even if the system contains hundreds of computed models. Here, we make use of a specialized grid service,

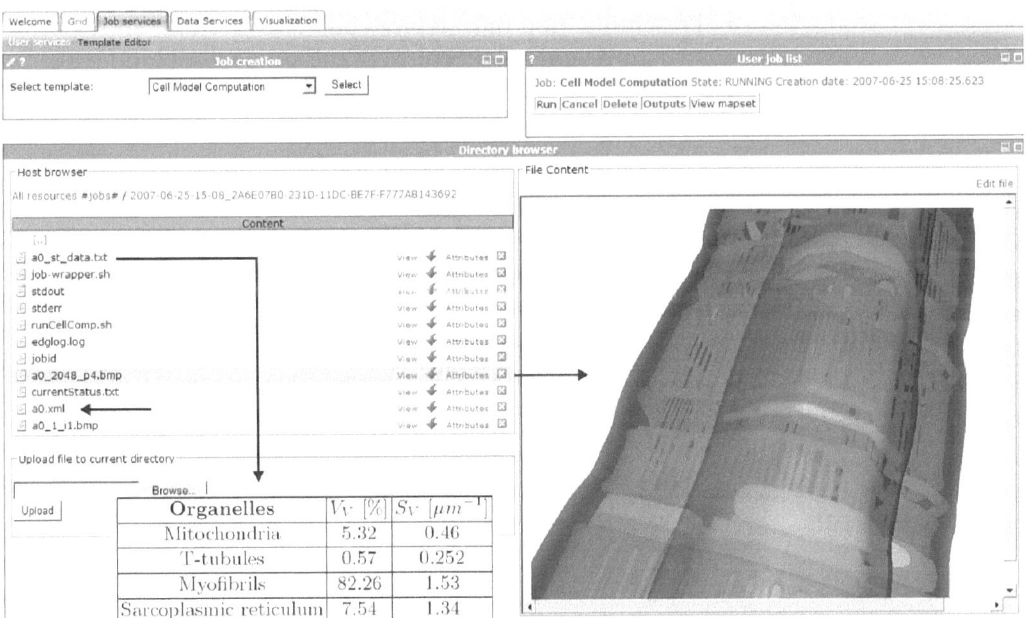

Fig. 11. A screenshot demonstrating the PSE-SBM portal GUI that enables to prepare, submit, and verify models. `File a0_st_data.txt` contains the evaluated VSDs for each organelle class (*the bent arrow*); file `a0.xml` includes XISL definition of the organelles (*left arrow*), and file `a0_2048_p4.bmp` is one of rendered images per the a0 model (*right arrow*). The remaining files, visible in the list on the left side, were generated automatically.

which is capable of registering cell models, their estimated VSDs, and additional parameters in the system. A user then enters a query containing descriptive data of interests and executes it against the system. Plausible models can be then directly downloaded to the user's local machine. The web interface provides users with download and all presented tools for sophisticated model utilization (polygonization, exporting to other formats, interactive rendering, etc.).

3.5. Model Applications

The modeling approach we presented here was stimulated by the need to verify the results of stereological measurements on muscle tissue, to develop a kind of "golden standard" that would be helpful for quantitative morphology. The final result promises more general and wider use worth of further exploration. Here, we summarize the most tempting and promising uses:

For biologists

- To test biological hypotheses: Is the architecture well understood? Can it provide for the observed images and the studied function? Are the stereological and morphometric data reliably measured by the selected method?
- Comparative analysis of cells and tissues in phylogeny and ontogeny, in health and disease, in stressed or relaxed state, etc.
- To assist research and teaching in biology as a tool for reporting and presenting complex results in a concise form.
- Presenting not only typical but also average and specific structures and architectures of the studied cells and tissues.
- Easy generation of numerous variants and views of studied objects.

For informaticians, theoreticians, and biophysicists

- Developing and testing new approaches for modeling and visualization of complex 3D scenes
- Extension by functional models related to real cell structures
- Development of principles for modeling of structures growing and adapting in time and space

The list is long and ambitious. We hope that it is also inspiring.

Acknowledgments

This work was made possible due to the support of APVT 51-31104, APVV-20-056105, and VEGA 2/6079/26. The authors are thankful to M. Ciglan for collaboration on grid application, and to A. Zahradníková for comments on the manuscript.

References

1. Zahradník, I., Parulek, J., and Sramek, P. (2004) Geometrical modelling of the ultrastructure of muscle cells. *Physiol Res* 53, 44P.
2. Novotová, M., Pavlovičová, M., Veksler, V.I., Ventura-Clapier, R., Zahradník, I. (2006) Ultrastructural remodeling of fast skeletal muscle fibers induced by invalidation of creatine kinase. *Am J Physiol Cell Physiol.* 291, C1279–1285.
3. Parulek, J. (2007) Problem solving environment for stereology based implicit modeling of muscle cells. Ph.D. thesis, Faculty of Mathematics, Physics and Informatics, Comenius University, Bratislava, Slovak Republic.
4. Parulek, J., Šrámek, M., and Zahradník, I. (2004) Geometrical modelling of muscle cells based on functional representation of polygons. *J WSCG* 12, 121–124.
5. Parulek, J., Zahradník, I., and Šrámek, M. (2004) A modelling tool for construction of 3d structure of muscle cells, in *Analysis of Biomedical Signals and Images. Proceedings of the 17th Biennial International EURASIP Conference BIOSIGNAL 2004* (Jan, J., Kozumplik, J., and Provazník, I., eds.), Vutium Press, Brno, pp. 267–269.
6. Parulek, J., Zahradník, I., Novotová, M., and Šrámek, M. (2006) Geometric modeling of muscle cells. *G.I.T. Imaging Microsc* 8, 58–59.
7. Parulek, J., Ciglán, M., Šimo, B., Šrámek, M., Hluchý, L., and Zahradník, I. (2007) Grid problem solving environment for stereology based modeling, in *OTM Confederated International Conferences, Part II* (Meersman, R., Tari, Z., eds.), Springer, Berlin, pp. 1417–1434.
8. Blender, http://www.blender.org,accessed on Decenber 14, 2007.
9. Truespace, http://www.caligari.com/, accessed on December 14, 2007.
10. Bloomenthal, J., Bajaj, Ch., Blinn, J., Cani-Gascuel, M.-P., Rockwood, A., Wyvill, B., and Wyvill, G. (1997) *Introduction to Implicit Surfaces.* Morgan Kaufman, San Francisco, CA.
11. Velho, L., Figueiredo, L. H., and Gomes, J. A. (1998) *Implicit Objects in Computer Graphics.* Springer, New York.
12. Blinn, J. (1982) A generalization of algebraic surface drawing. *ACM Trans Graph* 1, 235–256.
13. Wyvill, G., Mcpheeters, C., and Wyvill, B. (1986) Data structure for soft objects. *Vis Comput* 2, 227–234.
14. Nishimura, H., Hirai, M., Kavai, T., Kawata, T., Shirakawa, I., and Omura, K. (1985) Object modeling by distribution function and a method of image generation. *Trans IECE Jpn* J68-D, 718–725.
15. Bloomenthal, J. and Wyvill, B. (1990) Interactive techniques for implicit modeling, in *SI3D'90: Proceedings of the 1990 Symposium on Interactive 3D graphics*, ACM, New York, pp. 109–116.
16. Bloomenthal, J. and Shoemaker, S. (1991) Convolution surfaces, in *SIGGRAPH'91: Proceedings of the 18th Annual Conference on Computer Graphics and Interactive Techniques*, ACM, New York, pp. 251–256.
17. Czanner, S., Durikovic, R., and Inoue, H. (2001) Growth simulation of human embryo brain, in *SCCG'01: Proceedings of the 17th Spring conference on Computer graphics*, IEEE Computer Society, Washington, DC, p. 139.
18. Durikovic, R. and Czanner, S. (2002) Implicit surfaces for dynamic growth of digestive system, in *SMI'02: Proceedings of the Shape Modeling International 2002*, IEEE Computer Society, Washington, DC, p. 111.
19. Oeltze, S. and Preim, B. (2004) Visualization of anatomic tree structures with convolution surfaces, in *Joint EUROGRAPHICS–IEEE TCVG Symposium on Visualization* (Deussen, O., Hansen, C., Keim, D. A., and Saupe, D., eds.), The Eurographics Association, Aire-la-Ville, Switzerland, pp. 311–320.
20. Barr, A. (1981) Superquadrics and angle-preserving transformations. *IEEE Comput Graph Appl* 1, 11–23.
21. Pasko, A. A., Adzhiev, V., Sourin, A., and Savchenko, V. V. (1995) Function representation in geometric modeling: concepts, implementation and applications. *Vis Comput* 11, 429–446.
22. Wyvill, B. and van Overveld, K. (1997) Warping as a modelling tool for csg/implicit models, in *Proceedings of the International Conference on Shape Modeling and Applications, 1997*, IEEE Computer Society, Aizu, Japan, pp. 205–214.
23. Ricci, A. (1972) A constructive geometry for computer graphics. *Comput J* 16, 157–160.
24. Shapiro, V. (2007) Semi-analytic geometry with R-functions. *Acta Numerica* 16, 239–303.
25. Pasko, A. A. and Savchenko, V. V. (1994) Blending operations for the functionally based constructive geometry, in *Set-theoretic Solid Modeling: Techniques and Applications, CSG 94 Conference Proceedings*, Information Geometers, Winchester, UK, pp. 151–161.
26. Dekkers, D., van Overveld, K., and Golsteijn, R. (2004) Combining CSG modeling with soft

blending using Lipschitz-based implicit surfaces. *Vis Comput* 20, 380–391.

27. Muraki, S. (1991) Volumetric shape description of range data using blobby model, in *SIGGRAPH'91: Proceedings of the18th Annual Conference on Computer Graphics and Interactive Techniques*, ACM, New York, pp. 227–235.
28. Duchon, J. (1977) Splines minimizing rotation-invariant seminorms in Sobolev spaces, in *Lecture Notes in Mathematics, Vol. 571* (Schempp, W. and Zeller, K., eds.), Springer, Berlin, pp. 85–100.
29. Floater, M. S. and Iske, A. (1996) Multistep scattered data interpolation using compactly supported radial basis functions. *J Comput Appl Math* 73, 65–78.
30. Savchenko, V. V., Pasko, A. A., Okunev, O. G., and Kunii, T. L. (1995) Function representation of solids reconstructed from scattered surface points and contours. *Comput Graph Forum* 14, 181–188.
31. Turk, G. and O'Brien, J. F. (2002) Modelling with implicit surfaces that interpolate. *ACM Trans Graph* 21, 855–873.
32. Carr, J. C., Beatson, R. K., Cherrie, J. B., Mitchell, T. J., Fright, W. R., McCallum, B. C., and Evans, T. R. (2001) Reconstruction and representation of 3d objects with radial basis functions, in *SIGGRAPH'01: Proceedings of the 28th Annual Conference on Computer Graphics and Interactive Techniques*, ACM, New York, pp. 67–76.
33. Crampin, E. J., Halstead, M., Hunter, P., Nielsen, P., Noble, D., Smith, N., and Tawhai, M. (2004) Computational physiology and the physiome project. *Exp Physiol* 89, 1–26.
34. Prusinkiewicz, P. (1993) Modeling and visualisation of biological structures, in *Proceedings of Graphics Interface'93*, Toronto, ON, pp. 128–137.
35. Wyvill, B., Guy, A., and Galin, E. (1999) Extending the csg tree (warping, blending and boolean operations in an implicit surface modeling system). *Comput Graph Forum* 18, 149–158.
36. Witkin, A. P. and Heckbert, P. S. (1994) Using particles to sample and control implicit surfaces, in *SIGGRAPH'94: Proceedings of the 21st Annual Conference on Computer Graphics and Interactive Techniques*, ACM, New York, pp. 269–277.
37. Adzhiev, V., Cartwright, R., Fausett, E., Ossipov, A., Pasko, A., and Savchenko, V. (1999) Hyperfun project: A framework for collaborative multidimensional f-rep modeling, in *Proceedings of the Implicit Surfaces '99 EUROGRAPHICS/ACM SIGGRAPH Workshop*, New York, pp. 59–69.
38. Schroeder,W.J.,Martin,K.M.,andLorensen,W.E. (1996) The design and implementation of an object-oriented toolkit for 3D graphics and visualization, in *VIS'96: Proceedings of the Seventh conference on Visualization'96*, IEEE Computer Society, Los Alamitos, CA, pp. 93–100.
39. Lorensen, W. E. and Cline, H. E. (1987) Marching cubes: a high resolution 3D surface construction algorithm. *Comput Graph* 21, 163–169.
40. Bloomenthal, J. (1988) Polygonization of implicit surfaces. *Comput Aided Geom Des* 5, 341–355.
41. Povray. Povray–the persistence of vision ray tracer, http://www.povray.org/, accessed on October 17, 2007.
42. Parulek, J., Novotný, P., and Šrámek, M. (2006) XISL – a development tool for construction of implicit surfaces, in *SCCG'06: Proceedings of the 22nd Spring Conference on Computer Graphics*, Comenius University, Bratislava, pp. 128–135.
43. Pasko, A., Savchenko, A., and Savchenko, V. (1996) Polygon-to-function conversion for sweeping, in *The Eurographics/SIGGRAPH Workshop on Implicit Surfaces* (Hart, J. and van Overveld, K., eds.), Eurographics, Eindhoven, The Netherlands, pp. 163–171.
44. Sourin, A. I. and Pasko, A. A. (1996) Function representation for sweeping by a moving solid. *IEEE Transactions on Visualization and Computer Graphics* 2, 11–18.
45. Schmidt, R. and Wyvill, B. (2005) Generalized sweep templates for implicit modeling, in *GRAPHITE'05: Proceedings of the 3rd International Conference on Computer Graphics and Iinteractive Techniques*, ACM, New York, pp. 187–196.
46. Parulek, J. and Šrámek, M. (2007) Implicit modeling by metamorphosis of 2D shapes, in *SCCG'07: Proceedings of the 23rd Spring Conference on Computer Graphics*, Comenius University, Bratislava, pp. 227–234.
47. GeomCell http://cvs.ui.sav.sk/twiki/bin/view/EGEE/UserGuide, accessed on December 13, 2008.
48. PLY format: http://www.cc.gatech.edu/projects/large_models/ply.html, accessed on December 14, 2007.
49. Sramek, M., Dimitrov, L. I., Straka, M., and Cervenansky, M. (2004) The f3d tools for processing and visualization of volumetric data. J Med Inform Technol 7, MIP-71–MIP-79.
50. Elias, H., Henning, A. and Schwarz, D. E. (1971) Stereology: application to biomedical research. *Physiol Rev* 51, 158–196.

Chapter 11

Location Proteomics: Systematic Determination of Protein Subcellular Location

Justin Newberg, Juchang Hua, and Robert F. Murphy

Summary

Proteomics seeks the systematic and comprehensive understanding of all aspects of proteins, and location proteomics is the relatively new subfield of proteomics concerned with the location of proteins within cells. This review provides a guide to the widening selection of methods for studying location proteomics and integrating the results into systems biology. Automated and objective methods for determining protein subcellular location have been described based on extracting numerical features from fluorescence microscope images and applying machine learning approaches to them. Systems to recognize all major protein subcellular location patterns in both two-dimensional and three-dimensional HeLa cell images with high accuracy (over 95% and 98%, respectively) have been built. The feasibility of objectively grouping proteins into subcellular location families, and in the process of discovering new subcellular patterns, has been demonstrated using cluster analysis of images from a library of randomly tagged protein clones. Generative models can be built to effectively capture and communicate the patterns in these families. While automated methods for high-resolution determination of subcellular location are now available, the task of applying these methods to all expressed proteins in many different cell types under many conditions represents a very significant challenge.

Key words: Location proteomics, Subcellular location trees, Subcellular location features, Fluorescence microscopy, Pattern recognition, Cluster analysis, Generative models, CD-tagging, Systems biology.

1. Introduction

A critical aspect of the analysis of a proteome is the collection of detailed information about the subcellular location of all of its proteins. Since subcellular location can change during the cell cycle and in response to internal (mutation) or external (drugs, hormones, metabolites) effectors, the acquisition of sufficient

Ivan V. Maly (ed.) *Methods in Molecular Biology, Systems Biology, vol. 500*

DOI: 10.1007/978-1-59745-525-1_11

information for even a single protein can be challenging. Two strategies are possible: experimental *determination* and computational *prediction*.

The former approach involves assigning class labels to data using automated learning methods. Depending on the application, classes can take on different meaning. Typically in location proteomic studies, various proteins or organelles define classes. If the classes of the data samples are known (in other words, if the data are class-labeled), then supervised learning approaches can be used, wherein classifiers are trained to distinguish between the classes, and new data can be automatically labeled as belonging to these classes. If data is not class labeled, then unsupervised learning approaches can be used, typically to group data by similarity and to identify important clusters in a dataset. In location proteomics, these clusters can correspond to important protein or organelle patterns.

A range of approaches to predicting location from sequence have been described, including detection of targeting motifs, analysis of amino acid composition, and modeling based on sequence homology *(1–6)*. What is clear is that all subcellular location prediction systems suffer from deficiencies in the training data: a limited number of proteins with known locations *and* insufficiently detailed descriptions for those that have been determined. This is because raw experimental data are converted into words that describe location, and both the process of assigning words and the limitations of the words themselves create loss of information. This is true even when standardized terms such as the Cellular Component terms from the Genome Ontology *(7)* are used. (Of course, many determinations of location are done by microscopy at low magnification and therefore the resolution of the imaging becomes the limiting factor.) There is thus an urgent need to collect new protein subcellular location data with high resolution. We first consider approaches using visual assignment of location.

Such efforts can be characterized along three dimensions: whether or not the approach used involves a selective *screen* for a particular location, whether or not the proteins to be analyzed are chosen *randomly*, and whether or not the resolution of the determinations is at or near the limit of optical microscopy. Tate et al. *(8)* used a gene trap approach to screen for proteins localized in the nucleus of mouse embryonic stem cells. Rolls et al. *(9)* used a cDNA library fused with GFP to screen for proteins with nuclear envelope distributions. Similarly, Misawa et al. *(10)* used a GFP-fusion cDNA library to identify 25 proteins showing specific intracellular localization. In contrast, Simpson et al. *(11)* used N- and C-terminal GFP fusion of cDNAs to assign locations to more than 100 novel proteins in monkey Vero cells, while Jarvik et al. *(12)* used random genomic tagging (CD-tagging) to create more than 300 GFP-expressing cell clones and assign locations. Huh et al. *(13)* created a even larger library of 6,029 yeast strains with GFP-tagged ORFs (open reading frames) to characterize the localization of yeast proteins.

While the vast majority of studies of protein location using fluorescence microscopy have employed visual interpretation of the resulting images, there have been efforts to bring automation to this process *(14–23)*. These have been based on work over the past decade demonstrating not only that computational analysis can be used to recognize ***known*** subcellular location patterns *(24–30)* but also that the accuracies achieved are equal to, and in some cases better than, those of visual analysis *(17)*.

Images from many of these studies are publicly available. **Table 1** summarizes some of these and other studies and illustrates how they are different by design. In addition, Schubert et al. *(21)* have developed multiepitope ligand cartography, a robotically controlled immunofluorescence microscopy system that can capture as many as 100 distinct antibodies in the same

Table 1
Data collections relevant to location proteomics

Project	Species (cell type)	Number of proteins	Public access	Tagging method	2D/3D	Mag
Yeast GFP fusion localization database	Yeast	>4,000	yeastgfp.ucsf.edu	cDNA c-terminal GFP fusion	2D	100×
Human Protein Atlas	Human (>40 tissue types)	>6,000	proteinatlas.org	Immunohistochemical staining	2D	20×
CD-tagging database	Mouse 3T3	>100	cdtag.bio.cmu.edu	Internal GFP fusion	3D	60×
GFP-cDNA localization project	Human (HeLa) and monkey (Vero)	>1,000	gfp-cdna.embl.de	cDNA terminal GFP fusion	2D	63×
Protein subcellular location image database	Human (Hela) and mouse (3T3)	>100	pslid.cbi.cmu.edu	Immunofluorescence and genomic internal GFP fusion	2D/3D	100× + 60×
Cell centered database	Various	Various	ccdb.ucsd.edu	Various	2D/3D	60×–40,000×

image sample, but collections of images from this approach are not yet publicly available.

This review briefly covers the process of data collection for determination of subcellular location, followed by a more detailed discussion of a range of automated methods for analysis of the resulting images. The large scale application of these methods over the next few years will help to address the need for large sets of proteins with well-characterized locations, and this in turn will further aid development of future systems capable of modeling and predicting subcellular location.

2. Acquisition of Protein Subcellular Location Images

Perhaps the most common method for determining the subcellular location of a protein is to label the protein with a fluorescent probe and then to visualize the distribution of the protein within cells under a fluorescence microscope. We will limit our discussion to variations on this approach, and we will not consider alternatives such as cell fractionation followed by protein identification and quantitation. Such approaches have been described *(18, 20)* but are fundamentally limited by the resolution of the fractionation step.

A typical fluorescence microscope consists of a light source such as an arc lamp or laser. Light passes through an excitation filter that allows only a specific wavelength through. Next, a condenser focuses the light onto the sample. This excites fluorophores in the sample to emit higher wavelength light that passes through the objective and then an emission filter that removes any undesired wavelengths. Next, the emitted, filtered light hits the detector (a photomultiplier tube or CCD-camera) and is stored digitally as a grayscale image. Multiple filter sets and corresponding probes can be thus used to obtain multiple grayscale images, producing a multichannel image.

The various approaches to tagging a protein for fluorescence microscopy can basically be divided into those that tag native proteins with a fluorescent dye and those that modify the coding sequence of the protein to introduce a fluorescent group into the molecule (for review *see* **refs.** *9, 14)*.

Native proteins are most commonly tagged in situ using antibodies conjugated with a fluorescent dye, but fluorescent probes that can specifically bind to a protein, such as phalloidin binding to F-actin, are also used. However, these approaches cannot usually be applied to a living cell, since the cell membrane has to be made permeable for the probes to enter the cell; moreover, they also require antibodies or probes with appropriate specificity, which make them hard to apply on a proteome wide scale.

Significant efforts to apply immunolabeling at the proteomic level have been undertaken, notably by the Human Protein Atlas *(47)*.

Tagging of proteins by modifying their DNA sequence does not have the above disadvantages. This approach involves either modifying a coding sequence (cDNA) and then introducing this sequence into cells or modifying the genome sequence directly (in either a targeted or a random manner). One of the powerful random tagging techniques is CD tagging *(14)*. In this approach, the coding sequence of a green fluorescent protein (GFP) is inserted randomly into genomic DNA by a retroviral vector. Because the tagging happens to the genomic DNA, the modified protein keeps its original regulatory sequences and expression level. This is in contrast to cDNA modification, in which a constitutive, highly expressed promoter is usually used and thus the expression level of the protein is typically higher than normal. By repeatedly performing random tagging on cells of identical lineage, most of, or eventually all of the proteins within a given cell line can be tagged and have their subcellular locations determined.

3. Interpretation of Protein Subcellular Location Images

3.1. Subcellular Location Features

As mentioned earlier, systems for recognizing subcellular patterns in a number of cell types have been developed. The heart of each of these systems is a set of numerical features that quantitatively describe the subcellular location pattern in a fluorescence microscope image. These features, termed subcellular location features (SLFs), are designed to be insensitive to the position, rotation, and total intensity of a cell image *(29)*. The only requirement for the calculation of these SLFs is that each input image contain a single cell. This requirement can be met in multiple cell images by segmenting the images into single cell regions either manually or automatically, using approaches such as modified Voronoi tessellation *(28)*, watershed *(26, 27)*, levelset methods *(30)*, and graphical model methods *(29)*.

A specific nomenclature has been used to enable unambiguous references to the features used in a particular study. Sets of features are referred to using the prefix "SLF" followed by a set number. Individual features are referred to by the set name followed by a period and its index within the set. For example, SLF1 refers to the first set of features, and SLF1.2 refers to the second feature in this set. We briefly summarize the various types of SLFs below.

3.2. SLFs for 2D Images

Morphological features (SLF1.1–1.8). The high intensity blobs of pixels in fluorescence microscope images might be the first thing a cell biologist looks at when trying to resolve subcellular location patterns. Morphological features mainly describe the characteristics of these blobs, or *objects*. An object is defined as a group of touching (connected) pixels that are above a threshold (the threshold is determined automatically). Eight morphological features have been defined *(15)* to describe the number, size, and relative position of the objects.

Edge features (SLF1.9–1.13). The edge features are calculated by first finding edges in the fluorescence image. These edges can be thought of as consisting of positions that have low intensity in one direction and high intensity in the opposite direction. The number of above-threshold pixels that are along an edge, the total fluorescence of the edge pixels, and measures of the homogeneity with which edges are aligned in the image are especially useful for characterizing proteins whose patterns are not easily divided into objects (such as cytoskeletal proteins). Proteins showing a radiating (star-like) distribution (such as tubulin) have low edge homogeneity, while those showing aligned fibers (such as actin) have higher edge homogeneity *(15)*.

Geometric features (SLF1.14–1.16). The starting point for these features is determination of the convex hull of the cell, which is defined as the smallest convex set which surrounds all above threshold pixels. Three features have been defined using the convex hull: the fraction of the area of the convex hull that is occupied by above threshold pixels, the roundness of the convex hull, and the eccentricity of the convex hull *(15)*.

DNA features (SLF2.17–2.22). The central landmark in eukaryotic cells is the nucleus, and thus having a parallel image of the DNA distribution of a cell is quite valuable. When this is present, a set of features can be calculated to measure quantities such as how far on average protein objects are from the nucleus, and how much overlap exists between the protein and DNA distributions *(15)*.

Haralick texture features (SLF3.66–3.78). For patterns that are not easily decomposed into objects using thresholding, measures of image *texture* are often very useful. Texture features are calculated as various statistics defined by Haralick *(24)* that summarize the relative frequency with which one gray level appears adjacent to another one. Adjacency can be defined in the horizontal, vertical, and two diagonal directions in two-dimensional (2D) images. The texture features are averaged over these four directions to achieve rotational invariance. These features were first introduced for classification of cell patterns in the initial demonstration of the feasibility of automated subcellular pattern analysis *(25)*.

Zernike moment features (SLF3.17–3.65). Like the convex hull and texture features, the rationale behind using these moment features is to capture general information about the distribution of a protein in a rotationally invariant way. Because the Zernike moments are defined on the unit circle, a cell image is first mapped to the unit circle using polar coordinates, where the center of a cell is the origin of the unit circle. Then, the similarity between the transformed image and the Zernike polynomials are calculated by conjugation. By using the absolute value of the resulting moments, the features become rotation invariant *(25)*.

Skeleton features (SLF7.80–7.84). The goal behind these features is to characterize the shape of the objects found by thresholding. This is done by first obtaining the skeleton of each object by a recursive erosion operation on the edge. Each skeleton is then described by features such as its length and degree of branching, and these are averaged over all objects to give features of the cell as a whole *(17)*.

Daubechies 4 wavelet features (SLF15.145–15.174). The principle behind wavelet decomposition is to measure the response of an image to a filter (a wavelet) applied in the horizontal, vertical, and diagonal directions. Wavelet decomposition can be performed recursively, with each pass measuring the response of the filter at a lower frequency *(31)*. Thus the average energy (sum of squared intensities) at each level of decomposition of an image using a wavelet function provides (among other things) information on the frequency (size) distribution of fluorescent objects but without the need for thresholding.

Gabor texture features (SLF15.85–15.144). These features are calculated by convolving an image with a 2D Gabor filter and calculating the mean and standard deviation of the resulting image *(32)*. By using different parameters to generate the Gabor filter, a total of sixty Gabor texture features can be calculated *(33)*.

3.3. Classification of 2D Images

In a series of studies, the SLFs described above have been applied to a set of 2D HeLa cells images showing the distribution of nine proteins and a parallel DNA-binding probe *(15)*. The nine proteins that were labeled by immunofluorescence are located in the endoplasmic reticulum (the protein p63), the Golgi complex (the proteins giantin and gpp130), lysosomes (LAMP2), endosomes (transferrin receptor), mitochondria, nucleoli (nucleolin), and cytoskeleton (beta-tubulin and F-actin). These protein classes which represent the major organelles in a cell were combined with a DNA-stained nucleus class selected from the parallel DNA images to form a 10-class subcellular location dataset. Example images are shown in **Fig. 1**. To evaluate the performance of an automated classifier, 90% of the images in each class were used to train that classifier and then its accuracy was obtained by testing it with the remaining 10% of the images. The process was then repeated nine additional times using different training and testing sets under the constraint that each image appears in a test set only once (this approach is termed tenfold cross-validation), and

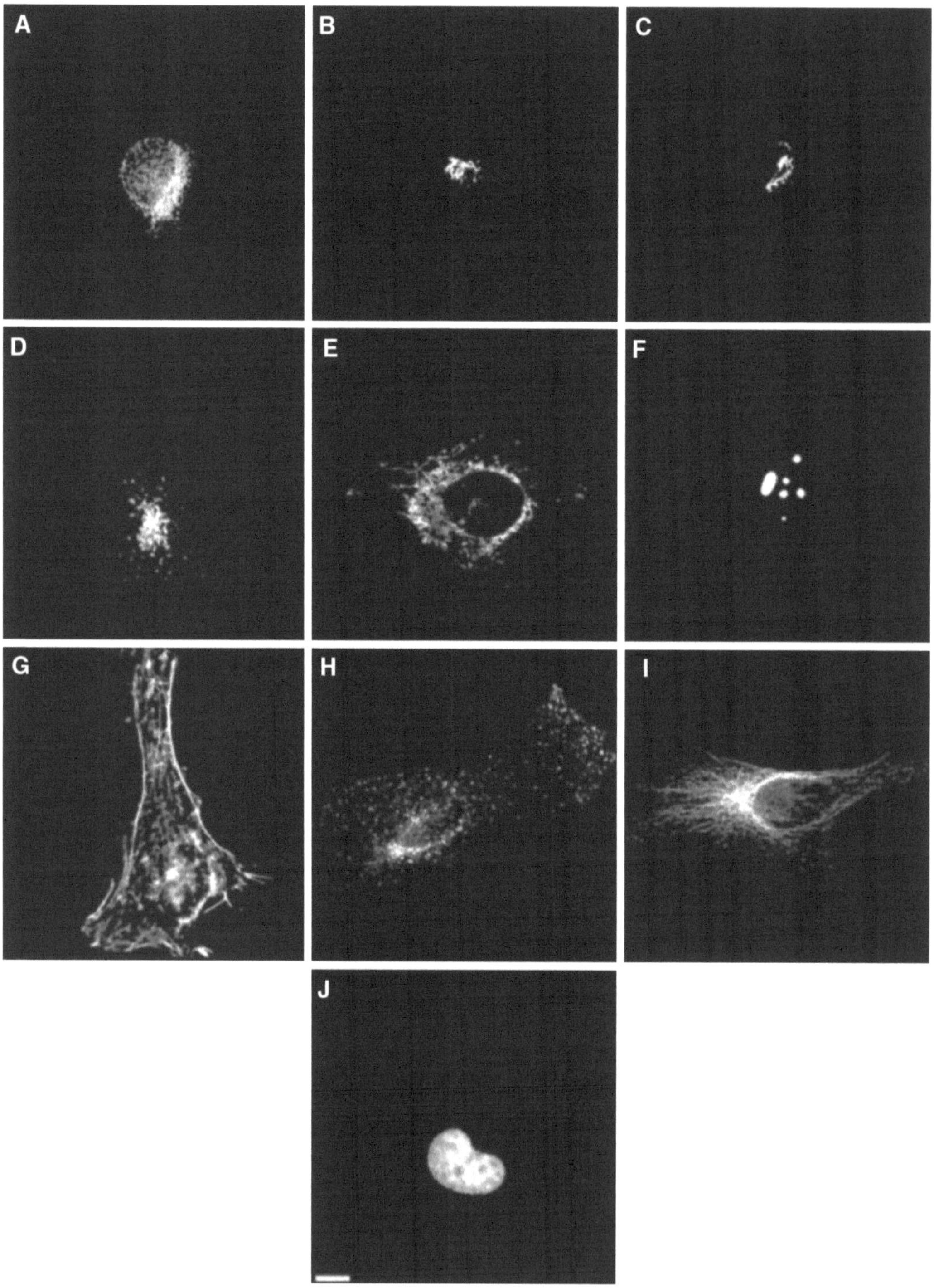

Fig. 1. Representative images of 2D HeLa dataset. These images have been preprocessed to remove background fluorescence and pixels below threshold. Images show the subcellular localization of (**A**) an ER protein, (**B**) Golgi protein giantin, (**C**) Golgi protein Gpp130, (**D**) lysosomal protein LAMP2, (**E**) a mitochondrial protein, (**F**) nucleolar protein nucleolin, (**G**) filamentous actin, (**H**) transferin receptor, (**I**) cytoskeleton protein tubulin, and (**J**) DNA. Scale bar = 10 μm. Reprinted from **ref.** *15* as allowed by Oxford University Press.

results from each repeat were averaged to get an overall classification accuracy. When the image set was first collected, an accuracy of 83% was obtained using a neural network classifier and a set of 37 SLFs *(15)*. Through the use of additional features and classifiers over the past few years, the accuracy on this dataset has risen to 92% for a majority-voting ensemble classifier using a set of 47 SLFs *(33)*. These results are shown in **Table 2**. Even better results have been obtained on this dataset using a multiresolution classification scheme that achieved an accuracy of 95% *(34)*. The automated systems are able to distinguish two Golgi proteins, GPP130 and giantin, which have been shown to be very hard to discriminate by visual inspection, as shown in **Table 3** *(17)*. A comparison of computer (**Table 2**) and human (**Table 3**) classifications is shown in **Fig. 2** *(35)*.

3.4. SLFs for 3D Images

Although most adherent cultured cells are very thin compared to their diameter in the plane of the substrate, a high resolution 2D image (which typically samples from only 0.5 to 1 μm in the axial direction) represents only a fraction of the compartments that are present in the three-dimensional (3D) cell. By taking 2D confocal microscope images at a series of depths within a cell, we can obtain a 3D image of a cell. Sampling in the axial direction is done typically every 0.5–2 μm, but depends on the microscope and the experimental design. Three types of 2D SLFs have

Table 2
Confusion matrix of 2D HeLa cell images using optimal majority-voting ensemble classifier with feature set SLF16

	DNA	ER	Gia	Gpp	Lam	Mit	Nuc	Act	TfR	Tub
DNA	**98.9**	1.2	0	0	0	0	0	0	0	0
ER	0	**96.5**	0	0	0	2.3	0	0	0	1.2
Gia	0	0	**90.8**	6.9	0	0	0	0	2.3	0
Gpp	0	0	14.1	**82.4**	0	0	2.4	0	1.2	0
Lam	0	0	1.2	0	**88.1**	1.2	0	0	9.5	0
Mit	0	2.7	0	0	0	**91.8**	0	0	2.7	2.7
Nuc	0	0	0	0	0	0	**98.8**	0	1.3	0
Act	0	0	0	0	0	0	0	**100**	0	0
TfR	0	1.1	0	0	12.1	2.2	0	1.1	**81.3**	2.2
Tub	1.1	2.2	0	0	0	1.1	0	0	1.1	**94.5**

The overall accuracy was 92.3%. Data from **ref.** *33*

Table 3
Confusion matrix of human classification of images from 2D HeLa dataset

	DNA	ER	Gia	Gpp	Lam	Mit	Nuc	Act	TfR	Tub
DNA	**100**	0	0	0	0	0	0	0	0	0
ER	0	**90**	0	0	3	6	0	0	0	0
Gia	0	0	**56**	36	3	3	0	0	0	0
Gpp	0	0	53	**43**	0	0	0	0	3	0
Lam	0	0	6	0	**73**	0	0	0	20	0
Mit	0	3	0	0	0	**96**	0	0	0	0
Nuc	0	0	0	0	0	0	**100**	0	0	0
Act	0	0	0	0	0	0	0	**100**	0	0
TfR	0	13	0	0	3	0	0	0	**83**	0
Tub	0	3	0	0	0	0	0	3	0	**93**

The overall accuracy was 83%. The major confusion came from the two Golgi protein, giantin and Gpp130, which were hard to distinguish by human inspection. Data from **ref.** *17*

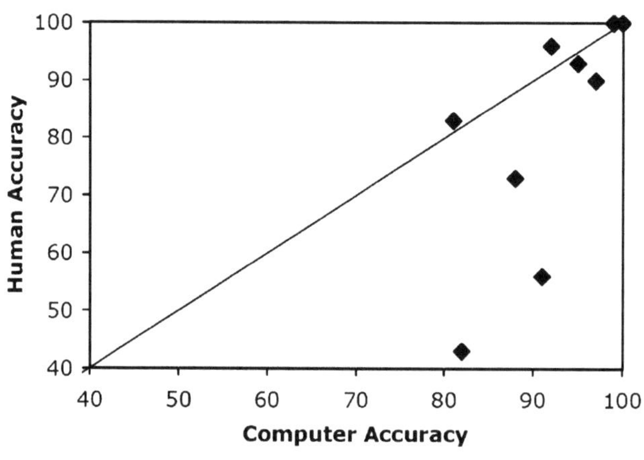

Fig. 2. Comparison of automated and visual classification of subcellular location patterns in 2D images of HeLa cells. Each *dark square* shows the classification accuracy of a specific pattern, while the *solid line* indicates equal performance between the two approaches. While six of the patterns are classified equally well by both, the computer performs significantly better on three of the patterns (two Golgi and one lysosomal). Reprinted from **ref.** *35* with permission (© 2004 IEEE).

been extended to three dimensions so that they can capture some information which is not available in 2D images. Results from automated classification using these 3D SLFs show improvement over the 2D SLF. Brief descriptions of these 3D features are presented below.

Morphological features (SLF9.1–9.28). The 3D morphological features are direct extensions of their 2D counterparts. Objects are found in 3D and size is replaced by volume. Moreover, distance features are decomposed into two components, one situated in the plane of the image and the other axially through the stack. Similar to the case for 2D images, a few 3D features (SLF9.9–9.14) can be defined relative to a parallel DNA image *(27)*.

Edge features (SLF11.15–11.16). The number of pixels along the edges and the total fluorescence of these pixels are calculated on every slice of the 3D images and then summed up. The fractions of these two values over the entire 3D image are used as 3D edge features *(16)*.

Haralick texture features (SLF11.17–11.42). The Haralick texture features can be extended to 3D images by considering all 13 directions in which a pixel can be considered adjacent to its neighbor pixels in 3D space (rather than the four directions in 2D space). The average value and the range of the 13 texture statistics over all 13 directions are used, yielding 26 features. Haralick texture features require a choice of image resolution and gray level bit depth to optimize the performance of recognizing patterns. Experiments revealed that 0.4 μm per pixel resolution and 256 (8 bit) gray levels were the best combination for recognizing subcellular patterns in the 3D HeLa image dataset described below *(36)*.

3.5. Classification of 3D Images

The 3D SLFs have been applied to a set of 3D HeLa images of the same nine proteins as in the 2D HeLa image collection *(27)*. A three-laser confocal microscope was used to record images of cells labeled simultaneously with three different probes (the images were collected in the Center for Biologic Imaging at the University of Pittsburgh with the kind assistance and support of Dr. Simon Watkins). In addition to probes for one of the nine targeted patterns, propidium iodide was used to stain DNA (after RNAse treatment), and a third probe was used to label total cell protein. The image of this third tag was used in combination with the DNA image to automatically segment images into single cell regions *(27)*.

The first evaluation of automated classification of this dataset used 28 morphological features, including 14 features which depend on the parallel DNA image. By using a neural network classifier, an overall accuracy of 91% was achieved *(27)*. To determine how well classification could be performed without using a parallel DNA image, a new feature set SLF14 was created with 14 DNA-independent morphological images, two edge features, and 26 Haralick texture features. An overall accuracy of 98% was achieved using features selected from this set *(36)* as shown in **Table 4**. The results are nearly perfect, and the extension from 2D to 3D significantly increases the ability to distinguish the two Golgi proteins, Gpp130 and Giantin.

3.6. Clustering of Subcellular Location Images

The classification results described above have shown the ability of the SLFs to distinguish major subcellular location patterns with a classifier trained on class-labeled images. This is supervised learning, in which the protein or location classes are known at the outset. In contrast, unsupervised learning tries to find an optimal way of dividing *unlabeled* images into distinct groups or clusters. In location proteomics, clustering methods are used to find the major subcellular location pattern groups for all proteins across a proteome or large dataset. An optimal clustering on the location patterns of proteomes (finding subcellular location families) can offer a fundamental framework for assigning locations to proteins. Such a framework is useful for many reasons, one of which is because it can be used to automatically generate an ontology that effectively describes protein locations, and another of which is that each pattern (family) is tied to the images that defined it.

Table 4
Confusion matrix of 3D HeLa images using neural network classifier with seven features selected from SLF17

	DNA	ER	Gia	Gpp	Lam	Mit	Nuc	Act	TfR	Tub
DNA	**98**	2	0	0	0	0	0	0	0	0
ER	0	**100**	0	0	0	0	0	0	0	0
Gia	0	0	**100**	0	0	0	0	0	0	0
Gpp	0	0	0	**96**	4	0	0	0	0	0
Lam	0	0	0	4	**95**	0	0	0	0	2
Mit	0	0	2	0	0	**96**	0	2	0	0
Nuc	0	0	0	0	0	0	**100**	0	0	0
Act	0	0	0	0	0	0	0	**100**	0	0
TfR	0	0	0	0	2	0	0	0	**96**	2
Tub	0	2	0	0	0	0	0	0	0	**98**

The overall accuracy was 98%. Data from **ref.** *36*

There are many different clustering algorithms and most of them require a similarity (or distance) function, which defines the way to calculate the similarity (or dissimilarity) of images in the feature space. Two well-known distance functions are Euclidean distance and Mahalanobis distance. Instead of unscaled Euclidean distances, which calculate the straight line distance in feature space between two images, standardized (or *z*-scored) Euclidean distances, which are Euclidean distances calculated after normalizing each feature to zero mean and unit variance, can be used. The Mahalanobis distance takes into account the correlation between features by scaling the distance with the covariance. Standardized Euclidean distance was shown to empirically produce the best agreement among different clustering algorithms applied to subcellular location images *(19)*.

k-Means clustering is a well-known centroid-based algorithm. Each data point is grouped into one of *k* clusters whose centroid is closest to it in the feature space. A centroid of a cluster is defined as the average feature vector of all the data points in that cluster. The starting centroids of the *k* clusters are randomly chosen from the data points or randomly generated. When a new data point is clustered into a certain cluster, the cluster centroid is updated accordingly. The process is repeated over all data points a few times until all the clusters converge.

To determine into how many groups the data should be clustered, an Akaike Information Content (AIC) score can be calculated for many values of *k*, the number of clusters. AIC measures the log-likelihood of the model penalized by the number of parameters of the model. A clustering result with small *k* and small variance of each cluster will have a relatively low AIC score, which means the clustering result is good. By varying *k* and comparing the AIC scores, an optimal *k* can be found *(37)*. Bayesian Information Criterion can be used in place of AIC.

Unlike the k-means clustering algorithm, hierarchical clustering does not depend on the choice of the number of clusters. Initially, each of the data points is a cluster. The distances of all of the clusters are calculated pairwise and the closest two clusters are joined together. This is repeated until all are joined. The result of hierarchical clustering shows how the clusters converge to fewer but larger clusters. A dendrogram is usually used to show the result of hierarchical clustering. A dendrogram generated from the SLFs of fluorescence microscope images has been termed a "subcellular location tree (SLT)" *(16)*. A SLT tells us how close the subcellular location pattern of one protein is to that of another protein. In order to increase the robustness of hierarchical clustering, consensus methods can be used *(19)*. In consensus clustering, a random half of images from each protein is used to build a hierarchical tree.

This is repeated and a consensus tree is built to show branches that are conserved *(38)*.

A third clustering approach is based on the confusion matrix generated by a classifier. This approach starts with training a classifier to discriminate all different proteins regardless of the possibility that some proteins may share the same location pattern. If two proteins actually do share a same location pattern, the classifier will not be able to tell them apart, which will then be shown in the confusion matrix as a large number in off-diagonal cells. By merging such confused proteins into a group, we can finally combine proteins which share a location pattern and obtain clusters which can be well separated by the classifier *(19)*.

As described before, the CD-tagging technique has been used to introduce an internal GFP domain in randomly targeted proteins in mouse 3T3 cells and to prepare a large library of subcellular location images *(12)*. 3D images have been collected for these clones using spinning disk confocal microscopy. The consensus clustering based on k-means algorithm divided 90 proteins into 17 groups, which represent the major location patterns distinguishable by the current 3D SLFs. A SLT was also generated on the same dataset. The proteins assigned to the same branch of the SLT often visually appear to display similar patterns. On the other hand, the proteins with distinct location patterns are well separated. This SLT (shown in **Fig. 3**) and the representative images of each leaf are available online at http://murphylab.web.cmu.edu/services/PSLID/ *(19)*. The whole process of building such a consensus SLT is automated and objective. The tree shows the major subcellular location patterns which are distinguishable in a collection of 90 different proteins in 3T3 cells as well as the hierarchical relations among these patterns. This clustering method is very promising to reveal the framework of protein subcellular location families when a complete image collection is available for all the proteins in a given cell type. Recently, images of 188 randomly tagged clones have been clustered into 35 distinct location clusters *(23)*. In addition to being used to group proteins by their location patterns, clustering of images has been used to group drugs by their effects upon subcellular patterns *(39)*.

3.7. Multiple Cell Image Analysis

Thus far we have discussed analysis of independent single cell regions. However, most fluorescence micrographs contain multiple cells per image field, and there is useful information in the spatial distribution of cells. Moreover, these cells may be expressing extracellular proteins of interest, and may be influencing each other (through things like cell division, hormonal signaling, or mechanical coupling).

There are various approaches to dealing with multicell images. The simplest are applied to images containing only one

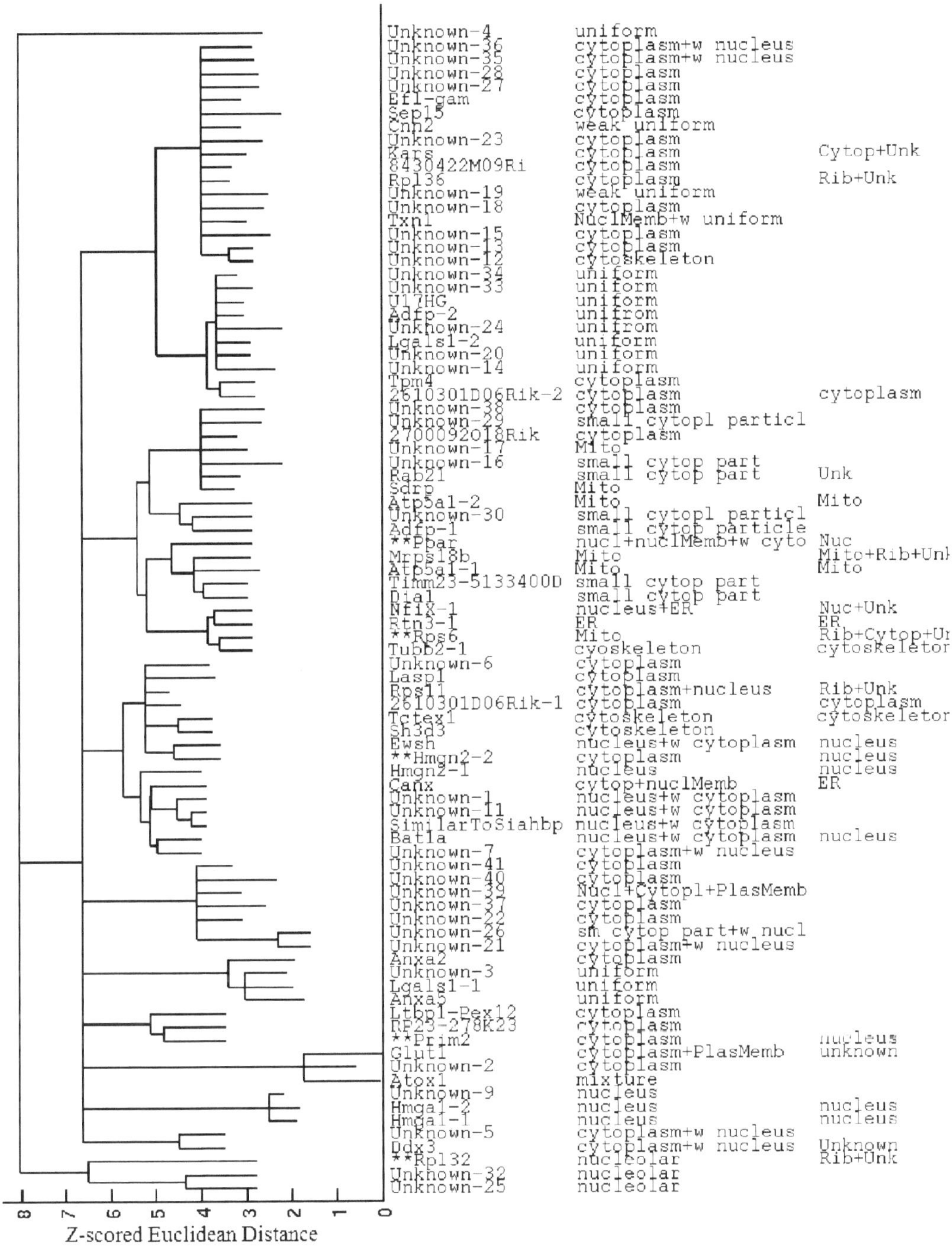

Fig. 3. A consensus subcellular location tree generated from the 3D 3T3. image dataset. The SLF11 feature set and standard (*z*-scored) Euclidean distance were used. The columns on the *right* of the tree show the protein name (if known), human observation of subcellular location, and subcellular location inferred from Gene Ontology (GO) annotations. Proteins marked with a *double asterisk* have significantly different locations between the description of human observation and the inference from GO annotation. Reprinted from **ref.** *19* under the terms of the Creative Commons License.

pattern in all of the cells. In one approach, field level features, which are independent of the number and rotation of cells in the image, are used to train classifiers. Huang and Murphy *(40)* showed that such features could be used to give a 95% accuracy. Their work was done using a modified version of the 2D HeLa images described above, where they used multiple single cell regions to synthesize multicell images containing anywhere from 2 to 6 cells. Following this work, Newberg and Murphy *(41)* showed that field level features combined with voting classification schemes can be used to effectively analyze protein patterns across human tissues. They trained a system that could distinguish between eight major organelle patterns with an 83% accuracy; this became 97% when only the most confident classification assignments were considered.

In another approach, information from surrounding cells is used to influence the classifier assignments for a local cell region in the image. This approach thus involves segmentation as a first step. If the image contains a homogeneous pattern (that is, all of the cells express the same protein pattern), simple voting methods can be used. These involve segmenting images, using SLFs in the classification of single cell regions, and then simply assigned the most common class label in the multicell image to all regions in that image. When multicell images contain more than one pattern (i.e., one group of cells expressing a tubulin pattern and another expressing a nuclear pattern), more complex voting schemes are needed. Chen and Murphy *(42)* showed that a graphical models approach can effectively deal with inhomogeneous data. This works by allowing close cell regions to have more influence than further away regions when deciding upon a class label for that region. Distances can be measured in both the physical space (where regions lie in an image) and feature space. Using synthetic multicell data (generated from the 2D HeLa image set), they were able to achieve greater than 90% accuracy in images containing up to four different types of patterns. This initial approach has been significantly improved and extended in subsequent work *(43, 44)*.

3.8. Object Type Recognition and Generative Models

The aforementioned methods consider protein subcellular location patterns at the level of each cell (or group of cells) and do not capture any information about the individual components of the cellular pattern. When they are applied to a new mixture pattern which combines the components from several different basic patterns (i.e., the location pattern of a protein which exists in different organelles or compartments), the cell level recognition methods tend to either generate a new location group (clustering) or simply be confused (classification). A more desirable result, however, might be a quantitative breakdown of how basic patterns compose the new mixed pattern *(45)*. To this end, an object-based method was developed wherein object types are learned from several class-labeled images, and then they are used to recognize a new image pattern based on this pattern's object type composition. In this two-stage learning problem, first objects are extracted from known image classes and the object types are learned by clustering on object features, termed subcellular object features (SOFs). Note that objects in an image are defined as a group of connected pixels that are above some threshold. In the second stage, features, which describe the object type composition as well as the relative positions of these objects, are extracted from new mixture patterns. These in turn can be used to train classifiers to recognize the new patterns *(45)*.

This two-stage method has been applied to the previously described 2D HeLa dataset, which consists of ten different subcellular location classes. AIC-based k-means clustering on the extracted SOFs indicated that there were 19 unique object types in the images. Next, from each image sample, 11 SOFs and two SLFs were extracted for each of the 19 object types. A classifier was trained using a subset of these features to distinguish between the ten classes. Classification accuracy using cross-validation was 75%, and when the two Golgi apparatus proteins were merged, the accuracy increased to 82%. These results indicate that the SOFs and object types are informative for describing the protein patterns *(45)*.

The utility of these features and object types is that they can be used to characterize mixture patterns. Zhao et al. *(45)* demonstrated this using an unmixing approach to decompose mixture patterns into components of fundamental patterns. A linear regression method was first applied. It assumes that the features of a mixture pattern are linear combinations of the features of fundamental patterns. The coefficients (weights) of each fundamental pattern can be solved from linear equations. However, even in fundamental patterns, the fractions of each object type are not fixed. They vary from cell to cell. In a second unmixing approach, multinomial distributions were used to model the object type components of fundamental patterns and the fundamental pattern components of mixture patterns. The parameters of the model were then solved by the maximum likelihood method.

The object-type-based pattern recognition enables systems to recognize patterns composed of a mixture of components (object types) of the basic patterns. The learned object types can potentially be used to describe new subcellular location patterns or subtle protein location changes that might occur when cells are treated with drugs. More importantly, the recognition of the object types makes it possible to build generative models for protein location patterns. Zhao and Murphy *(46)* defined a method that uses a three part model, with a nuclear, cell boundary, and protein component. Each component is learned separately, and the protein model uses object types at its core. In addition to capturing a subcellular pattern, the models also capture the variance of the pattern between images. Thus, these generative models can be used to create sets of images. The power of these generative models is that they, unlike conventional microscopy which only allows for a few proteins to be specifically imaged at a time, potentially allow for the creation of images that contain as many data channels as there are proteins in a proteome, and thus, these models are expected to become an essential tool for location proteomics and systems biology.

References

1. Nakai, K. (2000) Protein sorting signals and prediction of subcellular localization. *Adv. Protein Chem.* 54, 277–344.
2. Park, K. J. and Kanehisa, M. (2003) Prediction of protein subcellular locations by support vector machines using compositions of amino acids and amino acid pairs. *Bioinformatics* 19, 1656–1663.
3. Guda, C., Fahy, E., and Subramaniam, S. (2004) MITOPRED: A genome-scale method for prediction of nucleus-encoded mitochondrial proteins. *Bioinformatics* 20, 1785–1794.
4. Lu, Z., Szafron, D., Greiner, R., Lu, P., Wishart, D. S., Poulin, B., Anvik, J., Macdonell, C., and Eisner, R. (2004) Predicting subcellular localization of proteins using machine-learned classifiers. *Bioinformatics* 20, 547–556.
5. Chou, K. C., and Shen, H. B. (2006) Hum-PLoc: A novel ensemble classifier for predicting human protein subcellular localization. *Biochem. Biophys. Res. Commun.* 347, 150–157.
6. Yu, C. S., Chen, Y. C., Lu, C. H., and Hwang, J. K. (2006) Prediction of protein subcellular localization. *Proteins* 64, 643–651.
7. Harris, M.A., Clark, J., Ireland, A., Lomax, J., Ashburner, M., Foulger, R., Eilbeck, K., Lewis, S., Marshall, B., Mungall, C., Richter, J., Rubin, G.M., Blake, J.A., Bult, C., Dolan, M., Drabkin, H., Eppig, J.T., Hill, D.P., Ni, L., Ringwald, M., Balakrishnan, R., Cherry, J.M., Christie, K.R., Costanzo, M.C., Dwight, S.S., Engel, S., Fisk, D.G., Hirschman, J.E., Hong, E.L., Nash, R.S., Sethuraman, A., Theesfeld, C.L., Botstein, D., Dolinski, K., Feierbach, B., Berardini, T., Mundodi, S., Rhee, S.Y., Apweiler, R., Barrell, D., Camon, E., Dimmer, E., Lee, V., Chisholm, R., Gaudet, P., Kibbe, W., Kishore, R., Schwarz, E.M., Sternberg, P., Gwinn, M., Hannick, L., Wortman, J., Berriman, M., Wood, V., de la Cruz, N., Tonellato, P., Jaiswal, P., Seigfried, T., and White, R. (2004) The Gene Ontology (GO) database and informatics resource. *Nucleic Acids Res.* 32, D258–D261.
8. Tate, P., Lee, M., Tweedie, S., Skarnes, W. C., and Bickmore, W. A. (1998) Capturing novel mouse genes encoding chromosomal and other nuclear proteins. *J. Cell Sci.* 111, 2575–2585.
9. Rolls, M. M., Stein, P. A., Taylor, S. S., Ha, E., McKeon, F., and Rapoport, T. A. (1999) A visual screen of a GFP-fusion library identifies a new type of nuclear envelope membrane protein. *J. Cell Biol.* 146, 29–44.
10. Misawa, K., Nosaka, T., Morita, S., Kaneko, A., Nakahata, T., Asano, S., and Kitamura, T. (2000) A method to identify cDNAs based on localization of green fluorescent protein fusion products. *Proc. Natl Acad. Sci. USA* 97, 3062–3066.
11. Simpson, J. C., Wellenreuther, R., Poustka, A., Pepperkok, R., and Wiemann, S. (2000) Systematic subcellular localization of novel proteins identified by large-scale cDNA sequencing. *EMBO Rep.* 1, 287–292.
12. Jarvik, J. W., Fisher, G. W., Shi, C., Hennen, L., Hauser, C., Adler, S., and Berget, P. B. (2002) In vivo functional proteomics: Mammalian genome annotation using CD-tagging. *BioTechniques* 33, 852–867.
13. Huh, W.-K., Falvo, J. V., Gerke, L. C., Carroll, A. S., Howson, R. W., Weissman, J. S., and O'Shea, E. K. (2003) Global analysis of protein localization in budding yeast. *Nature* 425, 686–691.

14. Jarvik, J. W., Adler, S. A., Telmer, C. A., Subramaniam, V., and Lopez, A. J. (1996) CD-Tagging: A new approach to gene and protein discovery and analysis. *BioTechniques* 20, 896–904.
15. Boland, M. V. and Murphy, R. F. (2001) A neural network classifier capable of recognizing the patterns of all major subcellular structures in fluorescence microscope images of HeLa cells. *Bioinformatics* 17, 1213–1223.
16. Chen, X., Velliste, M., Weinstein, S., Jarvik, J. W., and Murphy, R. F. (2003) Location proteomics – Building subcellular location trees from high resolution 3D fluorescence microscope images of randomly-tagged proteins. *Proc. SPIE* 4962, 298–306.
17. Murphy, R. F., Velliste, M., and Porreca, G. (2003) Robust numerical features for description and classification of subcellular location patterns in fluorescence microscope images. *J. VLSI Sig. Proc.* 35, 311–321.
18. Jiang, X. S., Zhou, H., Zhang, L., Sheng, Q. H., Li, S. J., Li, L., Hao, P., Li, Y. X., Xia, Q. C., Wu, J. R., and Zeng, R. (2004) A high-throughput approach for subcellular proteome: Identification of rat liver proteins using subcellular fractionation coupled with two-dimensional liquid chromatography tandem mass spectrometry and bioinformatic analysis. *Mol. Cell. Proteomics* 3, 441–455.
19. Chen, X. and Murphy, R. F. (2005) Objective clustering of proteins based on subcellular location patterns. *J. Biomed. Biotechnol.* 2005, 87–95.
20. Drahos, K. L., Tran, H. C., Kiri, A. N., Lan, W., McRorie, D. K., and Horn, M. J. (2005) Comparison of Golgi apparatus and endoplasmic reticulum proteins from livers of juvenile and aged rats using a novel technique for separation and enrichment of organelles. *J. Biomol. Tech.* 16, 347–355.
21. Schubert, W., Bonnekoh, B., Pmmer, A. J., Philipsen, L., Bockelmann, R., Malykh, Y., Gollnick, H., Friedenberger, M., Bode, M., and Dress, A. W. M. (2006) Analyzing proteome topology and function by automated multi-dimensional fluorescence microscopy. *Nat. Biotechnol.* 24, 1270–1278.
22. Sigal, A., Milo, R., Cohen, A., Geva-Zatorsky, N., Klein, Y., Alaluf, I., Swerdlin, N., Perzov, N., Danon, T., Liron, Y., Raveh, T., Carpenter, A. E., Lahav, G., and Alon, U. (2006) Dynamic proteomics in individual human cells uncovers widespread cell-cycle dependence of nuclear proteins. *Nat. Methods* 3, 525–531.
23. Garcia Osuna, E., Hua, J., Bateman, N., Zhao, T., Berget, P., and Murphy, R. (2007) Large-scale automated analysis of location patterns in randomly tagged 3T3 cells. *Ann. Biomed. Eng.* 35, 1081–1087.
24. Haralick, R., Shanmugam, K., and Dinstein, I (1973) Textural features for image classification. *IEEE Trans. Systems Man Cybernet.* SMS-3, 610–621.
25. Boland, M. V., Markey, M. K., and Murphy, R. F. (1998) Automated recognition of patterns characteristic of subcellular structures in fluorescence microscopy images. *Cytometry* 33, 366–375.
26. Adiga, P. S. and Chaudhuri, B. B. (2000) Region based techniques for segmentation of volumetric histo-pathological images. *Comput. Methods Programs Biomed.* 61, 23–47.
27. Velliste, M. and Murphy, R.F. (2002) Automated determination of protein subcellular locations from 3D fluorescence microscope images. *Proceedings of the 2002 IEEE International Symposium on Biomedical Imaging,* 867–870.
28. Jones, T.R., Carpenter, A.E., and Golland, P. (2005) Voronoi-based segmentation of cells on image manifolds. *ICCV Workshop on Computer Vision for Biomedical Image Applications,* 535–543.
29. Chen, S.-C., Zhao, T., Gordon , G.J., and Murphy, R.F. (2006) A novel graphical model approach to segmenting cell images. *Proceedings of the IEEE Symposium on Computational Intelligence in Bioinformatics and Computational Biology,* 1–8.
30. Coulot, L., Kirschner, H., Chebira, A., Moura, J.M.F., Kovacevic, J., Osuna, E.G., and Murphy, R.F. (2006) Topology preserving STACS segmentation of protein subcellular location images. *Proceedings of the 2006 IEEE International Symposium on Biomedical Imaging,* 566–569.
31. Daubechies, I. (1988) Orthonormal bases of compactly supported wavelets. *Commun. Pure Appl. Math.* 41, 909–996.
32. Daugman, J. D. (1988) Complete discrete 2-D Gabor transforms by neural networks for image analysis and compression. *IEEE Trans. Acoustics Speech Sig. Proc.* 36, 1169–1179.
33. Huang, K. and Murphy, R. F. (2004) Boosting accuracy of automated classification of fluorescence microscope images for location proteomics. *BMC Bioinform.* 5, 78.
34. Chebira, A., Barbotin, Y., Jackson, C., Merryman, T., Srinivasa, G., Murphy, R. F., and Kovacevic, J. (2007) A multiresolution approach to automated classification of protein subcellular location images. *BMC Bioinform.* 8, 210.
35. Murphy, R.F. (2004) Automated interpretation of subcellular location patterns. 2004

IEEE International Symposium on Biomedical Imaging, 53–56.

36. Chen, X. and Murphy, R. F. (2004) Robust classification of subcellular location patterns in high resolution 3D fluorescence microscope images. *Proceedings of the 26th Annual International Conference of the IEEE Engineering in Medicine and Biology Society*, 1632–1635.
37. Ichimura, N. (1997) Robust clustering based on a maximum-likelihood method for estimating a suitable number of clusters. *Syst. Comput. Jpn* 28, 10–23.
38. Thorley, J. L. and Page, R. M. (2000) RadCon: Phylogenetic tree comparison and consensus. *Bioinformatics* 16, 486–487.
39. Perlman, Z. E., Slack, M. D., Feng, Y., Mitchison, T. J., Wu, L. F., and Altschuler, S. J. (2004) Multidimensional drug profiling by automated microscopy. *Science* 306, 1194–1198.
40. Huang, K. and Murphy, R.F. (2004) Automated classification of subcellular patterns in multicell images without segmentation into single cells. *Proceedings of the 2004 IEEE International Symposium on Biomedical Imaging*, 1139–1142.
41. Newberg, J. Y. and Murphy, R. F. (2008) A framework for the automated analysis of subcellular patterns in human protein atlas images. *J. Proteome Res.* 7, 2300–2308.
42. Chen, S.-C. and Murphy, R. F. (2006) A graphical model approach to automated classification of protein subcellular location patterns in multi-cell images. *BMC Bioinform.* 7, 90.
43. Chen, S.-C., Gordon, G., and Murphy, R.F. (2006) A novel approximate inference approach to automated classification of protein subcellular location patterns in multi-cell images. *Proceedings of the 2006 IEEE International Symposium on Biomedical Imaging*, 558–561.
44. Chen, S.-C., Gordon, G. J., and Murphy, R. F. (2008) Graphical models for structured classification, with an application to interpreting images of protein subcellular location patterns. *J. Mach. Learning Res.* 9, 651–682.
45. Zhao, T., Velliste, M., Boland, M. V., and Murphy, R. F. (2005) Object type recognition for automated analysis of protein subcellular location. *IEEE Trans. Image Process.* 14, 1351–1359.
46. Zhao, T. and Murphy, R. F. (2007) Automated learning of generative models for subcellular location: Building blocks for systems biology. *Cytometry Part A* **71A**, 978–990.
47. Uhlen et al. (2005) A human protein atlas for normal and cancer tissues based on antibody proteomics. *Mol. Cell Proteomics*, 4, 1920–1932.

Part IV

Methods for Larger-Scale Systems Analysis

Chapter 12

Model-Based Global Analysis of Heterogeneous Experimental Data Using *gfit*

Mikhail K. Levin, Manju M. Hingorani, Raquell M. Holmes, Smita S. Patel, and John H. Carson

Summary

Regression analysis is indispensible for quantitative understanding of biological systems and for developing accurate computational models. By applying regression analysis, one can validate models and quantify components of the system, including ones that cannot be observed directly. Global (simultaneous) analysis of all experimental data available for the system produces the most informative results. To quantify components of a complex system, the dataset needs to contain experiments of different types performed under a broad range of conditions. However, heterogeneity of such datasets complicates implementation of the global analysis. Computational models continuously evolve to include new knowledge and to account for novel experimental data, creating the demand for flexible and efficient analysis procedures. To address these problems, we have developed *gfit* software to globally analyze many types of experiments, to validate computational models, and to extract maximum information from the available experimental data.

Key words: Regression analysis, Computational model, Curve fitting, *MATLAB*, Computer simulation, Least-squares.

1. Introduction

Computational models play increasingly important roles in biology. Constructing a model that accurately represents the mechanism of a system, reliably simulates its behavior, and has well-defined parameter values is the ultimate goal of many research projects. Models are used for interpreting experimental observations, testing hypotheses, integrating knowledge, discovering components responsible for certain behavior, designing more informative

Ivan V. Maly (ed.) *Methods in Molecular Biology, Systems Biology, vol. 500*

DOI: 10.1007/978-1-59745-525-1_12

experiments, and making quantitative predictions *(1)*. Remarkably, computational models act both as tools for studying biology and as representations of the resulting knowledge. Indeed, quantitative mechanistic information incorporated into a model allows it to make predictions outside the domain of existing observations.

The focus of this chapter is on understanding experimental data and extracting useful information from it. The role of a model in this process is to postulate a relationship between conditions of experiments and the observed results. Using regression analysis, different models can be tested for their ability to explain the experimental observations, and their parameters can be estimated. Thus, regression analysis ties together models and data, validating the former and extracting information from the latter *(2, 3)*.

Unfortunately, practical application of this procedure to biological systems can be complicated. As will be shown in this chapter, even relatively simple models may contain too many parameters to estimate based on a single experiment of any type. Therefore, to test whether the model is consistent with the data and to determine its parameters, data from multiple experiments need to be analyzed globally, while applying all known constraints to the values of parameters *(4)*.

In this chapter, we discuss the challenges associated with practical application of regression analysis to biological systems. The problems we describe are exacerbated in complex models and experimental designs, and thus are especially frustrating for quantitative biologists. We describe our software, *gfit*, which helps to overcome these problems and illustrate its utility with three biological systems of increasing complexity.

1.1. Regression Analysis

Regression analysis includes a range of methods for establishing a model that accurately represents a system and makes accurate predictions of its behavior. The specific tasks include searching for optimal parameter values, testing whether the model agrees with experimental data, estimating parameter confidence intervals, testing whether more experimental data are needed, detecting outlier points, and selecting the preferred model from two possible ones. In regression analysis, model $\boldsymbol{F}$ is defined as a quantitative relationship between experimental measurements (dependent variables) $\boldsymbol{Y}$ and experiment conditions (independent variables) $\boldsymbol{C}$

$$Y = F(C, x) + \varepsilon \quad (1)$$

where $\boldsymbol{x}$ is a vector of model parameters (variables affecting behavior of the system that cannot be controlled or directly observed during experiment), and ε is a set of measurement errors (*see* **Note 1**) *(5)*.

Goodness of fit, the closeness of model simulations to the measurements, is quantified by objective function $S(\boldsymbol{x})$. The most commonly used objective function is a sum of squared residuals (*see* **Note 2**),

$$S(\boldsymbol{x}) = \sum_{i=1}^{N} \left[Y_i - F(C_i, \boldsymbol{x}) \right]^2 \tag{2}$$

or, in case of nonuniformly distributed ε, a weighted sum of squared residuals

$$S(\boldsymbol{x}) = \sum_{i=1}^{N} \left[\frac{Y_i - F(C_i, \boldsymbol{x})}{\sigma_i} \right]^2, \text{ where } \sigma_i \text{ is a standard deviation of } \varepsilon_i \tag{3}$$

Curve fitting is a problem of finding parameters $\boldsymbol{x}$ that produce the best fit, that is minimize the objective function:

$$\min_{x} S(\boldsymbol{x}). \tag{4}$$

Curve fitting is an optimization problem, performed by optimization engines. Many tasks of regression analysis are based on curve fitting.

1.2. Applying Regression Analysis to Experimental Data

One common obstacle to broader application of regression analysis to biological problems is failure of many models to directly simulate the experimentally observed variable. For example, a typical system model may simulate concentrations of reacting species, values that are rarely observed in an experiment directly. One way of addressing this discrepancy is to convert measured values into the type simulated by the model. However, such conversions often introduce statistical errors and are not always possible. The better solution is to simulate exactly the same value type as measured in the experiment. To achieve that, separate experiment models may be required. Experiment models use the system model to simulate the system's response to manipulations and the experimentally measured signal (*see* **Fig. 1**). The approach of separating system models and experiment models is used in Virtual Cell software *(6)*.

A curve fitting procedure for a heterogeneous dataset can be quite complex and require extensive communication between its entities, i.e., model, optimization engine, experiment conditions, measurements, parameters, and constraints (*see* **Fig. 2A**). Before a search for optimal parameter values can begin, the data for each experiment has to be examined:

- To determine which variables need to be simulated and their sizes
- To check that the data required for the simulation has been provided

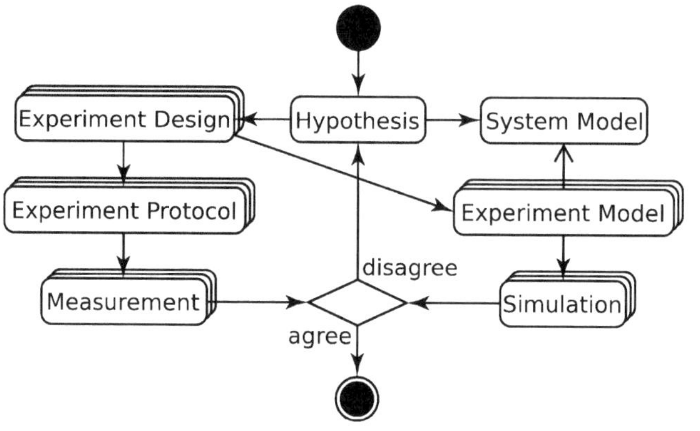

Fig. 1. Application of scientific method to quantitative biology. Mechanistic Hypothesis about a biological system leads to a System Model, a quantitative description of system components and their interactions. To test the Hypothesis, the system is treated in a controlled manner and its behavior is measured. The ideas, assumptions, and, possibly, hypotheses involved in the treatment are parts of Experiment Design. The Measurement obtained by following the Experiment Protocol is compared with Simulation. To make the comparison meaningful, all Simulations need to have same physical meaning and dimensionality as their corresponding Measurements. Therefore, an Experiment Model, an in silico counterpart of Experiment Protocol, is derived from each Experiment Design. Experiment Models interacting with the System Model produce Simulations that are quantitatively compared with the Measurements.

- To check against constraints on variable dimensions and values imposed by the model
- To determine what parameters can be estimated and to choose their starting values

Once the data have been examined, the optimization procedure can be initiated by passing a vector of starting parameter values to the optimization engine. Depending on the engine type, parameter constraints can be also provided. The engine conducts optimization by repeatedly changing parameters and recalculating the objective function on the basis of experimental measurements and simulations. To simulate each experiment, the input data for the model has to be assembled from applicable optimization parameters and experiment conditions. The input data also have to be checked against the constraints, since not all of them can be enforced by optimization engines. After simulating all experiments, the appropriate objective function can be computed and used by optimization engine to determine the direction of the search.

Curve fitting procedure follows complicated rules that depend on the computational model, experimental data, and optimization engine. In addition, parameter constraints need to reflect various considerations related to the research project. These factors make the analysis procedure not only complex, but also highly variable, making design and maintenance of project-specific software prohibitively expensive. Fortunately, the patterns of data flow during regression analysis are largely independent of the system under investigation. This fact allowed us to design software that solves the analysis problem generally and for any model type.

1.3. Design of gfit

The purpose of *gfit* is connecting models with various types of experimental data. First, it simplifies the model's task of directly simulating experimentally observable variables. Second, during regression analysis, *gfit* maintains communications between the analysis components, acting as a mediator (*see* **Fig. 2B**). Third, by defining standard application interfaces for models, optimization engines, objective functions, and other entities, it facilitates customization of the analysis procedure.

Of all components, application interfaces of models represent the biggest problem. Almost every step of regression analysis procedure depends on what information is required and produced by the model. Yet, every model has different inputs and outputs. To be able to perform regression analysis with any kind of computational model, *gfit* uses a metadata approach. Any model used by *gfit* is expected to have an attached Model Description (*see* **Note 3**) defining its inputs and outputs as sets of variables (*see* **Note 4**). More information about Model Descriptions is provided later in this chapter. Once the rules for performing simulations with the model are known, the analysis process becomes more straightforward and independent of the model type (**Fig. 3**).

Regression analysis is a complicated process with many pitfalls. *gfit* strives to provide information that can help researchers avoid mistakes related to the analysis. In the protocols that follow, the reader will build simple models and use the existing models and experimental data for parameter estimation.

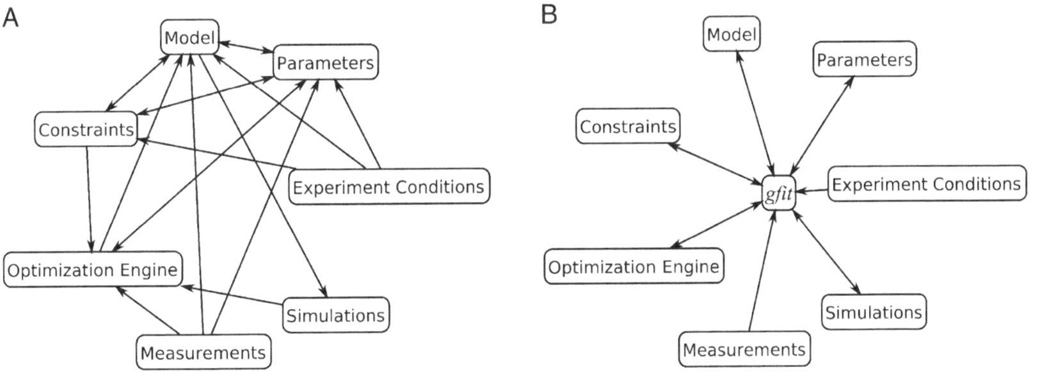

Fig. 2. Components of regression analysis. *Arrows* indicate information flow between components. (**A**) Analysis procedure requires extensive interactions between components. (**B**) To streamline the procedure, *gfit* mediates all interactions between components.

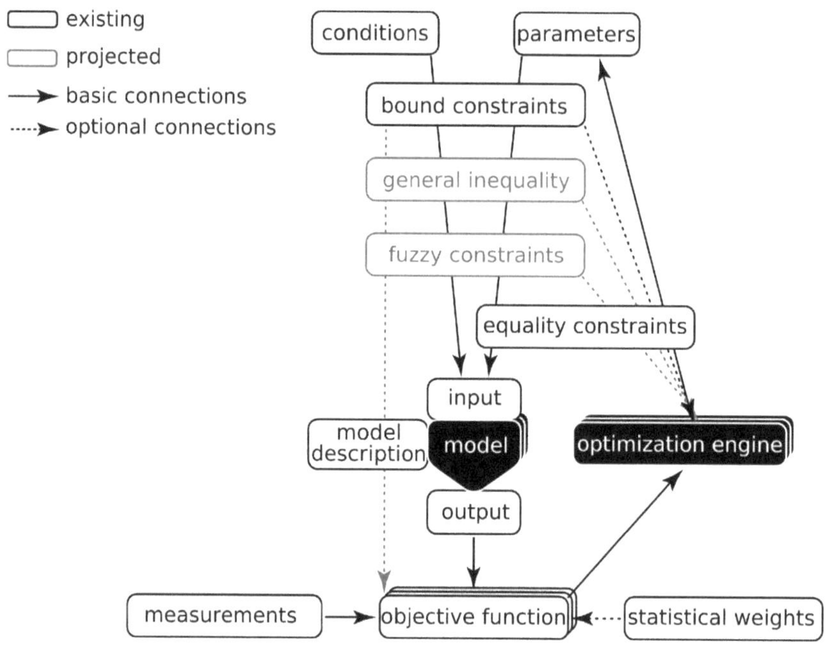

Fig. 3. Flow of information through regression analysis components. Communications between the components are controlled by *gfit* according to Model Description. During simulation of each experiment, independent variables from the experiment conditions and parameters are tested against constraints and combined into model input data. The input received by the model is guaranteed to be valid and to contain sufficient information for the simulation. Combining conditions with parameters keeps the model agnostic about the purpose of the simulation. Size of variables in model output depends on the input. The dependence is defined in Model Description and used by *gfit* to provide the input that will result in model output directly comparable with the measured variables for that experiment.

2. Materials

2.1. Software Requirements

1. Version 6.5 or later of *MATLAB* (Mathworks, Natick, MA), a common science and engineering computing software, is required for running simulations.
2. *MATLAB Optimization Toolbox* (Mathworks) is required for regression analysis.

2.2. Installation of gfit

1. Download the latest version of *gfit* from `http://gfit.sourceforge.net`. The zip-archive contains `gfit.jar` library and other files required for interaction of *gfit* with *MATLAB*.
2. Unzip the file to a convenient location on your hard disk. For this chapter we will assume location `C:/`. Folder `C:/Mgfit` will be created.
3. Start *MATLAB*.
4. Change *MATLAB*'s current directory to `C:/Mgfit`.
5. To start installation, type `mgfit` in *MATLAB*'s command line and press **Enter**.

6. Respond *Yes* to the query about adding `C:/Mgfit` to *MATLAB*'s path.
7. Restart *MATLAB* if requested.
8. After installation, the same command, `mgfit`, will bring up *gfit* user interface window.

2.3. Installation of rsys Library

rsys library is used for solving ODEs for mass-action reaction systems, as described in **Subheading 3.3**.

1. Download the latest version of *rsys* for your operating system from `http://gfit.sourceforge.net`.
2. Unzip the file and put the library file anywhere on the *MATLAB* path. For example, to `C:/Mgfit` folder.

2.4. Data and Models

A zip archive containing all model and data files mentioned in this chapter can be downloaded from `http://gfit.sourceforge.net`. The data files included are in tab/newline-delimited format. These files can be opened in a text editor, but it is more convenient to view them in a spreadsheet. Please check the `readme.txt` file for the most current information.

3. Using *gfit*: Examples

3.1 Simple Model Example: Equilibrium Binding

In this section, we will create a model for equilibrium binding of a protein, *E*, to a ligand, *L*, (**Eq. 5**) and use it for analysis of experimental data. This analysis is quite simple and can be accomplished with many existing programs (including commonly used spreadsheet applications). We will use it to illustrate the principles of data analysis and model validation used by *gfit*, and later apply them to more interesting examples.

$$E + L \underset{}{\overset{K_D}{\rightleftharpoons}} EL, \text{where } K_{\mathrm{D}} = \frac{[E][L]}{[EL]} \tag{5}$$

If only total concentrations of *E* and *L* are known, $E_T = [E] + [EL]$ and $[L_T] = [L] + [EL]$, equilibrium concentration of the complex can be written as

$$[EL] = \frac{K_{\mathrm{D}} + E_{\mathrm{T}} + L_{\mathrm{T}} - \sqrt{(K_{\mathrm{D}} + E_{\mathrm{T}} + L_{\mathrm{T}})^2 - 4E_{\mathrm{T}}L_{\mathrm{T}}}}{2} \tag{6}$$

3.1.1. Create Standalone Model

1. Open *MATLAB* editor by typing `edit` in the command line. Editor window will appear.
2. Add the code shown in **Listing 1** and save the file as `eq_binding.m` in `C:/Mgfit/Models` folder.

Listing 1 *MATLAB* function simulating binding equilibrium

```
1  function signal = eq_binding(Et, Lt, Kd)
2  %simulate equilibrium binding E + L <=> EL
3  EL = ( Kd + Et + Lt - sqrt( (Kd + Et + Lt).^2 - 4 * Et .* Lt ) )/2;
4  signal = EL;
```

Steps 1 and 2 create a *MATLAB* m-file function. Line 1 contains *MATLAB* keyword `function` followed by the name of the output variable *signal*, followed by the name of the function, *eq_binding*, and a list of input variables (*Et, Lt, Kd*). Line 2 is a comment, marked in *MATLAB* by the % character. Line 3 performs the calculation according to **Eq. 6**. Line 4 assigns the calculated, [EL], to the output variable.

3.1.2. Perform Simulation

The model we created has three input variables and one output variable. To perform a simulation, we need to supply values for input variables and store the result in the output variable. Variables in *MATLAB* (and in *gfit*) can contain a single value or arrays of numbers. The arrays may have 0, 1, or many dimensions. 0D array, scalar, stores a single number; 1D array, stores a vector of numbers; 2D array contains a matrix, etc. Models can take advantage of this fact. For example, our model can accept vectors of *Et* and *Lt* concentration and simulate an entire titration curve in one call. We will use the model in **Listing 1** to simulate different experiments.

3.1.3. Simulate Titration Curves

1. Simulate and plot a titration curve for *Lt* changing from 0 to 50

   ```
   Lt = 0:50; EL = eq_binding(1, Lt, 5);
   plot(Lt, EL)
   ```

2. Simulate and plot a titration curve for *Et* changing from 0 to 20

   ```
   Et = 1:20; EL = eq_binding(Et, 10, 5);
   plot(Et, EL)
   ```

   The model is flexible in that it can simulate experiments with different numbers of measurements (20 or 50) and with different combinations of variables that change and remain constant. However, if we did not know how input variables were used inside the model, it would be easy to call the model with illegal variables. For example, the following command will produce an error, because the size of *Et* vector is not equal to the size of *Lt* vector.

   ```
   Et = 1:20; Lt = 0:50; EL = eq_binding
   (Et, Lt, 5);
   ```

As well, mistakenly supplying negative values for the input parameters will produce a warning and a meaningless simulation result. Although we can keep track of correct variable sizes and values when manually executing a simple model, this task becomes tedious if many different experiments have to be simulated by a complex model. To create a connection between experimental data and a model that is practically useful, there has to be a method for automatically tracking the requirements of the model and for reconciling these requirements with existing experimental data. *gfit* learns about the requirement of the model and its expected output by reading the associated user generated Model Description (*see* **Note 3**).

3.1.4. Create gfit Model Description

1. Insert lines 3–13 into previously created `eq_binding.m`, as shown in **Listing 2**.
2. Add variable *signal* to the list of inputs.

   MATLAB m-files with a tag as on line 3 are recognized by *gfit* as models. Lines 3–13 are occupied by Model Description. Note that each of the lines starts from character %, a comment in *MATLAB* language. Therefore, Model Description is ignored by *MATLAB* and the meaning of the original program is not changed. Model Description is used by *gfit* to ensure that the model always receives legitimate input variables and to interpret experimental data that belongs to the model.

The information in Model Description is organized in a tabular fashion. Any number of tables can be present. Tables in Model

Listing 2. *gfit* model for binding equilibrium.

```
1  function signal = eq_binding(Et, Lt, Kd, signal)
2  %simulate equilibrium binding E + L <=> EL
3  %<gfitModelDescription version="100">
4  %variable  type         minVal
5  % Et       free         0
6  % Lt       independent  0
7  % Kd       para         0
8  % signal   dependent    ()
9  %
10 %variable  size  plotVs
11 % Et       Lt    ()
12 % signal   Lt    Lt
13 %</gfitModelDescription>
14 EL = ( Kd + Et + Lt - sqrt( (Kd + Et + Lt).^2 - 4 * Et .* Lt ) )/2;
15 signal = EL;
```

Description are space/tab/newline-delimited. Rows in each table should contain same number of elements. However, if an element needs to be empty (as in *minVal of signal* variable), an empty pair of brackets ((), [], '', or "") can be used. Brackets should also surround multiword elements.

The first table (lines 4–8) should list all model I/O variables in the same order as in the input list of the function. The first column of every table contains variable names. The second column of the first table defines variable types. Type *free* for variable *Et* means that its value could either be known precisely and supplied with the experiment conditions, or not known and appear as one of the optimization parameters. Variable *Lt is independent*, meaning that its value should always be known exactly and appear in the experimental data. Variable *Kd* has *para* type and is sought as an optimization parameter. Variable *signal* has *dependent* type and therefore is expected to be calculated by the model.

The Model Description is also used to set bound limits on the input variables. Column three of the first table, (*minVal*) specifies the minimum value of zero for all variables except *signal*, which is *dependent* and cannot have its value constrained. This has been included in the first table but can also be placed in a separate table.

In this example, the size of variables is set in the second table of Model Description (lines 10–12). The length of *Et* variable vector is set equal to the length of *Lt* as required by the model. It also sets the length of *signal* produced by the model to have the length of *Lt*. Since the size of the output variable must be known by *gfit*, the latter expression guarantees that the results are equal to a known length. The table also states (line 12, in the third column-*plotVs*) that *signal* should be plotted versus *Lt*.

3.1.5. Import Data into gfit

1. Start *gfit* by typing `mgfit` in *MATLAB* command line. *gfit* window will appear.
2. Select `eq_binding.m` model by choosing menu **Model →Pick Model**. The name of the model will appear in the model field.
3. Arrange data for two experiments in a spreadsheet application as shown in **Fig. 4A**. Names of the experiments should appear in the top line of the data block. First experiment *Titration one* contains two variables *Lt* and *Et* having equal size. Second experiment, *Titration two*, contains only the independent variable, *Lt*, required by the model. Note that variables *Lt* in experiments one and two have different sizes.
4. To transfer data to *gfit*, select both experiments in the spreadsheet and copy it into the clipboard. Names of experiments should appear in the top row of the selected block.

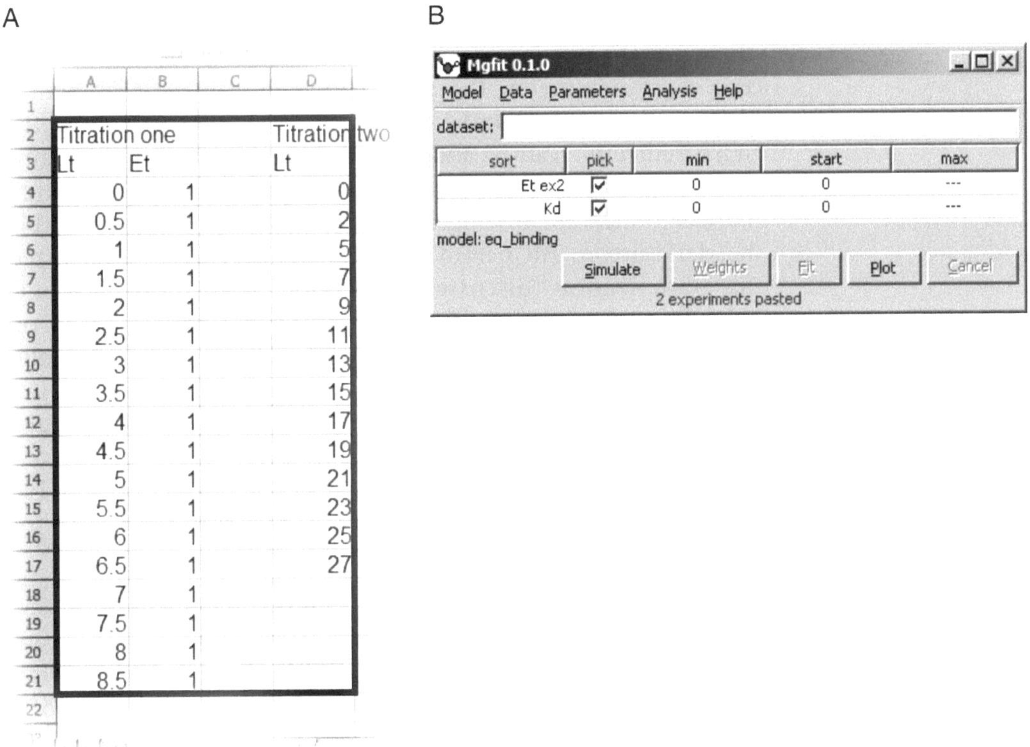

Fig. 4. *gfit* imports experimental data from a spreadsheet. (**A**) Spreadsheet containing two experiments–*Titration one* and *Titration two*. *Black rectangle* shows the cells to be selected, copied, and imported to *gfit*. (**B**) Once the data are imported, *gfit* user interface shows model parameters.

5. In *gfit* window, choose menu **Data→Paste-add Data**. *gfit* recognizes tab/newline-delimited data stored in clipboard and checks it against the requirements of Model Description to assure that it can be used with the current model. If there is an inconsistency between data and the requirements of the Model Description, an error message will appear.

Once the data is acquired by *gfit*, it generates parameters for all *para* variables and for every *free* model variable in the model that is missing in the experimental data (**Fig. 4B**). Parameters are generated based on the input variables defined in Model Description.

During simulation of an experiment, input variables draw their values either from the supplied experimental data, or from current values of parameters. In *gfit*, parameters and input variables have a many-to-many relationship. When simulating different experiments, a variable containing an array of numbers can collect its values from many parameters. One parameter can be also connected to multiple variables as long as the variables have the same physical units (discussed below).

Linkage of parameters to different variables can be adjusted through *gfit* parameter table in the user interface (**Fig. 4B**). For every parameter, the table shows its name, optimization flag (**pick**), low bound constraint (smallest value allowed), start (current) value, and upper bound constraint. Parameters in the table can be sorted by several criteria. To switch between different sorting methods, click **sort** parameter table header button. All parameters can be selected or deselected for optimization (discussed below) by clicking on **pick** button. Bound constraints for each parameter are set based on the variable's constraints in the Model Description. The valid interval of constraints can be only reduced through the user interface. For example, minimal value for K_D can be changed to 1.0, but not to −1.0.

3.1.6. Simulate with gfit

1. Set parameter *Et ex2* to 15.0 and parameter *Kd* to 1.
2. To perform simulation click on **Simulate** button.
3. Click on button **Plot**. A plot will appear.

The model we created can be conveniently used for simulating equilibrium concentration of *EL* complex under different experiment conditions. Unfortunately, concentration of a complex is seldom measured directly in an experiment. In the simplest case, experimentally observed values are proportional to [*EL*]. To avoid transformation of the data (even a linear one) prior to analysis, the model needs to simulate the measured signal.

3.1.7. Refine Model

1. To simulate the signal observed in a real experiment (e.g., binding induced change of fluorescence), introduce two more variables, signal gain, *gain*, and signal background, *c*. Assume the *signal* to be proportional to [EL] with an offset.
2. Add more columns to the first table of Model Description to set default starting values of parameters, physical units, and human-readable descriptions of model variables. Resulting model is shown in **Listing 3**.

These changes make this model more flexible because it can now be used for any experiment that measures a value proportional to the equilibrium concentration of the complex. Parameters by default take more reasonable starting values. Human readable variable descriptions appear when mouse cursor hovers over a name in the parameter table. Units prevent mixing incompatible variables in the same parameter.

3.1.8. Fit Data

1. From the data archive, open file `eq_binding_data_fig5.txt` in a text editor. The file contains data for one experiment, select and copy its entire contents.

Listing 3. Updated *gfit* model for binding equilibrium.

```
function signal = eq_binding(Et, Lt, Kd, signal, gain, c)
%binding equilibrium E + L <=> EL; simulate signal proportional to [EL]
%<gfitModelDescription version="100">
%variable type        minVal  startVal  unit  comment
% Et      ()          0       1         uM    'total concentration of E'
% Lt      independent 0       1         uM    'total concentration of L'
% Kd      para        0       0.1       uM    'dissociation constant'
% signal  dependent   ()      ()        ()    'complex formation'
% gain    para        ()      1         ()    'signal gain'
% c       para        ()      0         ()    'signal background'
%
%variable size plotVs
% Et      Lt   ()
% signal  Lt   Lt
%</gfitModelDescription>
EL = ( Kd + Et + Lt - sqrt( (Kd + Et + Lt).^2 - 4 * Et .* Lt ) )/2;
signal = c + gain * EL;
```

2. In *gfit* interface, select the updated model and import the data into *gfit*.
3. To view the imported data click **Plot**.
4. Click **Fit**. Optimization engine will search parameter space for the best fit, and will display the optimized values and their confidence intervals (**Fig. 5**) (*see* **Notes 5, 6,** and 7).
5. To view fitted data click **Plot**.

In this section, we have created a regular *MATLAB* m-file and added a *gfit* Model Description to it, which allowed us to connect it with experimental data to perform simulation and fitting. In the following sections, we will apply this technique to more complex biological systems.

3.2. More Complex Example: Equilibrium Binding to a Polymer

The model created in the previous section can be used for studying relatively simple systems where binding properties are characterized only by the dissociation constant. However, binding processes are often more complex and characterized by multiple parameters. Binding of proteins to discrete positions on a linear lattice (DNA,

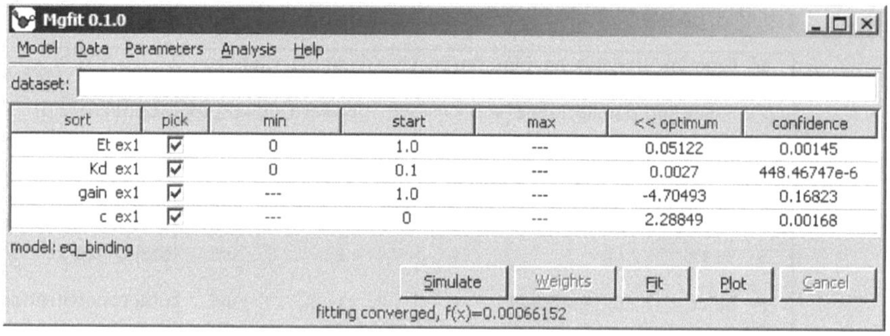

Fig. 5. Fitting the results of a fluorimetric titration experiment. The interface shows optimal values for each parameter and their confidence intervals.

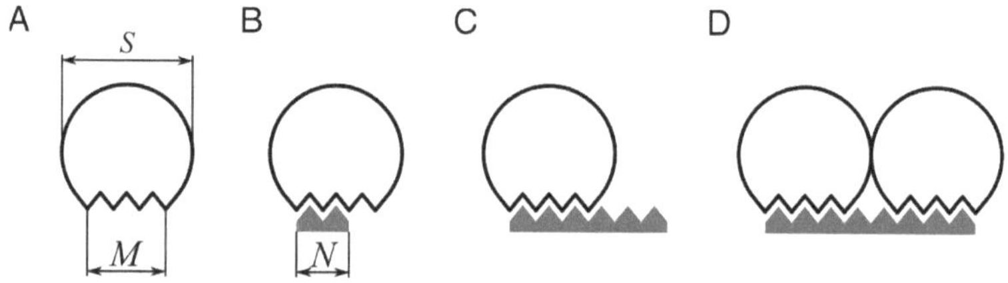

Fig. 6. Binding of proteins to lattices depends on their geometry. (**A**) Protein geometry parameters are minimal binding site, M, and occlusion site, S. (**B**) If lattice length, $N < M$, binding free energy is approximately proportional to N. (**C**) If $N > M$, the observed K_D is inversely proportional to the number of alternative binding configurations, $N - M + 1$. (**D**) If more than one protein can bind to a lattice molecule, $N \geq S + M$, binding configurations for all possible numbers of bound proteins have to be considered.

RNA, microtubules, and microfilaments) plays important roles in many biological processes. In this section, we will discuss binding of the helicase from hepatitis C virus to single-stranded (ss) DNA substrates. The experimental data were obtained by titrating a constant concentration of the helicase with oligonucleotides of different lengths while monitoring the reduction of intrinsic fluorescence of the helicase caused by ssDNA binding *(7)*. With this example we take regression analysis a step further and globally fit many titration curves to quantify helicase properties that are not apparent from any single experiment.

The model of equilibrium binding to a lattice, in addition to concentrations and the dissociation constant, uses parameters related to the geometry of the molecules, namely lattice length N, protein's minimal binding site (M, number of lattice units interacting with the protein), and protein's occlusion site (S, number of lattice units from which one protein molecule excludes the others) (**Fig. 6A**). The model assumes noncooperative and sequence-independent binding. Since the K_D observed in lattice binding experiments changes with the lattice length, the model uses a more fundamental microscopic dissociation constant, K_D^0, defined as the dissociation constant observed with lattice of length M. Thus, variables K_D^0, M, and S keep same values in all experiments.

The model calculates concentration of bound protein, E_B, from total protein concentration, E_T, total concentration of lattice, L_T, and N. Depending on relative values of M, S, and N, three cases can be considered. If lattice is shorter than the minimal binding site (**Fig. 6B**), $N < M$, and assuming equal contribution of each lattice unit to the binding free energy, the binding can be described by **Eq. 6**, where the observed dissociation constant

$$K_D = K_D^{0(N/M)} \xi^{(1-N/M)}, \text{ where } \xi \text{ is a concentration unit conversion factor.} \quad (7)$$

For longer lattices that can accommodate only one protein molecule (**Fig. 6C**), $M \le N < 2S$, the observed dissociation constant is inversely proportional to the number of distinct positions the protein can occupy.

$$K_D = \frac{K_D^0}{N - M + 1} \quad (8)$$

If multiple proteins can bind to a single lattice molecule, **Eq. 6** can no longer be used, and the extent of binding has to be calculated by numerically solving **Eq. 9** *(8)*. Alternatively, if the number of proteins bound to a lattice is large, a lower order equation can provide accurate results *(9)*.

$$\frac{E_B}{L_T} = \frac{\sum_{i=0}^{L_{max}} i\Omega_i \left(\frac{E_T - E_B}{K_D^0}\right)^i}{\sum_{i=0}^{L_{max}} \Omega_i \left(\frac{E_T - E_B}{K_D^0}\right)^i}, \quad \text{where } \Omega_i = \frac{(N - iS + i)!}{(N - iS)!\,i!}. \quad (9)$$

Although apparent dissociation constant, K_D, can be estimated on the basis of a single titration experiment, true K_D^0 cannot be determined without prior knowledge of M and S. All three parameters can be determined simultaneously by globally fitting titration curves for many ssDNA substrates of different lengths.

The model for lattice binding `lattice_binding_v1.m`, provided in the data archive, follows a similar pattern as previously created `eq_binding.m` model. Both models start with Model Description. We will now test whether the model is consistent with the observed results, estimate the parameters, and determine their confidence intervals.

3.2.1. Import Titration Data for Global Analysis

1. Start *gfit* and choose model `lattice_binding_v1.m`.
2. Open file `lattice_binding_data.txt` in a spreadsheet. The file contains data for nine titration experiments. As before, the top row contains only names of experiments with variable names and values appearing below (*see* **Note 8**).
3. Select and copy entire dataset. Make sure that selection starts at experiment name row and no values on the bottom are left out.
4. Choose menu **Data→Paste-add Data** to import data to *gfit*. The parameters generated for the dataset will appear. View data by clicking button **Plot**.

3.2.2. Fit the Data

1. Make sure that all parameters are selected for optimization and click button **Fit**. Because of the larger number of parameters and more involved computation, fitting may take a couple of minutes to complete.
2. Plot the fitted data. The fit does not match the data. Notice, however, that unlike the data, all fitted curves start from the same value. This happened because same value of *f0* parameter was used for all experiments.
3. Right-click on the name of *f0* parameter and choose menu **Separate Elements**. Parameter *f0* separated into nine parameters, one for each experiment. Also separate elements of parameter *fg*, because the gain of the signal is also known to vary between experiments.
4. Click button **Fit**.

Using this model and dataset, *gfit* is expected to produce a good fit at the first attempt (**Figs. 7** and **8**). However, this result is not typical. In our experience with other models, local optimization algorithms seldom find the global minimum in the first attempt. This highlights the importance of global optimization methods *(10, 11)*. Generally, finding a global minimum of a nonlinear problem is unattainable within finite time. Nevertheless, even if a good fit has been found, it is advisable to search the parameter space for alternative minima. Discovering distinct sets of parameters that produce "as good," or "nearly as good" fits is important to diagnose overparameterized models or insufficient amount of experimental observations.

The method currently used by *gfit* for exploring parameter space is random restart. This simple method is implemented as a "globalizer" on top of the existing local optimization engine. Random restart repeatedly reinitiates local optimization with a randomly chosen set of parameters. The new starting parameter values are picked from a uniform distribution for doubly-constrained parameters, from a truncated normal distribution for singly-constrained parameters, and from a normal distribution for the unconstrained ones. Random restart procedure is implemented without a defined termination condition. It is supposed to be interrupted by the user.

3.2.3. Search for Global Minimum

1. Choose menu **Analysis→Random Restart** (*see* **Note 9**).
2. Allow the program to perform a few hundred optimizations and interrupt it by clicking button **Cancel**. The best found parameter values appear in the right column of the parameter table.
3. To check the goodness of fit visually, copy best found parameter values to the start column by clicking the table header button **<<value**, click button **Simulate**, then button **Plot** (*see* **Note 10**).

More generally, the following rule-of-thumb procedure can be used to decide if the random restart search should be continued.

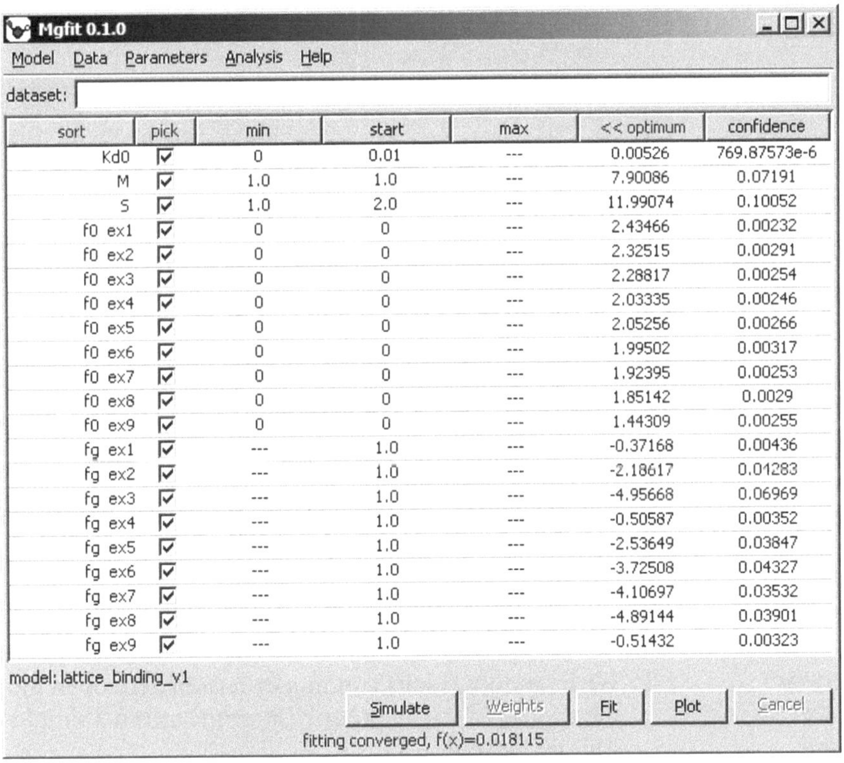

sort	pick	min	start	max	<< optimum	confidence
Kd0	☑	0	0.01	---	0.00526	769.87573e-6
M	☑	1.0	1.0	---	7.90086	0.07191
S	☑	1.0	2.0	---	11.99074	0.10052
f0 ex1	☑	0	0	---	2.43466	0.00232
f0 ex2	☑	0	0	---	2.32515	0.00291
f0 ex3	☑	0	0	---	2.28817	0.00254
f0 ex4	☑	0	0	---	2.03335	0.00246
f0 ex5	☑	0	0	---	2.05256	0.00266
f0 ex6	☑	0	0	---	1.99502	0.00317
f0 ex7	☑	0	0	---	1.92395	0.00253
f0 ex8	☑	0	0	---	1.85142	0.0029
f0 ex9	☑	0	0	---	1.44309	0.00255
fg ex1	☑	---	1.0	---	-0.37168	0.00436
fg ex2	☑	---	1.0	---	-2.18617	0.01283
fg ex3	☑	---	1.0	---	-4.95668	0.06969
fg ex4	☑	---	1.0	---	-0.50587	0.00352
fg ex5	☑	---	1.0	---	-2.53649	0.03847
fg ex6	☑	---	1.0	---	-3.72508	0.04327
fg ex7	☑	---	1.0	---	-4.10697	0.03532
fg ex8	☑	---	1.0	---	-4.89144	0.03901
fg ex9	☑	---	1.0	---	-0.51432	0.00323

Fig. 7. Results of global fitting of nine equilibrium titration experiments. The interface shows a table of 21 parameters. Three parameters, *Kd0*, *M*, and *S*, were globally applied to all experiments, while each of the others were used in one experiment only. For example, parameter *f0 ex3* was used in experiment 3 only. Optimization started with parameter values displayed in column **start,** and arrived to the optimal values displayed in column **optimum.** Column **confidence** shows parameter standard errors (68% confidence intervals) estimated using asymptotic method.

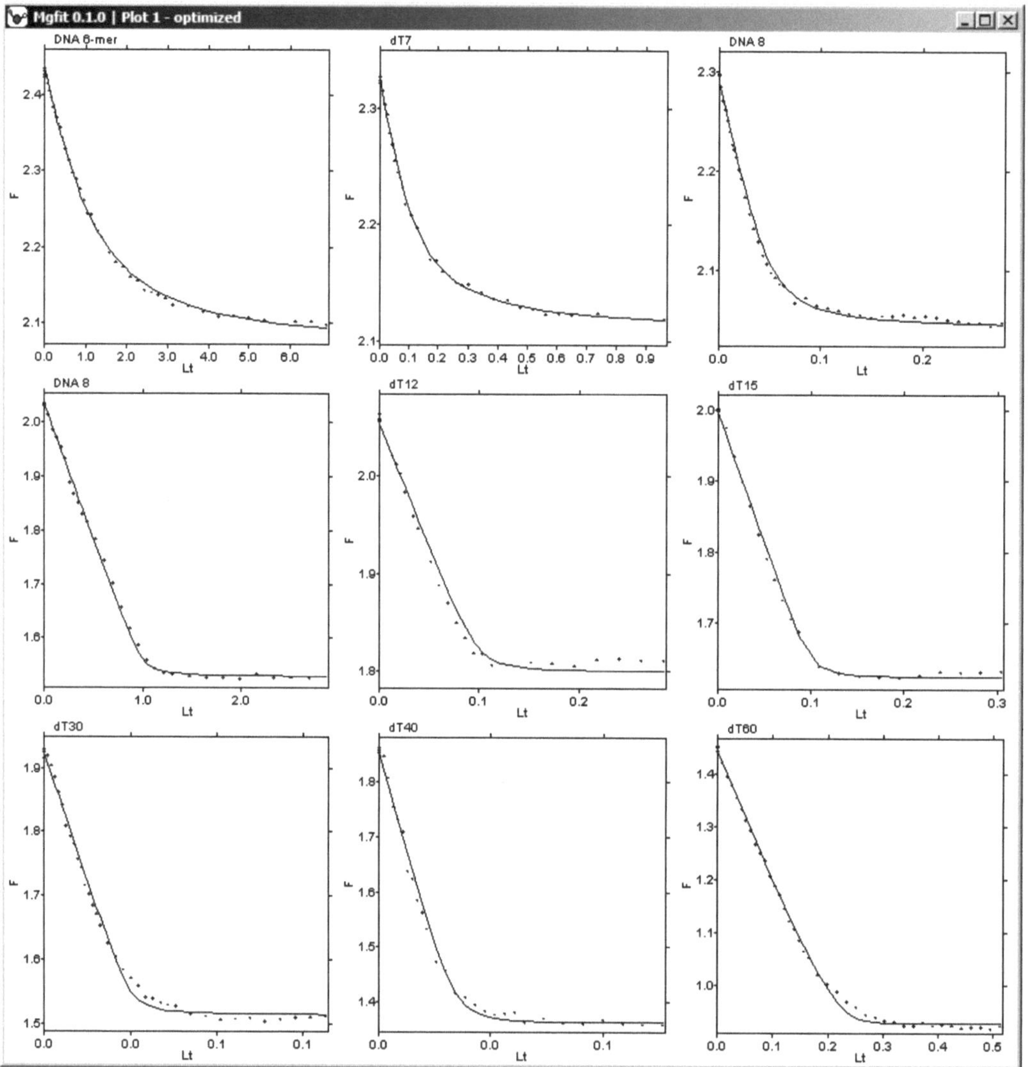

Fig. 8. Plots of nine globally fitted titration experiments. In each titration, increasing concentrations of ssDNA were added to the constant concentration of helicase. Concentrations of helicase as well as lengths of ssDNA were different in each titration. Binding to DNA reduced the intrinsic fluorescence of the helicase. Experimental measurements of fluorescence are shown as *dots*. Results of the simulation are shown as *solid lines*.

3.2.4. When Should Random Restart be Terminated?

1. With random restart running or terminated, open most recent file `C:/Mgfit/Temp/Optim_nnn.txt` in a simple text editor (*see* **Note 9**).
2. Select and copy entire contents of the file and paste it into a spreadsheet application. Now each row of the spreadsheet contains a set of optimized parameters, while the first column contains the values of objective function. Parameter sets appear in the order they were calculated.

3. Sort parameter sets by their objective function. The best fitting parameter set is now at the top row of the table.
4. Examine the better fitting parameter sets. Identify a group of better fitting sets with similar goodness of fit values starting from the top row.
   a If the group is small, the search is worth continuing.
   b If the group is large and at least some of the parameters have significantly different values, the data does not sufficiently constrain model parameters. Additional experiments may be needed.
   c If the group is large and parameter values are similar between different sets, the group is probably in the vicinity of the global minimum of the problem.

3.3. Advanced Example: Kinetic Mechanism of Clamp Assembly on DNA

In this section, we describe modeling and analysis of a kinetic pathway responsible for loading sliding clamp proteins onto DNA during replication initiation. The clamp, PCNA, encircles DNA and binds to DNA polymerase, conferring processivity to the replication complex. PCNA is loaded onto DNA by the heteropentameric clamp loader protein, RFC, in a process that involves ATP binding and hydrolysis and conformational changes of the proteins (**Fig. 9A**) *(12)*. Although the process has been extensively studied, understanding at the quantitative

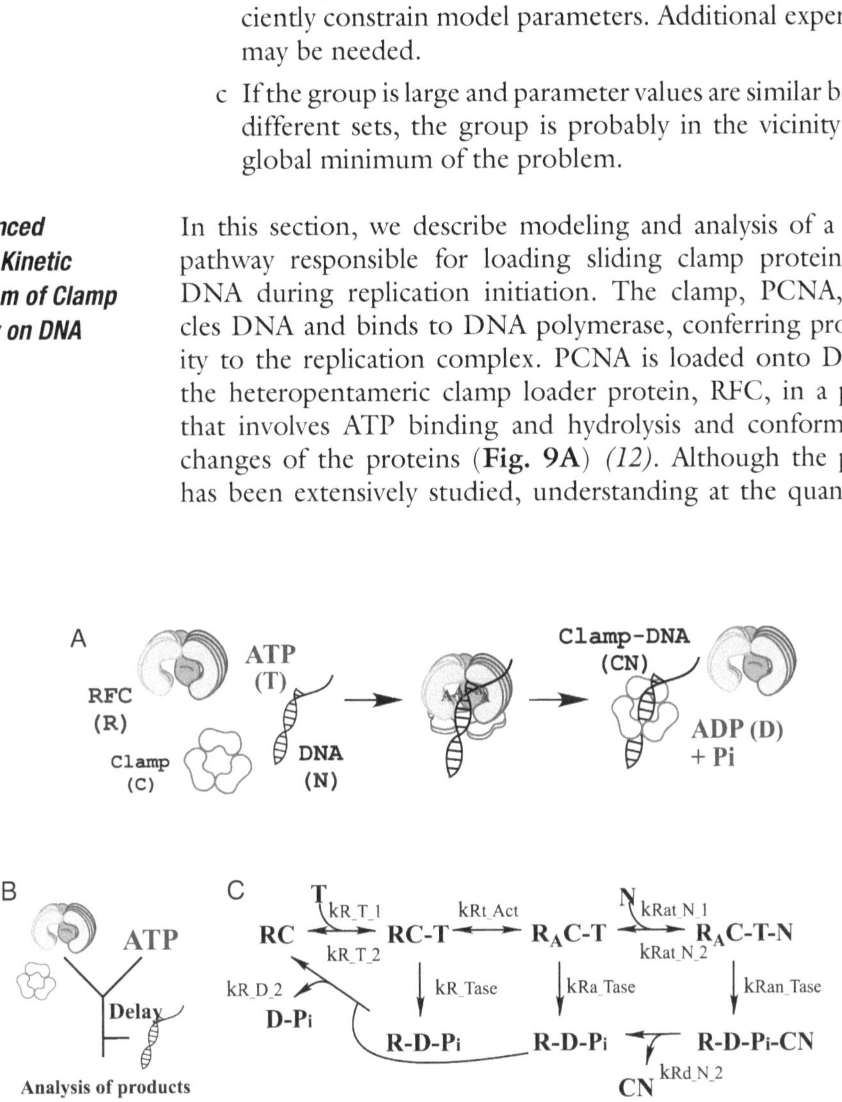

Fig. 9. Mechanism of clamp loading on DNA. (**A**) Clamp loader protein, RFC, catalyzes assembly of circular PCNA clamps onto primed DNA in a reaction driven by ATP binding and hydrolysis. (**B**) Design of ATPase experiments, with initial mixing of RFC and PCNA with ATP, followed by varying delay times, mixing with DNA, and measurement of product kinetics. (**C**) Simplified clamp loading reaction scheme. RFC (in the presence of PCNA; RC) binds ATP (T). The ternary complex undergoes an activation step (RA) before DNA (N) binding, rapid ATP hydrolysis and dissociation from the clamp-DNA complex (CN).

level is incomplete. Properties of individual species and reactions involved in this process are difficult to determine because almost any measurement technique is affected by a combination of many simultaneously occurring reactions. Computational modeling in conjunction with regression analysis provides a feasible solution to this complex problem.

Currently we are building, validating, and refining a mechanistic model of the process that can resolve many species and quantify the rates of individual reactions that may not be observed experimentally. A simplified reaction scheme, the first iteration of modeling process, is shown in **Fig. 9C**. The challenge of estimating the rates of many individual reactions can be addressed by monitoring the process from different perspectives. As noted earlier, each measurement is a function of many or all of the rates in the process; however, measurements from different perspectives are likely to be affected in distinct ways by individual reaction rates. Therefore, taken together, multiple measurements of presteady-state kinetics monitored by a few different methods can provide sufficient constraints to the parameters of the model.

Results of global regression analysis of data from two types of presteady-state experiments, one measuring ATP hydrolysis and the other phosphate (Pi) release by RFC protein, are shown in **Figs. 10** and **11**. ATP hydrolysis was monitored by a radiometric assay measuring formation of ^{32}P-ADP over time from ^{32}P-ATP, and Pi release was monitored by the change in fluorescence of a reporter, Pi Binding Protein, on binding the Pi released by RFC following ATP hydrolysis. Salient features of the experimental design in both cases are (**Fig. 9B**): (a) rapid mixing of a constant concentration of RFC (in the presence of excess PCNA clamp) with excess ATP; (b) incubation of RFC with ATP for varying times; (c) rapid mixing of the RFC, ATP, PCNA mix with excess DNA and measurement of product formation and release over time. Salient features of the model mechanism are (**Fig. 9C**): (a) RFC binding to ATP; (b) a proposed step that might account for the observed increase in ATPase activity with increasing RFC-ATP incubation time; (c) DNA binding to the RFC-ATP complex; (d) ATP hydrolysis; (e) release of ADP, Pi and clamp–DNA complex.

Simultaneous fitting of a large number of experiments requires efficient simulation. The most computationally intensive operation performed by RFC model is integration of ODE systems to simulate mass-action reactions. To accelerate this process, reaction kinetics was simulated using native *rsys* library. A combination of a flexible and user-friendly scripting language, such as *MATLAB*, with a native library for computationally intensive parts of simulation was found to be especially productive.

Given estimates of ATP and DNA-binding rate constants, the model produces a good fit for datasets from both ATP hydrolysis

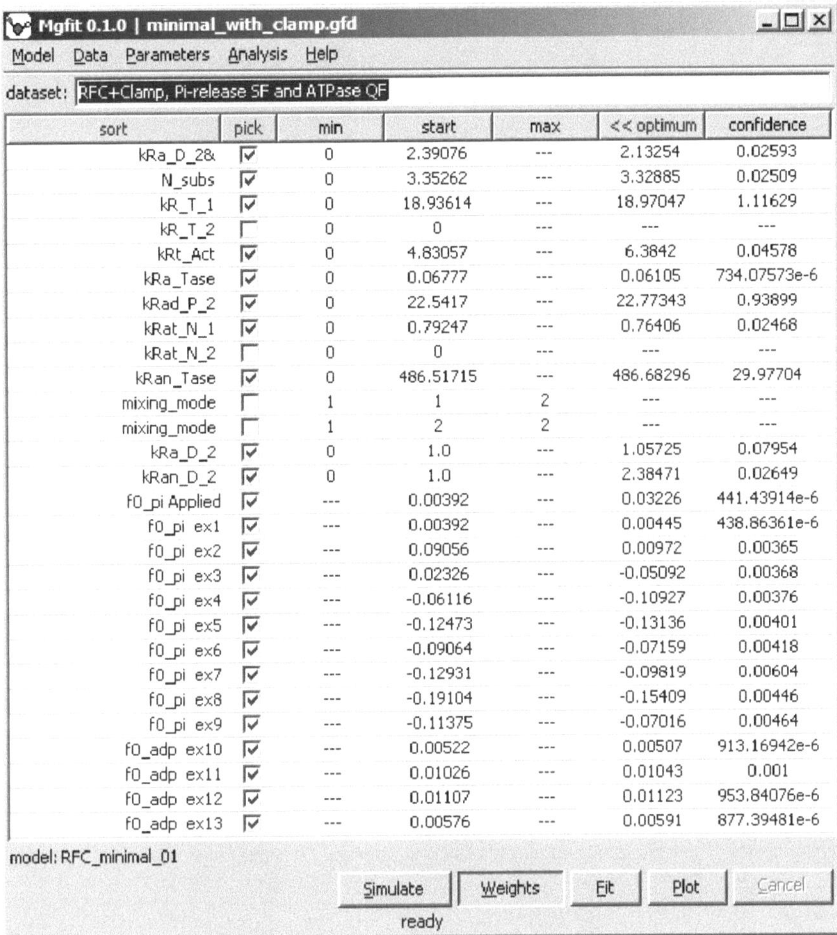
Mgfit 0.1.0 | minimal_with_clamp.gfd

Model Data Parameters Analysis Help

dataset: RFC+Clamp, Pi-release SF and ATPase QF

sort	pick	min	start	max	<< optimum	confidence
kRa_D_2&	☑	0	2.39076	---	2.13254	0.02593
N_subs	☑	0	3.35262	---	3.32885	0.02509
kR_T_1	☑	0	18.93614	---	18.97047	1.11629
kR_T_2	☐	0	0	---	---	---
kRt_Act	☑	0	4.83057	---	6.3842	0.04578
kRa_Tase	☑	0	0.06777	---	0.06105	734.07573e-6
kRad_P_2	☑	0	22.5417	---	22.77343	0.93899
kRat_N_1	☑	0	0.79247	---	0.76406	0.02468
kRat_N_2	☐	0	0	---	---	---
kRan_Tase	☑	0	486.51715	---	486.68296	29.97704
mixing_mode	☐	1	1	2	---	---
mixing_mode	☐	1	2	2	---	---
kRa_D_2	☑	0	1.0	---	1.05725	0.07954
kRan_D_2	☑	0	1.0	---	2.38471	0.02649
f0_pi Applied	☑	---	0.00392	---	0.03226	441.43914e-6
f0_pi ex1	☑	---	0.00392	---	0.00445	438.86361e-6
f0_pi ex2	☑	---	0.09056	---	0.00972	0.00365
f0_pi ex3	☑	---	0.02326	---	-0.05092	0.00368
f0_pi ex4	☑	---	-0.06116	---	-0.10927	0.00376
f0_pi ex5	☑	---	-0.12473	---	-0.13136	0.00401
f0_pi ex6	☑	---	-0.09064	---	-0.07159	0.00418
f0_pi ex7	☑	---	-0.12931	---	-0.09819	0.00604
f0_pi ex8	☑	---	-0.19104	---	-0.15409	0.00446
f0_pi ex9	☑	---	-0.11375	---	-0.07016	0.00464
f0_adp ex10	☑	---	0.00522	---	0.00507	913.16942e-6
f0_adp ex11	☑	---	0.01026	---	0.01043	0.001
f0_adp ex12	☑	---	0.01107	---	0.01123	953.84076e-6
f0_adp ex13	☑	---	0.00576	---	0.00591	877.39481e-6

model: RFC_minimal_01

Simulate Weights Fit Plot Cancel

ready

Fig. 10. Parameters of RFC clamp loader model. The values of parameters were obtained by global fitting of 22 presteady-state kinetic measurements.

as well as Pi release experiments in a single seamless operation. Confidence in the parameters obtained from the fits is increased by using the random restart method, as described in **Subheading 3.2.3**. A highlight of the findings is that global fitting of the ATPase data validates the proposed step between ATP binding and hydrolysis (**Fig. 9C**), and reveals that it is a relatively slow, and thus mechanistically important, step in the clamp assembly reaction (rate constant: *kRt_Act*). We can now formulate specific, testable hypotheses regarding the nature of this step; e.g., an ATP binding-driven conformational change in RFC that enables productive interactions with the clamp and DNA.

It is clear that the reaction depicted in **Fig. 9C** represents a simplified model of the clamp assembly reaction, in which the

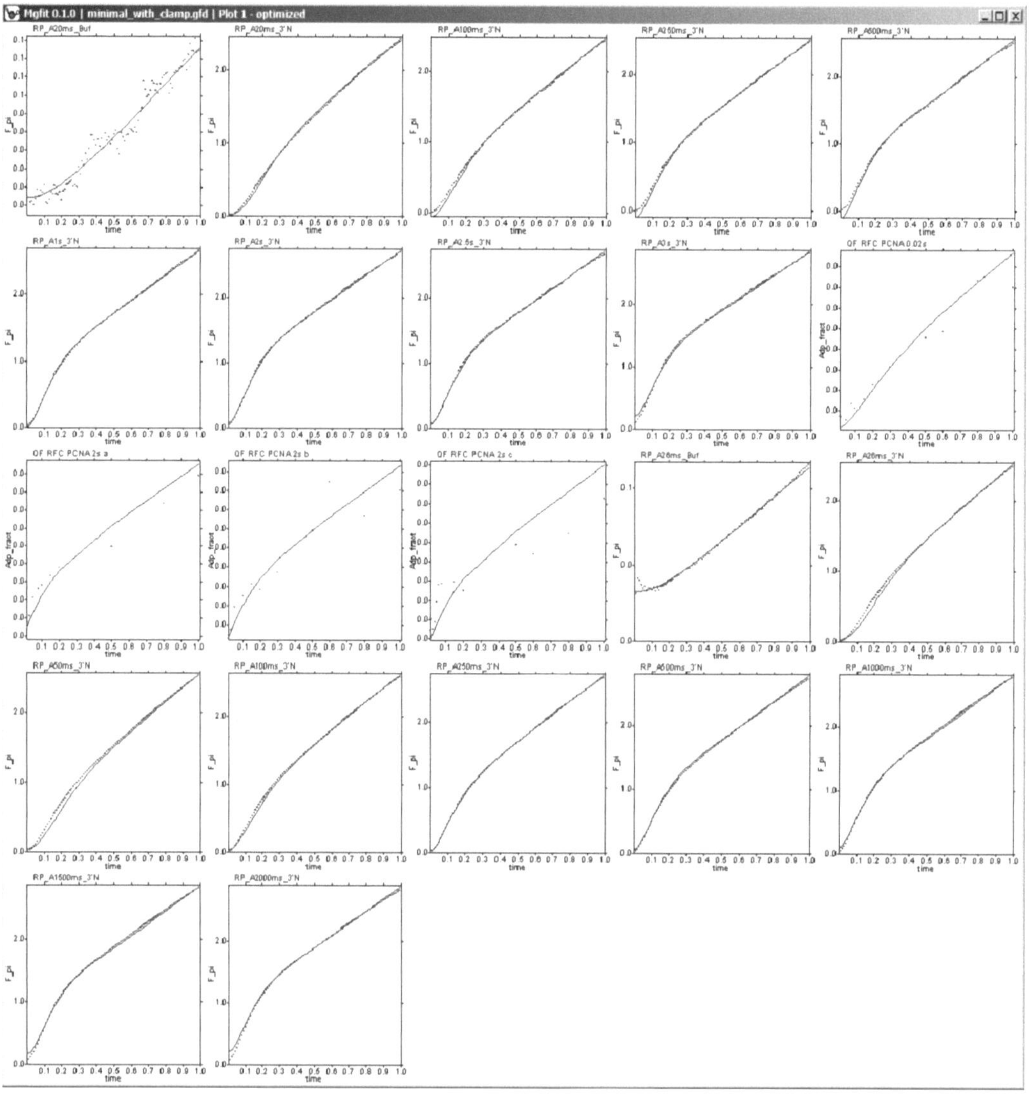

Fig. 11. Fitting of kinetic data by the clamp loader model. Depending on the measurement conditions, kinetics of Pi release and ADP production exhibits different extent of lag followed by a burst and approach to the steady state. The presence of multiple "features" in the kinetic curves facilitates constraining parameters of the model.

number of possible species and steps are restricted. Such limitations are often introduced in models to focus on specific experimentally measured parameters, and thereby increase confidence in the fit. For example, this model omits RFC binding to the clamp, since the parameters defining this step are unknown and are not measured explicitly in the ATPase experiments. Under such circumstances, however, it is entirely possible that goodness of fit to the data becomes unrelated to the quality of the model. *gfit* enables model-based global analysis of a variety of experiments to find parameter values that are consistent across the board. Therefore, a more comprehensive model of clamp assembly can be developed from the start, and then continually validated and refined by data input from experiments measuring clamp binding, clamp opening, DNA binding, clamp–DNA release, etc., leading to discovery of the preferred mechanism of clamp assembly on DNA.

4. Notes

1. In regression analysis literature, dependent variables may be referred to as response or observed variables; independent variables may be referred to as predictor or explanatory variables.
2. Residual is the difference between the experimental measurement and the simulation produced by the model.
3. Model Description is metadata attached to *gfit* models that defines their correct usage. It contains model name, version, general human-readable comments about the purpose of the model and its algorithm, and, most importantly, machine-readable descriptions of the model's input and output variables. For each variable, it specifies name, type, physical unit, dimensions, and a range of acceptable values. Variable dimensions are defined either as constants or in relationship to another variable dimension or index variable. Variables may change their size depending on experimental data and user input. Dimensions of each variable usually change in concert with dimensions of other variables.
4. Variable (in *gfit* context) is an array of elements (numbers) defined in Model Description. A variable may contain a single element (scalar variable, 0D), a vector of elements (1D), a matrix (2D), etc. Variables are used for storing information about an experiment, for passing data to the model and for receiving data simulated by the model. Depending on the Model Description, each variable dimension may be fixed, or vary individually or in concert with other variable dimensions. This property of variables increases flexibility of *gfit* models.
5. During calculations, all control elements of *gfit* interface are disabled with the exception of the button **Cancel**. Clicking this button prevents simulation of the next experiment. Simulation of the current experiment will not be aborted.
6. During fitting, information about each iteration is displayed in *MATLAB* command window. If fitting is aborted, the last values of parameters appear in the right column of parameter table.

7. Currently *gfit* uses an asymptotic method for determining confidence intervals of parameters. This method is known to be inaccurate for nonlinear models.
8. In spreadsheet-arranged data, a colon following a variable name indicates that the variable is scalar and its value should appear in the cell to the right of the name.
9. During a random restart run, starting parameter values and optimization results are accumulated in files `Start_`*nnn*`.txt` and `Optim_`*nnn*`.txt`, respectively, in folder `C:/Mgfit/Temp/`, where *nnn* are digits starting from `000` and incremented for each subsequent random restart search. The files contain tab/newline-delimited tables with each set of parameters occupying one row. The first row contains parameter names. In addition, the first column of `Optim_`*nnn*`.txt` file contains objective function values, $S(x)$. To check the progress of the search without interrupting it, open either of the files in a simple text editor, select, and copy its entire contents, and paste it into a spreadsheet application.
10. Column 6 of parameter table contains values produced by optimization (completed or interrupted). To be able to edit the values, to use them for simulation, or as starting values for fitting, copy them to column 4 by clicking table header button **<<value**.

Acknowledgments

Authors thank members of the Center for Cell Analysis and Modeling: Pavel Kraykivski, Igor Novak, Jim Schaff, and Boris Slepchenko for their advice and critical discussions and Les Loew for his support. Experimental data for clamp loader protein was provided thanks to Siying Chen. This work was supported in part by grants NS15190 (NIH), RR13186 (NIH), RR022232 (NIH) and RR022624 (NIH) to J.H.C; GM55310 (NIH) to S.S.P; GM64514-01 (NIH) and MCB 0448379 (NSF) to M.M.H.

References

1. Mogilner, A., Wollman, R., and Marshall, W. F. (2006) Quantitative modeling in cell biology: what is it good for? *Dev. Cell.* 11, 279–287.
2. Albeck, J. G., MacBeath, G., White, F. M., Sorger, P. K., Lauffenburger, D. A., and Gaudet, S. (2006) Collecting and organizing systematic sets of protein data. *Nat. Rev. Mol. Cell. Biol.* 7, 803–812.
3. Jaqaman, K. and Danuser, G. (2006) Linking data to models: data regression. *Nat. Rev. Mol. Cell. Biol.* 7, 813–819.
4. Beechem, J. M. (1992) Global analysis of biochemical and biophysical data. *Meth. Enzymol.* 210, 37–54.
5. Draper, N.R. and Smith, H. (1998) Applied Regression Analysis. Wiley, New York.

6. Slepchenko, B. M., Schaff, J. C., Macara, I., and Loew, L. M. (2003) Quantitative cell biology with the Virtual Cell. *Trends Cell. Biol.* 13, 570–576.
7. Levin, M. K. and Patel, S. S. (2002) Helicase from hepatitis C virus, energetics of DNA binding. *J. Biol. Chem.* 277, 29377–29385.
8. Epstein, I. R. (1978) Cooperative and non-cooperative binding of large ligands to a finite one-dimensional lattice. A model for ligand-oligonucleotide interactions. *Biophys. Chem.* 8, 327–339.
9. Tsodikov, O. V., Holbrook, J. A., Shkel, I. A., and Record, M. T., Jr. (2001) Analytic binding isotherms describing competitive interactions of a protein ligand with specific and nonspecific sites on the same DNA oligomer. *Biophys. J.* 81, 1960–1969.
10. Moles, C. G., Mendes, P., and Banga, J. R. (2003) Parameter estimation in biochemical pathways: a comparison of global optimization methods. *Genome Res.* 13, 2467–2474.
11. Banga, J. R. (2008) Optimization in computational systems biology. *BMC Syst. Biol.* 2, 47.
12. Johnson, A. and O Donnell, M. (2005) Cellular DNA replicases: components and dynamics at the replication fork. *Annu. Rev. Biochem.* 74, 283–315.

Chapter 13

Multicell Simulations of Development and Disease Using the CompuCell3D Simulation Environment

Maciej H. Swat, Susan D. Hester, Ariel I. Balter, Randy W. Heiland, Benjamin L. Zaitlen, and James A. Glazier

Summary

Mathematical modeling and computer simulation have become crucial to biological fields from genomics to ecology. However, multicell, tissue-level simulations of development and disease have lagged behind other areas because they are mathematically more complex and lack easy-to-use software tools that allow building and running *in silico* experiments without requiring in-depth knowledge of programming. This tutorial introduces Glazier–Graner–Hogeweg (GGH) multicell simulations and CompuCell3D, a simulation framework that allows users to build, test, and run GGH simulations.

Key words: Glazier–Graner–Hogeweg model, GGH, CompuCell3D, Mitosis, Cell growth, Cell sorting, Chemotaxis, Multicell modeling, Tissue-level modeling, Developmental biology, Computational biology.

1. Introduction

Most contemporary life scientists, whether theoretical or experimental, have relatively narrow disciplinary training. This specialization is partly a consequence of the speed of current progress in the life sciences and concomitant growth in the number of active researchers.

While the success of contemporary biology might lead naïve observers to conclude that our understanding is a simple superposition of achievements in the subfields composing life sciences, only rarely can we understand how a biological phenomenon operates by analyzing and understanding how its isolated components operate.

Ivan V. Maly (ed.) *Methods in Molecular Biology, Systems Biology, vol. 500*

DOI: 10.1007/978-1-59745-525-1_13

Just as knowing how transistors work is not sufficient to design and build a modern microprocessor, knowing the "function" of an enzyme does not suffice to design cells' biochemical networks or even to predict the phenotypic effect of knocking out specific genes.

Systems biology is a scientific discipline that studies complex interactions in biology, relying more on knowledge integration than on detailed studies of individual biological subsystems. Systems biologists often build mathematical models and computer simulations of living cells, tissues, organs, or even entire organisms to embody their understanding of this integration.

The last decade has seen fairly realistic simulations of single cells that can confirm or predict experimental findings. Because they are computationally expensive, they can simulate at most several cells at once. Even more detailed subcellular simulations can replicate some of the processes taking place inside individual cells. For example, Virtual Cell (http://www.nrcam.uchc.edu) supports microscopic simulations of intracellular dynamics to produce detailed replicas of individual cells, but can only simulate single cells or small cell clusters.

Simulations of tissues, organs, and organisms present a somewhat different challenge: how to simplify and adapt single cell simulations to apply them efficiently to study, *in silico*, ensembles of several million cells. To be useful, these simplified simulations should capture key cell-level behaviors, providing a phenomenological description of cell interactions without requiring prohibitively detailed molecular-level simulations of the internal state of each cell. While an understanding of cell biology, biochemistry, genetics, etc. is essential for building useful, predictive simulations, the hardest part of simulation building is identifying and quantitatively describing appropriate subsets of this knowledge. In the excitement of discovery, scientists often forget that modeling and simulation, by definition, require simplification of reality.

One choice is to ignore cells completely, e.g., Physiome *(1)* models tissues as continua with bulk mechanical properties and detailed molecular reaction networks, which is computationally efficient for describing dense tissues and noncellular materials like bone, extracellular matrix (ECM), fluids, and diffusing chemicals *(2, 3)*, but not for situations where cells reorganize or migrate.

Multicell simulations are useful for interpolating between single-cell and continuum-tissue extremes because cells provide a natural level of abstraction for simulation of tissues, organs, and organisms *(4)*. Treating cells phenomenologically reduces the millions of interactions of gene products to several behaviors: most cells can move, divide, die, differentiate, change shape, exert forces, secrete and absorb chemicals and electrical charges, and change their distribution of surface properties. The *Glazier–Graner–Hogeweg* (*GGH*) approach facilitates multi-scale simulations by defining spatially extended *generalized cells*, which can represent clusters of cells, single cells, subcompartments of single cells, or small subdomains of noncellular materials. This flexible definition allows tuning of the level of detail in a simulation from intracellular to continuum without switching simulation frameworks to examine the effect of changing the level of detail on a macroscopic outcome, e.g., by switching from a coupled ordinary differential equation (ODE) *Reaction-Kinetics* (RK) model of gene regulation to a Boolean description or from a simulation that includes subcellular structures to one that neglects them.

2. GGH Applications

Because it uses a regular cell lattice and regular field lattices, GGH simulations are often faster than equivalent *Finite Element* (FE) simulations operating at the same spatial granularity and level of modeling detail, permitting simulation of tens to hundreds of thousands of cells on lattices of up to $1{,}024^3$ pixels on a single processor. This speed, combined with the ability to add biological mechanisms via terms in the effective energy, permit GGH modeling of a wide variety of situations, including tumor growth *(5–9)*, gastrulation *(10–12)*, skin pigmentation *(13–16)*, neurospheres *(17)*, angiogenesis *(18–23)*, the immune system *(24, 25)*, yeast colony growth *(26, 27)*, *myxobacteria (28–31)*, stem-cell differentiation *(32, 33)*, *Dictyostelium discoideum (34–37)*, simulated evolution *(38–43)*, general developmental patterning *(14, 44)*, convergent extension *(45, 46)*, epidermal formation *(47)*, *Hydra* regeneration *(18, 49)*, plant growth, retinal patterning *(50, 51)*, wound healing *(47, 52, 53)*, biofilms *(54–57)*, and limb-bud development *(58, 59)*.

3. GGH Simulation Overview

All GGH simulations include a list of *objects*, a description of their *interactions* and *dynamics*, and appropriate *initial conditions*.

Objects in a GGH simulation are either generalized cells or *fields* in two dimensions (2D) or three dimensions (3D). Generalized cells are spatially extended objects (**Fig. 1**), which reside on a single *cell lattice* and may correspond to biological cells,

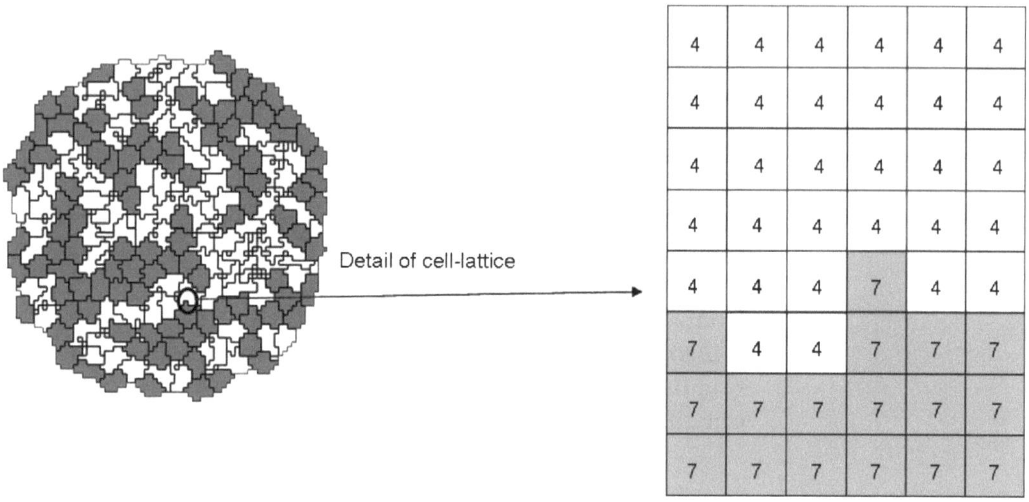

Fig. 1. Detail of a typical two-dimensional GGH cell-lattice configuration. Each domain represents a single spatially extended cell. The detail shows that each generalized cell is a set of cell-lattice sites (or pixels), $\vec{i}$, with a unique index, $\sigma(\vec{i})$, here 4 or 7. The *shade of gray* denotes the cell type, $\tau(\sigma(\vec{i}))$.

subcompartments of biological cells, or to portions of noncellular materials, e.g., ECM, fluids, solids, etc. *(8, 48, 60–72)*. We denote a lattice site or *pixel* by a vector of integers, $\vec{i}$, the *cell index* of the generalized cell occupying pixel $\vec{i}$ by $\sigma(\vec{i})$ and the *type* of the generalized cell $\sigma(\vec{i})$ by $\tau(\sigma(\vec{i}))$. Each generalized cell has a unique cell index and contains many pixels. Many generalized cells may share the same cell type. Generalized cells permit coarsening or refinement of simulations by increasing or decreasing the number of lattice sites per cell, grouping multiple cells into clusters or subdividing cells into variable numbers of *subcells* (subcellular compartments). Compartmental simulation permits detailed representation of phenomena like cell shape and polarity, force transduction, intracellular membranes and organelles, and cell-shape changes. For details on the use of subcells, which we do not discuss in this chapter, *see* **refs.** *27, 31, 73, 74*. Each generalized cell has an associated list of attributes, e.g., *cell type*, *surface area*, and *volume*, as well as more complex attributes describing a cell's state, biochemical interaction networks, etc. *Fields* are continuously variable concentrations, each of which resides on its own lattice. Fields can represent chemical diffusants, nondiffusing ECM, etc. Multiple fields can be combined to represent materials with textures, e.g., fibers.

Interaction descriptions and *dynamics* define how GGH objects behave both biologically and physically. Generalized-cell behaviors and interactions are embodied primarily in the *effective energy*, which determines a generalized cell's shape, motility, adhesion, and response to extracellular signals. The effective energy mixes true energies, such as cell–cell adhesion, with terms that mimic energies, e.g., the response of a cell to a chemotactic gradient of a field *(75)*. Adding *constraints* to the effective energy allows description of many other cell properties, including osmotic pressure, membrane area, etc. *(76–83)*.

The cell lattice evolves through attempts by generalized cells to move their boundaries in a caricature of cytoskeletally driven cell motility. These movements, called *index-copy attempts*, change the effective energy, and we accept or reject each attempt with a probability that depends on the resulting *change of the effective energy*, *H*, according to an *acceptance function*. Nonequilibrium statistical physics then shows that the cell lattice evolves to locally minimize the total effective energy. The classical GGH implements a modified version of a classical stochastic Monte-Carlo pattern-evolution dynamics, called *Metropolis dynamics with Boltzmann acceptance (84, 85)*. A *Monte Carlo Step* (*MCS*) consists of one index-copy attempt for each pixel in the cell lattice.

Auxiliary equations describe cells' absorption and secretion of chemical diffusants and extracellular materials (i.e., their interactions with fields), state changes within cells, mitosis, and cell death. These auxiliary equations can be complex, e.g., detailed RK descriptions of complex regulatory pathways. Usually, state changes affect generalized-cell behaviors by changing parameters in the terms in the effective energy (e.g., cell target volume or type or the surface density of particular cell-adhesion molecules).

Fields also evolve due to secretion, absorption, diffusion, reaction, and decay according to *partial differential equations* (PDEs). While complex coupled-PDE models are possible, most simulations require only secretion, absorption, diffusion, and decay, with all reactions described by ODEs running inside individual generalized cells. The movement of cells and variations in local diffusion constants (or diffusion tensors in anisotropic ECM) mean that diffusion occurs in an environment with moving boundary conditions and often with advection. These constraints rule out most sophisticated PDE solvers and have led to a general use of simple forward-Euler methods, which can tolerate them.

The *initial condition* specifies the initial configurations of the cell lattice, fields, a list of cells and their internal states related to auxiliary equations, and any other information required to completely describe the simulation.

3.1. Effective Energy

The core of GGH simulations is the *effective energy*, which describes cell behaviors and interactions.

One of the most important effective-energy terms describes cell adhesion. If cells did not stick to each other and to extracellular materials, complex life would not exist *(86)*. Adhesion provides a mechanism for building complex structures, as well as for holding them together once they have formed. The many families of adhesion molecules (CAMs, cadherins, etc.) allow embryos to control the relative adhesivities of their various cell types to each other and to the noncellular ECM surrounding them, and thus to define complex architectures in terms of the cell configurations which minimize the adhesion energy.

To represent variations in energy due to adhesion between cells of different types, we define a *boundary energy* that depends on $J(\tau(\sigma), \tau(\sigma'))$.

The *boundary energy per unit area* between two cells (σ, σ') of given types $(\tau(\sigma), \tau(\sigma'))$ at a *link* (the interface between two neighboring pixels):

$$H_{\text{boundary}} = \sum_{\substack{\vec{i},\vec{j} \\ \text{neighbors}}} J(\tau(\sigma(\vec{i})),\tau(\sigma(\vec{j})))\,(1-\delta(\sigma(\vec{i}),\sigma(\vec{j}))), \quad (1)$$

where the sum is over all neighboring pairs of lattice sites $\vec{i}$ and $\vec{j}$, and the boundary-energy coefficients are symmetric

$$J(\tau(\sigma),\tau(\sigma')) = J(\tau(\sigma'),\tau(\sigma)). \quad (2)$$

In addition to boundary energy, most simulations include multiple constraints on cell behavior. The use of constraints to describe behaviors comes from the physics of classical mechanics. In the GGH context we write *constraint energies* in a general *elastic* form:

$$H_{\text{constraint}} = \lambda\,(\text{value} - \text{target\_value})^2. \quad (3)$$

The constraint energy is zero if value = target_value (the constraint is *satisfied*) and grows as *value* diverges from *target_value*. The constraint is *elastic* because the exponent of 2 effectively creates an ideal spring pushing on the cells and driving them to satisfy the constraint. λ is the *spring constant* (a positive real number), which determines the *constraint strength*. Smaller values of λ allow the pattern to deviate more from the *equilibrium condition* (i.e., the condition satisfying the constraint). Because the constraint energy decreases smoothly to a minimum when the constraint is satisfied, the energy-minimizing dynamics used in the GGH automatically drives any configuration toward one that satisfies the constraint. However, because of the stochastic simulation method, the cell lattice need not satisfy the constraint exactly at any given time, resulting in random fluctuations. In addition, multiple constraints may conflict, leading to configurations which only partially satisfy some constraints.

Because biological cells have a given volume at any time, most GGH simulations employ a *volume constraint*, which restricts volume variations of generalized cells from their target volumes:

$$H_{\mathrm{vol}} = \sum_{\sigma} \lambda_{\mathrm{vol}}(\sigma)(v(\sigma) - V_{\mathrm{t}}(\sigma))^2, \tag{4}$$

where for cell σ, $\lambda_{\mathrm{vol}}(\sigma)$ denotes the *inverse compressibility* of the cell, $v(\sigma)$ is the number of pixels in the cell (its *volume*), and $V_{\mathrm{t}}(\sigma)$ is the cell's *target volume*. This constraint defines $P \equiv -2\lambda(v(\sigma) - V_{\mathrm{t}}(\sigma))$ as the *pressure* inside the cell. A cell with $v < V_{\mathrm{t}}$ has a positive internal pressure, while a cell with $v > V_{\mathrm{t}}$ has a negative internal pressure.

Since many cells have nearly fixed amounts of cell membrane, we often use a *surface-area constraint* of form:

$$H_{\mathrm{surf}} = \sum_{\sigma} \lambda_{\mathrm{surf}}(\sigma)(s(\sigma) - S_{\mathrm{t}}(\sigma))^2, \tag{5}$$

where $s(\sigma)$ is the surface area of cell σ, S_{t} is its target surface area, and $\lambda_{\mathrm{surf}}(\sigma)$ is its *inverse membrane compressibility*.[1]

Adding the boundary energy and volume constraint terms together (**Eqs. 1** and **4**), we obtain the *basic GGH effective energy*:

$$H_{\mathrm{GGH}} = \sum_{\substack{\vec{i},\vec{j} \\ \text{neighbors}}} J(\tau(\sigma(\vec{i})), \tau(\sigma(\vec{j})))\ (1 - \delta(\sigma(\vec{i}), \sigma(\vec{j}))) + \sum_{\sigma} \lambda_{\mathrm{vol}}(\sigma)(v(\sigma) - V_{\mathrm{t}}(\sigma))^2. \tag{6}$$

3.2. Dynamics

A GGH simulation consists of many attempts to copy cell indices between neighboring pixels. In CompuCell3D, the default dynamical algorithm is *modified Metropolis dynamics*. During each index-copy attempt, we select a *target* pixel, $\vec{i}$, randomly from the cell lattice, then randomly select one of its neighboring pixels, $\vec{i}'$, as a *source* pixel (note that the neighbor range may be greater than one). If they belong to the same generalized cell (i.e., if $\sigma(\vec{i}) = \sigma(\vec{i}')$) we do nothing. Otherwise, the cell containing the source pixel, $\sigma(\vec{i}')$, attempts to occupy the target pixel. Consequently, a successful index copy increases the volume of the *source* cell and decreases the volume of the *target* cell by one pixel (**Fig. 2**).

In the modified Metropolis algorithm we evaluate the change in the total effective energy due to the attempted index copy and accept the index-copy attempt with probability

$$P(\sigma(\vec{i}) \to \sigma(\vec{i}')) = \begin{cases} \exp(-\Delta H / T_{\mathrm{m}}) : \Delta H > 0; \\ 1 : \Delta H \le 0 \end{cases} \tag{7}$$

where T_{m} is a parameter representing the *effective cell motility* (we can think of T_{m} as the amplitude of cell-membrane fluctuations). **Equation 7** is the *Boltzmann acceptance function*. Users

[1] Because of lattice discretization and the option of defining long-range neighborhoods, the surface area of a cell scales in a non-Euclidian, lattice-dependent manner with cell volume (*see* **ref.** *61* on bubble growth).

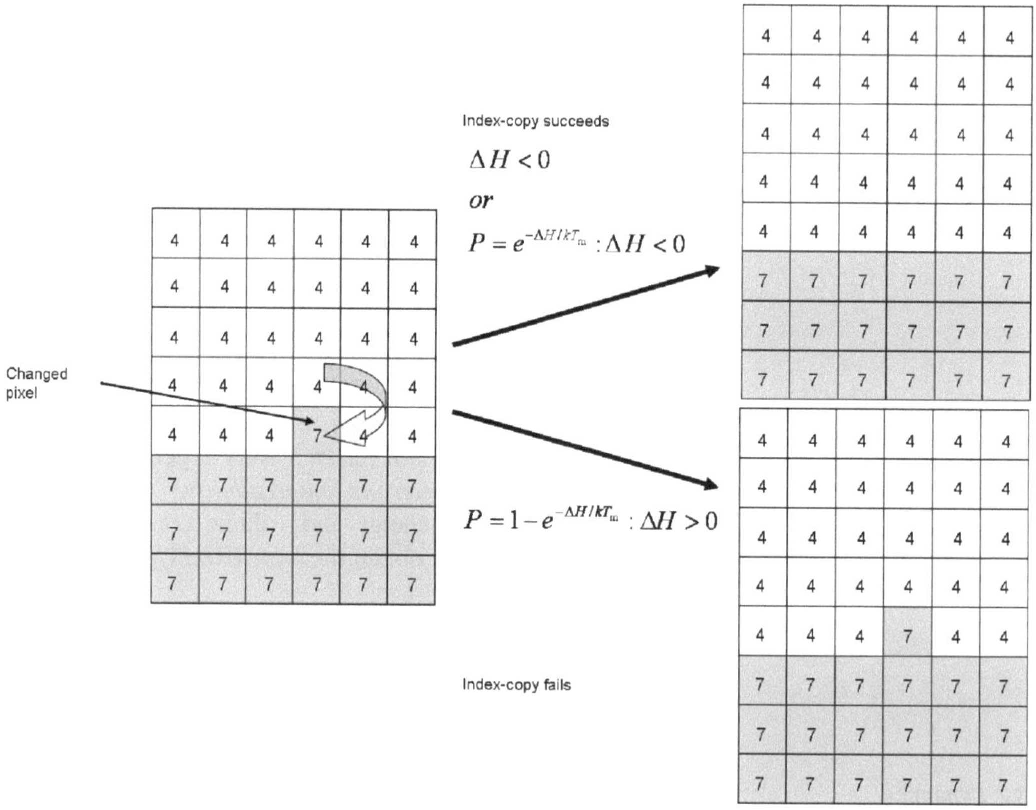

Fig. 2. GGH representation of an index-copy attempt for two cells on a 2D square lattice – The "*white*" pixel (source) attempts to replace the "*gray*" pixel (target). The probability of accepting the index copy is given by Eq. 7.

can define other acceptance functions in CompuCell3D. The conversion between MCS and experimental time depends on the average values of $\Delta H/T_m$. MCS and experimental time are proportional in biologically meaningful situations *(20, 87–89)*.

3.3. Algorithmic Implementation of Effective-Energy Calculations

Consider an effective energy consisting of boundary-energy and volume-constraint terms as in **Eq. 6**. After choosing the source ($\vec{i}'$) and destination ($\vec{i}$) pixels (the cell index of the source will overwrite the target pixel if the index copy is accepted), we calculate the change in the effective energy that would result from the copy. We evaluate the change in the boundary energy and volume constraint as follows. First, we visit the target pixel's neighbors ($\vec{i}''$). If the neighbor pixel belongs to a different generalized cell from the target pixel, i.e., when $\sigma(\vec{i}'') \neq \sigma(\vec{i})$ (see **Eq. 1**), we decrease ΔH by $J(\tau(\sigma(\vec{i})), \tau(\sigma(\vec{i}'')))$. If the neighbor belongs to a cell different from the source pixel ($\vec{i}'$) we increase ΔH by $J(\tau(\sigma(\vec{i}')), \tau(\sigma(\vec{i}'')))$.

The change in volume-constraint energy is evaluated according to

$$\begin{aligned}\Delta H_{\text{vol}} &= H_{\text{vol}}^{\text{new}} - H_{\text{vol}}^{\text{old}} \\ &= \lambda_{\text{vol}}[(v(\sigma(\vec{i}'))+1-V_{\text{t}}(\sigma(\vec{i}')))^2+(v(\sigma(\vec{i}))-1-V_{\text{t}}(\sigma(\vec{i})))^2] \\ &\quad -\lambda_{\text{vol}}[(v(\sigma(\vec{i}'))-V_{\text{t}}(\sigma(\vec{i}')))^2+(v(\sigma(\vec{i}))-V_{\text{t}}(\sigma(\vec{i})))^2] \\ &= \lambda_{\text{vol}}[\{1+2(v(\sigma(\vec{i}'))-V_{\text{t}}(\sigma(\vec{i}')))\}+\{1-2(v(\sigma(\vec{i}))-V_{\text{t}}(\sigma(\vec{i})))\}],\end{aligned} \tag{8}$$

where $v(\sigma(\vec{i}'))$ and $v(\sigma(\vec{i}))$ denote the volumes of the generalized cells containing the source and target pixels, respectively.

In this example, we could calculate the change in the effective energy locally, i.e., by visiting the neighbors of the target of the index copy. Most effective energies are quasi-local, allowing calculations of ΔH similar to those presented above (**Fig. 3**). The locality of the effective energy is crucial to the utility of the GGH approach. If we had to calculate the effective energy for the entire cell lattice for each index-copy attempt, the algorithm would be prohibitively slow.

For longer-range interactions we use the appropriate list of neighbors, as shown in **Fig. 4** and **Table 1**. Longer-range interactions are less anisotropic but result in slower simulations.

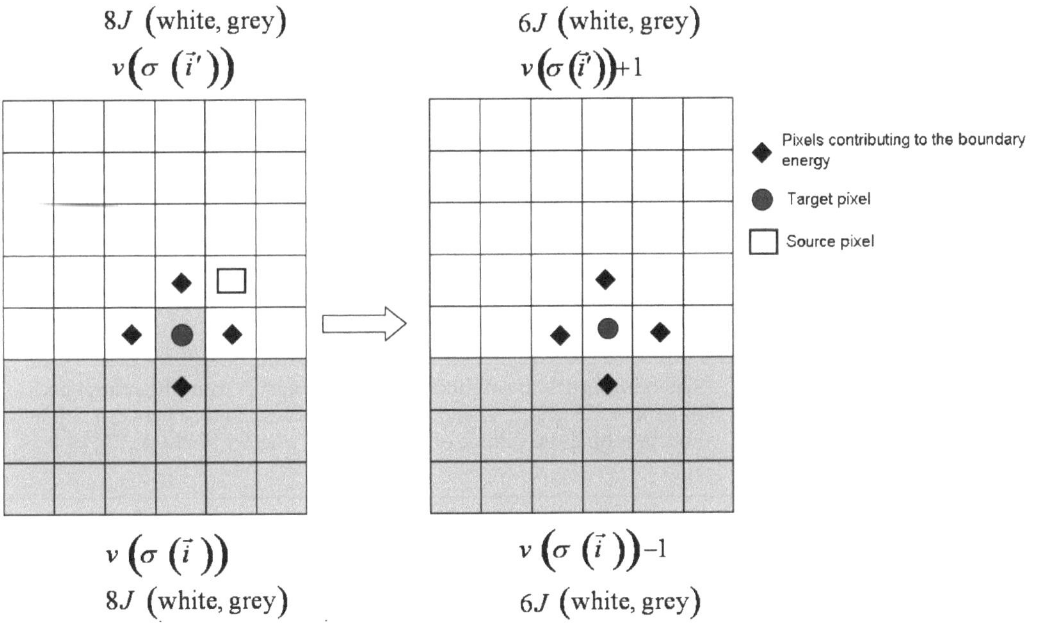

Fig. 3. Calculating changes in the boundary energy and the volume-constraint energy on a nearest-neighbor square lattice.

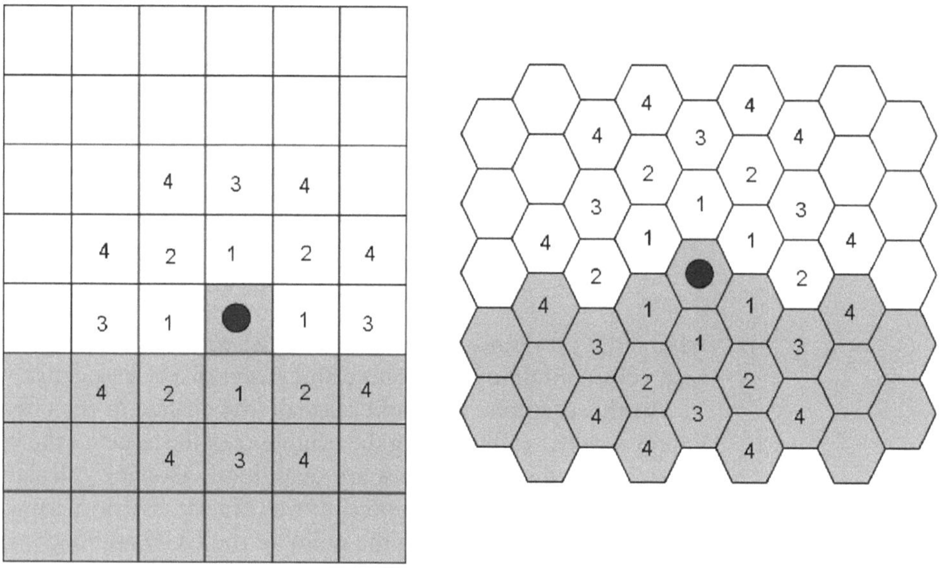

Fig. 4. Locations of *n*th nearest neighbors on a 2D square lattice and a 2D hexagonal lattice.

Table 1
Multiplicity and Euclidian distances of *n*th-nearest neighbors for 2D square and hexagonal lattices

	2D Square lattice		2D Hexagonal lattice	
Neighbor order	Number of neighbors	Euclidian distance	Number of Neighbors	Euclidian distance
1	4	1	6	$\sqrt{2/\sqrt{3}}$
2	4	$\sqrt{2}$	6	$\sqrt{6/\sqrt{3}}$
3	4	2	6	$\sqrt{8/\sqrt{3}}$
4	8	$\sqrt{5}$	12	$\sqrt{14/\sqrt{3}}$

The number of *n*th neighbors and their distances from the central pixel differ in a 3D lattice. CompuCell3D calculates distance between neighbors by setting the volume of a single pixel (whether in 2D or 3D) to 1.

4. CompuCell3D

One advantage of the GGH model over alternative techniques is its mathematical simplicity. We can implement fairly easily a computer program that initializes the cell lattice and fields, performs index copies, and displays the results. In the 15 years since the GGH model was developed, researchers have written numerous programs to run GGH simulations. Because all GGH implementations share the same logical structure and perform similar tasks, much of this coding effort has gone into rewriting code already developed by someone else. This redundancy leads to significant research overhead and unnecessary duplication of effort and makes model reproduction, sharing and validation needlessly cumbersome.

To overcome these problems, we developed CompuCell3D as a framework for GGH simulations *(90, 91)*. Unlike specialized research code, CompuCell3D is a *simulation environment* that allows researchers to rapidly build and run shareable GGH-based simulations. It greatly reduces the need to develop custom code and its adherence to open-source standards ensures that any such code is shareable.

CompuCell3D supports nonprogrammers by providing visualization tools, an *eXtensible Markup Language* (XML) interface for defining simulations, and the ability to extend the framework through specialized modules. The C+ computational kernel of CompuCell3D is also accessible using the open-source scripting language Python, allowing users to create complex simulations without programming in lower-level languages such as C or C+. Unlike typical research code, changing a simulation does not require recompiling CompuCell3D.

Users define simulations using *CompuCell3D XML (CC3DML) configuration files* and/or Python scripts. CompuCell3D reads and parses the CC3DML configuration file and uses it to define the basic simulation structure, then initializes appropriate Python services (if they are specified) and finally executes the underlying simulation algorithm.

CompuCell3D is modular: each module carries out a defined task. CompuCell3D terminology calls modules associated with index copies or index-copy attempts *plugins*. Some plugins calculate changes in effective energy, while others (*lattice monitors*) react to accepted index copies, e.g., by updating generalized cells' surface areas, volumes, or lists of neighbors. Plugins may depend on other plugins. For example, the `Volume` plugin (which calculates the volume–energy constraint in **Eq. 4**) depends on `VolumeTracker` (a lattice monitor), which, as its name suggests, tracks changes in generalized cells' volumes. When implicit plugin dependencies exist, CompuCell3D automatically loads and initializes dependent plugins. In addition to plugins, CompuCell3D defines modules called *steppables* which run either repeatedly after a defined intervals of Monte Carlo Steps or once at the beginning or end of the simulation. Steppables typically define initial conditions, alter cell states, update fields, or output intermediate results.

Figure 5 shows the relations among index-copy attempts, Monte Carlo Steps, steppables, and plugins.

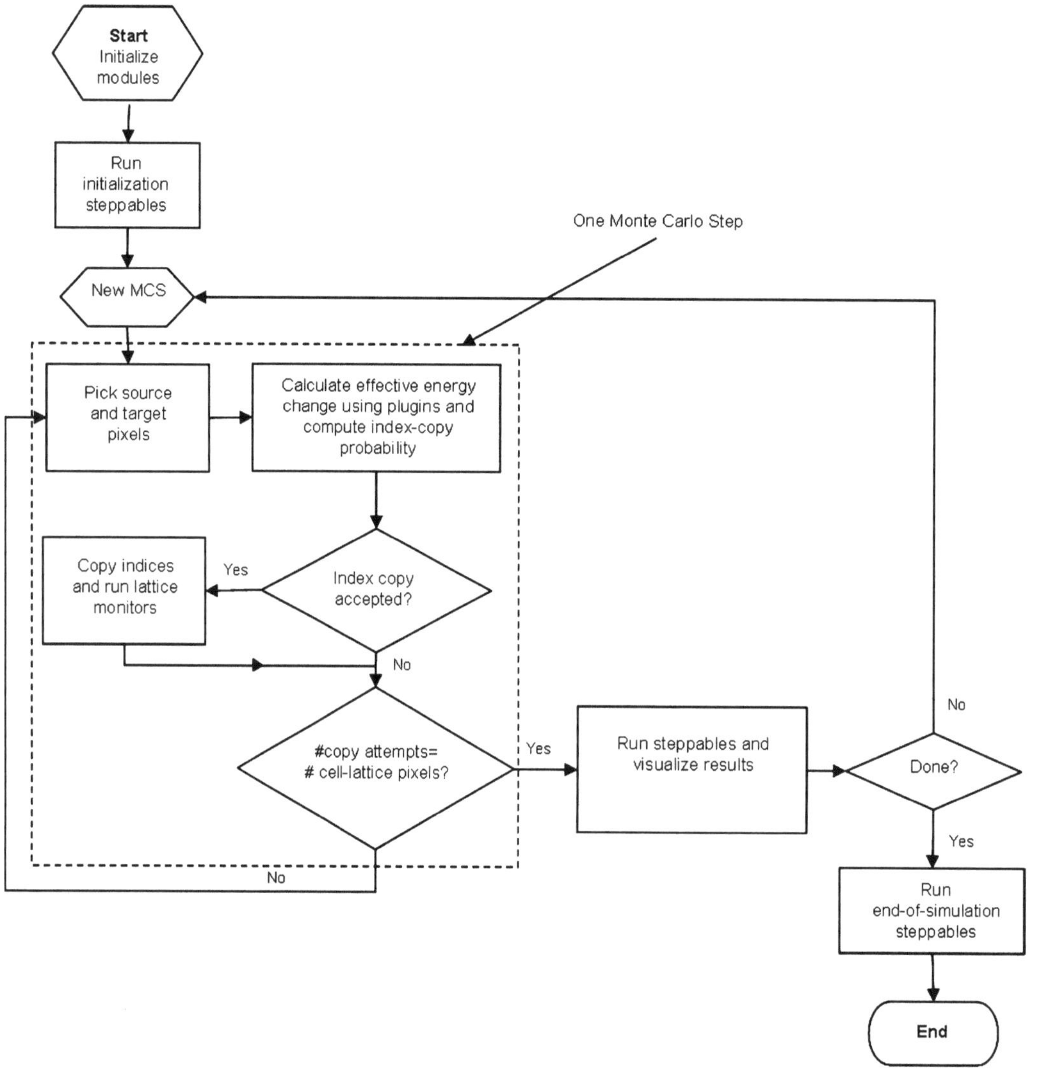

Fig. 5. Flow chart of the GGH algorithm as implemented in CompuCell3D. CompuCell3D includes a Graphical User Interface (GUI) and visualization tool, CompuCell Player (also referred to as Player). From Player the user opens a CC3DML configuration file and/or Python file and hits the "Play" button to run the simulation. Player allows users to define multiple 2D or 3D visualizations of generalized cells, fields or various vector plots while the simulation is running and save them automatically for postprocessing.

5. Building CC3DML-Based Simulations Using CompuCell3D

To show how to build simulations in CompuCell3D, the reminder of this chapter provides a series of examples of gradually increasing complexity. For each example we provide a brief explanation of the physical and/or biological background of the simulation and listings of the CC3DML configuration file and Python scripts, followed by a detailed explanation of their syntax and algorithms.

We can specify many simulations using only a simple CC3DML configuration file. We begin with three examples using only CC3DML to define simulations.

5.1. A Short Introduction to XML

XML is a text-based data-description language that allows standardized representations of data. XML syntax consists of lists of *elements*, each either contained between opening (<Tag>) and closing (</Tag>) tags:[2]

```
<Tag Attribute1="text1">ElementText</Tag>
```

or of form:

```
<Tag Attribute1="attribute_text1" Attribute2 ="
attribute_text2"/>
```

We will denote the <Tag>…</Tag> syntax as a <Tag> *tag pair*. The opening tag of an XML element may contain additional *attributes* characterizing the element. The content of the XML element (*ElementText* in the above example) and the values of its attributes (*text1, attribute_text1, attribute_text2*) are strings of characters. Computer programs that read XML may interpret these strings as other data types such as integers, Booleans, or floating point numbers. XML elements may be nested. The simple example below defines an element Cell with subelements (represented as nested XML elements) Nucleus and Membrane assigning the element Nucleus an attribute Size set to "10" and the element Membrane an attribute Area set to "20.5," and setting the value of the Membrane element to Expanding:

```
<Cell>
 <Nucleus Size="10"/>
 <Membrane Area="20.5">Expanding</Membrane>
</Cell>
```

Although XML parsers ignore indentation, all the listings presented in this chapter are block indented for better readability.

5.2. Grain-Growth Simulation

One of the simplest CompuCell3D simulations mimics crystalline grain growth, or *annealing*. Most simple metals are composed of microcrystals, or *grains*, each of which has a particular crystalline-lattice orientation. The atoms at the surfaces of these grains have a higher energy than those in the bulk because of their missing neighbors. We can characterize this excess energy

[2]In the text, we denote XML, CC3DML, and Python code using the Courier font. In listings presenting syntax, user-supplied variables are given in *italics*. Broken-out listings are boxed. Punctuation at the end of boxes is implicit.

as a *boundary energy*. Atoms in convex regions of a grain's surface have a higher energy than those in concave regions, in particular those in the concave face of an adjoining grain, because they have more missing neighbors. For this reason, an atom at a convex curved boundary can reduce its energy by "hopping" across the grain boundary to the concave side *(62)*. The movement of atoms moves the grain boundaries, lowering the net configuration energy through *annealing* or *coarsening*, with the net size of grains growing because of grain disappearance. Boundary motion may require thermal activation because atoms in the space between grains may have higher energy than atoms in grains. The effective energy driving grain growth is simply the boundary energy in **Eq. 1**.

In CompuCell3D, we can represent grains as generalized cells. CC3DML **Listing** 1 defines our grain-growth simulation.

Each CC3DML configuration file begins with the <CompuCell3D> tag and ends with the </CompuCell3D> tag. A CC3DML configuration file contains three sections in the following sequence: the *lattice section* (contained within the

Listing 1. CC3DML configuration file for 2D grain-growth simulation.

```
<CompuCell3D>
 <Potts>
  <Dimensions x=100" y="100" z="1"/>
  <Steps>10000</Steps>
  <Temperature>5</Temperature>
  <Boundary_y>Periodic</Boundary_y>
  <Boundary_x>Periodic</Boundary_x>
  <NeighborOrder>2</NeighborOrder>
 </Potts>

 <Plugin Name="CellType">
  <CellType TypeName="Medium" TypeId="0"/>
  <CellType TypeName="Grain" TypeId="1"/>
 </Plugin>

 <Plugin Name="Contact">
  <Energy Type1="Medium" Type2="Grain">0</Energy>
  <Energy Type1="Grain" Type2="Grain">5</Energy>
  <Energy Type1="Medium" Type2="Medium">0</Energy>
  <NeighborOrder>3</NeighborOrder>
 </Plugin>

 <Steppable Type="UniformInitializer">
  <Region>
   <BoxMin x="0" y="0" z="0"/>
   <BoxMax x="100" y="100" z="1"/>
   <Gap>0</Gap>
   <Width>5</Width>
   <Types>Grain</Types>
  </Region>
 </Steppable>

</CompuCell3D>
```

Lattice Section

Plugins Section

Steppables Section

<Potts> tag pair), the *plugins section*, and the *steppables section*. The lattice section defines global parameters for the simulation: cell-lattice and field-lattice dimensions (specified using the syntax <Dimensions x="*x_dim*" y="*y_dim*" z="*z_dim*"/>), the number of Monte Carlo Steps to run (defined within the <Steps> tag pair) the effective cell motility, T_m, (defined within the <Temperature> tag pair), and boundary conditions. The default boundary conditions are *no-flux*. However, in this simulation, we have changed them to be periodic along the *x* and *y* axes by assigning the value Periodic to the <Boundary_x> and <Boundary_y> tag pairs. The value set by the <NeighborOrder> tag pair defines the range over which source pixels are selected for index-copy attempts (*see* **Fig. 4** and **Table 1**).

The plugins section lists the plugins the simulation will use. The syntax for all plugins that require parameter specification is:

```
<Plugin Name="PluginName">
  <ParameterSpecification/>
</Plugin>
```

The CellType plugin uses the parameter syntax

```
<CellType  TypeName="Name"  TypeId="Integer
Number"/>
```

to map verbose generalized-cell-type names to numeric cell TypeIds for all generalized-cell types. It does not participate directly in index copies, but is used by other plugins for cell-type-to-cell-index mapping.

Note: Even though the grain-growth simulation fills the entire cell lattice with cells of type Grain, the current implementation of CompuCell3D requires that all simulations define the Medium cell type with TypeId 0. Medium is a special cell type with unconstrained volume and surface area that fills all cell-lattice pixels unoccupied by cells of other types.

The Contact plugin calculates changes in the boundary energy defined in **Eq. 1** for each index-copy attempt. The parameter syntax for the Contact plugin is:

```
<Energy Type1="TypeName1" Type2="TypeName1">
EnergyValue</Energy>
```

where *TypeName1* and *TypeName2* are the names of the cell types and *EnergyValue* is the boundary-energy coefficient, *J* (*TypeName1*, *TypeName2*), between cells of *TypeName1* and *TypeName2* (*see* **Eq. 1**). The <NeighborOrder> tag pair specifies the interaction range of the boundary energy. The default value of this parameter is 1.

The steppables section includes only the UniformInitializer steppable. All steppables have the following syntax:

```
<Steppable Type="SteppableName" Frequency="
  FrequencyMCS">
  <ParameterSpecification/>
</Steppable>
```

The `Frequency` attribute is optional and by default is 1 MCS.

CompuCell3D simulations require specification of initial condition. The simplest way to define the initial cell lattice is to use the built-in initializer steppables, which construct simple regions filled with generalized cells.

The `UniformInitializer` steppable in the grain-growth simulation defines one or more rectangular (parallelepiped in 3D) regions filled with generalized cells of user selectable types and sizes. We enclose each region definition within a `<Region>` tag pair. We use the `<BoxMin>` and `<BoxMax>` tags to describe the boundaries of the region, The `<Width>` tag pair defines the size of the square (cubical in 3D) generalized cells and the `<Gap>` tag pair creates space between neighboring cells. The `<Types>` tag pair lists the types of generalized cells. The grain-growth simulation uses only one cell type, `Grain`, but we can also initialize cells using types randomly chosen from the list, as in **Listing 2**.

Note: The coordinate values in the `BoxMax` element must be one more than the coordinates of the corresponding corner of the region to be filled. So to fill a square of side 10 beginning with pixel location *(5, 5)* we use the following region-boundary specification:

```
<BoxMin x="5" y="5" z="0"/>
<BoxMax x="16" y="16" z="1"/>
```

Listing the same type multiple times results in a proportionally higher fraction of generalized cells of that type. For example,

```
<Types>Condensing,Condensing,NonCondensing
</Types>
```

Listing 2. CC3DML code excerpt using the `UniformInitializer` steppable to initialize a rectangular region filled with 5 × 5 pixel generalized cells with randomly assigned cell types (either `Condensing` or `NonCondensing`).

```
<Steppable Type="UniformInitializer"
   <Region>
      <BoxMin x="10" y="10" z="0"/>
      <BoxMax x="90" y="90" z="1"/>
      <Gap>0</Gap>
      <Width>5</Width>
      <Types>Condensing,NonCondensing</Types>
   </Region>
</Steppable>
```

will allocate approximately 2/3 of the generalized cells to type `Condensing` and 1/3 to type `NonCondensing`. `Uniform-Initializer` allows specification of multiple regions. Each region is independent and can have its own cell sizes, types, and cell spacing. If the regions overlap, later-specified regions overwrite earlier-specified ones. If region specification does not cover the entire lattice, uninitialized pixels have type `Medium`.

Figure 6 shows sample output generated by the grain-growth simulation.

One advantage of GGH simulations compared to FE simulations is that going from 2D to 3D is easy. To run a 3D grain-

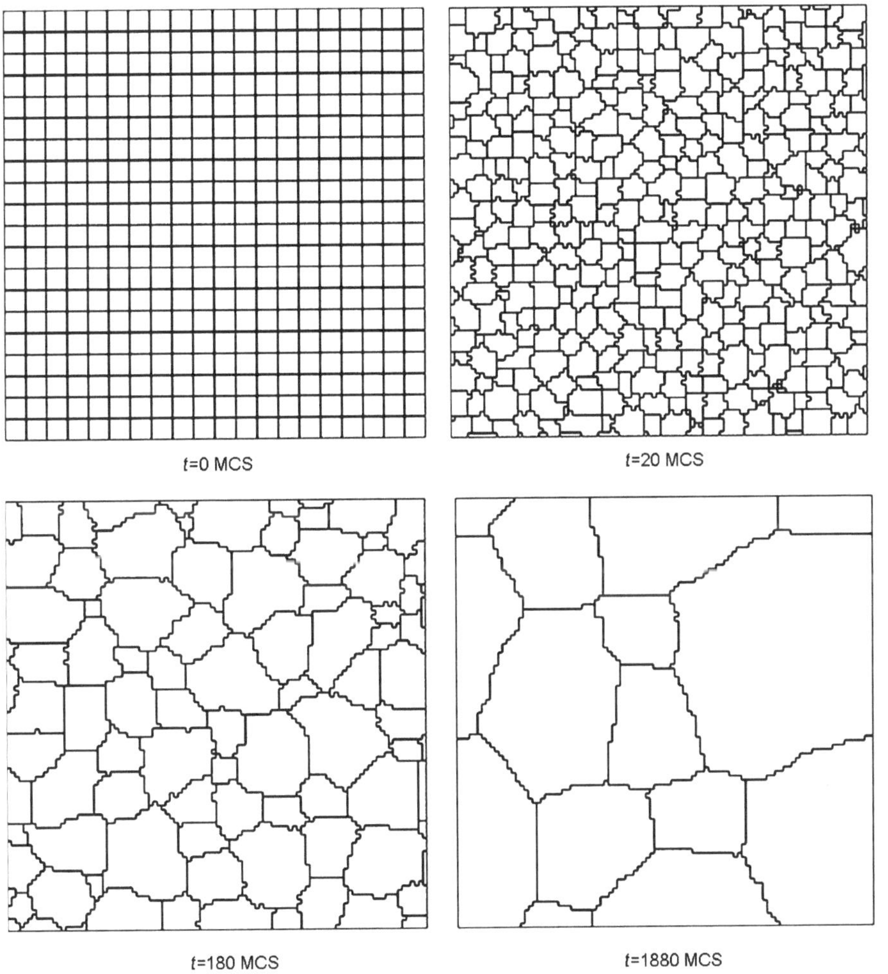

Fig. 6. Snapshots of the cell-lattice configuration for the grain-growth simulation on a 100 × 100 pixel third-neighbor square lattice, as specified in Listing 1. Boundary conditions are periodic in both directions.

growth simulation on a 100 × 100 × 100 lattice we only need to make the following replacements in **Listing** 1:

```
<Dimensions x="100" y="100" z="1"/> →
<Dimensions x="100" y="100" z="100"/>
```

and

```
<BoxMax x="100" y="100" z="1"/> → <BoxMax
x="100" y="100" z="100"/>
```

Grain growth simulations are particularly sensitive to lattice anisotropy, so running them on lower-anisotropy lattices is desirable. Longer-range lattices are less anisotropic but cause simulations to run more slowly. Fortunately, a hexagonal lattice of a given range is less anisotropic than a square lattice of the same range. To run the grain-growth simulation on a hexagonal lattice, we add `<LatticeType>Hexagonal</LatticeType>` to the lattice section in **Listing** 1 and change the two occurrences of

```
<NeighborOrder>3</NeighborOrder>
```

to

```
<NeighborOrder>1</NeighborOrder>
```

Figure 7 shows snapshots for this simulation.

Note: One inconvenience of the current implementation of CompuCell3D is that it does not automatically rescale parameter values when interaction range, lattice dimensionality or lattice type change. When changing these attributes, users must recalculate parameters to keep the underlying physics of the simulation the same.

CompuCell3D dramatically reduces the amount of code necessary to build and run a simulation. The grain-growth simulation took about 25 lines of CC3DML instead of 1,000 lines of C, C++ or Fortran.

5.3. Cell-Sorting Simulation

Cell sorting is an experimentally observed phenomenon in which cells with different adhesivities are randomly mixed and reaggregated. They can spontaneously sort to reestablish coherent homogenous domains *(92, 93)*. Sorting is a key mechanism in embryonic development.

The grain-growth simulation used only one type of generalized cell. Simulating sorting of two types of biological cell in an aggregate floating in solution is slightly more complex. **Listing** 3 shows a simple cell-sorting simulation. It is similar to **Listing** 1 with a few additional modules (shown in **bold**). The effective energy is that in **Eq. 6**.

The most significant departure from the lattice section in **Listing** 1 is that we omit the boundary condition specification and use default no-flux boundary conditions.

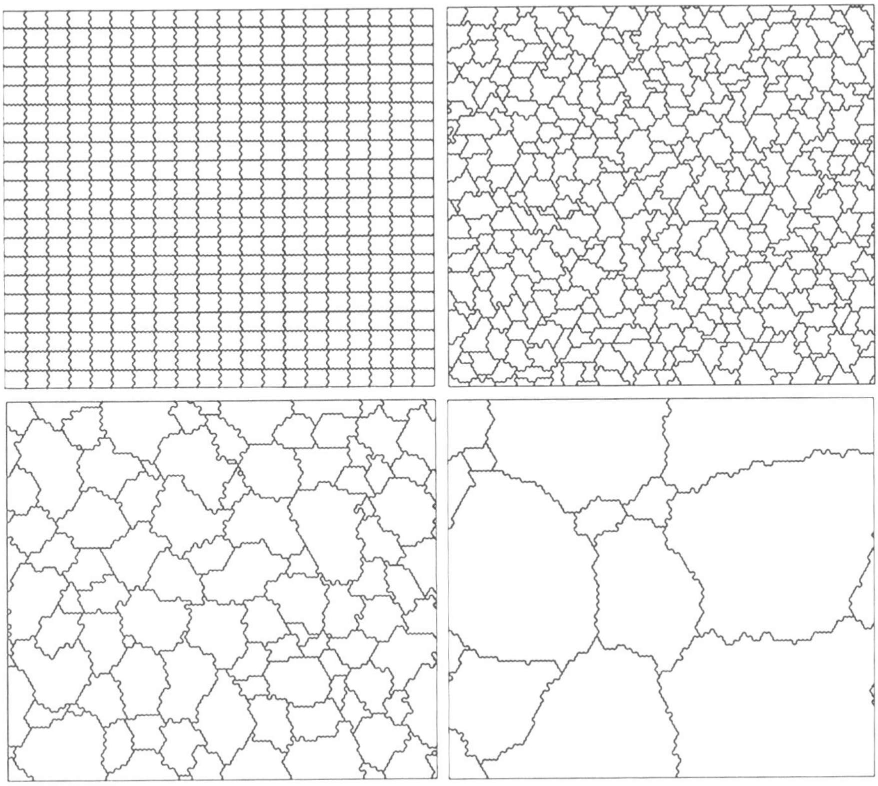

Fig. 7. Snapshots of the cell-lattice configuration for the grain-growth simulation on a 100 × 100 pixel first-neighbor hexagonal lattice as specified in Listing 1 with substitutions described in the text. The *x* and *y* length units in an hexagonal lattice differ, resulting in differing *x* and *y* dimensions for a cell lattice with an equal number of pixels in the *x* and *y* directions.

In the `CellType` plugin we introduce the two cell types, `Condensing` and `NonCondensing`, in place of `Grain`. In addition, we do not fill the lattice completely with `Condensing` and `NonCondensing` cells, so the interactions with `Medium` become important. The boundary-energy matrix in the `Contact` plugin thus requires entries for the additional cell-type pairs. The hierarchy of boundary energies listed results in cell sorting.

We also add the `Volume` plugin, which calculates the volume-constraint energy as given in **Eq. 4**. In this plugin the `<TargetVolume>` tag pair sets target volume $V_t = 25$ for both `Condensing` and `NonCondensing` cells; similarly, the `<LambdaVolume>` tag pair sets the constraint strength $\lambda_{vol} = 2.0$ for both cell types. We will see later how to define volume-constraint parameters for each cell type or each cell individually.

In the cell-sorting simulation we initialize the cell lattice using the `BlobInitializer` steppable, which specifies circular (or spherical in 3D) regions filled with square (or cubical in 3D) cells of user-defined size and types. The syntax is very similar to that for `UniformInitializer`.

Looking in detail at the syntax of `BlobInitializer` in **Listing 3**, the `<Radius>` tag pair defines the radius of a circular

Listing 3. CC3DML configuration file simulating cell sorting between `Condensing` and `NonCondensing` cell types. Highlighted text indicates modules absent in Listing 1. Notice how little modification of the grain-growth CC3DML configuration file this simulation requires.

```
<CompuCell3D>
 <Potts>
  <Dimensions x="100" y="100" z="1"/>
  <Steps>10000</Steps>
  <Temperature>10</Temperature>
  <NeighborOrder>2</NeighborOrder>
 </Potts>

 <Plugin Name="Volume">
  <TargetVolume>25</TargetVolume>
  <LambdaVolume>2.0</LambdaVolume>
 </Plugin>

 <Plugin Name="CellType">
  <CellType TypeName="Medium" TypeId="0"/>
  <CellType TypeName="Condensing" TypeId="1"/>
  <CellType TypeName="NonCondensing" TypeId="2"/>
 </Plugin>

 <Plugin Name="Contact">
  <Energy Type1="Medium" Type2="Medium">0</Energy>
  <Energy Type1="NonCondensing" Type2="NonCondensing">16</Energy>
  <Energy Type1="Condensing"    Type2="Condensing">2</Energy>
  <Energy Type1="NonCondensing" Type2="Condensing">11</Energy>
  <Energy Type1="NonCondensing" Type2="Medium">16</Energy>
  <Energy Type1="Condensing"    Type2="Medium">16</Energy>
  <NeighborOrder>2</NeighborOrder>
 </Plugin>

 <Steppable Type="BlobInitializer">
  <Region>
   <Gap>0</Gap>
   <Width>5</Width>
   <Radius>40</Radius>
   <Center x="50" y="50" z="0"/>
   <Types>Condensing,NonCondensing</Types>
  </Region>
 </Steppable>

</CompuCell3D>
```

(or spherical) domain of cells in pixels. The `<Center>` tag, with syntax `<Center x="x_position" y="y_position" z="z_position"/>`, defines the coordinates of the center of the domain. The remaining tags are the same as for `UniformInitializer`. As with `UniformInitializer`, we can define multiple regions. We can use both `UniformInitializer` and `BlobInitializer` in the same simulation. In the case of overlap, later-specified regions overwrite earlier ones.

We show snapshots of the cell-sorting simulation in **Fig. 8**. The less cohesive `NonCondensing` cells engulf the more cohesive `Condensing` cells, which cluster and form a single central domain. By changing the boundary energies, we can produce other cell-sorting patterns *(94, 95)*.

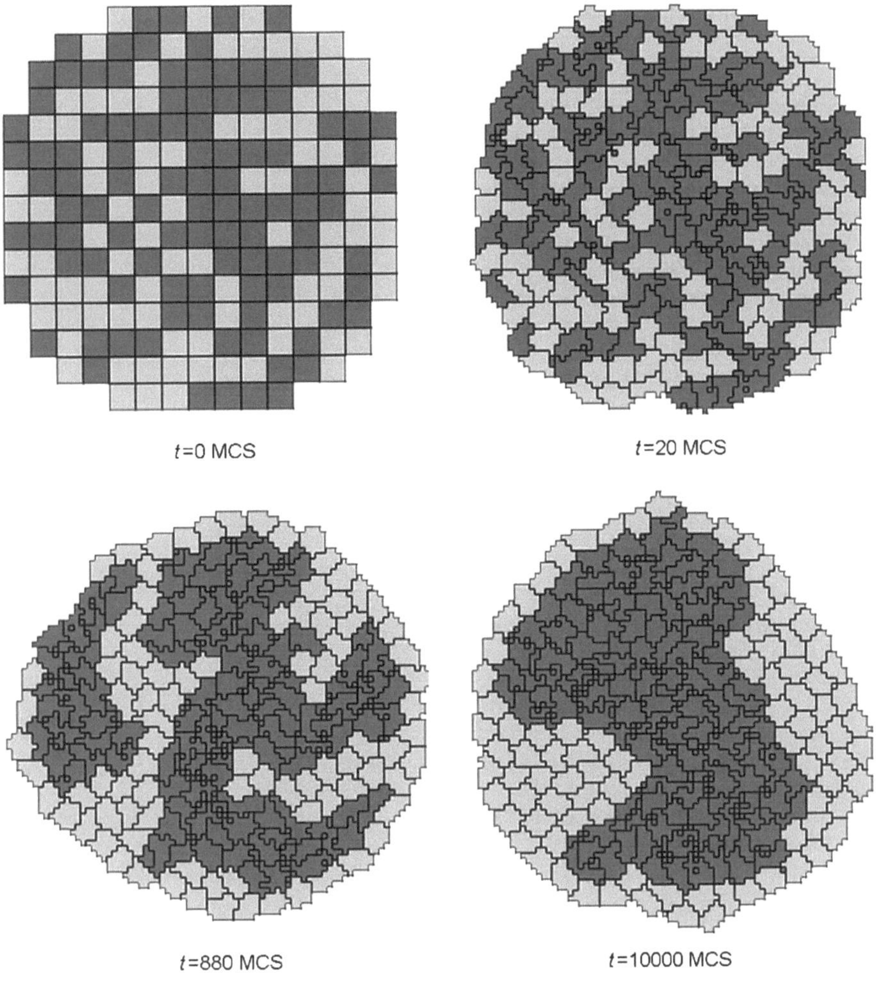

Fig. 8. Snapshots of the cell-lattice configurations for the cell-sorting simulation in Listing 3. The boundary-energy hierarchy drives NonCondensing (*light gray*) cells to surround Condensing (*dark gray*) cells. The *white* background denotes surrounding Medium.

5.4. Bacterium-and-Macrophage Simulation

In the two simulations we have presented so far, the cellular pattern develops without fields. Often, however, biological patterning mechanisms require us to introduce and evolve chemical fields and to have cells' behaviors depend on the fields. To illustrate the use of fields, we model the *in vitro* behavior of bacteria and macrophages in blood. In the famous experimental movie taken in the 1950s by David Rogers at Vanderbilt University, the macrophage appears to chase the bacterium, which seems to run away from the macrophage. We can model both behaviors using cell secretion of diffusible chemical signals and movement of the cells in response to the chemicals (*chemotaxis*): the bacterium secretes a signal (a *chemoattractant*) that attracts the macrophage and the macrophage secretes a signal (a *chemorepellant*) that repels the bacterium *(96)*.

Listing 4 shows the CC3DML configuration file for the bacterium-and-macrophage simulation.

Listing 4. CC3DML configuration file for the bacterium-and-macrophage simulation.

```
<CompuCell3D>
 <Potts>
  <Dimensions x="100" y="100" z="1"/>
  <Steps>100000</Steps>
  <Temperature>20</Temperature>
  <LatticeType>Hexagonal</LatticeType>
 </Potts>

 <Plugin Name="CellType">
  <CellType TypeName="Medium" TypeId="0"/>
  <CellType TypeName="Bacterium" TypeId="1" />
  <CellType TypeName="Macrophage" TypeId="2"/>
  <CellType TypeName="Red" TypeId="3"/>
  <CellType TypeName="Wall" TypeId="4" Freeze=""/>
 </Plugin>

 <Plugin Name="VolumeFlex">
  <VolumeEnergyParameters CellType="Macrophage" TargetVolume="150"
      LambdaVolume="15"/>
  <VolumeEnergyParameters CellType="Bacterium" TargetVolume="10"
     LambdaVolume="60"/>
  <VolumeEnergyParameters CellType="Red" TargetVolume="100"
     LambdaVolume="30"/>
 </Plugin>

 <Plugin Name="SurfaceFlex">
  <SurfaceEnergyParameters CellType="Macrophage" TargetSurface="50"
     LambdaSurface="30"/>
  <SurfaceEnergyParameters CellType="Bacterium" TargetSurface="10"
     LambdaSurface="4"/>
  <SurfaceEnergyParameters CellType="Red" TargetSurface="40"
     LambdaSurface="0"/>
 </Plugin>

 <Plugin Name="Contact">
  <Energy Type1="Medium" Type2="Medium">0</Energy>
  <Energy Type1="Macrophage" Type2="Macrophage">150</Energy>
  <Energy Type1="Macrophage" Type2="Medium">8</Energy>
  <Energy Type1="Bacterium" Type2="Bacterium">150</Energy>
  <Energy Type1="Bacterium" Type2="Macrophage">15</Energy>
  <Energy Type1="Bacterium" Type2="Medium">8</Energy>
  <Energy Type1="Wall" Type2="Wall">0</Energy>
  <Energy Type1="Wall" Type2="Medium">0</Energy>
  <Energy Type1="Wall" Type2="Bacterium">150</Energy>
  <Energy Type1="Wall" Type2="Macrophage">150</Energy>
  <Energy Type1="Wall" Type2="Red">150</Energy>
  <Energy Type1="Red" Type2="Red">150</Energy>
  <Energy Type1="Red" Type2="Medium">30</Energy>
  <Energy Type1="Red" Type2="Bacterium">150</Energy>
  <Energy Type1="Red" Type2="Macrophage">150</Energy>
  <NeighborOrder>2</NeighborOrder>
 </Plugin>

 <Plugin Name="Chemotaxis">
```

```
 <ChemicalField Source="FlexibleDiffusionSolverFE" Name="ATTR">
 <ChemotaxisByType Type="Macrophage" Lambda="1"/>
 </ChemicalField>

 <ChemicalField Source="FlexibleDiffusionSolverFE" Name="REP">
  <ChemotaxisByType Type="Bacterium" Lambda="-0.1"/>
 </ChemicalField>
</Plugin>

<Steppable Type="FlexibleDiffusionSolverFE">
 <DiffusionField>
  <DiffusionData>
   <FieldName>ATTR</FieldName>
   <DiffusionConstant>0.10</DiffusionConstant>
   <DecayConstant>0.00005</DecayConstant>
   <DoNotDiffuseTo>Wall</DoNotDiffuseTo>
   <DoNotDiffuseTo>Red</DoNotDiffuseTo>
  </DiffusionData>
   <SecretionData>
    <Secretion Type="Bacterium">200</Secretion>
   </SecretionData>
  </DiffusionField>

  <DiffusionField>
   <DiffusionData>
   <FieldName>REP</FieldName>
   <DiffusionConstant>0.10</DiffusionConstant>
   <DecayConstant>0.001</DecayConstant>
   <DoNotDiffuseTo>Wall</DoNotDiffuseTo>
   <DoNotDiffuseTo>Red</DoNotDiffuseTo>
  </DiffusionData>
  <SecretionData>
   <Secretion Type="Macrophage">200</Secretion>
  </SecretionData>
 </DiffusionField>
</Steppable>

<Steppable Type="PIFInitializer">
 <PIFName>bacterium_macrophage_2D_wall_v3.pif</PIFName>
</Steppable>

</CompuCell3D>
```

The simulation has five generalized-cell types: `Medium`, `Bacterium`, `Macrophage`, `Red` (blood) cells, and a surrounding `Wall`. It also has two diffusible fields, representing a chemoattractant, `ATTR`, and a chemorepellent, `REP`. Because the default boundary-energy between any generalized-cell type and the edge of the cell lattice is zero, we define a surrounding wall to prevent cells from sticking to the cell-lattice boundary. As in our previous simulations, we assign cell types using the `CellType` plugin. Note the new syntax in the line specifying the cell type making up the walls:

```
<CellType TypeName="Wall" TypeId="4" Freeze=""/>
```

The `Freeze=""` attribute excludes generalized cells of type `Wall` from participating in index copies, which makes the walls immobile.

We replace the `Volume` plugin with `VolumeFlex` and add the plugin `SurfaceFlex`. These plugins allow independent assignment of target values and constraint strengths in the volume-constraint and surface-constraint energies (**Eqs. 4** and **5**). These plugins require a line for each generalized-cell type, specifying the type name and the target volume (or target surface area), and λ_{vol} (or λ_{surf}) for that generalized-cell type:

```
<VolumeEnergyParameters         CellType="Name"
TargetVolume="Value" LambdaVolume="Value"/>
```

We implement the actual bacterium-macrophage "chasing" mechanism using the `Chemotaxis` plugin, which specifies how a generalized cell of a given type responds to a field. The `Chemotaxis` plugin biases a cell's motion up or down a field gradient by changing the calculated effective-energy change used in the acceptance function, **Eq. 7** by the addition of a term ΔH_{chem}. For a field $c(\vec{i})$:

$$\Delta H_{chem} = -\lambda_{chem}(c(\vec{i}) - c(\vec{i}\,')), \tag{9}$$

where $c(\vec{i})$ is the chemical field value at the index-copy target pixel, $c(\vec{i}\,')$ is the field at the index-copy source pixel, and λ_{chem} is the strength and direction of chemotaxis. If $\lambda_{chem} > 0$ and $c(\vec{i}) > c$ $c(\vec{i}\,')$, then ΔH_{chem} is negative, increasing the probability of accepting the index copy in **Eq. 7** (**Fig. 9**). The net effect is that the cell moves up the field gradient with a velocity $\sim\lambda_{chem}\vec{\nabla}c$ (i.e., the field describes a chemoattractant for the cell). If λ_{chem} is negative, the opposite occurs, and the cell will move down the field gradient (the chemo repellant for the cell). Plugins with more sophisticated ΔH_{chem} calculations (e.g., including response saturation) are available within CompuCell3D (see the *CompuCell3D User Guide http://www.compucell3d.org*).

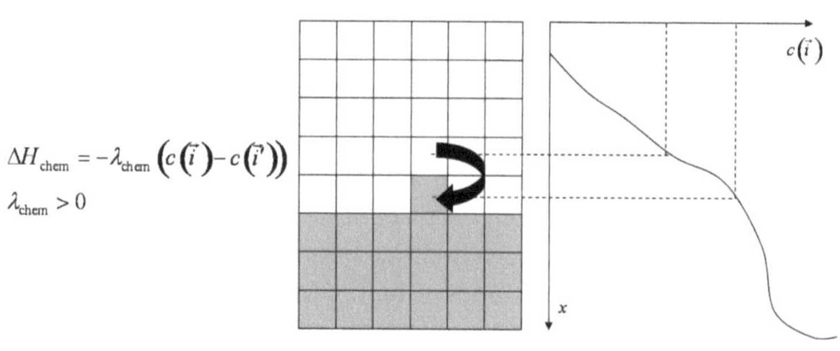

Fig. 9. Connecting a field to GGH dynamics using a chemotaxis-energy term. The difference in the value of the field c at the source ($\vec{i}\,'$) and target ($\vec{i}$) pixels changes the ΔH of the index-copy attempt. Here $c(\vec{i}) > c(\vec{i}\,')$ and $\lambda > 0$, so $\Delta H_{chem} < 0$, increasing the probability of accepting the index-copy attempt in Eq. 7.

In the Chemotaxis plugin we must identify the names of the fields, where the field information is stored, the list of the generalized-cell types that will respond to the fields, and the strength and direction of the response (Lambda = λ_{chem}). The information for each field is specified using the syntax

```
<ChemicalField Source="where field is stored"
Name="field name">
<ChemotaxisByType              Type="cell_type1"
Lambda="lambda1"/>
<ChemotaxisByType              Type="cell_type2"
Lambda="lambda1"/>
</ChemicalField>
```

In our current example, the first field, named ATTR, is stored in FlexibleDiffusionSolverFE. Macrophage cells are attracted to ATTR with $\lambda_{chem} = 1$. None of the other cell types responds to ATTR. Similarly, Bacterium cells are repelled by REP with $\lambda_{chem} = -0.1$.

Keep in mind that fields are *not* created within the Chemotaxis plugin, which only specifies how different cell types respond to the fields. We define and store the fields elsewhere. Here, we use the FlexibeDiffusionSolverFE steppable as the source of our fields. The FlexibleDiffusionSolverFE steppable is the main CompuCell3D tool for defining diffusing fields, which evolve according to the diffusion equation:

$$\frac{\partial c(\vec{i})}{\partial t} = D(\vec{i})\nabla^2 c(\vec{i}) - k(\vec{i})c(\vec{i}) + s(\vec{i}), \tag{10}$$

where $c(\vec{i})$ is the field concentration and $D(\vec{i})$, $k(\vec{i})$, and $s(\vec{i})$ denote the diffusion constant (in m^2/s), decay constant (in s^{-1}), and secretion rates (in concentration/s) of the field, respectively. $D(\vec{i})$, $k(\vec{i})$, and $s(\vec{i})$ may vary with position and cell-lattice configuration.

As in the Chemotaxis plugin, we may define the behaviors of multiple fields, enclosing each one within <DiffusionField> tag pairs. For each field, users provide values for the name of the field (using the <FieldName> tag pair), the diffusion constant (using the <DiffusionConstant> tag pair), and the decay constant (using the <DiffusionConstant> tag pair), all enclosed by the <DiffusionData> tag pair.

Note: Forward-Euler methods are numerically unstable for large diffusion constants, limiting the maximum nominal diffusion constant allowed in CompuCell3D simulations. However, by increasing the PDE-solver calling frequency, which reduces the effective time step, CompuCell3D can simulate arbitrarily large diffusion constants. For more information, see the *CompuCell3D User Guide*.

Each optional <DoNotDiffuseTo> tag pair, with syntax

```
<DoNotDiffuseTo>cell_type</DoNotDiffuseTo>
```

prevents the field from diffusing into field-lattice pixels where the corresponding cell-lattice pixel, $\vec{i}$, is occupied by a cell, $\sigma(\vec{i})$, of the specified type. In our case, chemical fields do not diffuse into the pixels occupied by Wall or Red cells. The optional <SecretionData> tag pair defines a subsection which identifies cell types that secrete or absorb the field and the rates of secretion:

```
<SecretionData>
<Secretion    Type="cell_type1">real_rate1</
Secretion>
<Secretion    Type="cell_type2">real_rate2</
Secretion>
<SecretionData>
```

A negative *rate* simulates absorption. In the bacterium and macrophage simulation, Bacterium cells secrete ATTR and Macrophage cells secrete REP.

We load the initial configuration for the bacterium-and-macrophage simulation using the PIFInitializer steppable. Many simulations require initial generalized-cell configurations that we cannot easily construct from primitive regions filled with cells using BlobInitializer and UniformInitializer. To allow maximum flexibility, CompuCell3D can read the initial cell-lattice configuration from Pixel Initialization Files (PIFs). A PIF is a text file that allows users to assign multiple rectangular (parallelepiped in 3D) pixel regions or single pixels to particular cells.

Each line in a PIF has the syntax

```
Cell_id Cell_type x_low x_high y_low y_high
z_low z_high
```

where *Cell_id* is a unique cell index. A PIF may have multiple, possibly non-adjacent, lines starting with the same *Cell_id*; all lines with the same *Cell_id* define pixels of the same generalized cell. The values *x_low*, *x_high*, *y_low*, *y_high*, *z_low*, and *z_high* define rectangles (parallelepipeds in 3D) of pixels belonging to the cell. In the case of overlapping pixels, a later line overwrites pixels defined by earlier lines. The following line describes a 6 × 6-pixel square cell with cell index 0 and type Amoeba:

```
0 Amoeba 10 15 10 15 0 0
```

If we save this line to the file "amoebae.pif," we can load it into a simulation using the following syntax:

```
<Steppable Type="PIFInitializer">
<PIFName>amoebae.pif</PIFName>
</Steppable>
```

Listing 5 illustrates how to construct arbitrary shapes using a PIF. Here we define two cells with indices 0 and 1, and cell types `Amoeba` and `Bacterium`, respectively. The main body of each cell is a 6 × 6 square to which we attach additional pixels.

All lines with the same cell index (first column) define a single cell.

Figure 10 shows the initial cell-lattice configuration specified in **Listing 5**.

Listing 6 shows the PIF for the bacterium-and-macrophage simulation.

In **Listing 4**, we read the cell-lattice configuration from the file "bacterium_macrophage_2D_wall_v3.pif" using the lines:

Listing 5. Simple PIF initializing two cells, one each of type Bacterium and Amoeba.

```
0 Amoeba 10 15 10 15 0 0

1 Bacterium 25 30 25 30 0 0

0 Amoeba 16 16 15 15 0 0

1 Bacterium 25 27 31 35 0 0
```

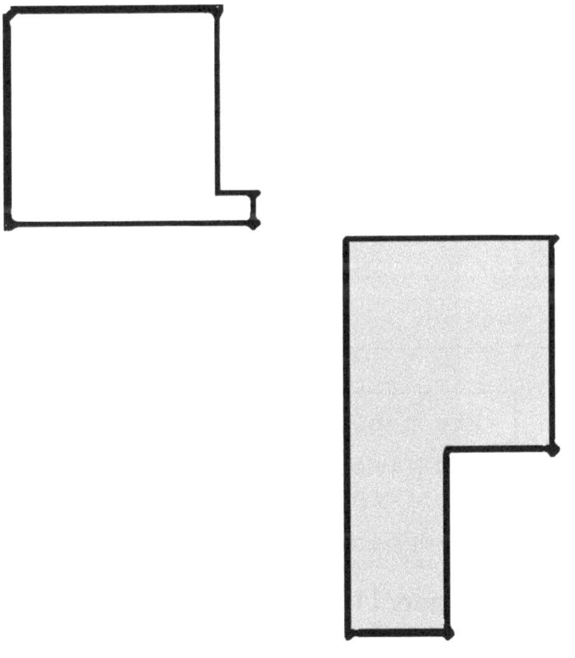

Fig. 10. Initial configuration of the cell lattice based on the PIF in Listing 5. In practice, because constructing complex PIFs by hand is cumbersome, we generally use custom-written scripts to generate the file directly, or convert images stored in graphical formats (e.g., gif, jpeg, png) from experiments or other programs.

Listing 6. PIF defining the initial cell-lattice configuration for the bacterium-and-macrophage simulation. The file is stored as "bacterium_macrophage_2D_wall_v3.pif".

```
0 Red 10 20 10 20 0 0
1 Red 10 20 40 50 0 0
2 Red 10 20 70 80 0 0
3 Red 40 50 0 10 0 0
4 Red 40 50 30 40 0 0
5 Red 40 50 60 70 0 0
6 Red 40 50 90 95 0 0
7 Red 70 80 10 20 0 0
8 Red 70 80 40 50 0 0
9 Red 70 80 70 80 0 0
10 Wall 0 99 0 1 0 0
10 Wall 98 99 0 99 0 0
10 Wall 0 99 98 99 0 0
10 Wall 0 1 0 99 0 0
11 Bacterium 5 5 5 5 0 0
12 Macrophage 35 35 35 35 0 0
13 Bacterium 65 65 65 65 0 0
14 Bacterium 65 65 5 5 0 0
15 Bacterium 5 5 65 65 0 0
16 Macrophage 75 75 95 95 0 0
17 Red 24 28 10 20 0 0
18 Red 24 28 40 50 0 0
19 Red 24 28 70 80 0 0
20 Red 40 50 14 20 0 0
21 Red 40 50 44 50 0 0
22 Red 40 50 74 80 0 0
23 Red 54 59 90 95 0 0
24 Red 70 80 24 28 0 0
25 Red 70 80 54 59 0 0
26 Red 70 80 84 90 0 0
27 Macrophage 10 10 95 95 0 0
```

```
<Steppable Type="PIFInitializer">
<PIFName>bacterium_macrophage_2D_wall_
v3.pif</PIFName>
</Steppable>
```

Figure 11 shows snapshots of the bacterium-and-macrophage simulation. By adjusting the properties and number of bacteria, macrophages and red blood cells and the diffusion properties of the chemical fields, we can build a surprisingly good reproduction of the experiment.

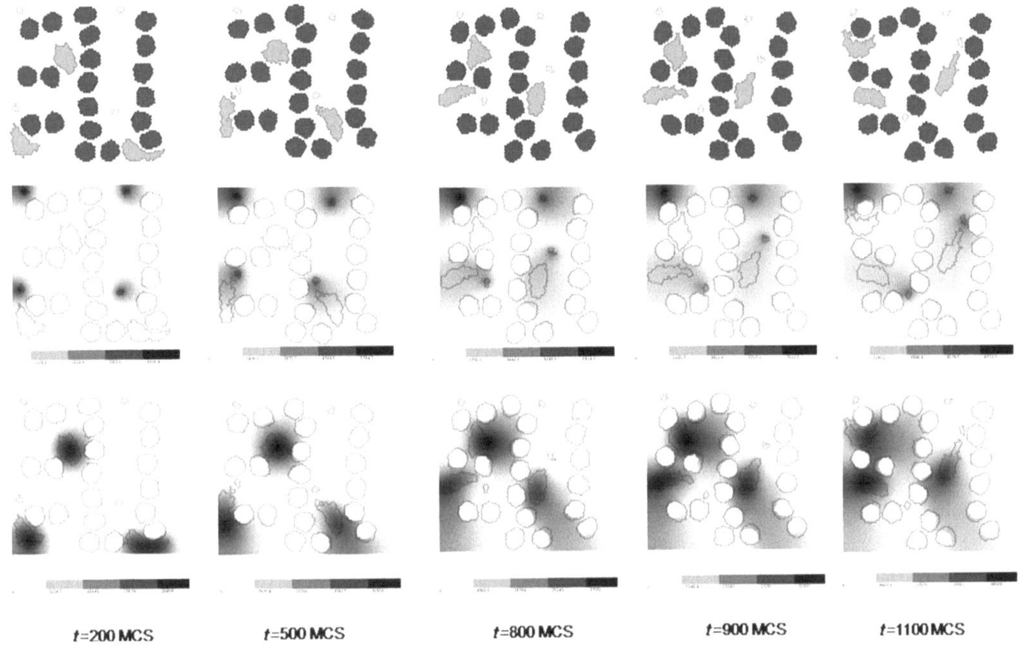

Fig. 11. Snapshots of the bacterium-and-macrophage simulation from **Listing 4** and the PIF in **Listing 6** saved in the file "bacterium_macrophage_2D_wall_v3.pif." The *upper row* shows the cell-lattice configuration with the Macrophages in *gray*, Bacteria in *white with black* borders, red blood cells in *dark gray*, and Medium in *white*. The *middle row* shows the concentration of the chemoattractant ATTR secreted by the Bacteria. The *bottom row* shows the concentration of the chemorepellant REPL secreted by the Macrophages. The bars at the *bottom* of the field images show the concentration scales (*white*, low concentration; *black*, high concentration).

6. Python Scripting

CC3DML is convenient for building simple simulations such as those we presented above. To describe more complex simulations, CompuCell3D allows users to write specialized, shareable modules in C/C++ (through the *CompuCell3D Application Programming Interface*, or *CC3D API*) or Python (through a Python-scripting interface). C and C++ modules have the advantage that they run at native speed. However, developing them requires knowledge of both C/C++ and the CC3D API, and their integration with CompuCell3D requires recompilation of the source code. Python module development is less complicated, since Python has simpler syntax than C/C++ and users can modify and extend a library of Python-module templates included with CompuCell3D. Moreover, Python modules do not require recompilation.

Tasks performed by CompuCell3D modules either relate to index-copy attempts (*plugins*) or run either once, at the beginning or end of a simulation, or once every several MCS (*steppables*). Tasks running every index-copy attempt, like effective-energy-term calculations, are the most frequently-called tasks in a GGH simulation and writing them in Python may slow down simulations. *Steppables* and lattice monitors are called less frequently and thus they are good candidates for Python implementation. Using Python scripts users can perform cell parameter adjustments that depend on the state of the simulation, *e.g.* simulating cell growth in response to a certain chemical, cell-type differentiation and changes in cell-cell adhesion, *etc.*

6.1. A Short Introduction to Python

Python is an object-oriented scripting language with all the syntactic constructs present in any modern programming language. Python supports popular flow-control statements such as `if-elif-else` conditional instructions and `for` and `while` loops. Unlike C/C++, Python does not use ";" to end lines or "{" and "}" to define code blocks. Instead, Python relies on indentation to define blocks of code. `if` statements, `for` or `while` loops and their subsections are created by a ":" and increasing the level of indentation. The end of a block is indicated by a decrease in the level of indentation. Python uses the "=" operator for assignments and "==" for checking equality between objects. For example, in the following code:

```
b=2
vif b==2:
    a=10
    for c in range(0,a):
        b=a+c
        print b
```

we indent the body of the `if` statement and the body of the inner `for` loop. The `for` loop is executed inside the `if` statement. `a=0` assigns the variable `a` a value of 10, while `b==2` is true if `b` has a value of 2. The `for` loop assigns the variable c values 0 through `a−1` and executes instructions inside the loop body.

As an object-oriented language, Python supports *classes*, *inheritance*, and *polymorphism*. Accessing *members* of *objects* uses the "." operator. For example, to access the real part of a complex number, we use the following code:

```
a=complex(2,3)
a=1.5+0.5j
print a.real
```

Here, `real` is a member of the Python class `complex`, which represents complex numbers. If the object has composite subobjects, we use the "." operator recursively:

```
object.subobject.member_of_subobject
```

Users may define Python objects without declaring their type. A single data structure such as a list or dictionary can store objects of multiple types. Python provides automatic memory management, which frees users from remembering to deallocate memory for objects that are no longer used.

Long source code lines can be carried over to the following line using the "\" character:

```
very_long_variable_name = \
very_long_variable_name * very_important_constant
```

Note: Double underscore "__" has a reserved meaning in Python and should not be confused with a single underscore "_".

We will present additional Python features in the subsequent sections and explain, step-by-step, some basic concepts of Python programming (for more on Python, see *Learning Python*, by Mark Lutz *(97)*). For more information on Python scripting in CompuCell3D, see our *Python Tutorials* and *CompuCell3D User Guide* (available from the CompuCell3D website, *http://www.compucell3d.org*).

6.2. Building Python-Based CompuCell3D Simulations

Python scripting allows users to augment their CC3DML configuration files with Python scripts or to code their entire simulations in Python (in which case the Python script looks very similar to the CC3DML script it replaces). **Listing** 7 shows the standard block of template code for running a Python script in conjunction with a CC3DML configuration file.

The `import sys` line provides access to standard functions and variables needed to manipulate the Python runtime environment. The next two lines

```
from os import environ
from os import getcwd
```

import `environ` and `getcwd` housekeeping functions into the current *namespace* (i.e., current script) and are included in all of our Python programs. In the next three lines,

```
import string
sys.path.append(environ["PYTHON_MODULE_\
PATH"])
import CompuCellSetup
```

we import the `string` module, which contains convenience functions for performing operations on strings of characters; set the search path for Python modules; and import the `CompuCell-`

Listing 7. Basic Python template to run a CompuCell3D simulation through a Python interpreter. Later examples will be based on this script.

```
import sys

from os import environ

from os import getcwd

import string

sys.path.append(environ["PYTHON_MODULE_PATH"])

import CompuCellSetup


sim,simthread = CompuCellSetup.getCoreSimulationObjects()


#Create extra player fields here or add attributes

CompuCellSetup.initializeSimulationObjects(sim,simthread)


#Add Python steppables here

steppableRegistry=CompuCellSetup.getSteppableRegistry()

CompuCellSetup.mainLoop(sim,simthread,steppableRegistry)
```

Setup module, which provides a set of convenience functions that simplify initialization of CompuCell3D simulations.

Next, we create and initialize the core CompuCell3D modules:

```
sim,simthread = CompuCellSetup.getCoreSimu\
lationObjects()
CompuCellSetup.initializeSimulationObjects\
(sim,simthread)
```

We then create a steppable *registry* (a Python *container* that stores steppables, i.e., a list of all steppables that the Python code can access) and pass it to the function that runs the simulation:

```
steppableRegistry=CompuCellSetup.getSteppable\
Registry()
CompuCellSetup.mainLoop(sim,simthread,\
steppableRegistry)
```

In **Subheading 6.3**, we extend this template to build a simple simulation.

6.3. Cell-Type-Oscillator Simulation

Suppose that we would like to add a caricature of oscillatory gene expression to our cell-sorting simulation (**Listing 3**) so that cells exchange types every 100 MCS. We will implement the changes to cell types using a Python steppable, since it occurs at intervals of 100 MCS.

Listing 8 shows the changes to the Python template in **Listing** 7 that are necessary to create the desired type switching (changes are shown in **bold**).

A CompuCell3D steppable is a *class* (a type of *object*) that holds the parameters and functions necessary for carrying out a task. Every steppable defines at least four functions: `__init__(self, _simulator, _frequency)`, `start(self)`, `step(self, mcs)`, and `finish(self)`.

Listing 8. Python script expanding the template code in Listing 7 into a simple `TypeSwitcherSteppable` steppable. The code illustrates dynamic modification of cell parameters using a Python script. Lines added to Listing 7 are shown in bold.

```
import sys
from os import environ
from os import getcwd
import string

sys.path.append(environ["PYTHON_MODULE_PATH"])

import CompuCellSetup
sim,simthread = CompuCellSetup.getCoreSimulationObjects()

from PySteppables import *
class TypeSwitcherSteppable(SteppablePy):
   def __init__(self,_simulator,_frequency=100):
      SteppablePy.__init__(self,_frequency)
      self.simulator=_simulator
      self.inventory=self.simulator.getPotts().getCellInventory()
      self.cellList=CellList(self.inventory)

   def step(self,mcs):
      for cell in self.cellList:
         if cell.type==1:
            cell.type=2
         elif (cell.type==2):
            cell.type=1
         else:
            print "Unknown type. In cellsort simulation there should\
            only be two types 1 and 2"

#Create extra player fields here or add attributes

CompuCellSetup.initializeSimulationObjects(sim,simthread)

#Add Python steppables here
steppableRegistry=CompuCellSetup.getSteppableRegistry()

typeSwitcherSteppable=TypeSwitcherSteppable(sim,100);
steppableRegistry.registerSteppable(typeSwitcherSteppable)

CompuCellSetup.mainLoop(sim,simthread,steppableRegistry)
```

CompuCell3D calls the `start(self)` function once at the beginning of the simulation before any index-copy attempts. It calls the `step(self, mcs)` function periodically after every `_frequency` MCS. It calls the `finish(self)` function once at the end of the simulation. **Listing 8** does not have explicit `start(self)` or `finish(self)` functions. Instead, the class definition:

```
class TypeSwitcherSteppable(SteppablePy):
```

causes the `TypeSwitcherSteppable` to inherit components of the `SteppablePy` class. `SteppablePy` contains default definitions of the `start(self)`, `step(self,mcs)`, and `finish(self)` functions. Inheritance reduces the length of the user-written Python code and ensures that the `TypeSwitcher-Steppable` object has all needed components. The line

```
from PySteppables import *
```

makes the content of the "PySteppables.py" file (or module) available in the current namespace. The `PySteppables` module includes the `SteppablePy` *base class.*

The `__init__` function is a *constructor* that accepts user-defined parameters and initializes a steppable object. Consider the `__init__` function of the `TypeSwitcherSteppable`:

```
def __init__(self,_simulator,_frequency=100):
  SteppablePy.__init__(self,_frequency)
  self.simulator=_simulator
  self.inventory=self.simulator.getPotts()\
  .getCellInventory()
  self.cellList=CellList(self.inventory)
```

In the `def` line, we pass the necessary parameters: `self` (which is used in Python to access class variables from within the class), `_simulator` (the main CompuCell3D kernel object which runs the simulation), and `_frequency` (which tells `steppableReg-istry` how often to run the steppable here, every 100 MCS). Next we call the constructor for the inheritance class, `SteppablePy`, as required by Python. The statement

```
self.simulator=_simulator
```

assigns to the class variable `self.simulator` a reference to the `_simulator` object, passed from the main script. We can think of a Python reference as a pointer variable that stores the address of some object but not a copy of the object itself. The last two lines construct a list of all generalized cells in the simulation, a *cell inventory*, which allows us to visit all of the cells with a simple `for` loop to perform various tasks. The cell inventory is a dynamic structure, i.e., it updates automatically when cells are created or destroyed during a simulation.

The section of the `TypeSwitcherSteppable` steppable that implements the cell-type switching is found in the `step(self, mcs)` function:

```
def step(self,mcs):
   for cell in self.cellList:
      if cell.type==1:
         cell.type=2
      elif (cell.type==2):
         cell.type=1
      else:
         print "Unknown type"
```

We use the cell inventory to iterate over all cells in the simulation and reassign their cell types between `cell.type` 1 and `cell.type` 2. If we encounter a `cell.type` that is neither 1 nor 2 (which we should not), we print an error message.

Once we have created a steppable (i.e., created an object of class `TypeSwitcherSteppable`) we must register it using the `registerSteppable` function from the `steppableRegistry` object:

```
typeSwitcherSteppable=TypeSwitcherSteppable\
(sim,100);
steppableRegistry.registerSteppable(typeSwi\
tcherSteppable)
```

CompuCell3D will not run unregistered steppables.

As we will see, much of the script is not specific to this example. We will recycle it with slight changes in later examples. **Figure 12** shows snapshots of the cell-type-oscillator simulation.

We mentioned earlier that users can run simulations without a CC3DML configuration file. **Listing 9** shows the cell-type-oscillator simulation written entirely in Python, with changes to Listing 8 shown in **bold**.

The `configureSimulation` function replaces the CC3DML file from Listing 3. After importing `CompuCellSetup` and `ElementCC3D` from `XMLUtils` module, we have access to functions and modules that provide all the functionality necessary to code the simulation in Python. The general Python syntax corresponding to the opening lines of each CC3DML block is:

```
nameElem=parentElement.ElementCC3D("Name"),
```

where "`Name`" refers to the name of the section in a CC3DML configuration file (*e.g.* `Compucell3D, Potts, Plugin, Steppable`). parentElement denotes CC3DML element containing element "`Name`". The rest of the block usually follows the syntax:

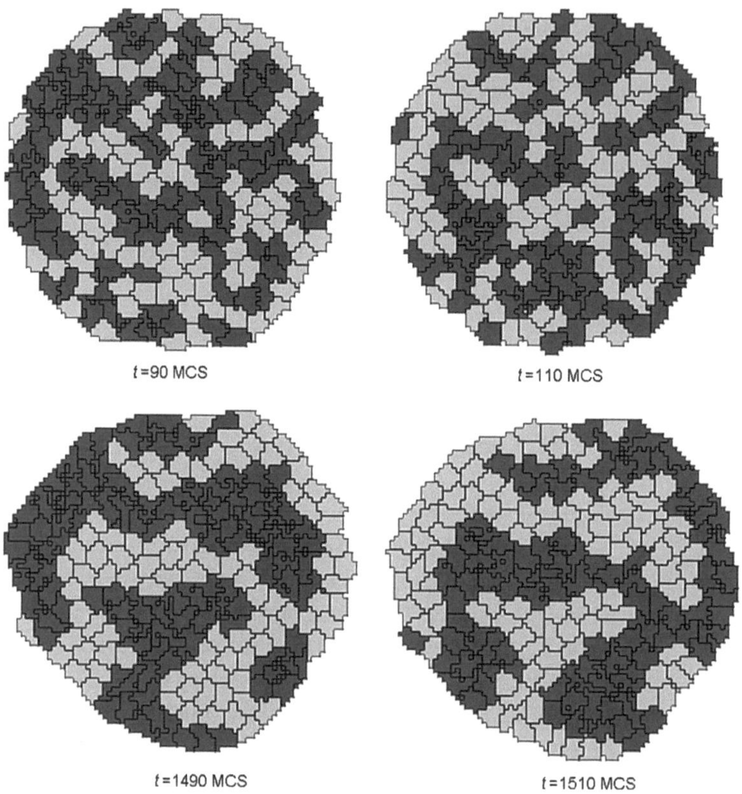

Fig. 12. Results of the Python cell-type-oscillator simulation using the TypeSwitcherSteppable steppable implemented in **Listing 8** in conjunction with the CC3DML cell-sorting simulation in **Listing 3**. Cells exchange types and corresponding adhesivities and colors every 100 MCS; i.e., between $t = 90$ MCS and $t = 110$ MCS and between $t = 1{,}490$ MCS and $t = 1{,}510$ MCS.

```
tagNameElem=parentElement.ElementCC3D\
("TagName",{attributes},value),
```

where "TagName" corresponds to the tag pair used to assign a value to a the parameter in a CC3DML file or, for values within subsections:

```
parentElement.ElementCC3D("SubSection",\
{attributes},value).
```

In case CC3DML element has only value but no attributes (*e.g.* <Temperature>10</Temperature>) we use the following syntax:

```
tagName=parentElement.ElementCC3D("TagName",\
{},value).
```

For CC3DML elements with attributes only and no values (*e.g.* <Dimensions x="100" y="100" z="1" />) the correct syntax is shown below:

```
tagName=parentElement.ElementCC3D("TagName",\
{attributes}).
```

Finally, for CC3DML elements with no attributes and no values (*e.g.* <Potts>) we use syntax of the form:

```
tagName= parentElement.ElementCC3D("TagName").
```

In the first block, corresponding to the `<Potts>` section of CC3DML code, we input the cell-lattice parameter values using rules and syntax described above:

```
potts.ElementCC3D("ParameterName",{attributes},\
value)
```

where `ParameterName` matches a parameter name used in the CC3DML lattice section.

Next we define the cell types using the syntax:

```
cellType.ElementCC3D("CellType",{"TypeName:\
type, "TypeId: id }).
```

The next section assigns contact energies between the cell types:

```
contact.ElementCC3D("Energy",{"Type1": type,\
"Type2": type },value).
```

We input the rest of the parameter values in a similar fashion, following the general syntax described above.

The examples in Listing 8 and Listing 9 define the `Type-SwitcherSteppable` class within the main Python script, but separating extension modules from the main script and using an `import` statement to refer to modules stored in external files is more practical because it ensures that each module can be used in multiple simulations without duplicating source code, and makes the scripts more readable and editable. We will follow this convention in our remaining examples.

Listing 9. Stand-alone Python cell-type-oscillator script containing an initial section that replaces the CC3DML configuration file from Listing 3. Lines added to Listing 8 are shown in **bold**.

```
def configureSimulation(sim):
   import CompuCellSetup
   from XMLUtils import ElementCC3D

   cc3d=ElementCC3D("CompuCell3D")
   potts=cc3d.ElementCC3D("Potts")
   potts.ElementCC3D("Dimensions",{"x":100,"y":100,"z":1})
   potts.ElementCC3D("Steps",{},1000)
   potts.ElementCC3D("Temperature",{},5)
   potts.ElementCC3D("NeighborOrder",{},2)

   cellType=cc3d.ElementCC3D("Plugin",{"Name":"CellType"})
   cellType.ElementCC3D("CellType",{"TypeName":"Medium",\
   "TypeId":"0"})
```

```
    cellType.ElementCC3D("CellType",{"TypeName":"Condensing",\
    "TypeId":"1"})
    cellType.ElementCC3D("CellType",{"TypeName":"NonCondensing",\
    "TypeId":"2"})

    volume=cc3d.ElementCC3D("Plugin",{"Name":"Volume"})
    volume.ElementCC3D("TargetVolume",{},25)
    volume.ElementCC3D("LambdaVolume",{},1.0)

    contact=cc3d.ElementCC3D("Plugin",{"Name":"Contact"})
    contact.ElementCC3D("Energy",{"Type1":"Medium",\
    "Type2":"Medium"},0)
    contact.ElementCC3D("Energy",{"Type1":"NonCondensing",\
    "Type2":"NonCondensing"},16)
    contact.ElementCC3D("Energy",{"Type1":"Condensing",\
    "Type2":"Condensing"},2)
    contact.ElementCC3D("Energy",{"Type1":"NonCondensing",\
    "Type2":"Condensing"},11)
    contact.ElementCC3D("Energy",{"Type1":"NonCondensing",\
    "Type2":"Medium"},16)
    contact.ElementCC3D("Energy",{"Type1":"Condensing",\
    "Type2":"Medium"},16)
    blobInitializer=cc3d.ElementCC3D("Steppable",\
    {"Type":"BlobInitializer"})
    blobInitializer.ElementCC3D("Gap",{},0)
    blobInitializer.ElementCC3D("Width",{},5)
    blobInitializer.ElementCC3D("Types",{},"Condensing")
    blobInitializer.ElementCC3D("Types",{},"NonCondensing")
    blobInitializer.ElementCC3D("Radius",{},40)

    CompuCellSetup.setSimulationXMLDescription(cc3d)
import sys
from os import environ
from os import getcwd
import string

sys.path.append(environ["PYTHON_MODULE_PATH"])

import CompuCellSetup
sim,simthread = CompuCellSetup.getCoreSimulationObjects()

configureSimulation(sim)

from PySteppables import *
class TypeSwitcherSteppable(SteppablePy):
    def __init__(self,_simulator,_frequency=100):
        SteppablePy.__init__(self,_frequency)
        self.simulator=_simulator
        self.inventory=self.simulator.getPotts().getCellInventory()
        self.cellList=CellList(self.inventory)

    def step(self,mcs):
        for cell in self.cellList:
            if cell.type==1:

                cell.type=2
            elif (cell.type==2):
                cell.type=1
            else:
                print "Unknown type. In cellsort simulation there should
only be two types 1 and 2"

#Create extra player fields here or add attributes
CompuCellSetup.initializeSimulationObjects(sim,simthread)

#Add Python steppables here
steppableRegistry=CompuCellSetup.getSteppableRegistry()

typeSwitcherSteppable=TypeSwitcherSteppable(sim,100);
steppableRegistry.registerSteppable(typeSwitcherSteppable)

CompuCellSetup.mainLoop(sim,simthread,steppableRegistry)

CompuCellSetup.mainLoop(sim,simthread,steppableRegistry)
```

6.4. Two-Dimensional Foam-Flow Simulation

CompuCell3D can simulate simple physical experiments with foams. Indeed, GGH techniques grew out of foam-simulation techniques *(73)*. Our next example shows how to use CC3DML and Python scripts to simulate quasi-2D foam flow.

The experimental apparatus (**Fig. 13**) consists of a channel created by two parallel rectangular glass plates separated by 5 mm, with the gap between their long sides sealed and that between their short sides open. A foam generator injects small, uniform-sized bubbles at one short end, pushing older bubbles toward the open end of the channel, creating a foam flow. The top glass plate has a hole through which we inject air. Bubbles passing under this point grow because of the air injected into them, forming characteristic patterns (**Fig. 14**) *(98)*.

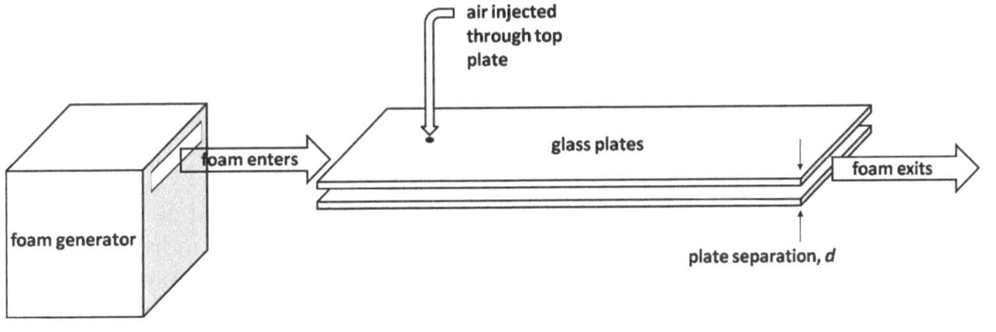

Fig. 13. Schematic of experiment for studying quasi-2D foam flow.

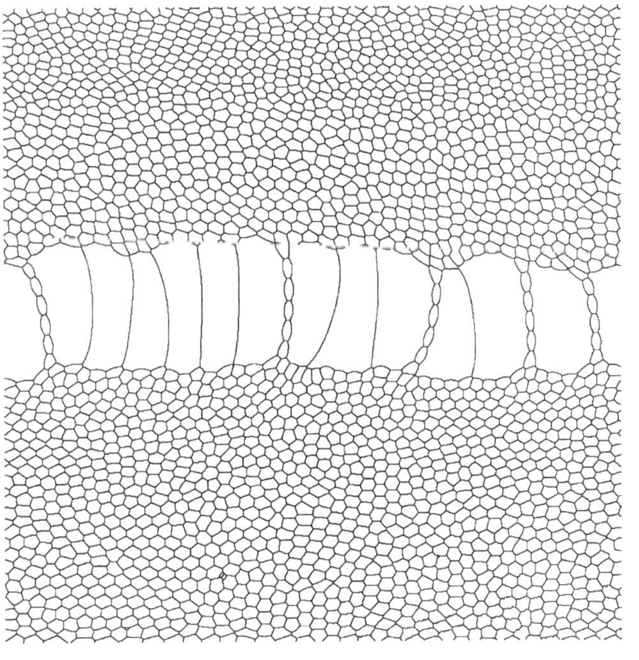

Fig. 14. Detail of processed experimental image of flowing quasi-2D bubbles. Image size is 15 cm × 15 cm.

Generalized cells will represent bubbles in this simulation. To simulate this experiment in CompuCell3D we need to write Python steppables that (1) create bubbles at one end of the channel, (2) inject air into the bubble which includes a given location (the identity of this bubble will change in time due to the flow), and (3) remove bubbles at the open end of the channel. We will store the source code in a file called "foamairSteppables.py". We will also need a main Python script to call these steppables appropriately.

We simulate bubble injection by creating generalized cells (bubbles) along the lattice edge corresponding to the left end of the channel (small-*x* values of the cell lattice). We simulate air injection into a bubble at the injection point by identifying the bubble currently at the injection point and increasing its target volume at a fixed rate. Removing a bubble from the simulation simply requires assigning it a target volume of zero once it comes close to the right end of the channel (large-*x* values of the cell lattice).

We first define a CC3DML configuration file for the foam-flow simulation (**Listing 10**). The CC3DML configuration file is simple: it initializes the `VolumeLocalFlex`, `CellType`, `Contact` and `CenterOfMass` plugins. We do not use a cell-lattice-initializer steppable, because all bubbles are created as the simulation runs. We use `VolumeLocalFlex` because individual bubbles will change their target volumes during the simulation. We also include the `CenterOfMass` plugin to track the changing centroids of each bubble.

Note: The `CenterOfMass` plugin in CompuCell3D actually calculates $\vec{x}_\sigma^{\mathrm{C}}$, the centroid of the generalized cell multiplied by volume of the cell:

$$\vec{x}_\sigma^{\mathrm{C}} = \sum_{\vec{i}} \vec{i}\,\delta(\sigma'(\vec{i}),\sigma), \tag{11}$$

so the actual centroid of the bubble is

$$\vec{x}_\sigma = \frac{\vec{x}_\sigma^{\mathrm{C}}}{v(\sigma)}. \tag{12}$$

The ability to track a generalized-cell's centroid is useful if we need to pick a single reference point in the cell. In this example we will remove bubbles whose centroids have *x*-coordinates greater than a cutoff value.

We will implement the Python script in four sections: (1) a main script (**Listing 11**), which runs every MCS and calls the steppables that (2) create bubbles at the left end of the cell lattice (`BubbleNucleator`, **Listing 12**), (3) enlarge the target volume of the bubble at the injector site (`AirInjector`, **Listing 13**) and (4) set the target volume of bubbles at the right end of the cell lattice to zero (`BubbleCellRemover`, **Listing 14**). We store classes (2–4) in a separate file called "foamairSteppables.py".

The main script in **Listing 11** builds on the template Python code in **Listing 7**; we show changes in **bold**. The line

Listing 10. CC3DML configuration file for the foam-flow simulation. This file initializes needed plugins but all of the interesting work is done in Python.

```
<CompuCell3D>
 <Potts>
  <Dimensions x="200" y="50" z="1"/>
  <Steps>10000</Steps>
  <Temperature>5</Temperature>
  <LatticeType>Hexagonal</LatticeType>
 </Potts>

 <Plugin Name="VolumeLocalFlex"/>

 <Plugin Name="CellType">
  <CellType TypeName="Medium" TypeId="0"/>
  <CellType TypeName="Foam"   TypeId="1"/>
 </Plugin>

 <Plugin Name="Contact">
  <Energy Type1="Medium" Type2="Medium">5</Energy>
  <Energy Type1="Foam"   Type2="Foam">5</Energy>
  <Energy Type1="Foam"   Type2="Medium">5</Energy>
  <NeighborOrder>3</NeighborOrder>
 </Plugin>

 <Plugin Name="CenterOfMass"/>

</CompuCell3D>
```

```
from foamairSteppables import BubbleNucleator
```

tells Python to look for the `BubbleNucleator` class in the file named "foamairSteppables.py". The line

```
bubbleNucleator=BubbleNucleator(sim, 20)
```

Listing 11. Main Python Script for foam-flow simulation. Changes to the template (Listing 7) are shown in **bold**.

```
import sys
from os import environ
import string
sys.path.append(environ["PYTHON_MODULE_PATH"])

import CompuCellSetup

sim,simthread = CompuCellSetup.getCoreSimulationObjects()

#Create extra player fields here
CompuCellSetup.initializeSimulationObjects(sim,simthread)

#Add Python steppables here
steppableRegistry=CompuCellSetup.getSteppableRegistry()

from foamairSteppables import BubbleNucleator
bubbleNucleator=BubbleNucleator(sim,20)
bubbleNucleator.setNumberOfNewBubbles(1)
bubbleNucleator.setInitialTargetVolume(25)
bubbleNucleator.setInitialLambdaVolume(2.0)
bubbleNucleator.setInitialCellType(1)
steppableRegistry.registerSteppable(bubbleNucleator)

from foamairSteppables import AirInjector
airInjector=AirInjector(sim,40)
airInjector.setVolumeIncrement(25)
airInjector.setInjectionPoint(50,25,0)
steppableRegistry.registerSteppable(airInjector)

from foamairSteppables import BubbleCellRemover
bubbleCellRemover=BubbleCellRemover(sim)
bubbleCellRemover.setCutoffValue(170)
steppableRegistry.registerSteppable(bubbleCellRemover)

CompuCellSetup.mainLoop(sim,simthread,steppableRegistry)
```

creates the steppable `BubbleNucleator` that will run every 20 MCS. The next few lines in this section pass the number of bubbles to create, which in our case is one:

Listing 12. Python code for the `BubbleNucleator` steppable, saved in the file "foamairSteppables.py." This module creates bubbles at points with random *y* coordinates and *x* coordinate of 3.

```
from CompuCell import Point3D
from random import randint

class BubbleNucleator(SteppablePy):
   def __init__(self,_simulator,_frequency=1):
      SteppablePy.__init__(self,_frequency)
      self.simulator=_simulator

   def start(self):
      self.Potts=self.simulator.getPotts()
      self.dim=self.Potts.getCellFieldG().getDim()

   def setNumberOfNewBubbles(self,_numNewBubbles):
         self.numNewBubbles=int(_numNewBubbles)

   def setInitialTargetVolume(self,_initTargetVolume):
      self.initTargetVolume=_initTargetVolume

   def setInitialLambdaVolume(self,_initLambdaVolume):
      self.initLambdaVolume=_initLambdaVolume

   def setInitialCellType(self,_initCellType):
      self.initCellType=_initCellType

   def createNewCell(self,pt):
      print "Nucleated bubble at ",pt
      cell=self.Potts.createCellG(pt)
      cell.targetVolume=self.initTargetVolume
      cell.type=self.initCellType
      cell.lambdaVolume=self.initLambdaVolume

   def nucleateBubble(self):
      pt=Point3D(0,0,0)
      pt.y=randint(0,self.dim.y-1)
      pt.x=3
      self.createNewCell(pt)

   def step(self,mcs):
      for i in xrange(self.numNewBubbles):
         self.nucleateBubble()
```

```
bubbleNucleator.setNumberOfNewBubbles(1)
```

the initial V_t for the new bubble, which is 25 pixels:

```
bubbleNucleator.setInitialTargetVolume(25)
```

the initial λ_{vol} for the bubble:

```
bubbleNucleator.setInitialLambdaVolume(2.0)
```

and the bubble's `type.id`:

```
bubbleNucleator.setInitialCellType(1)
```

Finally, we register the steppable:

```
steppableRegistry.registerSteppable(bubble \
Nucleator)
```

Listing 13. Python code for the `AirInjector` steppable which simulates air injection into the bubble currently occupying the cell-lattice pixel at location (*x,y,z*). Air injection begins after 5,000 MCS to allow the channel to partially fill with bubbles. The steppable is saved in file "foamairSteppables.py".

```
class AirInjector(SteppablePy):

   def __init__(self,_simulator,_frequency=1):

      SteppablePy.__init__(self,_frequency)

      self.simulator=_simulator

      self.Potts=self.simulator.getPotts()

      self.cellField=self.Potts.getCellFieldG()


   def start(self): pass


   def setInjectionPoint(self,_x,_y,_z):

      self.injectionPoint=CompuCell.Point3D(int(_x),int(_y),int(_z))


   def setVolumeIncrement(self,_increment):

      self.volumeIncrement=_increment


   def step(self,mcs):

      if mcs <5000:

         return

      cell=self.cellField.get(self.injectionPoint)

      if cell:

         cell.targetVolume+=self.volumeIncrement
```

The next group of lines repeats the process for the AirInjector steppable, reading it from the file "foamairSteppables.py":

```
from foamairSteppables import AirInjector
airInjector=AirInjector(sim, 40)
```

and increases V_t by 25:

```
airInjector.setVolumeIncrement(25)
```

Listing 14. Python code for the `BubbleCellRemover` steppable. This module removes cells once the *x*-coordinates of their centroids > `cutoffValue` by setting their target volumes to zero and increasing their λ_{vol} to 10,000. Like the other steppables in the foam-flow simulation, we save it in the file "foamairSteppables.py".

```
class BubbleCellRemover(SteppablePy):

   def __init__(self,_simulator,_frequency=1):

      SteppablePy.__init__(self,_frequency)

      self.simulator=_simulator

      self.inventory=self.simulator.getPotts().getCellInventory()

      self.cellList=CellList(self.inventory)


   def start(self):

      self.Potts=self.simulator.getPotts()

      self.dim=self.Potts.getCellFieldG().getDim()


   def setCutoffValue(self,_cutoffValue):

      self.cutoffValue=_cutoffValue


   def step(self,mcs):

      for cell in self.cellList:

         if cell:

            if int(cell.xCM/float(cell.volume))>self.cutoffValue:

               cell.targetVolume=0

               cell.lambdaVolume=10000
```

for the bubble occupying the pixel at the point (50, 25, 0) on the cell lattice:

```
airInjector.setInjectionPoint(50,25,0)
```

As before, the final line registers the steppable:

```
steppableRegistry.registerSteppable(airInjector)
```

The final new section reads the BubbleCellRemover steppable from the file "foamairSteppables.py":

```
from foamairSteppables import BubbleCellRemover
```

and invokes the steppable, telling it to run every MCS; note that we have omitted the number after `sim`:

```
bubbleCellRemover=BubbleCellRemover(sim)
```

Next we set 170 as the *x*-coordinate at which we will destroy bubbles:

```
bubbleCellRemover.setCutoffValue(170)
```

and, finally, register BubbleCellRemover:

```
steppableRegistry.registerSteppable(bubble\
CellRemover)
```

We must also write Python code to define the three steppables `BubbleNucleator`, `AirInjector`, and `BubbleCellRemover` and save them in the file "foamairSteppables.py".

Listing 12 shows the code for the `BubbleNucleator` steppable.

The first two lines import necessary modules, where the line

```
from CompuCell import Point3D
```

allows us to access points on the simulation cell lattice, and the line

```
from random import randint
```

allows us to generate random integers.

In the constructor of the `BubbleNucleator` steppable class we assign to the variable `self.simulator` a reference to the `simulator` object from the CompuCell3D kernel. In the `start(self)` function, we assign a reference to the `Potts` object from the CompuCell3D kernel to the variable `self.Potts`:

```
self.Potts=self.simulator.getPotts()
```

and assign the dimensions of the cell lattice to `self.dim`:

```
self.dim=self.Potts.getCellFieldG().getDim()
```

In addition to the four essential steppable member functions (`__init__(self, _simulator, _frequency)`, `start(self)`, `step(self, mcs)` and `finish(self)`), `BubbleNucleator` includes several functions, some of which set parameters and some of which perform necessary tasks. The functions `setNumberOfNewBubbles`, `setInitialTargetVolume`, and `setInitialLambdaVolume` accept the values passed from the main Python script in **Listing 11**.

The `CreateNewCell` function requires that we pass the coordinates of the point, `pt`, at which to create a new bubble:

```
def CreateNewCell (self,pt):
```

Then we use a built-in CompuCell3D function to add a new bubble at that location:

```
cell=self.Potts.createCellG(pt)
```

assigning the new cell a target volume V_t= `target volume`:

```
cell.targetVolume=self.initTargetVolume
```

type (τ = `type`):

```
cell.type=self.initCellType
```

and compressibility λ_{vol}= `lambda Volume`:

```
cell.lambdaVolume=initLambdaVolume
```

based on the values passed to the `BubbleNucleator` steppable from the main script.

The first three lines of the `nucleateBubble` function create a reference to a point on the cell lattice (`pt=Point3D(0,0,0)`), assign it a random *y*-coordinate between 0 and `y_dim-1`:

```
pt.y=randint(0,self.dim.y-1)
```

and an *x*-coordinate of 3:

```
pt.x=3
```

The line calls the `createNewCell` function and passes it the point (`pt`) at which to create the new bubble:

```
self.createNewCell(pt)
```

Finally, the `step(self,mcs)` function calls the `nucleateBubble` function `self.numNewBubbles` times per MCS.

Listing13 shows the code for the `AirInjector` steppable.

The first three lines of the `__init__(self,_simulator,_frequency)` function are identical to the same lines in the `BubbleNucleator` steppable (**Listing 12**). The final line of the function loads the cell-lattice parameters:

```
self.cellField=self.Potts.getCellFieldG()
```

The start(self) function in this steppable does not do anything:

```
def start(self): pass
```

The next two functions read the `injectionPoint` and `volumeIncrement` passed to the `AirInjector` steppable by the main Python script (**Listing 11**). The `step` function uses these values to identify the bubble at the injection site, `self.injectionPoint`:

```
cell=self.cellField.get(self.injection\
Point)
```

and then increments that bubble's target volume, V_t, by `self.volumeIncrement`:

```
if cell:
  cell.targetVolume+=self.volumeIncrement
```

Note the syntax:

```
if cell:
```

which we use to test whether a cell is `Medium` or not. `Medium` in CompuCell3D is assigned a `NULL` pointer, which, in Python, becomes a `None` object. Python evaluates the `None` object as `False` and other objects (in our case, bubbles) as `True`, so the task is only carried out on bubbles, not `Medium`.

In the first two lines of the `step(self,mcs)` function, we tell the function not to perform its task until 5,000 MCS have elapsed:

```
if mcs <5000:
  return
```

The 5,000-MCS delay allows the simulation to establish a uniform flow of small bubbles throughout a large portion of the cell lattice.

Finally, we define the `BubbleCellRemover` steppable (**Listing 14**).

At each MCS we scan the cell inventory looking for cells whose centroid has an *x*-coordinate close to the right end of the lattice and remove these cells from the simulation by setting their target volumes to zero and increasing λ_{vol} to 10,000.

The first two lines of the `__init__ (self,_simulator,_frequency)` function are identical to the corresponding lines in the BubbleNucleator and `AirInjector` steppables (**Listings 12** and **13**). In the third line of the function, we gain access to the generalized-cell inventory:

```
self.inventory=self.simulator.getPotts().\
getCellInventory()
```

and in the fourth line we make a list containing all of the generalized cells in the simulation:

```
self.cellList=CellList(self.inventory)
```

The `start(self)` function is identical to that of the BubbleNucleator steppable (**Listing 12**), and performs the same function.

The next function reads the `cutoffValue` for the *x*-coordinate that we passed to `BubbleCellRemover` from the main Python script (**Listing 11**):

```
setCutoffValue(self,_cutoffValue)
```

Finally, the `step(self, mcs)` function iterates through the cell inventory. We first check to make sure that the cell is not `Medium`:

```
if cell:
```

For each non-`Medium` cell, we test whether the *x*-coordinate of the cell's centroid is greater than the `cutoffValue`:

```
if int(cell.xCM/float(cell.volume))>self.cut-\
offValue:
```

and, if it is, set that cell's `targetVolume`, V_t, to zero:

```
cell.targetVolume=0
```

and its λ_{vol} to 10,000:

```
cell.lambdaVolume=10000
```

Running the CC3DML file from **Listing** 10 and the main Python script from **Listing** 11 (which loads the steppables in **Listings** 12–14 from the file "foamairSteppables.py") produces the snapshots shown in **Fig. 15**.

6.5. Diffusing-Field-Based Cell-Growth Simulation

One of the most frequent uses of Python scripting in CompuCell3D simulations is to modify cell behavior based on local field concentrations. To demonstrate this use, we incorporate stem-cell-like behavior into the cell-sorting simulation from **Listing 1**. This extension requires including relatively sophisticated interactions between cells and a diffusing chemical, FGF *(99)*.

We simulate a situation where `NonCondensing` cells secrete `FGF`, which diffuses freely through the cell lattice and obeys:

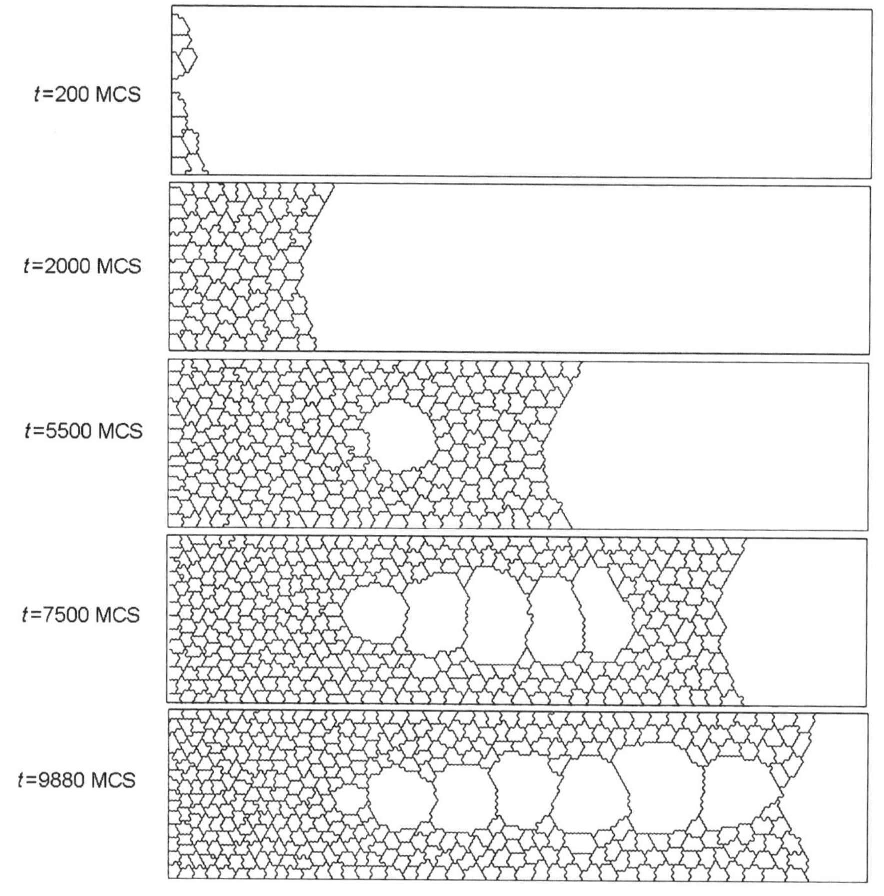

Fig. 15. Results of the foam-flow simulation on a 2D third-neighbor hexagonal lattice. Simulation code is given in Listings **10–14**.

$$\frac{\partial[\mathrm{FGF}](\vec{i})}{\partial t} = 0.10\nabla^2[\mathrm{FGF}](\vec{i}) + 0.05\delta(\tau(\sigma(\vec{i})), \mathit{NonCondensing}), \quad (13)$$

where [FGF] denotes the FGF concentration and Condensing cells respond to the field by growing at a constant rate proportional to the FGF concentration at their centroids:

$$\frac{\mathrm{d}V_t(\sigma)}{\mathrm{d}t} = 0.01[\mathrm{FGF}](\vec{x}_\sigma). \quad (14)$$

When they reach a threshold volume, the Condensing cells undergo mitosis. One of the resulting daughter cells remains a Condensing cell, while the other daughter cell has an equal probability of becoming either another Condensing cell or a DifferentiatedCondensing cell. DifferentiatedCondensing cells do not divide.

Each generalized cell in CompuCell3D has a default list of attributes, e.g. type, volume, surface (area), target volume, etc. However, CompuCell3D allows users to add cell attributes during execution of simulations. For example, in the current simulation, we will record data on each cell division in a list attached to each cell.

Note: Generalized cell attributes can be added using either C+ or Python. However, attributes added using Python are not accessible from C+ modules.

As in the foam-flow simulation, we divide the necessary simulation tasks among different Python modules (or classes) which we save in a file "cellsort_2D_field_modules.py" and call from the main Python script. We reuse elements of the CC3DML files we presented earlier to construct the CC3DML configuration file, presented in **Listing 15**.

The CC3DML code is a slightly extended version of the cell-sorting code in **Listing 3** plus the `FlexibleDiffusionSolverFE` discussed in the bacterium-and-macrophage simulation (*see* **Listing 4**). The initial cell lattice does not contain any `CondensingDifferentiated` cells. These cells appear only as the result of mitosis. We use the `VolumeLocalFlex` plugin to allow the target volume to vary individually for each cell, allowing cell growth as discussed in the foam-flow simulation. We manage the volume-constraint parameters using a Python script. The `CenterOfMass` plugin provides a reference point in each cell at which we measure the `FGF` concentration. We then adjust the cell's target volume accordingly.

To build this simulation in CompuCell3D we need to write several Python routines. We need (1) a steppable, `VolumeConstraintSteppable`, to initialize the volume-constraint parameters for each cell and to simulate cell growth by periodically increasing `Condensing` cells' target volumes in proportion to the `FGF` concentration at their centroids; (2) a plugin, `CellsortMitosis`, that runs the CompuCell3D mitosis algorithm when any cell reaches a threshold volume and then adjusts the parameters of the resulting parent and daughter cells, and also appends information about the time and type of cell division to a list attached to each cell; (3) a steppable, `MitosisDataPrinterSteppable`, that prints the cell-division information from the lists attached to each cell; (4) a class, `MitosisData`, which `MitosisDataPrinterSteppable` uses to extract and format the data it prints; and (5) a main Python script to call the steppables and the `CellsortMitosis` plugin appropriately. We store the source code for routines (1–4) in a separate file called "cellsort_2D_field_modules.py."

Listing 16 shows the main Python script for the diffusing-field-based cell-growth simulation, with changes to the template (**Listing 7**) shown in **bold**.

Listing 15. CC3DML code for the diffusing-field-based cell-growth simulation.

```
<CompuCell3D>
 <Potts>
   <Dimensions x="200" y="200" z="1"/>
   <Steps>10000</Steps>
   <Temperature>10</Temperature>
   <NeighborOrder>2</NeighborOrder>
 </Potts>

 <Plugin Name="VolumeLocalFlex"/>

 <Plugin Name="CellType">
  <CellType TypeName="Medium" TypeId="0"/>
  <CellType TypeName="Condensing" TypeId="1"/>
  <CellType TypeName="NonCondensing" TypeId="2"/>
  <CellType TypeName="CondensingDifferentiated" TypeId="3"/>
 </Plugin>

 <Plugin Name="Contact">
  <Energy Type1="Medium" Type2="Medium">0</Energy>
  <Energy Type1="NonCondensing" Type2="NonCondensing">16</Energy>
  <Energy Type1="Condensing"    Type2="Condensing">2</Energy>
  <Energy Type1="NonCondensing" Type2="Condensing">11</Energy>
  <Energy Type1="NonCondensing" Type2="Medium">16</Energy>
  <Energy Type1="Condensing"    Type2="Medium">16</Energy>
  <Energy Type1="CondensingDifferentiated"
      Type2="CondensingDifferentiated">2</Energy>
  <Energy Type1="CondensingDifferentiated"
      Type2="Condensing">2</Energy>
  <Energy Type1="CondensingDifferentiated"
      Type2="NonCondensing">11</Energy>
  <Energy Type1="CondensingDifferentiated" Type2="Medium">16</Energy>
  <NeighborOrder>2</NeighborOrder>
 </Plugin>

 <Plugin Name="CenterOfMass"/>
```

```
<Steppable Type="FlexibleDiffusionSolverFE">
 <DiffusionField>
  <DiffusionData>
   <FieldName>FGF</FieldName>
   <DiffusionConstant>0.10</DiffusionConstant>
   <DecayConstant>0.00005</DecayConstant>
  </DiffusionData>
  <SecretionData>
   <Secretion Type="NonCondensing">0.05</Secretion>
  </SecretionData>
 </DiffusionField>
</Steppable>

<Steppable Type="BlobInitializer">
 <Region>
  <Gap>0</Gap>
  <Width>5</Width>
  <Radius>40</Radius>
  <Center x="100" y="100" z="0"/>
  <Types>Condensing,NonCondensing</Types>
 </Region>
</Steppable>

</CompuCell3D>
```

The first change to the template code (**Listing 7**) is

```
pyAttributeAdder,listAdder=CompuCellSetup.\
attachListToCells(sim)
```

which instructs the CompuCell3D kernel to attach a Python-defined list to each cell when it creates it. This list serves as a generic container which can store any set of Python objects and hence any set of generalized-cell properties. In the current simulation, we use the list to store objects of the class `MitosisData`, which records the Monte Carlo Step at which each cell division involving the current cell, or its parent, happened, as well as the cell index and cell type of the parent and daughter cells.

Listing 16. Main Python script for the diffusing-field-based cell-growth simulation. Changes to the template code (Listing 7) shown in *bold*.

```
import sys
from os import environ
from os import getcwd
import string

sys.path.append(environ["PYTHON_MODULE_PATH"])

import CompuCellSetup

sim,simthread = CompuCellSetup.getCoreSimulationObjects()

#add additional attributes
pyAttributeAdder,listAdder=CompuCellSetup.attachListToCells(sim)

CompuCellSetup.initializeSimulationObjects(sim,simthread)

#notice importing CompuCell to main script has to be
#done after call to getCoreSimulationObjects()
import CompuCell
changeWatcherRegistry=CompuCellSetup.getChangeWatcherRegistry(sim)
stepperRegistry=CompuCellSetup.getStepperRegistry(sim)

from cellsort_2D_field_modules import CellsortMitosis
cellsortMitosis=CellsortMitosis(sim,changeWatcherRegistry,\
stepperRegistry)
cellsortMitosis.setDoublingVolume(50)

#Add Python steppables here
steppableRegistry=CompuCellSetup.getSteppableRegistry()

from cellsort_2D_field_modules import VolumeConstraintSteppable
volumeConstraint=VolumeConstraintSteppable(sim)
steppableRegistry.registerSteppable(volumeConstraint)

from cellsort_2D_field_modules import MitosisDataPrinterSteppable
mitosisDataPrinterSteppable=MitosisDataPrinterSteppable(sim)
steppableRegistry.registerSteppable(mitosisDataPrinterSteppable)

CompuCellSetup.mainLoop(sim,simthread,steppableRegistry)
```

Because one of our Python modules is a lattice monitor, rather than a steppable, we need to create `stepperRegistry` and `changeWatcherRegistry` objects, which store the two types of lattice monitors:

```
changeWatcherRegistry=CompuCellSetup.\
getChangeWatcherRegistry(sim)
stepperRegistry=CompuCellSetup.\
getStepper Registry(sim)
```

The `CellsortMitosis` plugin is a lattice monitor because it acts in response to certain index-copy events; it is invoked whenever a cell's volume reaches the threshold volume for mitosis. The following lines create the `CellsortMitosis` lattice monitor and register it with `stepperRegistry` and `changeWatcherRegistry`:

```
from cellsort_2D_field_modules import Cell\
sortMitosis
cellsortMitosis = CellsortMitosis(sim,change\
WatcherRegistry, stepperRegistry)
```

Because the base class inherited by `CellsortMitosis`, unlike our steppables, handles registration internally, we do not have to register `CellsortMitosis` explicitly. We can now set the threshold volume at which `Condensing` cells divide:

```
cellsortMitosis.setDoublingVolume(50)
```

Next we import the `VolumeConstraintSteppable` steppable, which initializes cells' target volumes and compressibilities at the beginning of the simulation and also implements chemical-dependent cell growth for `Condensing` cells, and register it:

```
from cellsort_2D_field_modules import Vol\
umeConstraintSteppable
volumeConstraint=VolumeConstraintSteppable(sim)
steppableRegistry.registerSteppable(volumeCo
nstraint)
```

Finally, we import, create and register the `MitosisDataPrinterSteppable` steppable, which prints the content of `MitosisData` objects for cells that have divided:

```
from cellsort_2D_field_modules import\
MitosisDataPrinterSteppable
mitosisDataPrinterSteppable=MitosisDataPrin\
terSteppable(sim)
steppableRegistry.registerSteppable(mitosis\
DataPrinterSteppable)
```

The number of `MitosisData` objects stored in each cell at any given Monte Carlo Step depends on cell type (`NonCondensing` cells do not divide, whereas `Condensing` cells can divide multiple times) and how often a given cell has divided.

Listing 17. Python code for the `VolumeConstraintSteppable`, saved in the file "cellsort_2D_field_modules.py," for the diffusing-field-based cell-growth simulation. The `VolumeConstraintSteppable` provides dynamic volume constraint parameters for each cell, which depend on the cell type and the chemical field concentration at the cell's centroid.

```
class VolumeConstraintSteppable(SteppablePy):

    def __init__(self,_simulator,_frequency=1):

        SteppablePy.__init__(self,_frequency)

        self.simulator=_simulator

        self.inventory=self.simulator.getPotts().getCellInventory()

        self.cellList=CellList(self.inventory)


    def start(self):

        for cell in self.cellList:

            cell.targetVolume=25

            cell.lambdaVolume=2.0

    def step(self,mcs):

        field=CompuCell.getConcentrationField(self.simulator,"FGF")

        comPt=CompuCell.Point3D()

        for cell in self.cellList:

            if cell.type==1: #Condensing cell

                comPt.x=int(round(cell.xCM/float(cell.volume)))

                comPt.y=int(round(cell.yCM/float(cell.volume)))

                comPt.z=int(round(cell.zCM/float(cell.volume)))

                concentration=field.get(comPt) # get concentration at comPt

                # and increase cell's target volume

                cell.targetVolume+=0.1*concentration
```

Moving on to the Python modules, we consider the `VolumeConstraintSteppable` steppable shown in **Listing 17**.

The `__init__` constructor looks very similar to the one in **Listing 14**, with the difference that we pass `_frequency=1` to update the cell volumes once per MCS. We also request the field-lattice dimensions and values from CompuCell3D:

```
self.dim=self.simulator.getPotts().get\
CellFieldG().getDim()
```

and specify that we will work with a field named `FGF`:

```
self.fieldName="FGF"
```

The script contains two functions: one that initializes the cells' volume-constraint parameters (`start(self)`) and one that updates them (`step(self, mcs)`).

The `start(self)` function executes only once, at the beginning of the simulation. It iterates over each cell (`for cell in self.cellList:`) and assigns the initial cells' `targetVolume` ($V_t(\sigma)$= 25 pixels) and `lambdaVolume` ($\lambda_{vol}(\sigma) = 2.0$) parameters as the `VolumeLocalFlex` plugin requires.

The first line of the `step(self, mcs)` function extracts a reference to the `FGF` concentration field defined using the `FlexibleDiffusionSolverFE` steppable in the CC3DML file (each field created in a CompuCell3D simulation is registered and accessible by both C+ and Python). The function then iterates over every cell in the simulation. If a `cell` is of `cell.type 1 (Condensing` – see the CC3DML configuration file, **Listing** 15), we calculate its centroid:

```
centerOfMassPoint.x=int(round(cell.xCM/\
float(cell.volume)))
centerOfMassPoint.y=int(round(cell.yCM/\
float(cell.volume)))
centerOfMassPoint.z=int(round(cell.zCM/\
float(cell.volume)))
```

and retrieve the `FGF` concentration at that point:

```
concentration=field.get(centerOfMassPoint)
```

We then increase the target volume of the cell by 0.01 times that concentration:

```
cell.targetVolume+=0.01*concentration
```

We must include the `CenterOfMass` plugin in the CC3DML code. Otherwise the centroid (`cell.xCM, cell.yCM, cell.zCM`) will have the default value (0,0,0).

Listing 18 shows the code for the `CellsortMitosis` plugin. The plugin divides the mitotic cell into two cells and adjusts both cells' attributes. It also initializes and appends `Mito-`

Listing 18. Python code for the `CellsortMitosis` plugin for the diffusing-field-based cell-growth simulation, saved in the file "cellsort_2D_field_modules.py." The plugin handles division of cells when they reach a threshold volume.

```
class VolumeConstraintSteppable(SteppablePy):
   def __init__(self,_simulator,_frequency=1):
      SteppablePy.__init__(self,_frequency)
      self.simulator=_simulator
      self.inventory=self.simulator.getPotts().getCellInventory()
      self.cellList=CellList(self.inventory)

   def start(self):
      for cell in self.cellList:
         cell.targetVolume=25
         cell.lambdaVolume=2.0
   def step(self,mcs):
      field=CompuCell.getConcentrationField(self.simulator,"FGF")
      comPt=CompuCell.Point3D()
      for cell in self.cellList:
         if cell.type==1: #Condensing cell
            comPt.x=int(round(cell.xCM/float(cell.volume)))
            comPt.y=int(round(cell.yCM/float(cell.volume)))
            comPt.z=int(round(cell.zCM/float(cell.volume)))
            concentration=field.get(comPt) # get concentration at comPt
            # and increase cell's target volume
            cell.targetVolume+=0.1*concentration
```

`sisData` objects to the original cell's (`self.parentCell`) and daughter cell's (`self.childCell`) attribute lists.

The second line of **Listing 18**:

```
from PyPluginsExamples import MitosisPyPlug\
inBase
```

lets us access the CompuCell3D base class `MitosisPyPlug-inBase`.

CellsortMitosis inherits the content of the MitosisPyPluginBase class. MitosisPyPluginBase internally accesses the CompuCell3D-provided Mitosis plugin, which is written in C++ and handles the technicalities of plugin initialization behind the scenes. The MitosisPyPluginBase class provides a simple-to-use interface to this plugin. To create a customized version of MitosisPyPluginBase, CellsortMitosis, we must call the constructor of MitosisPyPluginBase from the CellsortMitosis constructor:

```
MitosisPyPluginBase.__init__(self,_simulator,\
_changeWatcherRegistry,_stepperRegistry)
```

We also need to reimplement the function updateAttributes(self), which is called by MitosisPyPluginBase after mitosis takes place, to define the postdivision cells' parameters. The objects self.childCell and self.parentCell that appear in the function are initialized and managed by MitosisPyPluginBase. In the current simulation, after division we set V_t for the parent and daughter cells to half of the V_t of the parent just prior to cell division. λ_{vol} is left unchanged for the parent cell and the same value is assigned to the daughter cell:

```
self.parentCell.targetVolume=self.parentCell.\
volume/2.0
self.childCell.targetVolume=self.parentCell.\
targetVolume
self.childCell.lambdaVolume=self.parentCell.\
lambdaVolume
```

The cell type of one of the two daughter cells (childCell) is randomly chosen to be either Condensing (i.e., the same as the parent type) or CondensingDifferentiated, which we have defined to be cell.type 3 (**Listing 15**):

```
if (random()<0.5):
  self.childCell.type=self.parentCell.type
else:
  self.childCell.type=3
```

The parent cell remains `Condensing`. We now add a description of this cell division to the lists attached to each cell. First, we collect the data in a list called `mitData`:

```
mcs=self.simulator.getStep()
mitData=MitosisData(mcs,self.parentCell.\
id,self.parentCell.type,\
self.childCell.id,self.childCell.type)
```

then we access the lists attached to the two cells:

```
parentCellList=CompuCell.getPyAttrib(self.\
parentCell)
childCellList=CompuCell.getPyAttrib(self.\
childCell)
```

and append the new mitosis data to these lists:

```
parentCellList.append(mitData)
childCellList.append(mitData)
```

Listing 19. Python code for the `MitosisData` class for the diffusing-field-based cell-growth simulation, saved in the file "cellsort_2D_field_modules.py." `MitosisData` objects store information about cell divisions involving the parent and daughter cells.

```
class MitosisData:
   def __init__(self,_MCS,_parentId,_parentType,_offspringId,\
_offspringType):
      self.MCS=_MCS
      self.parentId=_parentId
      self.parentType=_parentType
      self.offspringId=_offspringId
      self.offspringType=_offspringType
   def __str__(self):
      return "Mitosis time="+str(self.MCS)+"\
      parentId="+str(self.parentId)+"\
      offspringId="+str(self.offspringId)
```

Listing 19 shows the Python code for the `MitosisData` class, which stores the data on the cell division that we append to the cells' attribute lists after each cell division.

In the constructor of `MitosisData`, we read in the time (in MCS) of the division, along with the parent and daughter cell indices and types. The `__str__(self)` convenience function returns an ASCII string representation of the time and cell indices only, to allow the Python `print` command to print out this information.

Listing 20 shows the Python code for the `MitosisDataPrinterSteppable` steppable, which prints the mitosis data to the user's screen.

The constructor is identical to that for the `VolumeConstraintSteppable` steppable (**Listing 17**). Within the `step(self,mcs)` function, we iterate over each cell (`for cell in self.cellList:`) and access the Python list attached to the cell (`mitDataList=CompuCell.getPyAttrib(cell)`).

Listing 20. The Python code for the `MitosisDataPrinter` steppable for the diffusing-field-based cell-growth simulation, saved in the file "cellsort_2D_field_modules.py." The steppable prints the cell-division history for dividing cells (*see* Fig. 18).

```
class MitosisDataPrinterSteppable(SteppablePy):

  def __init__(self,_simulator,_frequency=100):

    SteppablePy.__init__(self,_frequency)

    self.simulator=_simulator

    self.inventory=self.simulator.getPotts().getCellInventory()

    self.cellList=CellList(self.inventory)


  def step(self,mcs):

    for cell in self.cellList:

      mitDataList=CompuCell.getPyAttrib(cell)

      if len(mitDataList) > 0:

        print "MITOSIS DATA FOR CELL ID",cell.id

        for mitData in mitDataList:

          print mitData
```

If a given cell has undergone mitosis, then the list will have entries, and thus a nonzero length. If so, we print the MitosisData objects stored in the list:

```
if len(mitDataList) > 0:
  print "MITOSIS DATA FOR CELL ID",cell.id
  for mitData in mitDataList:
    print mitData
```

Figures 16 and **17** show snapshots of the diffusing-field-based cell-growth simulation. **Figure 18** shows a sample screen output of the cell-division history.

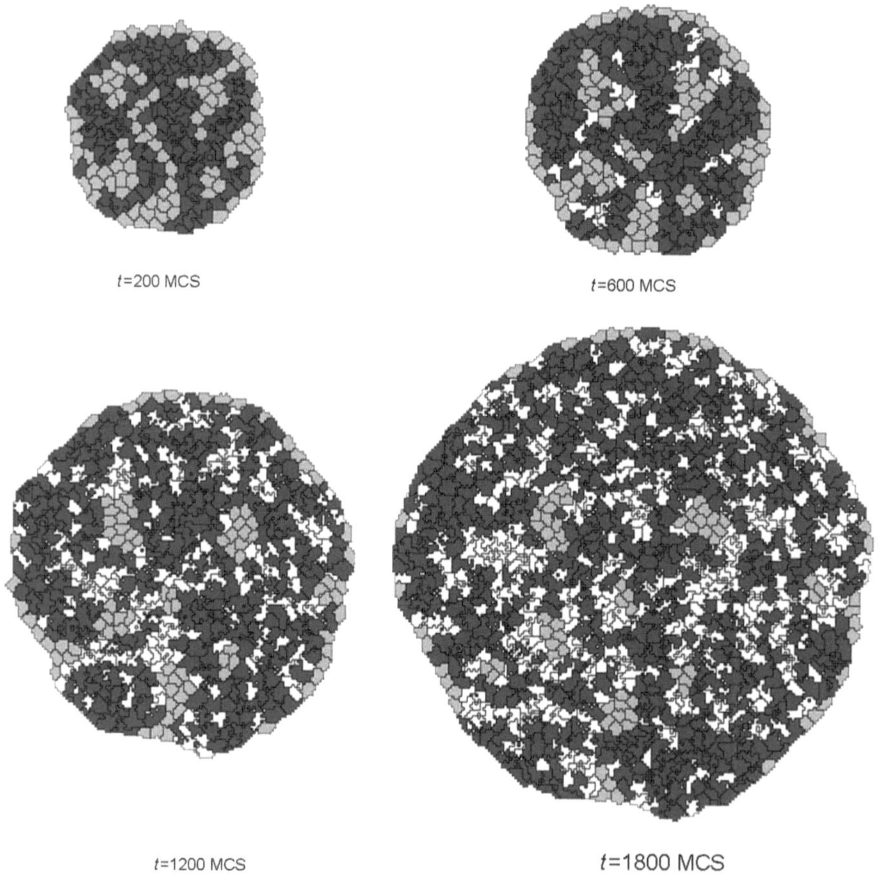

Fig. 16. Snapshots of the diffusing-field-based cell-growth simulation obtained by running the CC3DML file in Listing 15 in conjunction with the Python file in Listing 16. As the simulation progresses, NonCondensing cells (*light gray*) secrete diffusing chemical, FGF, which causes Condensing (*dark gray*) cells to proliferate. Some Condensing cells differentiate to CondensingDifferentiated (*white*) cells.

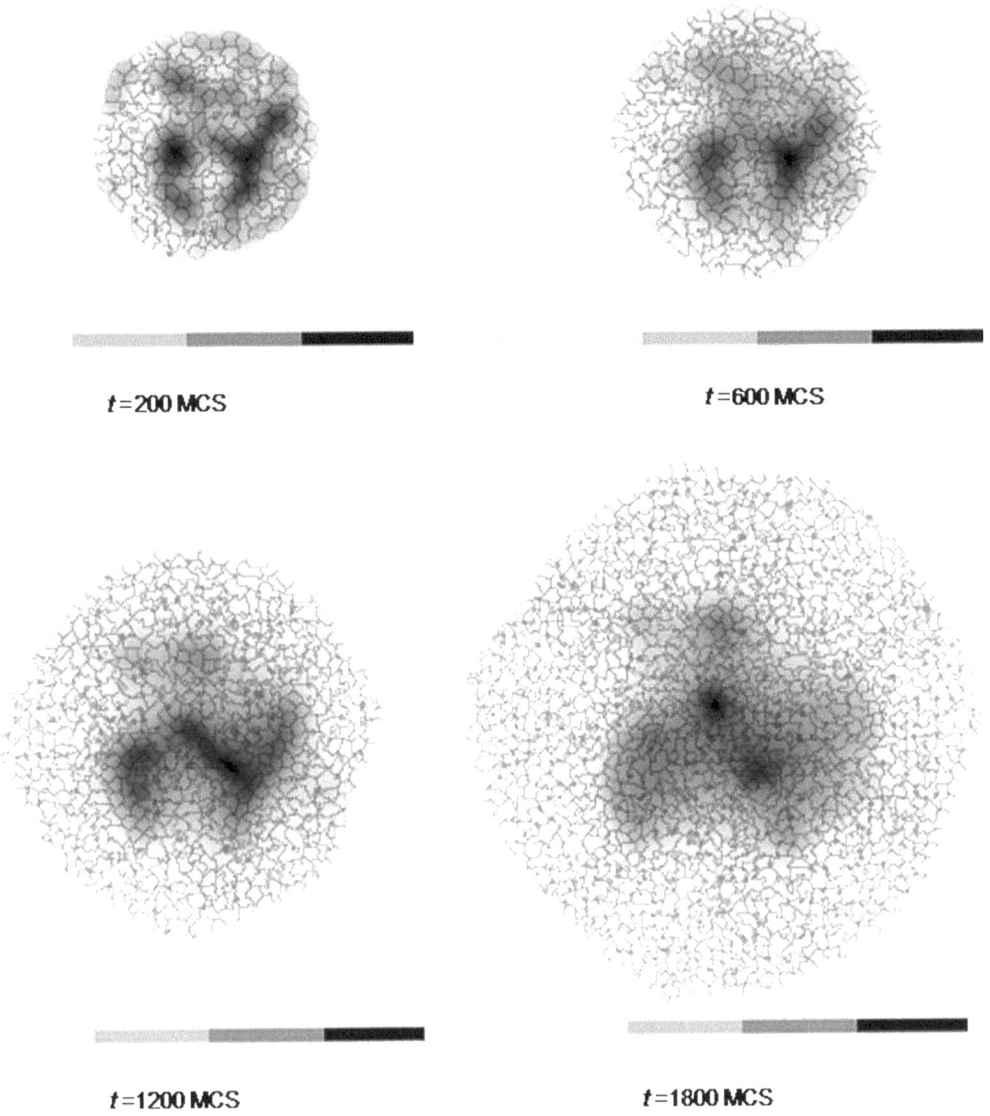

Fig. 17. Snapshots of FGF concentration in the diffusing-field-based cell-growth simulation obtained by running the CC3DML file in Listing 15 in conjunction with the Python files in **Listings 16–20**. The *bars* at the *bottom* of the field images show the concentration scales (*white*, low concentration; *black*, high concentration).

The diffusing-field-based cell-growth simulation includes concepts that extend easily to simulate biological phenomena that involve diffusants, cell growth, and mitosis, e.g., limb-bud development *(58, 59)*, tumor growth *(5–9)*, and *Drosophila* imaginal-disc development.

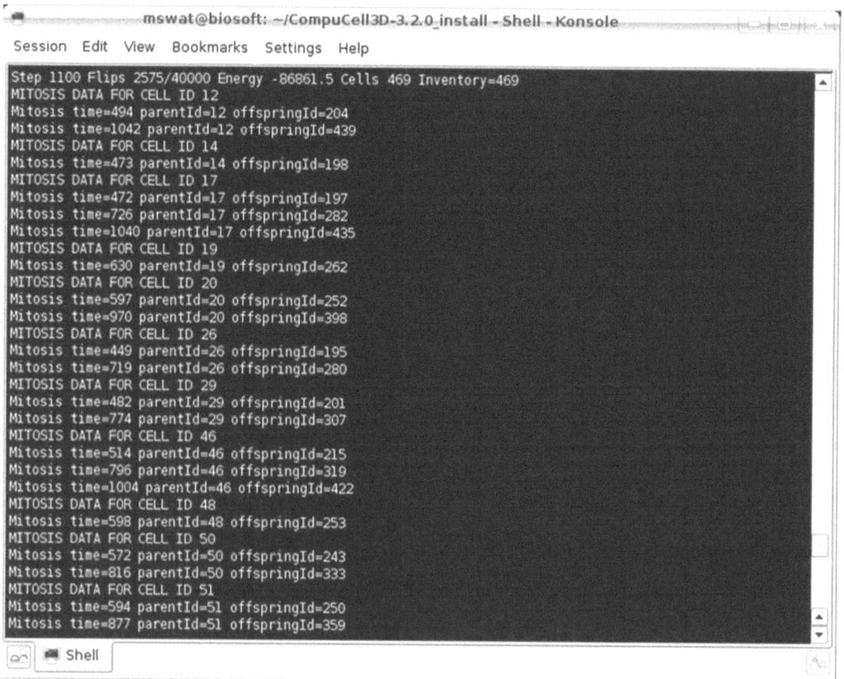

Fig. 18. Sample output from the MitosisDataPrinterSteppable steppable in **Listing 20**.

7. Conclusion

In most cases, building a complex CompuCell3D simulation requires writing Python modules, a main Python script, and a CC3DML configuration file. While the effort to write this code can be substantial, it is much less than that required to develop custom simulations in lower-level languages. Working from the substantial base of Python templates provided by CompuCell3D further streamlines simulation development. Python programs are fairly short, so simulations can be published in journal articles, greatly facilitating simulation validation, reuse, and adaptation. Finally, CompuCell3D's modular structure allows new Python modules to be reused from simulation to simulation. The CompuCell3D Web site, http://www.compucell3d.org, allows users to archive their modules and make them accessible to other users.

We hope the examples we have shown will convince readers to evaluate the suitability of GGH simulations using CompuCell3D for their research.

All the code examples presented in this chapter are available from http://www.compucell3d.org. They will be curated to ensure their correctness and compatibility with future versions of CompuCell3D.

Acknowledgments

We gratefully acknowledge support from the National Institutes of Health, National Institute of General Medical Sciences, grants 1R01 GM077138–01A1 and 1R01 GM076692-01, and the Office of Vice President for Research, the College of Arts and Sciences, the Pervasive Technologies Laboratories and the Biocomplexity Institute at Indiana University. Indiana University's University Information Technology Services provided time on their BigRed clusters for simulation execution. Early versions of CompuCell and CompuCell3D were developed at the University of Notre Dame by J.A.G., Dr. Mark Alber and Dr. Jesus Izaguirre and collaborators with the support of National Science Foundation, Division of Integrative Biology, grant IBN-00836563. Since the primary home of CompuCell3D moved to Indiana University in 2004, the Notre Dame team have continued to provide important support for its development.

References

1. Bassingthwaighte, J. B. (2000) Strategies for the Physiome project. *Ann. Biomed. Eng.* 28, 1043–1058.
2. Merks, R. M. H., Newman, S. A., and Glazier, J. A. (2004) Cell-oriented modeling of *in vitro* capillary development. *Lect. Notes Comp. Sci.* 3305, 425–434.
3. Turing, A. M. (1953) The chemical basis of morphogenesis. *Philos. Trans. R. Soc. B* 237, 37–72.
4. Merks, R. M. H. and Glazier, J. A. (2005) A cell-centered approach to developmental biology. *Phys. A* 352, 113–130.
5. Dormann, S. and Deutsch, A. (2002) Modeling of self-organized avascular tumor growth with a hybrid cellular automaton. *In Silico Biol.* 2, 1–14.
6. dos Reis, A. N., Mombach, J. C. M., Walter, M., and de Avila, L. F. (2003) The interplay between cell adhesion and environment rigidity in the morphology of tumors. *Phys. A* 322, 546–554.
7. Drasdo, D. and Hohme, S. (2003) Individual-based approaches to birth and death in avascular tumors. *Math. Comput. Model.* 37, 1163–1175.
8. Holm, E. A., Glazier, J. A., Srolovitz, D. J., and Grest, G. S. (1991) Effects of lattice anisotropy and temperature on domain growth in the two-dimensional Potts model. *Phys. Rev. A* 43, 2662–2669.
9. Turner, S. and Sherratt, J. A. (2002) Intercellular adhesion and cancer invasion: A discrete simulation using the extended Potts model. *J. Theor. Biol.* 216, 85–100.
10. Drasdo, D. and Forgacs, G. (2000) Modeling the interplay of generic and genetic mechanisms in cleavage, blastulation, and gastrulation. *Dev. Dynam.* 219, 182–191.
11. Drasdo, D., Kree, R., and McCaskill, J. S. (1995) Monte-Carlo approach to tissue-cell populations. *Phys. Rev. E* 52, 6635–6657.
12. Longo, D., Peirce, S. M., Skalak, T. C., Davidson, L., Marsden, M., and Dzamba, B. (2004) Multicellular computer simulation of morphogenesis: Blastocoel roof thinning and matrix assembly in *Xenopus laevis*. *Dev. Biol.* 271, 210–222.
13. Collier, J. R., Monk, N. A. M., Maini, P. K., and Lewis, J. H. (1996) Pattern formation by lateral inhibition with feedback: A mathematical model of Delta-Notch intercellular signaling. *J. Theor. Biol.* 183, 429–446.
14. Honda, H. and Mochizuki, A. (2002) Formation and maintenance of distinctive cell patterns by coexpression of membrane-bound ligands and their receptors. *Dev. Dynam.* 223, 180–192.
15. Moreira, J. and Deutsch, A. (2005) Pigment pattern formation in zebrafish during late larval stages: A model based on local interactions. *Dev. Dynam.* 232, 33–42.

16. Wearing, H. J., Owen, M. R., and Sherratt, J. A. (2000) Mathematical modelling of juxtacrine patterning. *Bull. Math. Biol.* 62, 293–320.
17. Zhdanov, V. P. and Kasemo, B. (2004) Simulation of the growth of neurospheres. *Europhys. Lett.* 68, 134–140.
18. Ambrosi, D., Gamba, A., and Serini, G. (2005) Cell directional persistence and chemotaxis in vascular morphogenesis. *Bull. Math. Biol.* 67, 195–195.
19. Gamba, A., Ambrosi, D., Coniglio, A., de Candia, A., di Talia, S., Giraudo, E., Serini, G., Preziosi, L., and Bussolino, F. (2003) Percolation, morphogenesis, and Burgers dynamics in blood vessels formation. *Phys. Rev. Lett.* 90, 118101.
20. Novak, B., Toth, A., Csikasz-Nagy, A., Gyorffy, B., Tyson, J. A., and Nasmyth, K. (1999) Finishing the cell cycle. *J. Theor. Biol.* 199, 223–233.
21. Peirce, S. M., van Gieson, E. J., and Skalak, T. C. (2004) Multicellular simulation predicts microvascular patterning and *in silico* tissue assembly. *FASEB J.* 18, 731–733.
22. Merks, R. M. H., Brodsky, S. V., Goligorksy, M. S., Newman, S. A., and Glazier, J. A. (2006) Cell elongation is key to *in silico* replication of *in vitro* vasculogenesis and subsequent remodeling. *Dev. Biol.* 289, 44–54.
23. Merks, R. M. H. and Glazier, J. A. (2005) Contact-inhibited chemotactic motility can drive both vasculogenesis and sprouting angiogenesis. *q-bio/0505033.*
24. Kesmir, C. and de Boer, R. J. (2003) A spatial model of germinal center reactions: Cellular adhesion based sorting of B cells results in efficient affinity maturation. *J. Theor. Biol.* 222, 9–22.
25. Meyer-Hermann, M., Deutsch, A., and Or-Guil, M. (2001) Recycling probability and dynamical properties of germinal center reactions. *J. Theor. Biol.* 210, 265–285.
26. Nguyen, B., Upadhyaya, A., van Oudenaarden, A., and Brenner, M. P. (2004) Elastic instability in growing yeast colonies. *Biophys. J.* 86, 2740–2747.
27. Walther, T., Reinsch, H., Grosse, A., Ostermann, K., Deutsch, A., and Bley, T. (2004) Mathematical modeling of regulatory mechanisms in yeast colony development. *J. Theor. Biol.* 229, 327–338.
28. Borner, U., Deutsch, A., Reichenbach, H., and Bar, M. (2002) Rippling patterns in aggregates of *myxobacteria* arise from cell–cell collisions. *Phys. Rev. Lett.* 89, 078101.
29. Bussemaker, H. J., Deutsch, A., and Geigant, E. (1997) Mean-field analysis of a dynamical phase transition in a cellular automaton model for collective motion. *Phys. Rev. Lett.* 78, 5018–5021.
30. Dormann, S., Deutsch, A., and Lawniczak, A. T. (2001) Fourier analysis of Turing-like pattern formation in cellular automaton models. *Future Gener. Comput. Syst.* 17, 901–909.
31. Börner, U., Deutsch, A., Reichenbach, H., and Bär, M. (2002) Rippling patterns in aggregates of myxobacteria arise from cell–cell collisions. *Phys. Rev. Lett.* 89, 078101.
32. Zhdanov, V. P. and Kasemo, B. (2004) Simulation of the growth and differentiation of stem cells on a heterogeneous scaffold. *Phys. Chem. Chem. Phys.* 6, 4347–4350.
33. Knewitz, M. A. and Mombach, J. C. (2006) Computer simulation of the influence of cellular adhesion on the morphology of the interface between tissues of proliferating and quiescent cells. *Comput. Biol. Med.* 36, 59–69.
34. Marée, A. F. M. and Hogeweg, P. (2001) How amoeboids self-organize into a fruiting body: Multicellular coordination in *Dictyostelium discoideum. Proc. Natl Acad. Sci. USA* 98, 3879–3883.
35. Marée, A. F. M. and Hogeweg, P. (2002) Modelling *Dictyostelium discoideum* morphogenesis: the culmination. *Bull. Math. Biol.* 64, 327–353.
36. Marée, A. F. M., Panfilov, A. V., and Hogeweg, P. (1999) Migration and thermotaxis of *Dictyostelium discoideum* slugs, a model study. *J. Theor. Biol.* 199, 297–309.
37. Savill, N. J. and Hogeweg, P. (1997) Modelling morphogenesis: From single cells to crawling slugs. *J. Theor. Biol.* 184, 229–235.
38. Hogeweg, P. (2000) Evolving mechanisms of morphogenesis: On the interplay between differential adhesion and cell differentiation. *J. Theor. Biol.* 203, 317–333.
39. Johnston, D. A. (1998) Thin animals. *J. Phys. A* 31, 9405–9417.
40. Groenenboom, M. A. and Hogeweg, P. (2002) Space and the persistence of male-killing endosymbionts in insect populations. *Proc. Biol. Sci.* 269, 2509–2518.
41. Groenenboom, M. A., Maree, A. F., and Hogeweg, P. (2005) The RNA silencing pathway: the bits and pieces that matter. *PLoS Comp. Biol.* 1, 155–165.
42. Kesmir, C., van Noort, V., de Boer, R. J., and Hogeweg, P. (2003) Bioinformatic analysis of functional differences between the immunoproteasome and the constitutive proteasome. *Immunogenetics* 55, 437–449.
43. Pagie, L. and Hogeweg, P. (2000) Individual- and population-based diversity in restriction-modification systems. *Bull. Math. Biol.* 62, 759–774.
44. Silva, H. S. and Martins, M. L. (2003) A cellular automata model for cell differentiation. *Phys. A* 322, 555–566.
45. Zajac, M., Jones, G. L., and Glazier, J. A. (2000) Model of convergent extension in animal morphogenesis. *Phys. Rev. Lett.* 85, 2022–2025.

46. Zajac, M., Jones, G. L., and Glazier, J. A. (2003) Simulating convergent extension by way of anisotropic differential adhesion. *J. Theor. Biol.* 222, 247–259.
47. Savill, N. J. and Sherratt, J. A. (2003) Control of epidermal stem cell clusters by Notch-mediated lateral induction. *Dev. Biol.* 258, 141–153.
48. Mombach, J. C. M., de Almeida, R. M. C., Thomas, G. L., Upadhyaya, A., and Glazier, J. A. (2001) Bursts and cavity formation in *Hydra* cells aggregates: Experiments and simulations. *Phys. A* 297, 495–508.
49. Rieu, J. P., Upadhyaya, A., Glazier, J. A., Ouchi, N. B., and Sawada, Y. (2000) Diffusion and deformations of single hydra cells in cellular aggregates. *Biophys. J.* 79, 1903–1914.
50. Mochizuki, A. (2002) Pattern formation of the cone mosaic in the zebrafish retina: A cell rearrangement model. *J. Theor. Biol.* 215, 345–361.
51. Takesue, A., Mochizuki, A., and Iwasa, Y. (1998) Cell-differentiation rules that generate regular mosaic patterns: Modelling motivated by cone mosaic formation in fish retina. *J. Theor. Biol.* 194, 575–586.
52. Dallon, J., Sherratt, J., Maini, P. K., and Ferguson, M. (2000) Biological implications of a discrete mathematical model for collagen deposition and alignment in dermal wound repair. *IMA J. Math. Appl. Med. Biol.* 17, 379–393.
53. Maini, P. K., Olsen, L., and Sherratt, J. A. (2002) Mathematical models for cell–matrix interactions during dermal wound healing. *Int. J. Bifurcat. Chaos* 12, 2021–2029.
54. Kreft, J. U., Picioreanu, C., Wimpenny, J. W. T., and van Loosdrecht, M. C. M. (2001) Individual-based modelling of biofilms. *Microbiology* 147, 2897–2912.
55. Picioreanu, C., van Loosdrecht, M. C. M., and Heijnen, J. J. (2001) Two-dimensional model of biofilm detachment caused by internal stress from liquid flow. *Biotechnol. Bioeng.* 72, 205–218.
56. van Loosdrecht, M. C. M., Heijnen, J. J., Eberl, H., Kreft, J., and Picioreanu, C. (2002) Mathematical modelling of biofilm structures. *Antonie Van Leeuwenhoek Int. J. General Mol. Microbiol.* 81, 245–256.
57. Pop awski, N. J., Shirinifard, A., Swat, M., and Glazier, J. A. (2008) Simulations of single-species bacterial-biofilm growth using the Glazier–Graner–Hogeweg model and the CompuCell3D modeling environment. *Math. Biosci. Eng.* 5, 355–388.
58. Chaturvedi, R., Huang, C., Izaguirre, J. A., Newman, S. A., Glazier, J. A., and Alber, M. S. (2004) A hybrid discrete-continuum model for 3-D skeletogenesis of the vertebrate limb. *Lect. Notes Comput. Sci.* 3305, 543–552.
59. Pop awski, N. J., Swat, M., Gens, J. S., and Glazier, J. A. (2007) Adhesion between cells, diffusion of growth factors, and elasticity of the AER produce the paddle shape of the chick limb. *Phys. A* 373, 521–532.
60. Glazier, J. A. and Weaire, D. (1992) The kinetics of cellular patterns. *J. Phys.: Condens. Matter* 4, 1867–1896.
61. Glazier, J. A. (1993) Grain growth in three dimensions depends on grain topology. *Phys. Rev. Lett.* 70, 2170–2173.
62. Glazier, J. A., Grest, G. S., and Anderson, M. P. (1990) Ideal two-dimensional grain growth, in Simulation and Theory of Evolving Microstructures (Anderson, M. P. and Rollett, A. D., eds.), The Minerals, Metals and Materials Society, Warrendale, PA, pp. 41–54.
63. Glazier, J. A., Anderson, M. P., and Grest, G. S. (1990) Coarsening in the two-dimensional soap froth and the large-Q Potts model: a detailed comparison. *Philos. Mag. B* 62, 615–637.
64. Grest, G. S., Glazier, J. A., Anderson, M. P., Holm, E. A., and Srolovitz, D. J. (1992) Coarsening in two-dimensional soap froths and the large-Q Potts model. *Mater. Res. Soc. Symp.* 237, 101–112.
65. Jiang, Y. and Glazier, J. A. (1996) Extended large-Q Potts model simulation of foam drainage. *Philos. Mag. Lett.* 74, 119–128.
66. Jiang, Y., Levine, H., and Glazier, J. A. (1998) Possible cooperation of differential adhesion and chemotaxis in mound formation of *Dictyostelium. Biophys. J.* 75, 2615–2625.
67. Jiang, Y., Mombach, J. C. M., and Glazier, J. A. (1995) Grain growth from homogeneous initial conditions: Anomalous grain growth and special scaling states. *Phys. Rev. E* 52, 3333–3336.
68. Jiang, Y., Swart, P. J., Saxena, A., Asipauskas, M., and Glazier, J. A. (1999) Hysteresis and avalanches in two-dimensional foam rheology simulations. *Phys. Rev. E* 59, 5819–5832.
69. Ling, S., Anderson, M. P., Grest, G. S., and Glazier, J. A. (1992) Comparison of soap froth and simulation of large-Q Potts model. *Mater. Sci. Forum* 94–96, 39–47.
70. Mombach, J. C. M. (2000) Universality of the threshold in the dynamics of biological cell sorting. *Phys. A* 276, 391–400.
71. Weaire, D. and Glazier, J. A. (1992) Modelling grain growth and soap froth coarsening: Past, present and future. *Mater. Sci. Forum* 94–96, 27–39.
72. Weaire, D., Bolton, F., Molho, P., and Glazier, J. A. (1991) Investigation of an elementary model for magnetic froth. *J. Phys.: Condens. Matter* 3, 2101–2113.

73. Glazer, J. A., Balter, A., and Pop awski, N. (2007) Magnetization to morphogenesis: A brief history of the Glazier–Graner–Hogeweg model, in *Single-Cell-Based Models in Biology and Medicine* (Anderson, A. R. A., Chaplain, M. A. J., and Rejniak, K. A., eds.), Birkhauser Verlag, Basel, pp. 79–106.
74. Walther, T., Reinsch, H., Ostermann, K., Deutsch, A., and Bley, T. (2005) Coordinated growth of yeast colonies: Experimental and mathematical analysis of possible regulatory mechanisms. *Eng. Life Sci.* 5, 115–133.
75. Keller, E. F. and Segel, L. A. (1971) Model for chemotaxis. *J. Theor. Biol.* 30, 225–234.
76. Glazier, J. A. and Upadhyaya, A. (1998) First steps towards a comprehensive model of tissues, or: A physicist looks at development, in Dynamical Networks in Physics and Biology: At the Frontier of Physics and Biology (Beysens, D. and Forgacs, G., eds.), EDP Sciences, Berlin, pp. 149–160.
77. Glazier, J. A. and Graner, F. (1993) Simulation of the differential adhesion driven rearrangement of biological cells. *Phys. Rev. E* 47, 2128–2154.
78. Glazier, J. A. (1993) Cellular patterns. *Bussei Kenkyu* 58, 608–612.
79. Glazier, J. A. (1996) Thermodynamics of cell sorting. *Bussei Kenkyu* 65, 691–700.
80. Glazier, J. A., Raphael, R. C., Graner, F., and Sawada, Y. (1995) The energetics of cell sorting in three dimensions, in Interplay of Genetic and Physical Processes in the Development of Biological Form (Beysens, D., Forgacs, G., and Gaill, F., eds.), World Scientific, Singapore, pp. 54–66.
81. Graner, F. and Glazier, J. A. (1992) Simulation of biological cell sorting using a 2-dimensional extended Potts model. *Phys. Rev. Lett.* 69, 2013–2016.
82. Mombach, J. C. M. and Glazier, J. A. (1996) Single cell motion in aggregates of embryonic cells. *Phys. Rev. Lett.* 76, 3032–3035.
83. Mombach,J.C.M.,Glazier,J.A.,Raphael,R.C., and Zajac, M. (1995) Quantitative comparison between differential adhesion models and cell sorting in the presence and absence of fluctuations. *Phys. Rev. Lett.* 75, 2244–2247.
84. Cipra, B. A. (1987) An introduction to the Ising-model. *Am. Math. Monthly* 94, 937–959.
85. Metropolis, N., Rosenbluth, A., Rosenbluth, M. N., Teller, A. H., and Teller, E. (1953) Equation of state calculations by fast computing machines. *J. Chem. Phys.* 21, 1087–1092.
86. Forgacs, G. and Newman, S. A. (2005). *Biological Physics of the Developing Embryo.* Cambridge University Press, Cambridge.
87. Alber, M. S., Kiskowski, M. A., Glazier, J. A., and Jiang, Y. (2002) On cellular automation approaches to modeling biological cells, in *Mathematical Systems Theory in Biology, Communication and Finance* (Rosenthal, J. and Gilliam, D. S., eds.), Springer, New York, NY, pp. 1–40.
88. Alber, M. S., Jiang, Y., and Kiskowski, M. A. (2004) Lattice gas cellular automation model for rippling and aggregation in *myxobacteria. Phys. D* 191, 343–358.
89. Upadhyaya, A., Rieu, J. P., Glazier, J. A., and Sawada, Y. (2001) Anomalous diffusion in two-dimensional *Hydra* cell aggregates. *Phys. A* 293, 549–558.
90. Cickovski, T., Aras, K., Alber, M. S., Izaguirre, J. A., Swat, M., Glazier, J. A., Merks, R. M. H., Glimm, T., Hentschel, H. G. E., and Newman, S. A. (2007) From genes to organisms via the cell: A problem-solving environment for multicellular development. *Comput. Sci. Eng.* 9, 50–60.
91. Izaguirre, J. A., Chaturvedi, R., Huang, C., Cickovski, T., Coffland, J., Thomas, G., Forgacs, G., Alber, M., Hentschel, G., Newman, S. A., and Glazier, J. A. (2004) CompuCell, a multi-model framework for simulation of morphogenesis. *Bioinformatics* 20, 1129–1137.
92. Armstrong, P. B. and Armstrong, M. T. (1984) A role for fibronectin in cell sorting out. *J. Cell Sci.* 69, 179–197.
93. Armstrong, P. B. and Parenti, D. (1972) Cell sorting in the presence of cytochalasin B. *J. Cell Sci.* 55, 542–553.
94. Glazier, J. A. and Graner, F. (1993) Simulation of the differential adhesion driven rearrangement of biological cells. *Phys. Rev. E* 47, 2128–2154.
95. Glazier, J. A. and Graner, F. (1992) Simulation of biological cell sorting using a two-dimensional extended Potts model. *Phys. Rev. Lett.* 69, 2013–2016.
96. Ward, P. A., Lepow, I. H., and Newman, L. J. (1968) Bacterial factors chemotactic for polymorphonuclear leukocytes. *Am. J. Pathol.* 52, 725–736.
97. Lutz, M. (1999) *Learning Python.* O'Reilly & Associates, Sebastopol, CA.
98. Balter, A. I., Glazier, J. A., and Perry, R. (2008) Probing soap-film friction with two-phase foam flow. *Philos. Mag. Lett.* 88, 679–691.
99. Dvorak, P., Dvorakova, D., and Hampl, A. (2006) Fibroblast growth factor signaling in embryonic and cancer stem cells. *FEBS Lett.* 580, 2869–2287.

Chapter 14

BioLogic: A Mathematical Modeling Framework for Immunologists

Shlomo Ta'asan and Rima Gandlin

Summary

The immune response to pathogens is a result of complex interactions among many cell types and a large number of molecular processes. As such it poses numerous challenges for modeling, simulation, and analysis. In this work we aim at addressing major issues regarding modeling of large biological systems with a special focus on the immune system. We address (1) the hierarchy in the system, from genes to organelles to cells to organs to organism, (2) the high variability due to experimentation, (3) the high variability among organisms, and (4) the need to bridge between immunologists/experimentalists and mathematicians/modelers. We provide an intuitive syntax to describe biological knowledge in terms of interactions (reactions) and objects (cells, organs, etc.) and illustrate how to use it in describing very complex systems. We describe the main elements of a simulation program that use that syntax to define models and to automatically simulate them. We restrict our discussion to modeling using logical network, although other modeling techniques, for example, differential equations and probabilistic/stochastic modeling, are also possible. Examples demonstrating the different features of the framework are given throughout the chapter.

Key words: BioLogic, Biological systems, Modeling, Simulation.

1. Introduction

Mathematical modeling and simulation technique are common in most areas of science and engineering. They assist in gaining deeper insights into physical phenomena; they help in estimating unknown parameters, and they help performing design tasks and more. Experimentalists in numerous areas such as materials science and engineering, civil engineering, aerospace engineering, etc. use mathematical models routinely. In biology, in contrast,

Ivan V. Maly (ed.) *Methods in Molecular Biology, Systems Biology, vol. 500*

DOI: 10.1007/978-1-59745-525-1_14

we encounter mostly statistical methods that are used to handle noisy measurements and for hypotheses testing. Mathematical modeling for gaining insight into how biological systems work and evolve is hardly used. Yet in biology the problems are more complex and the need for exploring mathematical techniques is essential. An example is an immune response to pathogens, which is a result of complex interactions among many cell types and a large number of molecular processes – a grand challenge for modeling simulation as well as analysis and interpretation.

Advances in experimental techniques in biology over the last decade allow the interrogation of systems at multiple levels yielding abundant data that call for new tools for its understanding. These experimental techniques include gene microarrays, high-dimensional flow cytometry, multiplex assays, and more. The information processing involved in an immune response is becoming progressively available through the use of these multiple techniques. Thus, advances in experimentation make it necessary to use modeling and simulation techniques for interpreting and analyzing these data due to their large scale.

This work aimed at answering some aspects of the grand challenge of modeling the immune system. Our study here is focused toward developing a general framework and toolbox for mathematical modeling of the immune system that is targeted toward immunologists. The rationale behind this is that the vast amount of information in immunology is known only to expert immunologists and it is best if we can supply them with a tool for modeling, rather than have a mathematician learn all the details of the immune system. Our framework is based on a few important observations. First, the large number of interactions and processing that take place in an immune response call for a new paradigm to capture the complex behavior of the system. The system is multiscale in nature, comprising genes, proteins, organelles, cells, and organs. Second, there is a high variability among genetically identical organisms in expression level of mRNA, receptors, and secreted cytokines. Similarly, reactions rates (on rate, off rate) are not known for most processes, and probably have high variability as well even among genetically identical organisms. Third, there is a large body of knowledge regarding molecular interactions in immune response, and this knowledge is rapidly growing, reflecting that a lot is still unknown.

Our modeling and simulation framework includes three components. The first is syntax for expressing experimental setups, data, and knowledge regarding states and interactions in the system. This part can be regarded as a bridge between immunologists and mathematicians as it is intuitive and can be understood easily by immunologists, yet it is formal enough so it defines the system precisely. This syntax also answers the challenge of the rapid growth in immunological knowledge since it allows for easy updates of hypothesis and knowledge. The second component of this framework is a database for storing such knowledge and information. This allows for reuse of previously defined objects or systems. The third component is a computer simulation environment that uses the aforementioned syntax to define realization of models and to simulate them. The framework addresses limitations in available measurements and in their high variability by using logical networks, in which molecular abundance and reaction rates are described qualitatively using just a few levels, denoted by 0, 1, 2, etc. The use of the logical network paradigm removes the need for detailed parameter estimation, such as association/dissociation constants.

Subheadings 2 and **3** are written for a potential user of this framework, a person with immunology background. Part of **Subheading 4** includes some mathematical arguments that can be skipped without hurting the ability to use the framework. The mathematical arguments are intended for readers with mathematical background who may question the relation of our approach to the classical approach using differential equations.

2. Mathematical Approaches and Models

The complex and dynamic nature of the immune system has stimulated many mathematicians and modelers to use mathematical modeling to gain an understanding of its functioning and regulation. Indeed several mathematical models of different aspects of the immune system have been developed *(1)*. Mathematical models have proved very useful in the study of some aspects of the dynamics of HIV infection and progression to AIDS *(2)*, particularly in relation to the development of novel treatment regimens *(3–5)* and the latent phase *(6)*. The study of T-cell activation and the cognate interaction with peptide/MHC complexes *(7, 8)* has benefited from the use of mathematical models *(9–11)*. They are more quantitative and as a result allow for a more precise and refined analysis of how the dynamics of receptor interaction leads to activation of the cells of the immune system *(12–14)*. Simple mathematical models were used to discuss the immune memory *(15)*.

None of the aforementioned approaches were designed to model the immune response in a biologically realistic manner. Computer-based approaches such as cellular automata or complex system modeling have been used in attempting to describe adequately such complex processes. Seiden and Celada have used cellular automata to model the immune system *(15–19)*. Further development in this direction was carried out in *(20–21)*, including parallel implementation and the capability to simulate both humoral and cellular responses. The package PARIMM is one of the most complete simulators of the immune system developed. These approaches are, however, too rigid in the sense that if the model needs to be updated due to new immunological data, it has to be done at the software level, which can be cumbersome and time consuming.

More flexible approaches, yet in different areas, have been developed to modeling biology. In these approaches a simulation environment is created, rather than a particular model. We cite a few examples only. In *(22)*, an automation-based semantic of temporal evolution of complex biochemical reactions is used starting from the representation of the system as given as a set of differential algebraic equations (DAE). Reasoning about the system is done using temporal logic. The software package for this approach is Simpathica/xssys. Another approach *(23)* for a rigorous formalism in modeling biological systems involves Petri nets. This is a mathematical formalism developed by computer scientists that allows biologists to focus on the content of their model rather than on the implementation. A software package UltraSAN is dedicated to this approach. In this approach one obtains probability distributions for molecular species; thus, it addresses low-abundance molecular species, where differential equations are not appropriate. Addressing the hierarchy in biological systems and building complex objects have been done *(24)*. This involves continuous system modeling, and it produces a self-contained independently executable model, which allows for multimodel multicomponent hierarchy. An attempt to create logic for biological systems *(25)* involves a language together with hybrid projection temporal logic of modeling, analyzing, and verifying biological systems, and deals with nontrivial mixture of discrete and continuous systems. That approach is too mathematical to be useful for immunologists.

In designing our framework we have focused on (1) modularity, (2) ease of understanding and use by immunologists, (3) a simple approach to modeling hierarchical structure, (4) addressing the high variability in experimental data, and (4) the insufficient information regarding reaction rates.

3. BioLogic: A Modeling Framework

Our framework describes the immune system structure and function by defining the space (which may include multiple compartments), the objects (which may be hierarchical), and the interaction rules among them. We begin by discussing objects and interactions without specifying concentration or cell count. This will make the presentation easier to follow. We adopt a free style in defining concepts instead of using precise mathematical definition in order to make the material accessible for nonmathematicians.

3.1. Syntax for Objects

Immunology deals with different types of objects such as cells and organs whose interactions are facilitated through molecular mediators. Our modeling approach will mimic this and will therefore be very intuitive for immunologists. At the abstract mathematical level we deal with **objects**, which may represent molecules, cells, organs, etc. However, we distinguish between simple objects and complex objects.

Simple objects are used to model biological entities whose internal structure is not needed for the specific modeling. Biologically these may include description cytokines, chemokines, genes, etc. However, in simplified models, even a cell may be modeled as a simple object as we see later. To allow the flexibility of modeling from the gene level all the way to the organism level using a simple syntax, we have introduced three types of simple objects, which we refer to by their biological significance: **genes**, **molecules**, and **transporters**. Genes can be either up *(1)* or down (0). Molecules and transporters can have arbitrary levels. Names for objects can include characters a–z, A–Z, numbers, and the symbols: - (minus) and _ (underscore). They cannot include other characters. Simple objects in our frameworks are the ones that perform actions (interactions).

Complex objects are containers for simple objects and additional complex objects, and the hierarchy is limitless. The syntaxes for describing complex objects are the square brackets [ ]. For example, the object [ ] is the empty object, i.e., it contains no simple objects and is of only mathematical interest (there is no biology for it to represent). We move to more interesting objects.

Example I: The object depicted in **Fig. 1** and having the syntax

X = [A,B,C, [D E]]

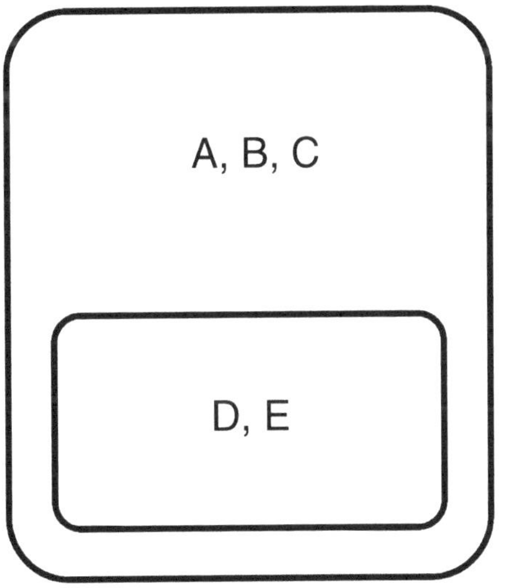

Fig. 1. A complex object.

is a complex object that contains three simple objects A, B, and C, and one complex object [D,E] consisting of two simple objects D and E.

Some terminology: An important concept is that of **parents** and **children** in these hierarchical structures. In the last example we may think of the object [D,E] as a child of the object that contains it, which we call the **parent**. We may need to refer to a **grandparent** later on, so keep this in mind as well. **Disjoint objects** are objects such that none contains the other. In the case of [[A,B],[C,[D,E]], the objects [A,B] and [C,[D,E]] are disjoint. The **innermost container** for an object, X, is the container, [], that contains X and no child of this container contains X. For example, in the object [[A,B],[C,[D,E]], the outermost brackets belong to the innermost container for both objects [A,B] and [C,[D,E]], but it is not the innermost container for [D,E] that is contained in a child, which contains both the simple object C and [D,E].

Example II. Suppose that we want to model the interaction between a macrophage (M) and a natural killer (NK) cell, but we do not want to go into description of signaling. This is an example in which we use simple objects of our framework to model complex biological objects. The model is defined as

Model = [M, NK],

which includes a description of structure but not of any interaction. We need to supply also a set of interaction rules that follow the biology and that would make this model interesting. As it is, it cannot evolve or perform any action. Before going to describing interaction we give some more examples.

Example III: A complex object describing a macrophage is shown in **Fig. 2**. In our syntax it is written as

```
Macrophage = [TLR4, TNF-R, IFNg-R, IL12-R,
                        [TLR3, TLR9, Phagosome,
                                [geneTNF, geneIL-6
                                ]
                        ]
             ].
```

We have used here indentation to show the beginning, [, and end,], of each compartment. In this object we distinguish three levels. The outer level contains the simple objects TLR4, TNF-R, IFNg-R, and IL12-R – all known receptors of a macrophage at its naive state. This outer level may be regarded as the membrane. In the next inner level we find TLR3, TLR9, and phagosome. In reality a phagosome is a complex object consisting of a membrane, receptors, and molecules in its interior; here, it is modeled as a simple object. The object containing TLR3, TLR9, and phagosome is the parent of the object [geneTNF, geneIL-6], while the top-level object containing TLR-4, etc., is the grandparent.

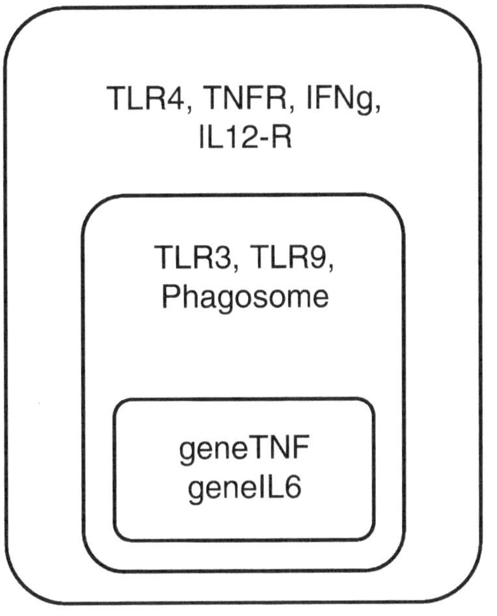

Fig. 2. A macrophage: graphical representation.

The object that contains the whole model is called the **space**. It may contain simple objects, complex objects, or both. It has a single copy of itself, and all interactions are within this object. In the last example **Model** is the parent object and thus defines the space for our simulation.

3.2. Syntax and Rules for Interactions

In the current version of our framework, interactions happen between simple objects only. We use the following syntax:

{reactants} {speed} {products},

where **{reactants}** and **{products}** contain lists of simple objects separated by the symbol +. For now, we will assume that speed is represented by the symbol ->. Later when we discuss logical variables we will extend this to include symbols such as ->>, ->>>, ->>>>, etc. We list a few examples of interactions (left) and their interpretation (right):

TNFaR + TNFa -> TNFa: TNFaR	*receptor ligand binding*
Caspase3 -> Apoptosis	*molecule initiating a process*
ProInflammatoryResponse ->TNFa	*process results in cytokine production*
Inflammation -> Cancer	*one complex process initiates another*

Note the use of names to represent real molecules and the results of their binding (first reaction); molecules that result in a complex biological process, such as apoptosis (second reaction); processes that result in production of cytokines (third reaction); a complex biological process that results in another complex biological process, both modeled as simple objects (fourth reaction).

The use of a hierarchical structure limits the possible interactions between simple objects. For example, simple objects that reside in disjoint objects may not be able to interact since they are meant to model physically different regions in the organism. We define the allowable interactions across compartments using the following rules:

Rule 1: Simple objects that belong to the same innermost container can interact.

Rule 2: Simple objects can interact with other simple objects of the parent's outermost level.

Rule 3: Simple objects in the outermost level of two disjoint objects that belong to the same innermost container can interact.

These rules are based on the intuitive idea of proximity in hierarchical structures. Rule 1 is obvious, implying that simple objects residing in different compartment may not be able to interact unless they obey rules 2 and 3. Rule 2 allows us to model signaling, as it mimics the situation in biology where molecules in the membrane can interact with molecules in the cytosol or molecules outside the cell. Rule 3 will allow us to model cell–cell interactions through surface molecules. We give an example to illustrate the rules.

Example IV: Consider the complex object that is shown in **Fig. 3**. The syntax for it is

Object = [[A,B,C, [D,E]], [F,G, [H, [K,L,M]]]].

In this example, A, B, and C are in the same container, and D and E are in another container, etc. From rule 1, the simple objects A, B, and C can interact; D and E can interact; F and G can interact; and K, L, and M can interact. From rule 2 we conclude that D and E can interact with A, B, and C but not with others; K, L, and M can interact with H. In addition, H can interact also with F and G. From rule 3 we have that A, B, and C can interact with F and G.

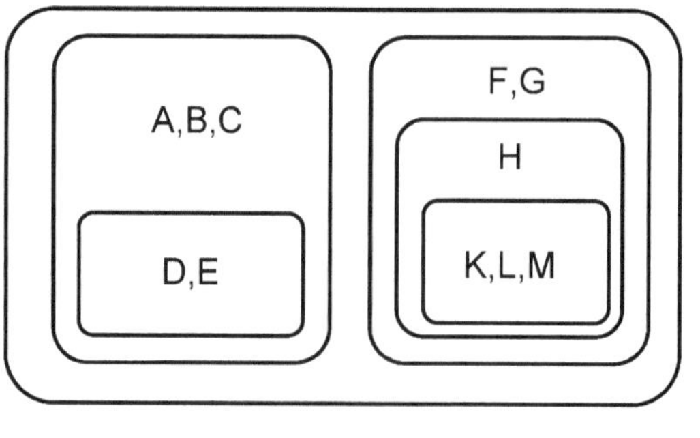

Fig. 3. A complex object containing two (children) complex objects.

Example V: Macrophage activation by bacteria. In this example we demonstrate the use of simple objects alone in modeling a biological scenario: the activation of a macrophage by bacteria. We start by defining the initial objects, bacteria (Bac), macrophage (M), and natural killer (NK),

Model = [Bac, M, NK],

together with all the possible interactions in the system,

R1: M + Bac -> paM + Bac + Bac
R2: paM -> paM + IL-12
R3: NK + IL-12 -> aNK
R4: aNK -> aNK + IFNg
R5: paM + IFNg -> aM
R6: aM + Bac -> aM

The first reaction, M + Bac -> paM + Bac + Bac, indicates that a macrophage (M), when interacting with bacteria (Bac), becomes a partially activated macrophage (paM), and the bacteria multiply (it appears twice on the right and only once on the left). The second reaction, paM -> paM + IL-12, describes the action of a partially active macrophage: it secretes IL-12. The fact that paM appears on both sides of the reaction means that the active macrophage does not change its state as a result of this secretion. Note that this is not the case in the first reaction: the naive macrophage (M) does not appear in the right hand side – it changed into a partially active macrophage (paM). The third reaction, NK + IL-12 -> aNK, means that a natural killer cell in the presence of IL-12 becomes activated, and the IL-12 is consumed. The fourth reaction, aNK -> aNK + IFNg, describes the action of an active natural killer: it secretes IFNg and does not change its state. The fifth reaction, paM + IFNg -> aM, indicates that a partially active macrophage (paM) becomes fully activated (aM) in the presence of IFNg. The last reaction, aM + Bac -> aM, describes the killing of bacteria by an active macrophage. Note that the object Bac appears on the left but not in the right-hand side of the reaction. This indicates that it disappears at the end of this reaction.

Our model can evolve only in accordance with the possible reactions available to it. The following is the evolution of the model, where we have indicated the reaction (R1–R6) that is responsible for each change:

[Bac, M, NK] (R1) ⇒ [Bac, paM, NK] (R2) ⇒ [Bac, paM, IL-12, NK] (R3) ⇒ [Bac, paM, IL-12, aNK] (R4) ⇒ [Bac, paM, aNK, IFNg] (R5) ⇒ [Bac, aNK, aM] (R6) ⇒ [aNK, aM].

In this sequence of events we assumed that there is a single macrophage, a single NK, and a single bacterium. In reality there is a population of each, and the reaction M + Bac -> aM + Bac does not change all macrophages but only those that encounter bacteria.

The sequence of states of the system will have the objects M, NK, etc. during the whole progression.

Syntax for transporter interactions. To allow for secretion of molecules and complex interaction such as phagocytosis, we introduced a special type of reactions, which we refer to as transporter reaction. It has the syntax

{transporter}:: {simpleObject} @ {origin} {speed} {simpleObject} @ {destination},

where **{transporter}** is a simple object, and **{origin}** and **{destination}** are one of the following keywords: SELF, PARENT, GPARENT (for grandparent). The origin and destination are with reference to the location of the transporter. The convention is to put the transporter in the innermost object involved, since the parent and grandparent are unique while an object may have several children. An example will clarify this. Consider again the case of a macrophage that was partially activated, by LPS, for example, and has produced IL-12 but not secreted it yet. The secretion can be facilitated using a transporter object, which is placed in the container of the IL-12. Let the object be

[[TLR4, [Tr, IL-12]]],

and the transporter reaction,

Tr :: IL-12 @ SELF -> IL-12 @ GPARENT.

This transporter reaction is read as "the transporter (Tr) takes (::) IL-12 molecules from its own compartment (SELF) and to the grandparent compartment (GPARENT)." It results in the following transformation of our object:

[[TLR4, [Tr, IL-12]]] ⇒ [IL-12 [TLR4, [Tr]]].

This example shows the reason for introducing GPARENT; the parent compartment here, containing the TLR4, is viewed as the membrane, and the secretion of the IL-12 needs to be done into the extracellular region.

The syntax introduced so far does not specify how to model the system. It describes only the structure and its logic. Certain things can happen while others cannot, etc. The actual dynamical modeling of systems described with the syntax explained here can be implemented in a variety of ways. One possibility is to use differential equations by translating all reactions into dynamical equations using the law of mass action. Another possibility is to use probabilistic models where each interaction happens only with a certain probability (also following the law of mass action). We will adopt a third approach that uses logical networks. We start with a brief description of what the logical variables are and how we add and subtract them, followed by their use in modeling interactions.

4. Implementation Using Logical Variables

As we have outlined in the introduction, a robust modeling approach to the immune system must address the high variability in molecular abundance and reaction rates. The system is robust and shows the same qualitative behavior even if the actual measurements differ from one organism to another. This suggest that instead of using real numbers to describe quantities such as molecular concentration, cell number, and reaction rates we might as well reduce the complexity by considering a few levels in each of these quantities. This is not a foreign idea to experimental immunologists. When discussing the results of an experiment they often use statements such as "the response was strong," "the number of cells was low," and "TNF-R level was high." These statements are not a use of an imprecise language by the immunologist. They reflect something very fundamental about the system; because of high variability across experiments and organisms we cannot use a more precise language.

4.1. Nonstandard Arithmetic

The mathematical concept of logical variables seems to fit the variability and the inability to be precise quite well. The simplest case is that of Boolean variables, which attain only two values, 0 and 1, referred to also as false and true, or in our case may represent high and low or fast and slow; the interpretation is up to us. A richer case that is more appropriate for modeling immunology would use a few such levels describing molecular abundance. For example, 0 would represent no expression, 1 – low expression, 2 – high expression, etc. Similarly, reaction rates will use the same numbers and 0 would mean slow rate, 1 – moderate, 2 – fast rate, etc.

Since molecular abundance may change during an immune response we may also need to define how to add and subtract logical variables. For example, suppose that we had two populations of the same cell type, and we have combined them. What is the size of the new combined population? If we follow our intuition from biology we know that if we add two small quantities we get a new quantity that is small, and by adding a small and a high quantity get a high quantity, etc. This is summarized in the following rules for addition.

None + None = None	$0 + 0 = 0$
Low + None = Low	$1 + 0 = 1$
Low + Low = Low	$1 + 1 = 1$
Low + High = High	$1 + 2 = 2$
High + High = High	$2 + 2 = 2$

The table on the right may look strange since we know that $1 + 1 = 2$; nevertheless in modeling with logical variables the aforementioned has to be kept in mind.

The problem with the arithmetic described earlier is that it does not allow for the following: Low + Low +... + Low = High, i.e., adding a small quantity many times results in a large quantity. To fix this we change the aforementioned arithmetic by making it probabilistic. That is, 1 + 1 = 1 most of the time, but with a small probability it is 2; for example, one of ten times it is 2, and in the other nine times it is 1. Of course, we do not want to determine in advance in which of the ten times it is 2, so we introduce a random element into our addition. The probabilities involved in the outcome of addition should be related to our interpretation of the logical variables in terms of real abundance. If we decide that 2 represents abundance that is 10 times larger than the abundance represented by 1, then the probability of 1 + 1 = 2 should be 0.1, and the probability of 1 + 1 = 1 should be 0.9. We can generalize this idea and consider the logical variables, 0,..., N, assuming that successive numbers represent abundances that differ by a factor of β (in the earlier example $\beta = 10$). If we also agree that the number 0 represents negligible amount and not really 0, it makes the addition more uniform, i.e., 0 + 0 + 0 +... + 0 = 1 in small probability. In summary, we will use the following general rule, using the notation $a \vee b \equiv \max(a,b)$ and $a \wedge b \equiv \min(a,b)$:

$$P(a + N = N) = 1 \qquad \text{for } a \leq N,$$

and

$$P(a + b = a \vee b) = 1 - \beta^{a \wedge b - a \vee b - 1} \qquad \text{for } a, b < N,$$

$$P(a + b = a \vee b + 1) = \beta^{a \wedge b - a \vee b - 1} \qquad \text{for } a, b < N,$$

where the symbol P stands for probability. A definition of subtraction follows from the relation $c - a = b$ being equivalent to $a + b = c$, and we have for given a and c with $a \leq c$,

$$P(c - a = b) = P(a + b = c) / \sum_x P(a + x = c),$$

where Σ_x denotes sum over all possible values of x. Note that with this definition $\Sigma_b P(c - a = b) = 1$, which is necessary for the definition to make sense. Note that we do not deal here with negative numbers; we did not define 1 − 2, for example. Our syntax uses the following symbols to distinguish among reaction rates: -> (0), ->> *(1)*, ->>> *(2)*, ->>>> *(3)*, etc.

To specify the number of objects of each type we use the brackets, (), and enclose in it both name and numbers. For example, (M,2) will represent macrophage at a high concentration.

4.2. Modeling of Interactions with Logical Variables

To motivate our approach for modeling interactions with logical variables, it is useful to keep in mind their intuitive relation to the real quantities. The most natural way to do it is to say that logical variables are proportional to the logarithm of the concentration (or number). We will obey the law of mass action translated into logical variables. We will explain it for three reaction types.

I. *Production at a constant rate*: -> A. The law of mass action here is simple, $dA/dt = c$, where c is some constant. Since our logical variable is $\ln(A)$ (natural logarithm of A) and not A itself, we need to construct an equation for $\ln(A)$. We will use natural logarithm, although any other base is suitable as well, and denote it by $\ln(A)$. Since $d \ln(A)/dt = (1/A)\, dA/dt$ we have $d \ln(A)/dt = c/A = c/e^{\ln(A)}$. This shows that as A increases, its rate of change decreases. This is quite intuitive.

II. *Unary interaction*: A -> B. Here, the law of mass action is simple as well, $dA/dt = -dB/dt = cA$. Translating this into logical variables that are proportional to $\ln(A)$ and $\ln(B)$ we have $d \ln(A)/dt = c$, $d \ln(B)/dt = -c$. That is per unit of time our logical variable in unary reaction changes by a fixed amount.

III. *Binary interaction*: $A + B$ -> C. This is slightly more complex. According to the law of mass action we have $dA/dt = -cAB$, $dB/dt = -cAB$, $dC/dt = cAB$. In translating these into log quantities we get $d \ln(A)/dt = -cB = -ce^{\ln(B)}$; $d \ln(B) = -cA = -ce^{\ln(A)}$; $d \ln(C) = cAB/C = c\, e^{\ln(A) + \ln(B) - \ln(C)}$. The quantities $\ln(A)$, $\ln(B)$, and $\ln(C)$ are the logical variables and their rate of change according to the earlier equation follows the law of mass action.

A small difficulty arises here since $\ln(A)$ may not be an integer, or may even be negative, and we need to define what exactly we do in these cases. The idea is simple: simulations are done with a given time step, Δt, and all changes are proportional to Δt. We accept only changes by integer values, so if a change is as a fractional part, we perform the fractional part in probability. For example, suppose that we need to make a change of 1.5, we make a change of 1, the integral part of 1.5, and we implement an additional change of ½ as a change of 1 but in probability ½.

5. Discussion

Our framework allows modeling of a biological system using different levels of resolutions. The coarsest description is in terms of the simple objects only and it is the closest to ordinary differential equation models. In this case there is usually one compartment only and all objects reside in it. This level of modeling is recommended as an initial step. It is done in order to gain an insight into the dynamical features of the system. Once the dominant aspects of the model are identified, a more refined model can be constructed. This process should continue as long as there are experimental data to back it up. When done in conjunction with experimental data, it can provide a feedback to experiments and in turn more experimental data to refine the model. The modeling framework described here allows us to combine elements that are represented crudely, using simple objects, together with elements that are represented with indefinite hierarchy. This allows accommodating existing gaps in knowledge. The modularity allows changing a model very easily; making updates as new information is becoming available is a straightforward task.

A web-based interface (http://www.math.cmu.edu/~shlomo/BioLogic.html) that implements the ideas described here, and gives additional details that have been omitted here, will be available by June 2008.

References

1. Morel, P. A. (1998) Mathematical modeling of immunological reactions. *Front. Biosci.* 3, d338–d347.
2. Blower, S., Schwartz, E. J., and Mills, J. (2003) Forecasting the future of HIV epidemics: the impact of antiretroviral therapies and imperfect vaccines. *AIDS Rev.* 5, 113–125.
3. Ho, D. D., Neumann, A. U., Perelson, A. S., Chen, W., Leonard, J. M., and Markowitz, M. M. (1995) Rapid turnover of plasma virions and CD4 lymphocytes in HIV-1 infection. *Nature* 373, 123–126.
4. Perelson, A. S., Essunger, P., Cao, Y., Vesanen, M., Hurley, A., Saksela, K., Markowitz, M., and Ho, D. D. (1997) Decay characteristics of HIV-1-infected compartments during combination therapy. *Nature* 387, 188–191.
5. Wei, X., Ghosh, S. K., Taylor, M. E., Johnson, V. A., Emini, E. A., Deutsch, P., Lifson, J. D., Bonhoeffer, S., Nowak, M. A., Hahn, B. H., Saag, M. S., and Shaw, G. M. (1995) Viral dynamics in human immunodeficiency virus type 1 infection. *Nature* 373, 117–122.
6. Nowak, M. A., May, R. M., and Sigmund, K. (1995) Immune responses against multiple epitopes. *J. Theor. Biol.* 175, 325–353.
7. Rabinowitz, J. D., Beeson, C., Lyons, D. S., Davis, M. M., and McConnell, H. M. (1996) Kinetic discrimination in T-cell activation. *Proc. Natl. Acad. Sci. USA* 93, 1401–1405.
8. Wülfing, C., Rabinowitz, J. D., Beeson, C., Sjaastad, M. D., McConnell, H. M., and Davis, M. M. (1997) Kinetics and extent of T cell activation as measured with the calcium signal. *J. Exp. Med.* 185, 1815–1825.
9. Agrawal, N. G. and Linderman, J. J. (1996) Mathematical modeling of helper T lymphocyte/antigen-presenting cell interactions: analysis of methods for modifying antigen processing and presentation. *J. Theor. Biol.* 182, 487–504.
10. De Boer, R. J. and Perelson, A. S. (1995) Towards a general function describing T cell proliferation. *J. Theor. Biol.* 175, 567–576.
11. McKeithan, T. W. (1995) Kinetic proofreading in T-cell receptor signal transduction. *Proc. Natl. Acad. Sci. USA* 92, 5042–5046.
12. Burke, M. A., Morel, B. F., Oriss, T. B., Bray, J., McCarthy, S. A., and Morel, P. A. (1997) Modeling the proliferative response of T cells to IL-2 and IL-4. *Cell. Immunol.* 178, 42–52.
13. Faeder, J. R., Hlavacek, W. S., Reischl, I., Blinov, M. L., Metzger, H., Redondo, A., Wofsy, C., and Goldstein, B. (2003) Investigation of early events in Fc epsilon RI-mediated signaling using a detailed mathematical model. *J. Immunol.* 170, 3769–3781.
14. Morel, B. F., Burke, M. A., Kalagnanam, J., McCarthy, S. A., Tweardy, D. J., and Morel, P. A. (1996) Making sense of the combined effect of interleukin-2 and interleukin-4 on lymphocytes using a mathematical model. *Bull. Math. Biol.* 58, 569–594.
15. Antia, R., Pilyugin, S. S., and Ahmed, R. (1998) Models of immune memory: on the role of cross-reactive stimulation, competition, and homeostasis in maintaining immune memory. *Proc. Natl. Acad. Sci. USA* 95, 14926–14931.
16. Celada, F. and Seiden, P. E. (1992) A computer model of cellular interactions in the immune system. *Immunol. Today* 13, 56–62.
17. Celada, F. and Seiden, P. E. (1996) Affinity maturation and hypermutation in a simulation of the humoral immune response. *Eur. J. Immunol.* 26, 1350–1358.
18. Morpurgo, D., Serenthà, R., Seiden, P. E., and Celada, F. (1995) Modelling thymic functions in a cellular automaton. *Int. Immunol.* 7, 505–516.
19. Seiden, P. E. and Celada, F. (1992) A model for simulating cognate recognition and response in the immune system. *J. Theor. Biol.* 158, 329–357.

20. Bernaschi, M., Succi, S., and Castiglione, F. (2000) Large-scale cellular automata simulations of the immune system response. *Phys. Rev. E Stat. Phys. Plasmas Fluids Relat. Interdiscip. Topics* 61, 1851–1854.
21. Bernaschi, M. and Castiglione, F. (2001) Design and implementation of an immune system simulator. *Comput. Biol. Med.* 31, 303–331.
22. Antoniotti, M., Policriti, A., Ugel, N., and Mishra, B. (2003) Model building and model checking for biochemical processes. *Cell Biochem. Biophys.* 38, 271–286.
23. Goss, P. J. and Peccoud, J. (1998) Quantitative modeling of stochastic systems in molecular biology by using stochastic Petri nets. *Proc. Natl. Acad. Sci. USA* 95, 6750–6755.
24. Hakman, M. and Groth, T. (1999) Object-oriented biomedical system modelling – the language. *Comput. Methods Programs Biomed.* 60, 153–181.
25. Duan, Z., Holcombe, M., and Bell, A. (2000) A logic for biological systems. *Biosystems* 55, 93–105.

Chapter 15

Dynamic Knowledge Representation Using Agent-Based Modeling: Ontology Instantiation and Verification of Conceptual Models

Gary An

Summary

The sheer volume of biomedical research threatens to overwhelm the capacity of individuals to effectively process this information. Adding to this challenge is the multiscale nature of both biological systems and the research community as a whole. Given this volume and rate of generation of biomedical information, the research community must develop methods for robust representation of knowledge in order for individuals, and the community as a whole, to "know what they know." Despite increasing emphasis on "data-driven" research, the fact remains that researchers guide their research using intuitively constructed conceptual models derived from knowledge extracted from publications, knowledge that is generally qualitatively expressed using natural language. Agent-based modeling (ABM) is a computational modeling method that is suited to translating the knowledge expressed in biomedical texts into dynamic representations of the conceptual models generated by researchers. The hierarchical object-class orientation of ABM maps well to biomedical ontological structures, facilitating the translation of ontologies into instantiated models. Furthermore, ABM is suited to producing the nonintuitive behaviors that often "break" conceptual models. Verification in this context is focused at determining the plausibility of a particular conceptual model, and qualitative knowledge representation is often sufficient for this goal. Thus, utilized in this fashion, ABM can provide a powerful adjunct to other computational methods within the research process, as well as providing a metamodeling framework to enhance the evolution of biomedical ontologies.

Key words: Agent-based modeling, Individual-based modeling, Mathematical models, Systems biology, Computational biology, Translational systems biology, Translational research, Knowledge representation, Biomedical ontology, Inflammation, Complexity, Complex systems, Metamodels, Model verification, Computer simulation.

Ivan V. Maly (ed.) *Methods in Molecular Biology, Systems Biology, vol. 500*

DOI: 10.1007/978-1-59745-525-1_15

1. Introduction: The Need for Dynamic Representation of Biomedical Knowledge

The biomedical research community today faces a challenge that has paradoxically arisen from its own success: as greater amounts of information become available at increasingly finer levels of biological mechanism it is also progressively difficult for individual researchers to effectively survey and integrate information even within their own area of expertise. While technology, via tools such as PubMED, the introduction of new publication formats like open-access journals, and the development of a whole slew of bioinformatics tools, has aided the distribution and availability of biomedical information, it still falls upon the individual researcher to concatenate that information into a conceptual mental model that represents that knowledge. These mental models guide the direction of their individual research and, in aggregate, they form the components of the evolving structure of community knowledge. However, the formal expression of mental models remains poorly defined, leading to limitations in the ability to share, critique, and evolve the knowledge represented in these conceptual models, particularly across disciplines. As a result it is increasingly difficult for both the individual researcher, and the community as a whole, to "know what it knows." Effective translational methodologies for knowledge representation need to move both "vertically" from the bench to the bedside, and be able to link "horizontally" across multiple researchers focused on different diseases. Information is generated by research endeavors at multiple scales and hierarchies of organization: gene ⇒ protein/enzyme ⇒ cell ⇒ tissue ⇒ organ ⇒ organism. The mirroring of these multiple levels in the organization of biomedical research has led to a disparate and compartmentalized research community and resulting organization of information. Recognition of this organizational challenge has led to extensive work in the area of developing biomedical ontological structures. These are classification systems, often hierarchical and "tree-like" in structure, to group biological objects together based on the rules of the particular ontology. However, while useful, these ontological structures are by and large static representations of knowledge, and do not help to address the "intuitive limit" in attempts to parse out cause and effect in complex, multiscale systems. The consequences of this intuitive limit are seen primarily in attempts to develop effective therapies for diseases resulting from disorders of internal regulatory processes, when the integration of knowledge requires crossing the multiple scales of organization (seen in **Fig. 1A–C**) to determine the organ and organism level consequences of molecular level manipulations *(1)*. Examples of such diseases are cancer, autoimmune disorders, and sepsis, all of which demonstrate complex, nonlinear behavior.

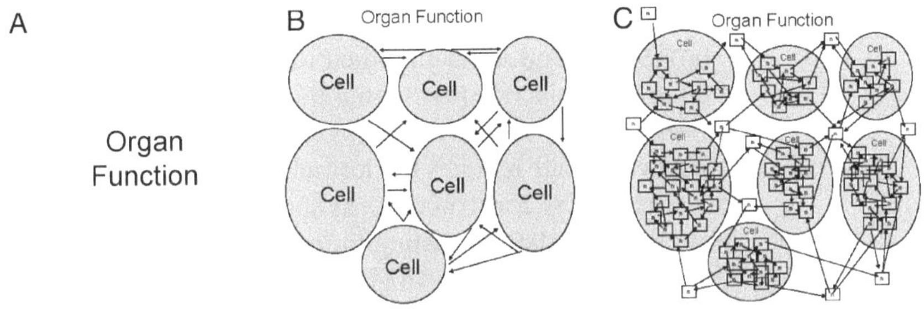

Fig. 1. Abstract demonstration of the expansion of information resulting from reductionist investigation of multiscale biological systems. (A) The highest level of clinically observed phenomenon at the organ level. (**B**) The mechanistic knowledge that organ function results from the interactions of multiple cells and types of cells. (**C**) What a conceptual mechanistic model would look like when a further finer grained level of resolution is used. This is where the overwhelming bulk of biomedical research is currently being conducted, particularly with respect to the search for drug candidates and mechanisms of disease. Note that the "indistinctness" of the last panel is intentional: attempts to "zoom in" on the figure may increase local clarity, but at the loss of being able to see the range of potential consequences to a particular manipulation. This figure is reproduced with the author's permission from **ref.** *1* under the terms of Creative Commons License.

1.1. A Possible Solution: Dynamic Knowledge Representation via Agent-Based Modeling

These limitations can be potentially overcome by developing methods of dynamically instantiating knowledge to allow researchers to express and evaluate conceptual models more effectively. Computer modeling can be seen as a means of dynamic knowledge representation to form a basis for formal means of testing, evaluating, and comparing what is currently known within the research community. To be able to "see" the consequences of a particular hypothesis structure/conceptual model, the formally represented knowledge is moved from a static depiction of relationships (as depicted in a flowchart or state diagram, similar to those seen in **Fig. 1C**) to a dynamic model in which the mechanistic consequences of each hypothesis can be observed and evaluated. This process can be termed Conceptual Model Verification: dynamic representation of a conceptual model is a means of its *verification*, analogous to model checking in computer science, i.e., does the model perform as expected based on its construction? It should be noted that *verification* in this context is distinct from *validation*, which can be considered the fidelity of a particular model to observed reality. For purposes of this discussion, *validation* is a process applied to the computational model, which in turn is used to *verify* the plausibility of a conceptual model.

Agent-based modeling (ABM) is an object-oriented computational modeling technique that is centered on the behaviors and interactions of the individual components of a system, and has been used to demonstrate the potential benefit of conceptual model verification *(2)*. ABM is a discrete event modeling system, meaning that the model cycles through a series of steps/loops/ticks during its execution. It has characteristics that make it well suited for creating aggregated modular multiscale models *(3, 4)*. ABM focuses on the rules and interactions between the individual components of a system, generating populations of those components and simulating their interactions in a "virtual world" to create an in silico experimental model *(2, 5–8)*. With its emphasis on parsing a system into groups or "classes" of system components ABM essentially requires the formulation of an ontological structure in order for its construction. As such, agent-based models are well suited to translating existing biomedical ontologies into a dynamic model. Furthermore, ABM rules are often expressed as conditional statements ("if–then" statements), which makes agent-based model suited to translating the hypotheses (expressed in natural language) that are generated from basic science research. There are three characteristics of ABM that deserve particular emphasis:

1. ABM is *spatial.* ABM has its origins in two-dimensional cellular automata, and as such many agent-based models are "grid-based." This spatial legacy makes ABM suited to representing structural relationships in a system under study. Nonmathematicians can model fairly complex topologies with greater ease and flexibility than may be possible with partial differential equations, leading to more intuitive knowledge translation into a model. The spatial nature of ABM also allows for modeling agents with "limited knowledge," i.e., input constrained by locality rules that determine its immediate environment. This property emphasizing local interactions also matches closely with the mechanisms of stimulus and response observed in biology.
2. ABM utilizes *parallelism.* This property of ABM sets it apart from other object-oriented modeling methods such as Petri nets or finite state machine models. In ABM each agent class has multiple instances within the model, forming a population of agents that interact in an emulated (usually) parallel processing environment. Within the execution of an agent-based model, heterogeneous individual agent behavior within a population of agents results in systemic dynamics that result in observable output that mirrors the behavior at the higher hierarchical level. A classic example of this is how relatively simple interaction rules among birds can lead to sophisticated flocking patterns.
3. ABM utilizes *stochasticity.* Many systems, particularly biological ones, include behaviors that appear to be random. "Appear to" is an important distinction, since what may appear to be random is actually deterministic from a mathematical standpoint. However, from a practical point of view, despite the fact that a particular system may follow the rules of deterministic chaos, at a higher-order observational level it is impossible to actually define the initial conditions from whence its behavior evolves. ABM addresses this issue via the generation of populations of agents. Once one is dealing with populations then it is possible to establish probabilities of a particular behavior for the population as a whole, and therefore also a probability

function for the behavior of a single agent. This probability function is incorporated into the agent's rules. When instantiated and run in parallel with other agents, each agent follows a particular trajectory of behavior as probabilities of its behavior rules "collapse" with each step of the model's run. In this fashion it is possible to generate a "population" of behavioral outputs from a single agent-based model, and move beyond the "behavior curves" seen in differential equation models toward "behavior spaces" more consistent with biological observation.

4. ABM reproduces *emergent properties.* Because of the parallelism, intrinsic stochasticity, and enforcement of locality resulting from its spatial architecture, a central hallmark of agent-based models is the fact that they generate systemic dynamics that often could not have been reasonably inferred from examination of the rules of the agents, resulting in so-called *emergent* behavior. To return to the example of the bird flock, superficial observation would seem to suggest the need for an overall "leader" to generate flock behavior, and therefore rules would seem to need to include a means of determining rules for flock-wide command and control communication. This, however, is not true; birds function via a series of locally constrained interaction rules and the flocking behavior emerges from the aggregate of these interactions. The capacity to generate emergent behavior is a vital advantage of using ABM for conceptual model verification, as it is often the paradoxical, nonintuitive nature of emergent behavior that "breaks" a conceptual model.

Although the use of ABM was pioneered in the areas of ecology, social science, and economics, it has been used to study biomedical processes such as sepsis *(2, 7)*, cancer *(4, 9)*, inflammatory cell trafficking *(10, 11)*, wound healing *(12)*, and intracellular structure and signaling *(13–15)*. In general, most biomedical ABM focuses on cells as the primary agent level (with notable exceptions from earlier **refs.** *13–15*). Cells are a natural agent level dictated by the organizational structure of biology, and from a knowledge translation standpoint, form a ready level of "encapsulated complexity" that can be addressed with relatively straightforward input–output rules. Furthermore, while the number of cells present in an organism is considerable, it is still magnitudes less than the number of molecules involved in intracellular signaling. Because of their spatial and structural relationships cellular populations are less amenable to the application of the mean field approximations and mass action kinetics that provide the basis for effective ordinary differential equation models. Equation-based modeling is generally considered to be the method of choice when dealing with interactions at the molecular level, where molecule populations can be considered to be well mixed and homogeneous. However, in circumstances where those approximations break down [such as in control of gene expression *(13)* or molecular crowding *(15)]*, ABM becomes a useful modeling option. The following sections will outline the steps in constructing an agent-based model, using as an example a previously published component agent-based model that represents relatively direct knowledge translation from an in vitro cell culture model to an agent-based model *(1)*.

2. Steps in the Development and Use of an Agent-Based Model: An Example Agent-Based Model of an In Vitro Model of Enterocyte Barrier

Many of the basic principles for developing a biomedical computational model are applicable to the construction and use of ABM. These steps typically involve (1) delineation of the system being modeled, (2) determination of the intended use of the model, and (3) the suitability of the modeling method in question to the answers to the aforementioned steps 1 and 2. As mentioned earlier, the structure of agent-based model facilitates its translational use for modeling both ontological structures and mechanistic information expressed in natural language, and as such is often more intuitive for nonmathematicians to grasp.

2.1. Example

An agent-based model based on an in vitro cell model is presented as an example of how knowledge generated from a basic science model/experiment can be effectively translated and dynamically represented. The ABM rule system focuses on particular molecular pathways in a specific cell type: tight junction protein metabolism and proinflammatory signaling as pertaining to gut epithelial barrier function seen in the enterocyte component of the gut. This model, then, will be called a gut epithelial barrier agent-based model (GEBABM). Calibration and validation follow the established pattern-oriented method well described for ABM *(2, 7, 16, 17)*. Pattern-oriented modeling suggests that models should be designed such that their properties and behaviors reflect those aspects of the system under study. Pattern-oriented modeling therefore consists of a "front-end" component: translating as directly as possible an accepted conceptual model of the mechanisms associated with enterocyte tight junction metabolism and inflammatory signaling, and a "back-end" component: comparing the behavior of the model with the in vitro reference model data.

2.2. What Is the Purpose of the Model?

This is a very basic, but often overlooked aspect of model construction. The question serves to remind us that models should not be created just because they can be; the justification of their development must be framed as to serve some particular purpose, even if that purpose is merely to demonstrate the capability of a particular type of model construction. Answering this question explicitly sets the groundwork for expectations with respect to interpretations of the model's output, and any conclusions that can be drawn from its behavior.

2.2.1. Example

In this case, the GEBABM is intended to serve two purposes. First, it is a method demonstration model to transparently illustrate the process of translating the basic science knowledge into an agent-based model. As such, it is a relatively direct and linear model, referenced to a tightly constrained in vitro preparation and therefore not expected or intended to vividly demonstrate the capacity of agent-based models to produce unexpected and paradoxical behavior. Second, the GEBABM is intended to be an example of a modular component in a cell-level, multiscale inflammatory modeling architecture *(18, 19)* (a more detailed description of this architecture is beyond the scope of this chapter).

2.3. What Is the Reference Model?

The reference model defines the informational basis of the agent-based model: its topology, the agents, and a starting point for identifying the literature-basis of the agent rules. Note that the reference model may be a particular experimental preparation (as is the case of the GEBABM) or an aggregated conceptual model in the mind of the researcher. If the latter is the case, then it is important to define as explicitly as possible the knowledge foundation of the conceptual model, particularly with respect to capabilities and limitations of the wet lab experiments that provide the basis of the conceptual model.

2.3.1. Example

The reference model for the GEBABM is a well-described human cultured enterocyte model (Caco-2) and its responses to inflammatory mediators including nitric oxide (NO) and a proinflammatory cytokine mix ("cytomix") that includes tumor necrosis factor (TNF), interleukin-1 (IL-1), and interferon-gamma (IFN-g) *(20–22)*. Integrating the information in these publications results in a conceptual model where enterocyte tight junction (TJ) proteins are involved in the integrity of gut epithelial barrier function, and where the production and localization of TJ proteins are impaired in a proinflammatory cytokine milieu.

2.4. What Is the Topology of the Model?

As mentioned earlier, many agent-based models are based on two-dimensional grids in which the edges "wrap" to form toruses. Other potential topologies include three-dimensional, cube-based structures and various network structures (such as scale-free, giant component or small-world configurations). In general, two-dimensional grids are sufficient to represent systems in which there is primarily one plane of agent interactions (along the surface of the grid), though it is possible to model multiple "layers" of data at a particular grid square [akin to the data structures of Geographical Information Systems (GIS)].

2.4.1. Example

The in vitro reference model consists of a monolayer of Caco-2 gut epithelial cells grown in a well with a chamber above the monolayer representing the luminal aspect of the enterocytes, and the chamber below representing the tissue interaction side of the enterocyte. Therefore, the GEBABM is modeled with a two-dimensional grid, with three "layers" per grid space: a central layer that holds the gut epithelial agent, one layer representing the apical extracellular space (from which the diffusate originates), and another layer representing the basal extracellular space (into which the diffusate flows if there is permeability failure).

2.5. What Are the Agents?

This is often the critical question and decision when constructing an agent-based model. Agents need to be a well-circumscribed group of components that can be treated as input–ouput devices (essentially finite state machines) following similar (if not identical) state-transition rules. The state of an agent is determined by a series of state variables internal to the agent, which are then modified based on external state variables in some spatially defined interaction environment for the agent. Therefore, selection of an agent level necessarily leads to some "compression of complexity" of the internal workings of the agent; an assumption implies that the informational basis of the input–output rules is valid irrespective of the particular mechanisms internal to the agent (the "black box" phenomenon). This is a critical point in determining the agent level. In general, the intended use of the model, vis-a-vis a planned intervention or particular targeted mechanism for study, will determine the resolution or *granularity* of the agent-based model. At the granularity chosen there needs to be a fairly certain linear approximation of mechanistic causality: i.e., how certain are you that state variable *a* goes to state variable *a*` with mechanism *b*? Explicit delineation of the granularity of the model and the corresponding assumptions are critical in avoiding *petito principii*, or "programming the proof." This can manifest as either treating the agent as a "black box" and focusing purely on its response as an input–ouput object, or, more commonly, with some degree of abstraction with respect to the progression of its internal state variables, such as by abstracting signaling and synthetic pathways.

2.5.1. Example

The GEBABM includes a single agent class that represents Caco-2 gut epithelial cells.

2.6. What Are the Agent Rules? Knowledge Translation and "Front-End" Validation

Once the agent level has been selected, attention is then turned to examination of the literature concerning the potential mechanisms to be modeled, and determining an interaction scheme between those mechanisms. It is often useful to express the latter goal in the form of a state or influence diagram that can be used to guide the actual coding of the agent-based model. The determination of the agent rules forms the primary translational step in ABM.

A qualitative approach is recommended at the outset in order to maintain a clear mapping between the basis of the reference/ conceptual model and the ABM computer code; too great an attention to specific details with respect to kinetic rate constants (for instance) in the initial translation phase can prove to be overwhelming in terms of how complicated the model appears to need to be. Therefore, it is useful to classify processes into relatively general groups of magnitude. For instance, with respect to determining the "time" it takes for a particular process it is usually sufficient to classify processes as "very fast (order of seconds)," "fast (order of minutes)," slow (order of hours)," and "very slow (order of days to weeks)." It should be acknowledged that this grouping is highly subjective; the overall subjectivity of the divisions is less important than (1) consistency within a particular agent-based model, (2) an awareness of the assumptions implicit upon the choice of the level of granularity, and (3) making this explicit and transparent when communicating the model.

2.6.1. Example

The GEBABM models the metabolism of TJ proteins, occludin, claudin-1, ZO-1, and ZO-3, involved in barrier function and their intersection with inflammatory signaling pathways. Activation of nuclear factor kappa-B (NF-kB) by proinflammatory cytokines leads to activation of inducible nitric oxide synthetase (iNOS). The nitric oxide (NO) produced inhibits synthesis of occludin, ZO-1, and ZO-3, while increasing production of claudin-1. Furthermore, NO impairs localization to the cell wall of synthesized occludin, claudin-1, and ZO-1. This appears to be due to the interference of NO with *N*-ethylmaleimide-sensitive factor (NSF), a molecule needed for localization of TJ proteins to the cell membrane *(23)*. These effects are seen with administration of both exogenous NO and intrinsic production of NO via the cytomix-NF-kB-iNOS pathway. These papers go on to investigate the effects of certain blocking agents. Addition of a NO scavenger *(22)* eliminates the effects of exogenous NO and cytomix. Administration of ethyl pyruvate *(20)* and nicotinamide adenine dinucleotide (NAD+) *(21)* both thought to inhibit NF-kB also attenuate the effects of cytomix. Data points for levels of NO, TJ protein expression and permeability were at 12, 24, and 48 h in all the experiments. **Figure 2** demonstrates a graphical representation of the general control logic underlying the agent rule systems based on the knowledge translated from the following references *(20–23)*.

2.7. Putting It Together: Programming the Model

There are two primary options when it comes time to write the code for an agent-based model: (1) a stand-alone program can be written in a basic computer language, such as C, or (2) a program can be written using an established ABM toolkit. Option 2 has the advantage that many of the programming underpinnings of ABM,

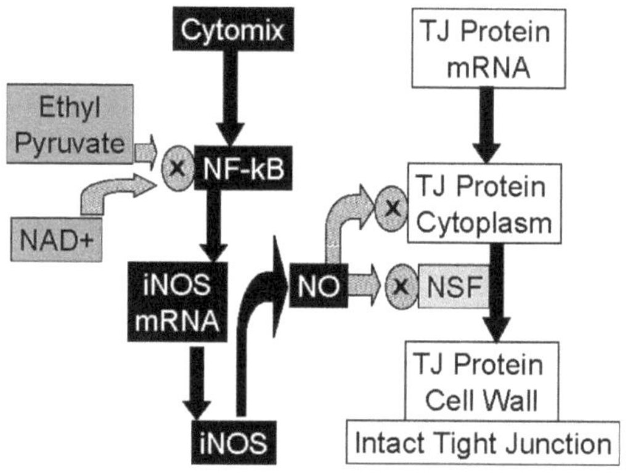

Fig. 2. Graphical representation of the control logic extracted from the basic science *(20,22,23)* on gut epithelial barrier function. General flowchart of the components and mechanisms of TJ protein synthesis and localization, the effects of proinflammatory stimulation, and the effects of interventions with ethyl pyruvate and NAD+. All labeled boxes correspond to agent or environment state variables within the GEBABM. In the actual code of the GEBABM there are distinct pathways for the different TJ proteins (not shown here for clarity purposes). This figure is reproduced with the author's permission from **ref.** *1* under the terms of Creative Commons License.

such as object-class definition, emulated parallelization, creation of a graphical user interface, and data collection tools, are not trivial programming tasks, and having these issues preaddressed in an established ABM toolkit allows a researcher to focus on the modeling aspect of the project rather than on the programming aspect. A list of available ABM toolkits/modeling environments can be seen in **Subheading 4**. Since ABM is a discrete event modeling system and the program progresses in a stepwise fashion, there must be a selection of the base time interval for each step. This selection is based upon the qualitative process-time course determined in the previous **Subheading 2.5**. Mechanisms to be translated into agent rules are broken into steps based on the duration of those mechanisms, and further translated into program code. It is important to keep in mind that a particular code block will run sequentially (even in an emulated parallel environment); therefore, the order or *schedule* of process events needs to free of inadvertent internal feedback loops. For instance, if a particular agent has a rule where it produces an external state variable that in turn affects the agent's subsequent production of that same external state variable, then placing the production code at the beginning of the code block will lead to an artifactual enhancement of any forward feedback effects of that particular rule.

2.7.1. Example

The GEBABM was constructed using the freeware software toolkit Netlogo *(24)*. The entire model, along with extensive documentation, is available on the Netlogo Community Models Website (http://ccl.northwestern.edu/netlogo/models/community/Shock2004_Gut_Epithelial_Barrier). The code for the model is included in **Subheading 3.5**. The GEBABM is a two-dimensional square grid, 21 × 21 cells, in each of which there is a gut epithelial cell agent ("epi-cell"). The size of this grid was arbitrarily chosen. A screenshot of the GEBABM during an experimental run can be seen in **Fig. 3**. Each epi-cell has eight immediate neighbors, and at each contact point there is a simulated tight junction (TJ). The integrity of the TJ requires both epi-cells opposite to have adequate production and localization of TJ proteins. The epi-cell agent class contains variables that represent the precursors, cytoplasmic levels, and cell wall levels of the TJ proteins, as well as intracellular levels of activated NF-kB and iNOS mRNA. Furthermore, there are "milieu" variables that represent NO, cytomix, and the diffusate. Algorithmic commands were written for the synthesis of TJ proteins as well as the pathway

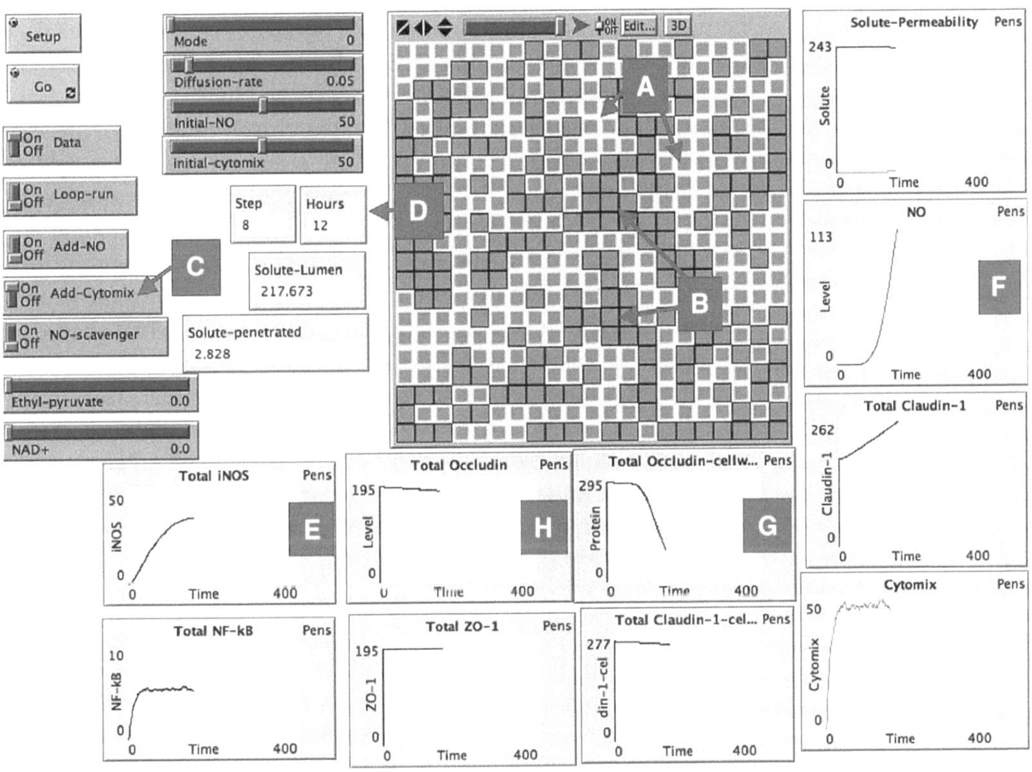

Fig. 3. Screen shot of the graphical user interface of the GEBABM. Control buttons are on the left; graphical output of the simulation is in the center. Graphs of variables corresponding to levels of mediators and tight junction proteins are at the bottom and right. In the graphical output Caco-2 agents are seen as *squares*; those with intact tight junctions are bordered in light color (letter A); those with failed tight junctions are bordered in dark color (letter B). This particular run is with the addition of cytomix (letter C), seen after 12 h of incubation (letter D). The heterogeneous pattern of tight junction failure can be seen in the graphical output. Levels of Caco-2 iNOS activation can be seen in graph's letter E, and produced nitric oxide (NO) can be seen in graph's letter F. Of note, the total amount of tight junction protein occludin does decrease slightly (graph's letter G), but the amount of occludin localized in the cell membrane drops much more rapidly (graph's letter H), reflecting the impairment of occludin transport due to NO interference with NSF and subsequent loss of tight junction integrity. This figure is reproduced with the author's permission from **ref.** *1* under the terms of Creative Commons License.

for NO induction (*see* **Subheading 3.5**). The time courses for these processes are primarily in the minutes to hours range, and therefore the program iterates with each step representing 5 min of simulated time. Since the reference data sets extend out to 48 h of observation, the simulation runs will terminate at that period of simulated time.

2.8. Calibration

Once the general control logic of the rule systems has been extracted from the reference texts, then the specifics of the rule algorithms need to be determined. Thus far, there has been a primarily qualitative translation of the mechanistic hypotheses derived from the reference literature into an abstracted influence diagram (**Fig. 2**) and then into conditional statements within the computer code (*see* **Subheading 3.5**). Running the agent-based model at this point will generate a set of behaviors that may have some qualitative utility, but in all likelihood will not be able to be matched to experimental data. Therefore, the qualitative representation of the modeled mechanisms must be calibrated to existing experimental reference data to produce, at least, a semi-quantitative model that can be more closely linked to the real world. This is done by "tuning" the stepwise rules that update the various state variables, usually via the addition of various constants and/or adjusting the algebraic relationships between the variables. Since these rules can be considered as computational equivalents to difference equations, changing a constant is akin to changing the slope of a particular kinetic curve, while changing the algebraic relationship from summation to multiplication will change the kinetic curve from linear to exponential. Care must be taken, however, not to change the actual variables associated with each rule in order to get a better "fit." Doing so constitutes rewriting the underlying knowledge representation of the agent-based model to match the observable; resorting to this in order to match real world observables is to deny the verity of the conceptual model being represented. Calibration, then, is done to establish the fidelity of baseline behavior of the model compared with the real-world data in order for additional interventions to be simulated. Inability to effectively calibrate a model at this point suggests an intrinsic flaw in the underlying conceptual model. A common challenge to calibration is the lack of sufficient experimental data against which to "fit" the model; not enough reference points exist to refine the agent-based model to a particular level of confidence. In these cases one must fall back upon qualitative interpretation of the agent-based model's behavior, making sure to be explicit with respect to that limitation when conclusions are drawn and communicated. Another challenge to calibration, which is actually more common in equation-based models, is overfitting the model to data, leading to "brittle" models with little applicability to additional conditions. This problem is relatively rare with

ABM, since due to the parallelism of agent-based models the direct predictability of the effect of rule adjustment to model output is less direct.

2.8.1. Example

Calibration of the GEBABM was done at three command points each with a different data set. The first calibration was for the basal diffusion rate. The diffusion coefficient in the unperturbed system was adjusted to match the rate of diffusion in the reference data set at times 12, 24, and 48 h. This established the baseline control permeability. The second calibration was done to reproduce the levels of administered cytomix and NO. The reference data sets were the levels of measured NO in both the exogenous NO donor arm and the cytomix administration arm (as seen in **Fig. 1** from **ref.** *22*). Calibration occurred by modifying the coefficients of the NO induction pathway algorithm. The third calibration was done with respect to the TJ protein synthesis/breakdown algorithms. Steady state TJ protein levels were established using the inhibition data extrapolated from the western blot results from **ref.** *22*. In silico experiments were run using these interventions with data points at 12, 24, and 48 h as per the reference papers. Data collection looked at permeability reflecting TJ integrity, levels of TJ proteins, and localization of TJ proteins. The results of the calibration runs of the GEBABM can be seen in **Figs. 4** and **5**. Note that the values of the in silico experiments are unitless, but the results qualitatively mirror the reference data set. Both of these figures include runs with exogenous NO, cytomix, and cytomix in the presence of a NO scavenger. The NO scavenger was simply modeled by reducing the level of the NO milieu variable after production. **Figure 4** demonstrates the calibrated levels of NO production, while **Fig. 5** demonstrates the permeability calibration results. These figures essentially reproduce the data generated in **ref.** *22*. These three levels of calibration established the baseline GEBABM. Note that this includes the GEBABM perturbed with both NO and cytomix. The next step is to perform "back-end" validation through the simulation of additional experimental interventions – ethyl pyruvate and NAD +.

2.9. "Back-End" Validation: Making Predictions with In Silico Experiments

By now it should be evident that agent-based models are relatively "complex" models, in so much that they derive a great deal of their behavior through parallel interactions that result in behaviors and dynamics that often cannot be completely described via a series of equations that can be subjected to formal mathematical analysis or proof. There is a paradox in that the need to build increasingly complex models to effectively represent complex systems results in models that may be too complex for analysis, or even to formally validate. With respect to ABM, pattern-oriented modeling has been proposed as a solution to this problem *(16, 17)*. Pattern-oriented

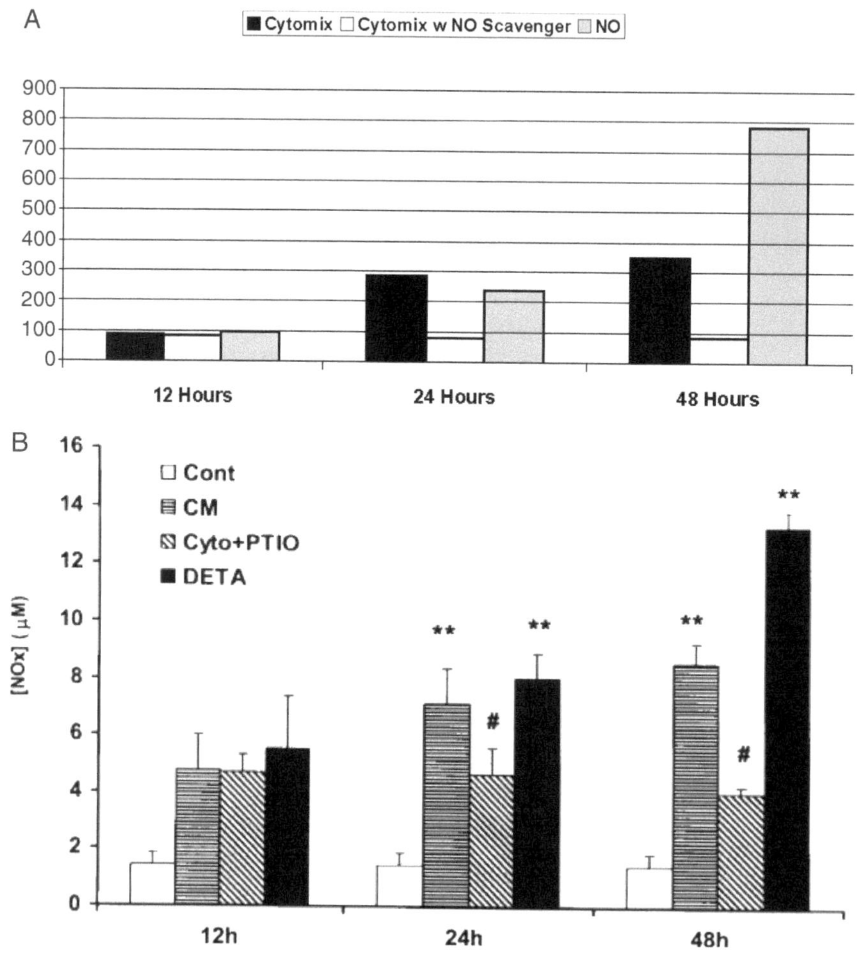

Fig. 4. Simulated nitrogen oxide (NO) production and response to NO scavenger. (**A**) Calibration data are seen in the black bars (cytomix) and the gray bars (NO) with respect to simulation rules for NO production. The NO data match the literature-reported levels of exogenous NO added in the experiments from **ref.** *22* in order to establish baseline responses of the epi-cell agent's TJ protein synthesis/localization algorithms and link them to the permeability data seen in the corresponding bars in **Fig. 5**. The Cytomix bars in panel (**A**) are used to calibrate the iNOS-NO production algorithms within the epi-cell agents. The middle data set (bars = cytomix + NO scavenger) shows the effect of exogenous NO reduction/elimination on the generated levels of NO in the face of cytomix. Panel (**B**) shows the literature-reported data from the upper portion of **Fig. 1** from **ref.** 22 (reproduced with permission from Lippincott Williams & Wilkins, © 2003). This demonstrates levels of NO in the control ("Cont"), cytomix of proinflammatory cytokines ("CM"), cytomix with the addition of a NO scavenger ("Cyto + PTIO") and with a free NO donor ("DETA"). Panel (**A**) of this figure is reproduced with the author's permission from **ref.** *1* under the terms of Common Creative License.

modeling has already been utilized in both the transparent translation of biomedical knowledge into agent rules ("front-end" validation) and in the calibration process in order to "tune" the model. However, the true test of the validity of a model is its ability to *predict* a behavior that has not already been used in the construction of the model. This is accomplished by performing "in silico" experiments in a fashion similar to performing experiments in the basic science lab: a particular intervention is planned based on a par-

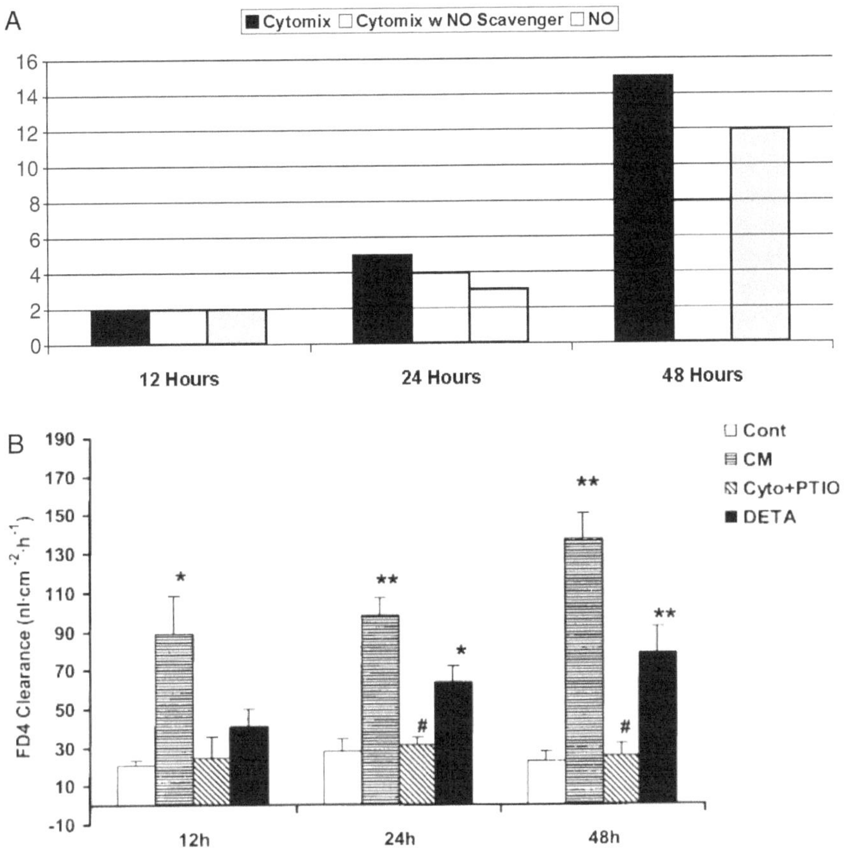

Fig. 5. Simulated permeability to NO, cytomix, and cytomix + NO scavenger. (**A**) Graph of calibration data of the permeability effects of NO and cytomix, representing the diffusion rate through a failed epithelial barrier and the effect of NO on the algorithms for epi-cell TJ protein synthesis/localization. As with **Fig. 4**, the black bars (cytomix) and gray bars (exogenous NO) are the calibration arms. This graph can be compared with panel (**B**), which is the lower panel of **Fig. 1** in **ref.** *22* (reproduced with permission from Lippincott Williams & Wilkins, © 2003). Panel (**B**) demonstrates permeability under the following conditions: control ("Cont"), cytomix of proinflammatory cytokines ("CM"), cytomix with the addition of a NO scavenger ("Cyto + PTIO") and with a free NO donor ("DETA"). Panel (**A**) of this figure is reproduced with the author's permission from **ref.** *1* under the terms of Creative Commons License

ticular mechanism of action within the experimental model, and the behavior of the model after the intervention is examined to see if a significant difference in the model's behavior arises. Agent-based models, as relatively direct translations of basic science hypotheses, can be treated as experimental templates in a similar fashion. Programming a particular intervention and its effect on the existing code represents the translation of a particular conceptual model of how that mechanism will affect the system as a whole. Notably, it does not (necessarily) address issues as to whether the physical compound that would be used in the real-world lab actually does what it is intended to do; rather the agent-based model tests the conceptual basis or justification for why that intervention should work. Therefore, ABM can be considered a "test of proof-of-concept" for a particular experiment. Matching the output of a proposed mechanism of intervention instantiated in the agent-based model with the results of a similarly designed wet lab experiment is thus evidence of the predictive capacity and robustness of the agent-based model, and enhances its claim toward being a valid model.

2.9.1. Example

As mentioned earlier, the reference papers *(20, 21, 25)* suggest that administration of both ethyl pyruvate *(20)* and nicotinamide adenine dinucleotide (NAD+) *(21)* inhibits NF-kB as a mechanism for their attenuation of the effects of cytomix. Of note, neither of these compounds or their presumptive effects was included in the development of the GEBABM. For the in silico experiments both NAD+ and ethyl pyruvate were modeled using their presumptive mechanisms of NF-kB inhibition by their insertion as negative influences in the NO induction pathway algorithm. No further modifications were done to the internal metabolism algorithms of the epi-cell class. For the complete code *see* **Subheading 3.5**.

In-silico experiments were run using these interventions with data points at 12, 24, and 48 h as per the reference papers. Data collection looked at permeability reflecting TJ integrity, levels of TJ proteins, and localization of TJ proteins. The results of these in silico interventions on the GEBABM can be seen in **Figs. 6–8**. Again, note that the values of the in silico experiments are unitless, but the results qualitatively mirror the reference data set. **Figure 6** demonstrates the effects of ethyl pyruvate and NAD+ on permeability, with the data in **Fig. 5** representing the control arm. The reference data for the effect of these interventions on the permeability changes with cytomix administration can be seen in **Fig. 1** from **ref.** *20* with ethyl pyruvate at 1.0-mM dose, and **Fig. 1A** from **ref.** *21* with NAD+ at 100-mcM dose. **Figures** 7 and **8** reproduce the results seen extrapolated from the western blot data on the effect of ethyl pyruvate and NAD+ administration on TJ proteins, specifically ZO-1 and occludin (**Fig. 6** from **ref.** *20* and **Fig. 2** from **ref.** *21*). ZO-1 is significantly decreased at 48 h, while occludin starts to drop at 24 h with the cytomix and continues to decrease at 48 h, but has a profile more similar to ZO-1 when run with the exogenous NO only. The simulation of adding both ethyl pyruvate and NAD+ obviated the effects of both exogenous NO and cytomix on both ZO-1 and occludin.

3. Notes

Many of the points described later have been alluded to in the preceding text. However, the following sections provide summaries of the key aspects of ABM with respect to design and utilization.

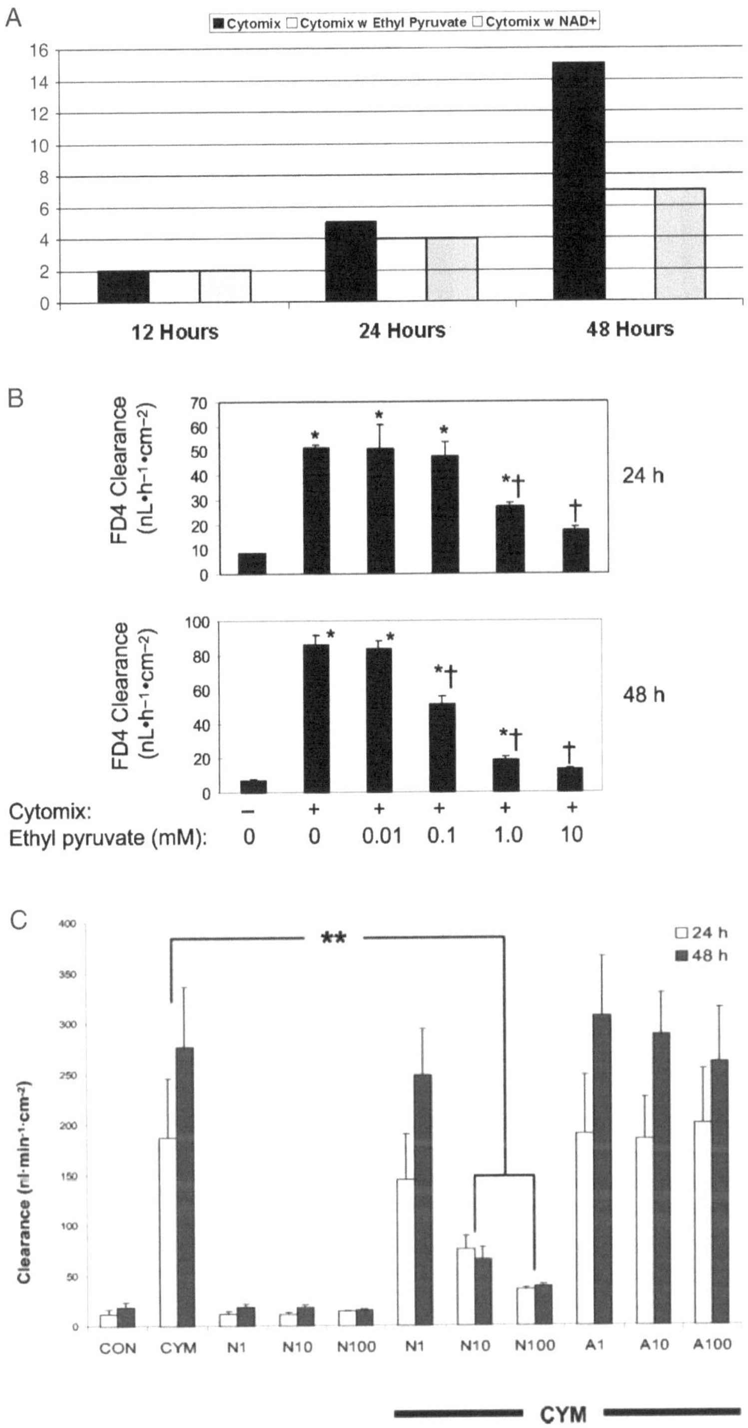

Fig. 6. Simulated permeability effects of ethyl pyruvate and NAD+ compared to literature-reported experimental experiments. Graph (**A**) demonstrating the effects of simulated addition of ethyl pyruvate and NAD+ on the proinflammatory algorithms within the epi-cell agents. Both of these substances interfere with NF-kB localization, and therefore are "upstream" from the iNOS-NO pathways as represented in those rules. This graph can be compared to panel (**B**): **Fig. 1** from **ref.** *20* with ethyl pyruvate at 1.0-mM dose, and panel (**C**): **Fig. 1a** from **ref.** *21* with NAD+ at 100-mcM dose ("CYM" = addition of cytomix). Panel (**A**) of this figure is reproduced with the author's permission from **ref.** *1* under the terms of Creative Commons License, and panels (**B**, **C**) are reproduced with permission from the American Society for Pharmacology and Experimental Therapeutics, © 2003.

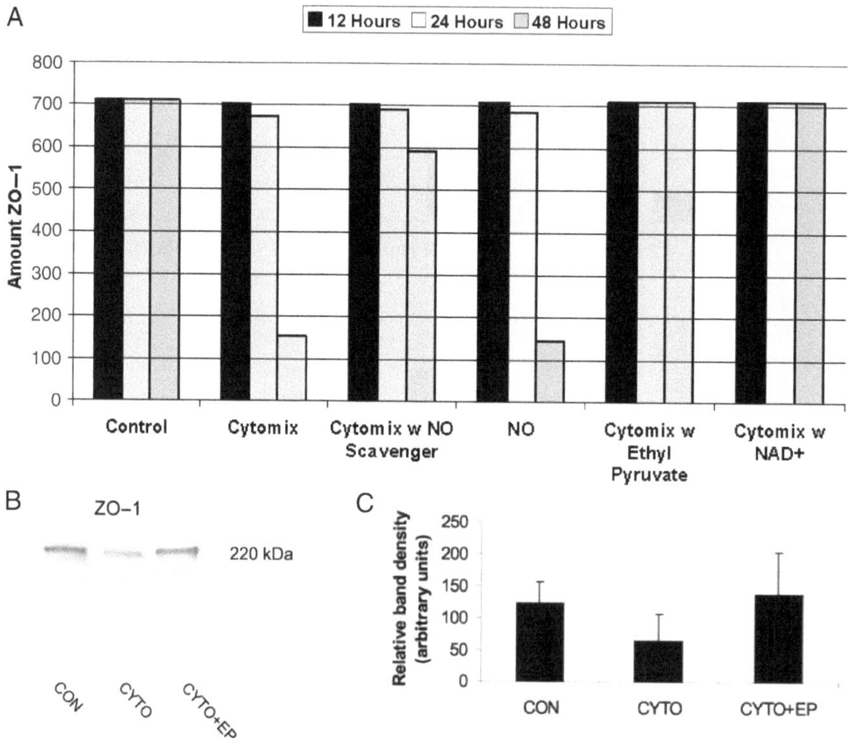

Fig. 7. Simulated levels of ZO-1 expression. (**A**) Graph demonstrating the levels of simulated ZO-1 expression in control, exogenous NO, cytomix, cytomix with NO scavenger, cytomix with ethyl pyruvate, and cytomix with NAD+ at 12, 24, and 48 h. Compare with (**B**): **Fig.** 6a from **ref.** *20* (reproduced with permission from the American Society for Pharmacology and Experimental Therapeutics, © 2003). Also compare with (**C**): data extrapolated from western blot analysis seen in **Fig. 2** from **ref.** *21*. Panel (**A**) of this figure is reproduced with the author's permission from ref. *1* under the terms of Creative Commons License.

3.1. Deciding When to Use ABM: Strengths and Weaknesses of ABM

The strengths of ABM as a modeling method have been emphasized in the preceding text: intuitive structure facilitating knowledge translation and representation, intrinsic management of spatial issues, the ability to capture complex behavior, and similarity in behavior and output to traditional wet lab experiments. These benefits would seem to suggest a fairly generous application of ABM in the biomedical arena. However, there are significant limitations to ABM, as with all modeling methods. These include high computational requirements for large-scale models, inability to "formally" analyze the inner workings of the model, difficulty in calibration due to the nonlinear relationships between agent rules and behavior, and difficulty in matching a specific run of a model's evolving conditions with a real-world reference (such as the case of attempting to predict the outcome of a specific patient). In the discussion of when one should use ABM it may be more useful to determine those instances where ABM does not suit the modeling problem at hand and the limitations listed aforementioned factor into that determination. These instances

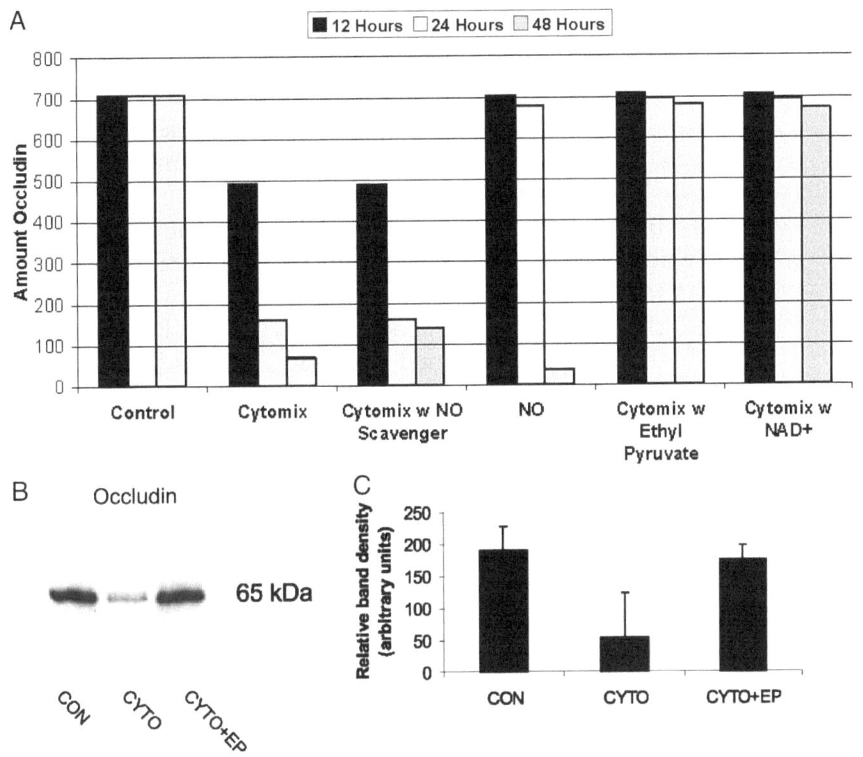

Fig. 8. Simulated level of occludin expression. (**A**) Graph demonstrating the levels of simulated occludin expression in control, exogenous NO, cytomix, cytomix with NO scavenger, cytomix with ethyl pyruvate, and cytomix with NAD+ at 12, 24, and 48 h. Compare with (**B**): **Fig. 6b** from **ref.** *20* (reproduced with permission from the American Society for Pharmacology and Experimental Therapeutics, © 2003). Also compare with (**C**): data extrapolated from western blot analysis seen in **Fig. 2** from **ref.** *21*. Panel (**A**) of this figure is reproduced with the author's permission from **ref.** *1* under terms of the Creative Commons License.

can mostly be expressed in terms of the suitability of using equation-based modeling, which remains the default method of mathematical modeling of dynamic system behavior:

1. Ordinary differential equation (ODE) modeling is preferable if the system can be characterized by well-mixed compartments/populations. In these situations mean field approximations will hold, and mass action kinetics approaches can be utilized.
2. Equation-based modeling is preferable if it appears possible to derive "formal" insights into the system's behavior. Systems whose behavior can be characterized with relatively simple order ordinary differential equations may be analyzed mathematically, leading to more comprehensive and general understanding of their dynamics.
3. Equation-based modeling is preferable if the development and calibration reference data already exists in equation form. This is particularly true when calibration is to be performed using an optimization algorithm. Again, in this situation the

nonlinearities between rule modification and model behavior make this extremely challenging.

4. Equation-based modeling is preferable when the number of agents needed to be modeled is extremely high. The computational demands of ABM when the number of agents reaches millions and billions are prohibitive.

In general, equation-based modeling remains the initial approach to modeling dynamic systems (including biological ones) and is well suited to modeling processes such as molecular kinetics, biochemical reactions, and gross physiologic behavior. However, if the criteria mentioned earlier do not hold, ABM can offer an advantageous modeling approach.

3.2. Construction Pitfalls

Most of the construction pitfalls have been mentioned earlier in **Subheading 2.1** (identifying the purpose of the agent-based model), **Subheading 2.2** (linking the agent-based model to its reference experimental/conceptual model), **Subheading 2.5** (agent level/class selection), and **Subheading 2.6** (agent rule determination). When trying to address these challenges, it is recommended to keep two goals in mind:

1. Minimize assumptions. It is important to remember the old computer programming adage: "Garbage in, garbage out." Agent rules should be tied as closely as possible to causal mechanisms defined by experiment. When generating state or influence diagrams, each individual step should be identified, and if there is not sufficient data or an intermediate step is unknown, this should be explicitly noted (see later). Recognize that each assumption made raises the risk of *petito principii.*

2. Be explicit, in both defining rules and their underlying justification and assumptions involved. Not only does this avoid the programming issues noted earlier, it also rigorously forces an objective assessment of a particular conceptual model. Many researchers are unpleasantly surprised when concepts that are taken as "given" are dissected in this fashion. This process is, of itself, beneficial in determining where the next set of wet lab experiments may need to be done. But it is also critical in being able to assess why a model has been "broken." An explicit representation of the underpinnings of a model provides a guide to where adjustments need to be made for improvement in the next generation. Finally, explicitness in model description is a vital component of being able to effectively communicate a model to other researchers. Transparent explicitness facilitates the evaluation, understanding, and eventual acceptance of a particular model in the research community at large.

3.3. Interpretation Pitfalls

Perhaps the greatest danger in interpreting the behavior of an agent-based model (or any model, for that matter) is assuming that the model represents some sort of objective truth. In other words, a model that appears to match the behavior seen in the real world is possibly only one of many *plausible* models that fit the data. Therein lies the key to using ABM as a test of the verity of a conceptual model: only a negative result can provide definitive information, i.e., the conceptual model is incorrect, whereas a positive result can only suggest that the conceptual model *may* be correct. The goal of model interpretation, then, is to develop models that, when broken, can provide some insight as to why they are broken, and the process of model utilization is to sequentially generate models, break them through falsification, and use that information to generate the next model.

3.4. Suggested Applications

Given the strengths and weaknesses of ABM as a method, it is possible to give some general suggestions for situations in which ABM can be effectively utilized. In addition to situations that do not suit equation-based modeling (*see* **Subheading 3.2**), the following areas deserve mention:

1. ABM is suited to modeling the behavior of cellular populations. As mentioned earlier, biology has provided a natural agent level in cells. Cells exhibit many of the preferable characteristics of a good agent level: there are readily defined classes; their aggregate behavior is relatively accessible for measurement; their behavior can be characterized in population-derived probability functions, and a great many conceptual models of the biological behavior are derived at the cellular level. Cells occupy the "middle ground" that appears to be an optimal resolution for effective modeling of biomedical systems *(8, 15, 16, 26, 27)*, and are suited as a translational level at the center of multiscale models *(27–29)*. Therefore, projects that involve representing cellular populations are suited to modeling with ABM.
2. *Multiscale problems.* Two aspects of ABM facilitate multiscale modeling. The first is their ability to generate emergent behavior, and therefore translate the complexity of structure and mechanism at a lower level into the behavior at a higher one. Second, agent-based models are intrinsically modular, so much that they can be organized and combined based on their spatial architecture and can communicate via commonalities in their constituent agent classes. For an example, see the multiscale, multiorgan architectures described in *(18, 19)*, of which the GEBABM is a component. However, it bears noting that due to computational limitations "pure" agent-based models of true multiscale processes are unlikely. Rather, drawing on the suitability of certain methods to modeling certain levels of biological organization *(28)* these multiscale models will almost certainly be "hybrid" models incorporating multiple different modeling methods *(15, 30)*.
3. *Biomedical ontology representation and evolution.* One of the most exciting possible applications for ABM architecture is

in being able to map directly to biomedical ontologies in an automated fashion, and thereby enhance the use, communication, and evolution of biomedical ontologies. It is in this area that meta-ABM methods hold great promise. Meta-ABM methods are attempts to create a general modeling interface that consists of representing the "essential" components of an agent-based model and allowing the metarepresentation to be instantiated via a series of different ABM toolkits. There is a marked similarity between this goal and the concept of a general-purpose biomedical ontology: both need to be general and robust enough to fit yet unspecified systems, but need to be specific enough to be of practical use. The likelihood is that a perfectly general ontology does not exist; however, as with the scientific process in general the important goal is to develop a developmental structure that facilitates an evolutionary process. Since the structure of ABM maps so well to ontological structure it would seem a natural fit for evolving meta-ABM architectures and their ability to instantiate biomedical ontologies to provide a means of conceptual model verification that will form the basis for the selection pressure on competing ontologies within the biomedical community.

3.5. Code for the GEBABM

The following code is in Netlogo language, a programming language specific to Netlogo. Of note, copying this code directly into a blank Netlogo file will not result in a functioning model, as Interface control commands are expressed directly through the graphical user interface. This code is made available so that interested individuals can examine the specific logic in the model and see the relationship between the code and the interpreted knowledge from the reference papers. This is done in the interest of complete transparency, as the interpretation of the knowledge into the code is inevitably subjective to some degree, and this degree of transparency is necessary for the appropriate evaluation and acceptance (either yea or nay) of the validity of the modeling assumptions. The entire model can be accessed and downloaded from http://ccl.northwestern.edu/netlogo/models/community/Shock2004_Gut_Epithelial_Barrier.

4. Software Resources

It is recommended to use existing general purpose ABM toolkits for at least the initial attempts in constructing a biomedical agent-based model. Even if the researcher/modeler is an experienced computer programmer and wants to write a special purpose agent-based model, developing a familiarity with existing methods of implementing an agent-based model will aid in addressing some specific programming challenges involved (i.e., object class definition, emulated parallel actions, spatial topography development, etc). One of those toolkits, Netlogo *(24)*, was used to create the GEBABM. Netlogo was originally designed to teach primary and secondary school students the dynamics of complex systems such as bird flocking, fish schooling, traffic, and ant colony behavior, and as such is very amenable to novices to computer programming. It has subsequently evolved into a very powerful modeling environment, particularly for developing the qualitative/semiquantitative knowledge translation models described in this chapter. Netlogo is freely available for download at http://ccl.northwestern.edu/netlogo/, and is available for Windows, Macintosh, and Linux. The GEBABM itself is available for download at http://ccl.northwestern.edu/netlogo/models/community/Shock2004_Gut_Epithelial_Barrier. Netlogo is also closely related to another introductory ABM toolkit called Starlogo, which is available for download at http://education.mit.edu/starlogo/.

Perhaps the prototypical ABM development tool is Swarm, an open source ABM platform originally developed at the Santa Fe Institute. Information on Swarm, and the ABM community in general, is available at http://www.swarm.org/wiki. Swarm spawned a series of open source ABM toolkits, such as Repast (http://repast.sourceforge.net/index.html), JAS (http://jaslibrary.sourceforge.net/index.html), Ascape (http://ascape.sourceforge.net/), and Mason (http://cs.gmu.edu/~eclab/projects/mason/), all of which have their own active development communities (a more comprehensive list of resources can be found on the Swarmwiki site). Finally, there is a strong trend toward "metamodeling" in the ABM community, reflected by the recent release of meta-ABM (http://www.metascapeabm.com/), and reflected in an active research and development community that can be followed at http://www.openabm.org/site/.

References

1. An, G. (2008) Introduction of an agent-based multi-scale modular architecture for dynamic knowledge representation of acute inflammation. *Theor. Biol. Med. Model.* 5, 11.
2. An, G. (2004) In silico experiments of existing and hypothetical cytokine-directed clinical trials using agent-based modeling. *Crit. Care Med.* 32, 2050–2060.
3. An, G. (2006) Concepts for developing a collaborative in silico model of the acute inflammatory response using agent-based modeling. *J. Crit. Care* 21, 105–110; discussion 110–111.
4. Zhang, L., Athale, C. A., and Deisboeck, T. S. (2007) Development of a three-dimensional multiscale agent-based tumor model: simulating gene–protein interaction profiles, cell phenotypes and multicellular patterns in brain cancer. *J. Theor. Biol.* 244, 96–107.
5. Bonabeau, E. (2002) Agent-based modeling: methods and techniques for simulating human systems. *Proc. Natl. Acad. Sci. USA* 99 Suppl 3, 7280–7287.
6. Bankes, S. C. (2002) Agent-based modeling: a revolution? *Proc. Natl. Acad. Sci. USA* 99 Suppl 3, 7199–7200.

7. An, G. (2001) Agent-based computer simulation and sirs: building a bridge between basic science and clinical trials. *Shock* 16, 266–273.
8. Thorne, B. C., Bailey, A. M., and Peirce, S. M. (2007) Combining experiments with multi-cell agent-based modeling to study biological tissue patterning. *Brief. Bioinform.* 8, 245–257.
9. Mansury, Y., Diggory, M., and Deisboeck, T. S. (2006) Evolutionary game theory in an agent-based brain tumor model: exploring the 'genotype–phenotype' link. *J. Theor. Biol.* 238, 146–156.
10. Bailey, A. M., Thorne, B. C., and Peirce, S. M. (2007) Multi-cell agent-based simulation of the microvasculature to study the dynamics of circulating inflammatory cell trafficking. *Ann. Biomed. Eng.* 35, 916–936.
11. Tang, J., Ley, K. F., and Hunt, C. A. (2007) Dynamics of in silico leukocyte rolling, activation, and adhesion. *BMC Syst. Biol.* 1, 14.
12. Walker, D. C., Hill, G., Wood, S. M., Smallwood, R. H., and Southgate, J. (2004) Agent-based computational modeling of wounded epithelial cell monolayers. *IEEE Trans. Nanobiosci.* 3, 153–163.
13. Pogson, M., Smallwood, R., Qwarnstrom, E., and Holcombe, M. (2006) Formal agent-based modelling of intracellular chemical interactions. *Biosystems* 85, 37–45.
14. Broderick, G., Ru'aini, M., Chan, E., and Ellison, M. J. (2005) A life-like virtual cell membrane using discrete automata. *In Silico Biol.* 5, 163–178.
15. Ridgway, D., Broderick, G., and Ellison, M. J. (2006) Accommodating space, time and randomness in network simulation. *Curr. Opin. Biotechnol.* 17, 493–498.
16. Grimm, V., Revilla, E., Berger, U., Jeltsch, F., Mooij, W., Railsback, S., Thulke, H.-H., Weiner, J., and Wiegand, T. (2005) Pattern-oriented modeling of agent-based complex systems: lessons from ecology. *Science* 310, 987–991.
17. Grimm, V. and Railsback, S. F. (2005) Individual-Based Modeling and Ecology. Princeton University Press, Princeton, NJ.
18. An, G. (2005) Multi-hierarchical agent-based modeling of the inflammatory aspects of the gut. *J. Crit. Care* 20, 383.
19. An, G. (2006) Integrative modeling of inflammation and organ function using agent based modeling. *Shock* 26, 2.
20. Sappington, P. L., Han, X., Yang, R., Delude, R. L., and Fink, M. P. (2003) Ethyl pyruvate ameliorates intestinal epithelial barrier dysfunction in endotoxemic mice and immunostimulated caco-2 enterocytic monolayers. *J. Pharmacol. Exp. Ther.* 304, 464–476.
21. Han, X., Uchiyama, T., Sappington, P. L., Yaguchi, A., Yang, R., Fink, M. P., and Delude, R. L. (2003) NAD+ ameliorates inflammation-induced epithelial barrier dysfunction in cultured enterocytes and mouse ileal mucosa. *J. Pharmacol. Exp. Ther.* 307, 443–449.
22. Han, X., Fink, M. P., and Delude, R. L. (2003) Proinflammatory cytokines cause NO*-dependent and -independent changes in expression and localization of tight junction proteins in intestinal epithelial cells. *Shock* 19, 229–237.
23. Matsushita, K., Morrell, C. N., Cambien, B., Yang, S. X., Yamakuchi, M., Bao, C., Hara, M. R., Quick, R. A., Cao, W., O'Rourke, B., Lowenstein, J. M., Pevsner, J., Wagner, D. D., and Lowenstein, C. J. (2003) Nitric oxide regulates exocytosis by *S*-nitrosylation of *N*-ethylmaleimide-sensitive factor. *Cell* 115, 139–150.
24. Wilensky, U. (1999) *NetLogo: http://ccl.northwestern.edu/netlogo.* Center for Connected Learning and Computer-Based Modeling of Northwestern University, Evanston, IL.
25. Han, Y., Englert, J. A., Yang, R., Delude, R. L., and Fink, M. P. (2005) Ethyl pyruvate inhibits nuclear factor-kappaB-dependent signaling by directly targeting p65. *J. Pharmacol. Exp. Ther.* 312, 1097–1105.
26. Tang, J., Hunt, C. A., Mellein, J., and Ley, K. (2004) Simulating leukocyte-venule interactions – a novel agent-oriented approach. *Conf. Proc. IEEE Eng. Med. Biol. Soc.* 7, 4978–4981.
27. Hunt, C. A., Ropella, G. E., Yan, L., Hung, D. Y., and Roberts, M. S. (2006), Physiologically based synthetic models of hepatic disposition. *J. Pharmacokinet. Pharmacodyn.* 33, 737–772.
28. Kirschner, D. E., Chang, S. T., Riggs, T. W., Perry, N., and Linderman, J. J. (2007) Toward a multiscale model of antigen presentation in immunity. *Immunol. Rev.* 216, 93–118.
29. Yan, L., Hunt, C. A., Ropella, G. E., and Roberts, M. S. (2004) In silico representation of the liver-connecting function to anatomy, physiology and heterogeneous microenvironments. *Conf. Proc. IEEE Eng. Med. Biol. Soc.* 2, 853–856.
30. Wakeland, W., Macovsky, L., and An, G. (2007) A hybrid simulation for studying the acute inflammatory response. *Proc. 2007 Spring Simulat. Multiconf. (Agent Directed Simulation Symposium)* 1, 39–46.

Chapter 16

Systems Biology of Microbial Communities

Ali Navid, Cheol-Min Ghim, Andrew T. Fenley, Sooyeon Yoon, Sungmin Lee, and Eivind Almaas

Summary

Microbes exist naturally in a wide range of environments in communities where their interactions are significant, spanning the extremes of high acidity and high temperature environments to soil and the ocean. We present a practical discussion of three different approaches for modeling microbial communities: rate equations, individual-based modeling, and population dynamics. We illustrate the approaches with detailed examples. Each approach is best fit to different levels of system representation, and they have different needs for detailed biological input. Thus, this set of approaches is able to address the operation and function of microbial communities on a wide range of organizational levels.

Key words: Microbial community, Rate equation, Agent-based modeling, Population dynamics, Quorum sensing, Biofilm.

1. Introduction

Microorganisms contribute a considerable fraction of the living biomass on Earth. While traditional studies of microbes have been based on the isolation and laboratory cultivation of pure species, relatively little is known about an estimated >99% of environmental microbes due to their difficulty of cultivation under standard laboratory conditions. In fact, the vast majority of microbes naturally only occurs and thrives when in microbial *communities*: there is frequently a synergistic partitioning of metabolic function between different microbial species *(1)*. The recent development of techniques to probe microorganisms in their natural environments, such as metagenomic sequencing, has uncovered an unanticipated

Ivan V. Maly (ed.) *Methods in Molecular Biology, Systems Biology, vol. 500*

DOI: 10.1007/978-1-59745-525-1_16

level of phylogenetic diversity and valuable insights into lifestyle and metabolic capabilities of microbial communities occupying a broad range of environmental niches *(2–4)*.

The function and operation of microbial communities has received significant interest with the introduction of these new technologies. There is also a growing realization that microbes contribute extensively to important environmental questions such as carbon sequestration and nitrogen cycling. It has been proposed that microbes and microbial communities may provide novel avenues for the degradation of lignocellulosic material and, thus, the generation of biofuels. Recently, new findings indicate that the activity and composition of microbial communities in, e.g., the intestine is of direct relevance to human obesity *(5)*, and revisiting the activity of pathogens, such as *Vibrio cholerae*, from the community context has revealed surprising insight with immediate consequences for generating clean drinking water *(6)*. Questions related to the function and interaction of microbial consortia has therefore taken a place of prominence in the current science literature.

In this chapter, we will address three methods that have proven useful in modeling the behavior of microbial communities. These methods have different requirements for the level of detail needed to model a multicellular microbial system. The first method we will discuss is based on representing a microbe by rate equations, requiring the highest level of detail. Not surprisingly, this approach has been limited in applicability due to the lack of measured kinetic parameters. However, it seems plausible that this drawback will be significantly tempered in the near future. We will then describe individual-based approaches (often called agent-based modeling (ABM)) capable of simulating the interaction of multiple microbes with a relatively narrow set of variables. While the focus is still on the individual microbes, this method is capable of addressing the spatial aggregation of large populations. We will complete this chapter with a discussion of population dynamics modeling, a method for which the species is the focal point. This class of approaches has become well known through the Lotka–Volterra representation of a predator–prey system *(7, 8)*.

2. Rate-Equation Models

2.1. Background

For good reasons, all developed genome-level models of microbial metabolism are based on the assumption that the system is at steady state (*see* Chapter “Flux-Balance Analysis: Interrogating Genome-Scale Metabolic Networks”). Although steady-state models (SSM) have shown great utility for assessing the metabolic capabilities of an organism, they ignore a number of crucial details needed to attain greater insights into the dynamics of a cell. For example:

- After an environmental or genetic perturbation, SSM only characterize the new steady state. SSM do not calculate how long it will take for the system to reach the new steady state and visited intermediary states.
- SSM ignore enzymatic capacity and thus cannot identify rate-limiting steps and metabolic bottlenecks.
- SSM do not account for the concentration of intermediates and thus cannot predict deleterious buildup of toxic metabolites.

Development of genome-scale kinetic models can overcome these failings; however, currently such undertakings are impractical. In order to develop a kinetic model of cellular metabolism, we must account for the time-dependent changes in metabolite concentrations. This requires the knowledge of a large number of kinetic parameters. Unfortunately, while recently developed analytical tools have accelerated genetic and proteomic analyses immensely, measurements of enzymatic kinetic parameters are still tedious and time consuming.

Kinetic models are usually developed only for well-studied pathways, such as central carbon metabolism in *Escherichia coli (9)*, urea cycle in *Rattus norvegicus (10)*, and glycolysis in a variety of organisms ranging from single cell organisms such as *Saccharomyces cerevisiae* (e.g. **refs.** *11–13*, for a review *see* **ref.** *14)* and *Trypanosoma brucei (15, 16)* to cells from organs such as skeletal muscle *(17)* and pancreatic b-cells *(18)*. Despite their limited metabolic scope, these models have been invaluable in enhancing our understanding of the complex collective dynamics of cellular groupings.

2.2. Theory and Methodology

Perhaps the most important question that one should consider prior to developing a kinetic model is: "How detailed should the model be?" The answer to this question is directly related to other questions that have to be answered early in the modeling process. For example:

- What kinetic parameters are available?
- Is it possible to bypass or generalize certain details of a pathway and still develop a sufficiently predictive model (*see* **ref.** *19)*?
- Which reactions are reversible and which are irreversible?
- Which metabolites can be transported across the cellular membrane? Are these processes passive or active (i.e., require energy expenditure)? Are they facilitated by (transporters) chaperones or are they nonfacilitated?
- What is the volume and surface area of a cell, and should the model account for changes to these physical characteristics?

These questions can be answered through a thorough examination of the available literature, and searching through databases such as BRENDA (http://www.brenda-enzymes.info). These answers will also determine how the dynamics of metabolic reactions are formulated.

Developing kinetic models of metabolic pathways involves writing the concentration changes of each metabolite as ordinary differential equations (ODEs). For example, given the two metabolic reactions:

$$A + B \xleftrightarrow{Q} C \quad \text{and} \quad C \xrightarrow{k_2} D,$$

where Q is the equilibrium constant equal to the ratio of forward and reverse reaction coefficients (k_1/k_{-1}), the change in concentration of the metabolites is written as:

$$\frac{\mathrm{d}[A]}{\mathrm{d}t} = -k_1[A][B] + k_{-1}[C]$$
$$\frac{\mathrm{d}[B]}{\mathrm{d}t} = -k_1[A][B] + k_{-1}[C]$$
$$\frac{\mathrm{d}[C]}{\mathrm{d}t} = k_1[A][B] - (k_{-1} + k_2)[C]$$
$$\frac{\mathrm{d}[D]}{\mathrm{d}t} = k_2[C].$$

By thus writing and solving similar ODEs for all metabolites, we may monitor the dynamic changes that occur in a microbe.

2.3. Examples

2.3.1. Coupling of Glycolytic Oscillations in *Saccharomyces cerevisiae*

One of the most studied problems of cellular nonlinear dynamics has been the coupling and synchronization of metabolic oscillators, such as baker's (or brewer's) yeast, *S. cerevisiae*. The initial reports of glycolytic oscillation with a frequency of several minutes in cell-free extracts of yeasts date back to 1964 *(20–22)*. Prolonged oscillations in biochemical systems require that at least one of the reactions obey nonlinear kinetics. Thus, it is not surprising that asynchronous *(23, 24)* and even chaotic *(25–27)* dynamics have been proposed. A large number of theoretical studies have examined oscillatory behavior in glycolysis, particularly in yeasts (for a review *see* **ref.** *28)*. The majority of theoretical studies involve the coupling of only a few metabolic pathways *(24, 29, 30)*, and the interaction is mediated through a common extracellular pool of metabolites that can be imported to and/or exported from different cells. In the case of *S. cerevisiae* suspensions, acetaldehyde (Acld) has been identified as the primary coupling metabolite *(31)*.

Glycolysis Oscillations and Synchronization

The model of glycolysis in yeast *(32)* was designed with the criterion that it should describe the observed experimental observations *(31, 33)*. A schematic of the modeled system is presented in **Fig. 1**. The characteristics of the model are:

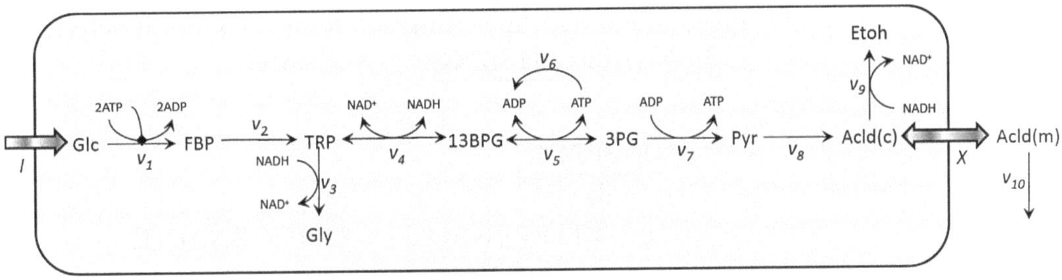

Fig. 1. Schematic diagram of anaerobic glycolysis. *Glc* Glucose; *TRP* Triose-phosphates; v_1 Hexokinase and PFK; v_2 Aldolase; v_3 Glycerol-3-phosphate dehydrogenase and glycerol kinase; v_4 GAPDH; v_5 PGK; v_6 ATPase; v_7 Phosphoglycerate mutase, enolase, and pyruvate kinase; v_8 Pyruvate decarboxylase; v_9 Alcohol dehydrogenase; v_{10} Degradation of Acld.

- Several chemical steps are lumped together, such as reactions catalyzed by hexokinase and phosphofructokinase (PFK) (v_1) and multistep conversions of dihydroxyacetone phosphate to glycerol (Gly) (v_3) and 3-phosphoglycerate (3PG) to pyruvate (Pyr) (v_7).
- Simulation is for anaerobic conditions with ethanol (Etoh) as the major product.
- Concentrations of Gly and Etoh are considered constant (reservoir). Import of glucose and export of Acld are the only modeled extracellular fluxes: Import of glucose (I) is assumed constant, and transport of Acld (X) is modeled as passive diffusion dependent on the concentration gradient of Acld across the membrane:

$$X = \frac{AJ}{V}([\text{Acld}(c)] - [\text{Acld}(m)]),$$

where A and V are the surface area and volume of the cell, respectively. J is the coefficient of permeability of the cellular membrane for Acld. c and m denote cytosolic and medium concentrations.

- Consumption of ATP by the cell is accounted for by an ATPase reaction. Intracellular pools of adenine nucleotides (ATP and ADP) and nicotinamide adenine dinucleotides (NAD^+ and NADH) are conserved:

$$[\text{ATP}] + [\text{ADP}] = A_{\text{Total}} \text{ and } [\text{NAD}^+] + [\text{NADH}] = N_{\text{Total}}.$$

- All reactions are considered irreversible, except for glyceraldehyde 3-phosphate dehydrogenase (GAPDH) (v_4) and phosphoglycerate kinase (PGK) (v_5).
- Reactions catalyzed by GAPDH and PGK are near equilibrium (Q_{GAPDH} = 0.0056, Q_{PGK} = 3,225, *(34)*), justifying a

quasi-steady-state approximation for 1,3-bisphosphoglycerate (13BPG), thus $\frac{d[13BPG]}{dt} = 0.$ And since,

$$\frac{d[13BPG]}{dt} = v_4 - v_5,$$
$$v_4 = k_4[TRP][NAD^+] - k_{-4}[13BPG][NADH],$$
$$v_5 = k_5[13BPG][ADP] - k_{-5}[3PG][ATP],$$

we can write the equation for concentration of 13BPG as

$$[13BPG] = \frac{k_4[TRP][NAD^+] + k_{-5}[3PG][ATP]}{k_{-4}[NADH] + k_5[ADP]}.$$

Thus, the reaction equation for v_4 and v_5 becomes

$$v_{4\&5} = \frac{k_4 k_5[TRP](N_{Total} - [NADH])(A_{Total} - [ATP]) - k_{-4}k_{-5}[3PG][ATP][NADH]}{k_{-4}[NADH] + k_5(A_{Total} - [ATP])}.$$

- Simple rate laws are used for all enzymatic reactions (*see* **Table 1**).
- The only regulatory behavior that is accounted for is the inhibitory effect of ATP on the hexokinase-PFK reaction (v_1) using K and n as the inhibition constant and cooperativity coefficient for ATP, respectively

$$f(ATP) = \left[1 + \left(\frac{[ATP]}{K}\right)^n\right]^{-1}.$$

Metabolites are distributed homogenously in the cytosolic and external medium.

Transduction of Oscillations

The model can be used to study the mechanism of intracellular propagation of nonlinear dynamics. It is reasonable to assume that when nonlinear dynamics are transmitted down the main backbone of the glycolytic pathway, the amplitude of substrates should be greater than that of the products it produces *(33)*, i.e., each enzymatic step dampens the oscillations. Not surprisingly, a series of simulations has shown that oscillations in glycolysis can be transmitted throughout the cell via the cofactors ADP and NAD.

As a follow up, Wolf and coworkers *(32)* examined whether it is possible for cells to synchronize their oscillating dynamics if oscillations are not propagated through the backbone of the glycolytic pathway. To this end, cells of yeast with identical kinetic capabilities, but different concentrations of metabolites, were coupled together via a shared extracellular Acld pool. In **Fig. 2**, we have simulated the coupled dynamics of three such cells. As it can be seen in **Fig. 2a**, the cells oscillate at the same frequency but with different amplitudes and phases. Gradually, the phase

Table 1
List of differential equations for a simplified model of glycolysis (*32*)

Model differential equations
$\frac{d[Glc]}{dt} = I - k_1[ATP][Glc]f(ATP)$
$\frac{d[FBP]}{dt} = k_1[ATP][Glc]f(ATP) - k_2[FBP]$
$\frac{d[TRP]}{dt} = 2k_2[FBP] - v_{4\&5} - k_3[TRP][NADH]$
$\frac{d[3PG]}{dt} = v_{4\&5} - k_7[3PG]\left(A_{Total} - [ATP]\right)$
$\frac{d[Pyr]}{dt} = k_7[3PG] - k_8[Pyr]$
$\frac{d[Acld(c)]}{dt} = k_8[Pyr] - k_9[NADH][Acld(c) - X$
$\frac{d[Acld(m)]}{dt} = \frac{V}{V_m}X - k_{10}[Acld(m)]$
$\frac{d[ATP]}{dt} = v_{4\&5} - 2k_1[ATP][Glc]f(ATP) - k_6[ATP] + k_7[3PG]\left(A_{Total} - [ATP]\right)$
$\frac{d[NADH]}{dt} = v_{4\&5} - k_3[TRP][NADH] - k_9[NADH][Acld(c)]$

V_m, Volume of the medium

shift disappeared, and in less than 20 min the dynamics of the cells were completely synchronized (**Fig. 2b**).

2.3.2. Quorum Sensing

Many bacteria synchronize the activation of particular functions by communicating their local cell density to each other through autoinducer (AI) molecules, an effect called "quorum sensing" *(35)*. As the cell population increases, the AIs accumulate in the surroundings, eventually reaching a critical concentration causing the differential expression of certain sets of genes, e.g., genes involved in bioluminescence. Experiments have shown that several necessary processes to bacterial colonization and virulence

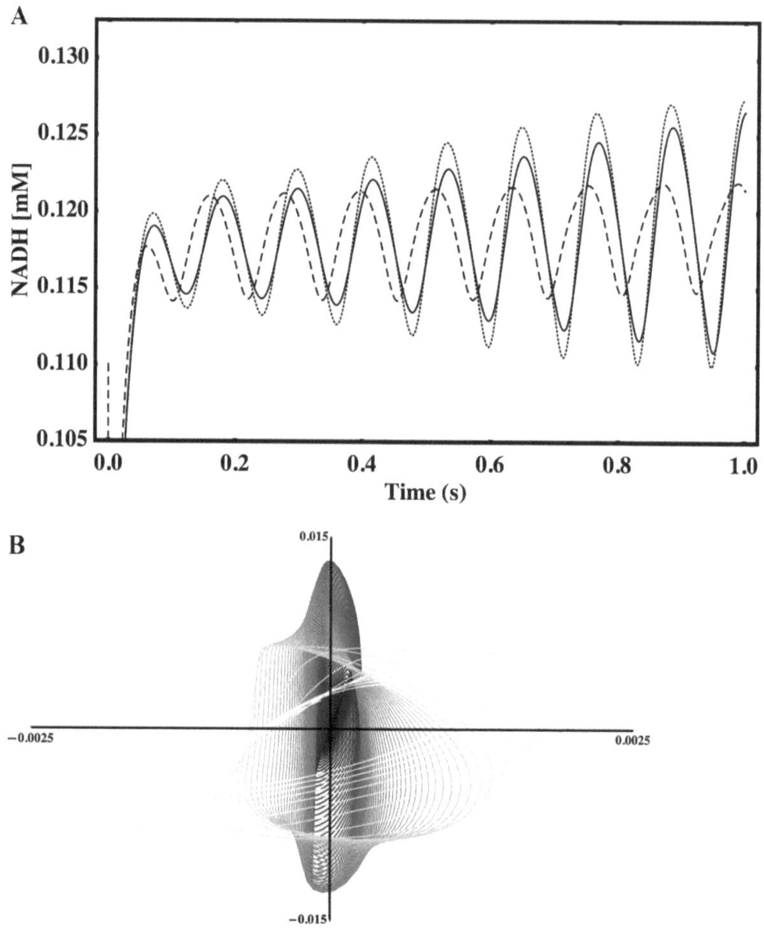

Fig. 2. Coupled dynamics of glycolysis in three cells with identical kinetic capabilities but different starting metabolite concentrations. (**A**) Oscillating concentration of NADH. The oscillation frequency is the same for all three cells while the starting amplitudes (Am) and phases differ ($Am_A > Am_B > Am_C$). (**B**) Amplitude differences in NADH oscillations between two pairs of cells (A–B and B–C). Time course is represented by ***shading*** (early = white, late = black).

such as biofilm formation, bioluminescence, type III secretion, and secretion of virulence factors are regulated via quorum sensing *(36–41)*.

The machinery of the gene regulatory networks that produce the AIs, detect the AIs, and respond to the changes in AI concentration differs depending on the bacteria. *Vibrio harveyi* and *V. cholerae* use sensors at the membrane to track changes in the AI concentration *(38, 42–45)*, while the AIs diffuse through the membrane and form a complex with a particular protein necessary for gene activation in *Vibrio fischeri* and *Pseudomonas aeruginosa (35, 36, 46–49)*. The latter type of quorum sensing bacteria will be the focus of this instructional example.

Quorum Sensing Model for *Vibrio fischeri*

We present a model of quorum sensing in *V. fischeri* introduced by James et al. *(50)*. *V. fischeri* is a gram-negative bioluminescent marine bacterium that uses acyl-homoserine lactones as its AIs to directly control the luminescence (*lux*) operon *(51–54)*. The model tracks the concentration of AIs (A), the concentration of the protein LuxR that the AI forms a complex with (R), and the concentration of the AI–protein complex (C) (*see* **Fig.** 3). The first interaction, AI forming a complex with LuxR, is described by the binding rate constant k_1, and the complex can break apart with dissociation rate constant k_2, giving the reactions:

$$A + R \xrightarrow{k_1} C \text{ and } C \xrightarrow{k_2} A + R.$$

The resulting three differential equations for A, R, and C are

$$\frac{dA}{dt} = k_2 C - k_1 AR, \frac{dR}{dt} = k_2 C - k_1 AR, \frac{dC}{dt} = k_1 AR - k_2 C.$$

Through binding to the *lux* box, the complex (C) is responsible for promoting the production of the *lux* operons, which include the genes responsible for light production, *luxCDABEG*, the gene responsible for producing the AI, *luxI*, and the gene that translates into the protein AI complexes with *luxR*. When the concentration of the complex is low (high), the *lux* box is predominantly unoccupied (occupied). This is accounted for by including a term

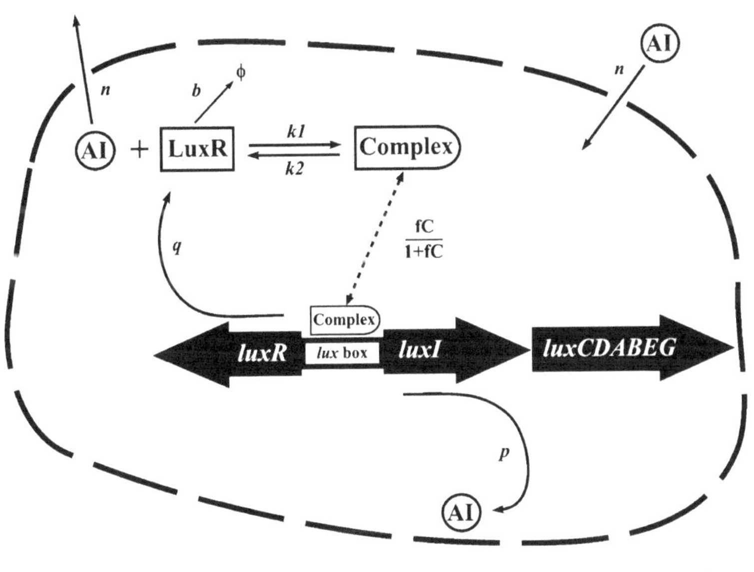

Fig. 3. Schematic of quorum sensing network used by *Vibrio fischeri* to regulate luminescence. AI (A) binds the protein LuxR (R) to form complex (C) with a forward rate of k_1 and a dissociation rate of k_2, and diffusion of AI through cellular membrane with constant n. LuxR is degraded at a rate b. The C complex occupies the *lux* box proportional to $fC/(1 + fC)$ and promotes the transcription of *luxR*, *luxI*, and *luxCDABEG* with rate q. AI is produced at rate p from LuxI.

$$lux\text{ box occupancy} = \frac{fC}{1+fC},$$

where f is a proportionality constant.

Since the complex promotes the transcription of *luxI* and *luxR*, the rates of transcription are proportional to the *lux* box occupancy time. The model does not explicitly include the translation step of *luxI* and *luxR* into LuxI and LuxR, or the direct synthesis of the AI from LuxI. Instead, it is assumes these reactions to be integrated into an additional proportionality constant times the occupancy of the *lux* box.

$$\text{LuxR synthesis rate} = q\frac{fC}{1+fC}, \quad \text{AI synthesis rate} = p\frac{fC}{1+fC}.$$

The differential equations for A and R are thus updated to be

$$\frac{dA}{dt} = k_2C + p\frac{fC}{1+fC} - k_1AR,$$

$$\frac{dR}{dt} = k_2C + q\frac{fC}{1+fC} - k_1AR,$$

$$\frac{dC}{dt} = k_1AR - k_2C.$$

The LuxR protein concentration is naturally reduced via enzymatic degradation and cellular volume changes from cell replication at a rate proportional to the LuxR concentration, and the chemically stable AIs freely diffuse through the cell membrane into the surrounding environment at a rate assumed to be proportional to the cellular AI concentration:

Degradation rate of LuxR= bR, Diffusion rate of AI=nA

Including this effect, the differential equations for A, R, and C then become:

$$\frac{dA}{dt} = k_2C + p\frac{fC}{1+fC} - k_1AR - nA,$$

$$\frac{dR}{dt} = k_2C + q\frac{fC}{1+fC} - k_1AR - bR,$$

$$\frac{dC}{dt} = k_1AR - k_2C.$$

Finally, an external concentration of AI (A_{ex}) generated by a colony of bacteria can be added to the model by including a forward rate of diffusion of AI proportional to the external concentration. This only modifies the equation for A by adding nA_{ex}:

$$\frac{\mathrm{d}A}{\mathrm{d}t} = k_2 C + p\frac{fC}{1+fC} - k_1 AR - n(A - A_{ex}).$$

Model Analysis

To illustrate how the system of three coupled differential equations can exhibit quorum sensing behavior, we solve for the time series solutions of the differential equations for two values of A_{ex} (A_{ex} = 1 nM and A_{ex} = 50 nM; see figure captions for parameter choices). Note that the model formulation in itself does not dictate a particular choice of units. The low value of A_{ex} corresponds to the low cell-density limit where the external concentration of AIs from surrounding bacteria is minimal. **Figure 4** shows the cellular concentrations of AI (dashed line) and the LuxR–AI complex (sold line). Both molecules are given initial concentrations of 1 nM. The system quickly reaches steady-state conditions where the internal AI concentration matches the external one, and the concentration of the LuxR–AI complex drops to a minimal value. Since the LuxR–AI complex is responsible for activating luminescence, this situation corresponds to a dark colony. Upon increasing the concentration of external AI (corresponding to high cell density), the LuxR–AI complex is able to reach a considerably larger steady-state concentration. Since the threshold concentration of the LuxR–AI complex necessary for light production is not known, the results in **Fig. 5** serve as an illustration of the cell's response to a large increase in external AI concentration.

Using the model for quorum sensing in *V. fischeri* proposed by James et al. *(50)*, it is clear that this relatively simple set of coupled differential equations is capable of exhibiting a quorum sensing-like response when the concentration of external AI is changed. Other models exist that include more interactions in the genetic regulatory network *(55, 56)*. There are also models of *P. aeruginosa*, a similar quorum-sensing bacteria to *V. fischeri* *(57, 58)*.

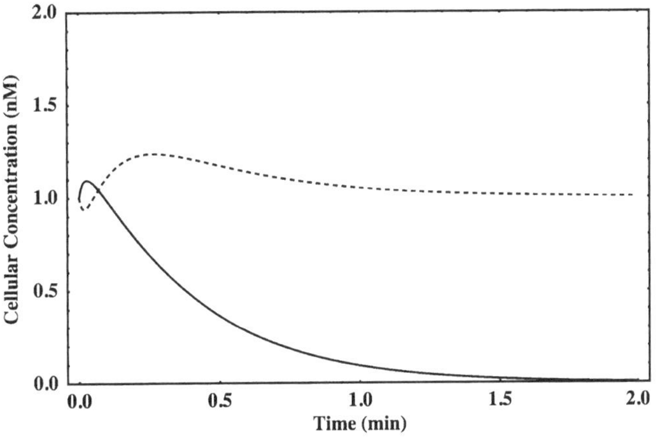

Fig. 4. Low cell density response of LuxR–AI complex (*solid line*) and AI (*dotted line*) concentrations. Starting concentrations are 1 nM for LuxR–AI, AI, and AI_{ext}. The system quickly reaches its steady-state values where LuxR–AI complex concentration is minimal and the AI concentration matches AI_{ext}. The other parameters are k_1 = 25 nM^{-1} min^{-1}, k_2 = 10 min^{-1}, n = 10 min^{-1}, b = 10 min^{-1}, p = 5 nM/min, q = 2.5 nM/min, and f = 0.25 nM^{-1}.

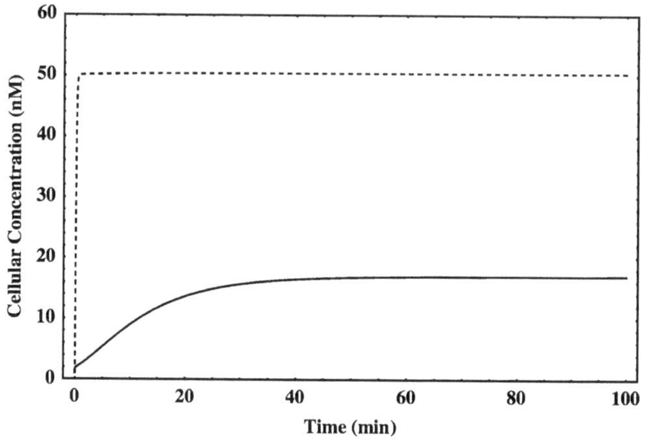

Fig. 5. High cell density response of LuxR–AI complex (*solid line*) and AI (*dotted line*). Starting concentrations are 1 nM for LuxR–AI, AI, and 50 nM for AI_{ext}. The system quickly reaches its steady-state values where LuxR–AI complex concentration can initiate light production and the AI concentration matches AI_{ext}. The other parameters are as in **Fig. 4**

2.4. Tools

ODEs can easily be solved by general scientific and engineering software such as Matlab (Mathworks, Natick, MA, http://www.mathworks.com) and Mathematica (Wolfram Research, Champaign, IL, http://www.wolfram.com). Many programs have been developed primarily to facilitate the modeling of dynamical systems:

- Virtual cell (http://www.vcell.org)
- E Cell (http://www.e-cell.org)
- CellDesigner (http://www.celldesigner.org)
- Karyote (biodynamics.indiana.edu/CellModeling)
- MathSBML (http://www.sbml.org/Software/MathSBML).

There are also a number of databases, such as http://www.siliconcell.net, where metabolic models are stored.

3. Individual-Based Modeling

3.1. Background

The history of individual-based modeling, also often called agent-based modeling (ABM), goes back to the late 1940s and early 1950s work by John von Neumann where he invented cellular automata (CA). CA are most frequently simulated on finite grids, and the state of a grid cell's neighbors is used to determine its state for the next time step. In a simple one-dimensional example, only two states (0 or 1) are available per cell and only nearest neighbors, and the CA update rules would then determine for which of the $2^3 = 8$ possible states (the cell itself and its two nearest neighbors) a cell would change its value. An example set of rules could be that a cell should only switch state if both of the neighbors are in an opposite state (majority rule): 101→111, and 010→000.

Individual-based models (IbMs) were suggested in the 1980s as a possible method for studying social systems on a computer. Differently from the CA, the IbMs are typically not occupying all available grid cells and, in fact, need not be based on a grid at all. However, similarly to the CA, each entity carries with it a predestined set of rules that it acts upon after polling its local environment. Due to the rapid increase in computational power for desktop PCs, IbMs started receiving serious attention in the 1990s *(59–61)*.

In the following, we will enlist the IbM framework to model microbial communities, and the agents will represent individual cells, being either bacteria, archaea, or single-cell eukaryotes. Contrasting the IbM framework with that of the rate equation approach, we quickly see that the chasm in representation can be bridged. For instance, one can imagine that the internal rule-set for an agent is based on monitoring the output of a set of rate equations, such as growth, internal ATP concentration, or autoinducer concentration in quorum sensing. However, the computational cost of including a highly detailed internal description should be measured carefully against the feasible number of simultaneous agents and the duration of the simulation.

3.2. Theory and Methodology

When a system is comprised of many agents whose interactions generate system-level dynamics that cannot be explained by their individual properties (emergent behavior), individual-based modeling is well suited for simulating the system function. Typically, IbMs of microbial communities are simulated on two-dimensional or three-dimensional grids where a single entity occupies a grid cell. Before taking on the task of designing or implementing an IbM, it is necessary to clearly define the contents and scope of the project. Important questions to clarify include:

- How many species will exist in the system?
- Will the microbes be allowed to move?
- What will be the inputs and outputs of each microbe?
- How much will a microbe eat before it divides?
- After cell division, how will the two cells be placed?
- Which boundary conditions will be chosen (e.g., hard walls, nutrient reservoir)?
- Which metabolic strategies, e.g., dormant or growing maximally, may be used?
- How will the microbes interact; through competition for nutrients or through more direct channels, e.g., quorum sensing, physical contact, or production of toxins?

Additionally, it is necessary to decide how nutrients and other chemicals will move in the system, as well as the shape and function of the system boundaries. In the modeling of biofilms, nutrient levels are sometimes chosen to be fixed along one of the system boundaries to simulate the presence of a reservoir, while a different boundary is chosen to be impermeable to both nutrients and cells, emulating a hard surface such as a wall.

The basis for any IbM is the set of "behavioral" rules that each microbe may follow. For every time increment, each microbe is visited and taken through the list of possible rules. In simple cases the rule set is deterministic: whenever the local conditions are identical, a given outcome is repeated. For more sophisticated models, the microbe may choose among the available strategies with a probability that depends on past history, the local environment, or both. While implemented behavioral rules frequently have been discrete in nature, this is not a requirement of the modeling approach. For instance, a common choice in calculating the growth of a microbe from one time-point to the next is to increment an "energy storage" variable with a fixed amount. However, one may alternatively describe the growth (rate) using Michaelis–Menten, or even double-saturation kinetics *(62)*.

It is in the selection of behavioral rules that IbM intersects with game theory. In simple IbMs, the rule set only allows for interactions through the use of nutrients or occupation of space (e.g., a microbe is not allowed to grow when adjacent grid cells are occupied). However, microbes may cooperate or compete through the production of chemical signals (quorum sensing) and toxins *(63)*. It is relatively straightforward to include a wide variety of competitive or cooperative behaviors in the behavioral rules. For instance, we can generate a class of cooperative microbes simply by lowering their possible growth rate while they produce a beneficial byproduct, such as extracellular polymeric substance (EPS) or a molecule that aids the function of a different microbial species. The competing behavioral class of "cheaters" will be allowed to avoid this burden (e.g., no EPS production) and can grow at the maximal rate. In such a scenario, it is possible either for the cooperators or for the cheaters to have the highest fitness, depending on the growth conditions and the structure of the environment *(62, 64)*.

When designing an IbM, it is also necessary to carefully consider how the nutrients are distributed. In the simplest models, nutrient concentrations are chosen to be constant, while more complex realizations include discretized differential equations for the diffusive nutrient transport. These hybrid methods, combining IbM dynamics for the microbes with differential equations for the nutrients, have given highly detailed insights into the dynamics of biofilms (see **ref.** *65* for an example of three-dimensional simulation). In these approaches, it is beneficial to utilize the difference in time scales between diffusion (fast process) and microbial activities (slow process) such as growth. The following two-step iterative process is frequently used (1) calculate the quasi-steady-state solution for the diffusive molecules and (2) use the identified local concentrations as input for the microbial IbM dynamics. Assuming that both microbial locations and their uptake and production rates are fixed, we may easily find the steady-state solution of the diffusion equations of, e.g., oxygen, glucose, and an autoinducer. Note that the microbes may act as both sinks (consumption of nutrient) and sources (production of signaling molecules).

Alternatively, we may consider the nutrients and other chemicals as discrete particles that conduct independent random walks. For instance, the nutrient particles may move with equal probability to an adjacent site, and multiple nutrient particles are allowed to occupy the same grid site. In this representation, the effective diffusion coefficient is determined by the number of steps in the walk. Fluid flow may be incorporated by biasing the direction of the random walk. Note that one must conduct the random walk step for all particles before updating the microbial states.

3.3. Example

In a simple, deterministic two-dimensional system where the only interaction between the microbes is competition over nutrients and available space, the rule set is:

1. Nutrient uptake:
   (a) If amount of nutrient $E > e$ is available in current and adjacent grid cells, eat amount e. Add to internal energy storage: $w \rightarrow w + e$ (and appropriately subtract from E).
   (b) If not, maintenance cost $m < e$ is deducted: $w \rightarrow w - m$
2. Duplication or sporulation:
   (a) If at least one adjacent grid cell is empty and internal energy storage $w > W$ (the duplication threshold) generate copy and set $w \rightarrow (w - W)/2$ in both microbes.
   (b) If internal energy storage $w < T$, the sporulation threshold, microbe is inactive until nutrient level in current grid cell is $E > e$.

Naturally, we choose $T \ll W$. In this simple example, we are inhibiting the movement of nutrient particles, similar to microbial growth on an agar plate. By allowing for the movement of nutrients, either as a random walk of discrete particles or by differential equations (diffusion), this simple IbM can be changed to describe biofilm growth in a liquid medium. Typical initial conditions start from either a single or multiple identical microbes in the middle of the grid or along a boundary. Multiple species are simply included by, e.g., changing the uptake amount from being a global constant e, to become species-dependent e_s.

We can create cooperative behavior by modifying, e.g., behavioral rule 1.a as follows:

(1.a) If amount of nutrient $E > e$ is available in current and adjacent grid cells and majority of adjacent grid cells are occupied, eat amount $e' = e-d$ ($d > 0$). If majority of adjacent grid cells are empty, eat $e' = e$. Add to internal energy storage: $w \rightarrow w + e'$ (and appropriately subtract from E).

This straightforward rule change forces microbes to behave altruistically by taking less of the nutrients when in a dense neighborhood, and thus, improve sharing of resources.

Figure 6 shows a snapshot of a biofilm simulation of two species competing over the same food source. In addition to rules 1 and 2, we have included nutrient diffusion using the random walk approach and cellular death instead of sporulation. It is not surprising that the fast growing species (dark gray) is dominating over the slower growing species (light gray) in the major bloom: the further away from the bottom layer (the wall) an individual is, the more nutrients are available and it can grow faster.

3.4. Tools

Several consortia have made available general-purpose IbM models. The most popular open-source implementations are Swarm (http://www.swarm.org) and Netlogo (http://ccl.northwestern.edu/netlogo). A listing of available IbM software packages is available at http://www.swarm.org/index.php?title=Tools_for_Agent-Based_Modelling. Programs specifically tailored to microbial communities include BacSim *(59)*, which is based on the Swarm toolkit, and BacLAB *(66)*.

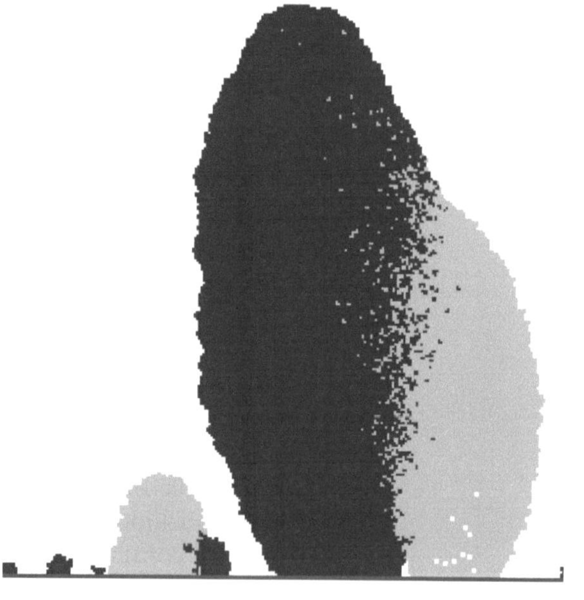

Fig. 6. Simulated biofilm of two competing species growing on an impermeable boundary. Substrate gradients are generated by random walks of discrete nutrient packets. Fast growing (*dark gray*) microbes dominate over slower growing ones (*light gray*).

4. Population Dynamics

4.1. Background

Population dynamics in community-level modeling comprises a coarse-grained approach compared to the two previous sections, where the focus has shifted from individual microbe to the species as basic unit. Population-level interactions between different species (macroscopic) can naturally be considered as effective per-capita rates resulting from the interacting individuals (microscopic). Thus, population interactions naturally arise from shared ecological niches and diverse metabolic capabilities of the constituent microbes.

A conventional way of classifying pair-wise population interactions is based on their effects on growth (*see* **Table 2**). The presence of one species may be beneficial [+], detrimental [−], or neutral [0] to the other. In fact, all possible combinations of effects are observed in nature, both the symmetric (reciprocal) interactions of mutualism [++] and competition [−−], and the asymmetric cases of ammensalism [0−], commensalism [0+], predator–prey or parasitism [+−]. However, this scheme does not reflect the microscopic origin of interactions. Simple abstractions of an interaction may be insufficient to quantitative analyses, and it is important to carefully consider the microscopic origin of interactions. We also note that these classification schemes constitute an idealization: In practice, the behavior of a mixed community is likely the combination of multiple interactions, often with opposing effects. A situation that is common to microbial communities consists of two (or more) species in a mixed population that compete for the same nutrient source while, at the same time, being physiologically coupled in a commensal way.

Table 2
Overview of species-level interaction classes in population-based modeling

Interaction mode	Reciprocity	Cell–cell direct contact	Sign of interactions
Mutualism	Y	N	[++]
Competition	Y	N	[−−]
Commensalism	N	N	[+0]
Ammensalism	N	N	[−0]
Predator–prey, parasitism	N	Y	[+−]

Thus, we should not expect that the resulting dynamics will be predictable by "effective" interaction models where the complex (competing) interactions are combined into one average contribution.

Population-level descriptions provide insights that are otherwise overlooked in microscopic studies. Microbial communities from compost, the bovine rumen, acid mine drainage, and hot springs are just a few among recently studied systems that will benefit from quantitative modeling.

4.2. Theory and Methodology

4.2.1. Lotka–Volterra Model and its Deterministic Variations

Since the early modeling of the predator–prey ecosystem, the Lotka–Volterra (LV) model *(7, 8)* has been the de facto standard template for modeling mixed populations. Though LV had originally aimed at modeling the specific case of predator–prey system, its current usage has been expanded past the predator–prey setting to include positive interactions. In its simplest version, the population size of a prey (n_1) and its predator (n_2) satisfy the following set of nonlinear differential equations

$$\frac{d}{dt}n_1(t) = \alpha n_1 - \beta n_1 n_2; \frac{\mathrm{d}}{dt}n_2(t) = \gamma n_1 n_2 - \delta n_2.$$

Here α and δ are the growth and decay rates for the prey and predator populations, unaffected by the negative (predation) interspecies interaction. The coefficients b and g represents the strength of the detrimental and the beneficial effects on prey and predator population owing to the predation. Due to the particular functional form of these equations, the Jacobian of this system has purely imaginary eigenvalues, regardless of the parameter combinations. Consequently, the two-species LV system has sustained oscillatory behavior with a characteristic frequency of $\sqrt{\alpha\delta} / 2\pi$.

The exponential growth of prey population has been a target for modifications. The original LV assumes no resource limits, which oftentimes is unrealistic. To include the resource-mediated intraspecies competition, we require a negative term that would counterbalance exponential growth. Thus introduced is the logistic growth rate, $\alpha n(1-n/k)$ where K is the carrying capacity of the ecosystem for the species involved. The modified LV with the logistic growth with finite carrying capacity for the prey population is now

$$\frac{\mathrm{d}}{\mathrm{d}t}n_1(t) = \alpha n_1\left(1 - \frac{n_1}{K}\right) - \beta n_1 n_2; \frac{\mathrm{d}}{\mathrm{d}t}n_2(t) = \gamma n_1 n_2 - \delta n_2,$$

which has the two nontrivial (excluding $n_1 = n_2 = 0$) steady states (**Fig. 7**)

$$(n_1, n_2) = (K, 0) \text{ or } \left(\frac{\delta}{\gamma}, \frac{\alpha}{\beta}\left(1 - \frac{\delta}{\gamma K}\right)\right).$$

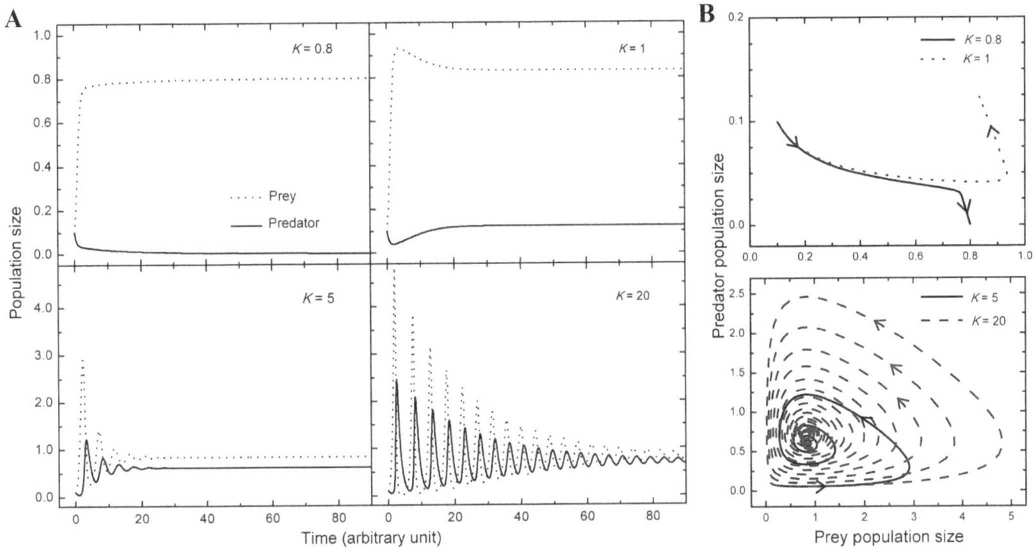

Fig. 7. Competitive Lotka–Volterra (LV) dynamics. (**A**) Time evolution of the population size from LV with logistic growth modification. All the systems start with $n_1(0) = n_2(0) = 0.1$ (arbitrary units) and the time scale is set in units of 1/d (~predator's lifespan). Rate parameters a = 2.3, b = 3.1, g = 1.2, and the carrying capacity K is varied from 0.8 to 20 ($K_c = 0.833$). The mixed population state is stable for $K > K_c$. (**B**) Trajectories in $n_1 - n_2$ space shows the attractor for different carrying capacities.

The first solution corresponds to predator extinction and prey proliferation, which is stable as long as $K < K_c \equiv \delta/\gamma$ (the extinction threshold). Stable population coexistence (second solution) is possible only when $K > K_c$. Linear stability analysis further shows that coexistence is either a stable node or a focus, and no oscillatory behavior is expected unless the carrying capacity diverges *(67)*.

We may generalize the LV population model (which we will refer to as GLV) to include competitive interactions among species by adding an extra, negative term following the spirit of mass-action:

$$\frac{\mathrm{d}}{\mathrm{d}t} n_i(t) = n_i \left(\alpha_i - \sum_{j=1}^{d} A_{ij} n_j \right); \quad i = 1, 2, \ldots, d,$$

where d is the total number of interacting species. The diagonal elements $A_{ii} > 0$ can be identified (to a multiplicative constant) with the inverse of the carrying capacity of species i. The off-diagonal elements $A_{ij} > 0$ represent the strength of j's negative effect on i, which is related to the distance between the two species in niche space.

Finally, a unified scheme for the community interactions is obtained by removing the positivity constraint on the off-diagonal elements A_{ij} in GLV. The majority of studies on mutualistic interactions have been using this representation as a template framework. However, all eigenvalues of the interaction matrix must have positive real parts for the system to be stable. High-diversity communities tend to become unstable as the interaction network becomes more complex, reminiscent of the work by Robert May in the 1970s *(68, 69)*. Recent studies have revisited this problem and found potential positive effects of complexity: High-diversity, stable LV systems arise if the interaction network evolves flexibility through adaptive behavior *(70, 71)*.

4.2.2. Effects of Spatial Heterogeneity

In general, microbial populations are spatially heterogeneous and not well-stirred "bioreactors" as assumed in the original LV work. Even marine microbes aggregate in the search for food using chemotaxis. We may introduce spatial structure into the deterministic framework by using an embedding space, where the individuals move around in the search for food and shelter. Now, interaction effects are no longer instantaneous but must propagate across the space, leading to time delays that stabilizes the community *(72, 73)*. A natural extension of LV to allow for the random movement of cells is by way of diffusion terms, turning the LV into the coupled partial differential equations

$$\begin{cases} \dfrac{\partial n_1(\boldsymbol{x},t)}{\partial t} = D_1\nabla^2 n_1(\mathbf{x},t) + \alpha n_1(\mathbf{x},t) - \beta n_1(\mathbf{x},t) n_2(\mathbf{x},t), \\ \dfrac{\partial n_2(\boldsymbol{x},t)}{\partial t} = D_2\nabla^2 n_2(\mathbf{x},t) + \gamma n_1(\mathbf{x},t) n_2(\mathbf{x},t) - \delta n_2(\mathbf{x},t), \end{cases}$$

where D_i is the diffusion coefficient of species *i*. For the case of two-species competition, this coupled reaction-diffusion system is known to contain propagating wave-front solutions in one dimension, of the form $n_i(t) = f(x - v_i t)$ that interpolate between the two steady states identified above. Convergence to the steady state monotonically or with oscillations depends on the choice of rate parameters *(72)*.

4.2.3. Stochastic Modeling

Randomness is a defining character of population processes, often diverting the dynamics from deterministic predictions. Depending on the origin of the "noise," population stochasticity may be classified by the following categories:

- Within-individual variability
- Cell-to-cell variability and age structure
- Spatial heterogeneity
- Temporal fluctuation of environment

The first two categories stem from the random timing of birth–death events and the discrete nature of individuals. These factors play a lesser role as the population grows in size, but may still have significant local effects. In fact, local extinctions commonly occur in nature, which is consistent with observations in stochastic simulations. The latter two categories are extrinsic in origin and can be described in terms of quenched or annealed noise. Note that, contrary to intrinsic noise, there is no constraint on the noise amplitude or temporal correlations. Overall, the different sources of noise work together in real ecosystems, and interesting behaviors emerges from their combinatorial effects *(74, 75)*. Given a noninteracting single population with a discrete phenotypic distribution, the time evolution of species n_i can be described by the following matrix equation

$$\frac{\mathrm{d}}{\mathrm{d}t}n_i(t) = r_i(E(t))n_i(t) + \sum_j T_{ij}(E(t))n_j(t),$$

where $E(t)$ is the random discrete variable representing the environment at time t, $r_i(E(t))$ is the environment-dependent fitness of phenotype i, and the matrix elements $T_{ij}(E(t))$ are transition probabilities for an individual to switch from the jth to the ith phenotype. Note that T_{ij} is a Laplacian matrix ($T_{ii} = -\sum_{j\neq i} T_{ij}$), and, on average, the loss term $T_{ii}n_i$ balances transitions to all other states. Interestingly, maximal growth occurs when the phenotypic switching rate is similar to that of environmental fluctuations. If environmental changes are slow or mostly predictable, random switching between states outperforms responsive switching, where the organism uses sensors to identify the optimal state of operation and transition probabilities to nonoptimal states are consequently set to zero.

4.3. Examples

4.3.1. Marine Phage Community

Recent work on marine phage communities demonstrates how the general framework of LV can be improved, and the importance of investigating microscopic origins of population growth. Hoffmann and colleagues *(76)* studied the interaction of marine phages (predator) and their host microbes (prey) by modeling the multispecies community as a simple predator–prey model. This can be justified since the phage–host interaction is highly specific and the dominant microbial species effectively is representative of the overall community *(77)*.

The key observation from this approach is that the observed cooperativity is caused by spatiotemporally nonuniform nutrient condition ascribed to a colloid-type organic detritus called "marine snow." The marine snow enhances aggregation of microbes and their predators, generating a positive feedback loop. The clustering around discrete food sources leads to locally high concentrations of lysed host cells that further attract more predators. The consequence is a superlinear dependence of predation rate in the phage population, represented as a quadratic dependence on the phage density:

$$\frac{d}{\mathrm{d}t}n_1(t) = \alpha n_1 - \beta n_1 n_2^2; \frac{\mathrm{d}}{\mathrm{d}t}n_2(t) = \gamma n_1 n_2^2 - \delta n_2^2,$$

where n_1 represents the microbial population and n_2 the phages. In order to preserve the oscillatory behavior of the predator–prey model, it is necessary to keep the phage-degradation term quadratic in the phage density. The population dynamics predicted by this model follow experimental data closely.

4.3.2. Identification of Unknown Species Interactions

Intra- and interspecies interactions among microbes are mainly responsible for the ripening process in spreadable cheeses. A recent study used the population dynamics approach to identify interactions in a spreadable cheese bacteria–eukaryote community composed of six bacteria and three yeast species *(78)*. The bacterial population behavior could be grouped into two quasispecies, resulting in a five-species model system. Using the GLV formulation as a starting point, entries in the interaction matrix *A* were selected to give simulated population dynamics that agreed with measurements. The identified possible realizations of *A* were further narrowed down through a species-removal study: A single quasispecies was removed at a time, and population dynamics for the remaining species were measured. As a result, the web of interaction between the five groups could be identified (*see* **Fig. 8**). Considering the experimental difficulties in resolving interspecies interactions in strongly interacting communities, the GLV modeling approach provides a useful first step.

4.4. Tools

4.4.1. General-Purpose ODE Solver

The SBML ODE solver library (SOSlib) is a programming library for formula representation to construct ODE systems, their Jacobian matrix, a parameter dependency matrix and other derivatives in the Systems Biology Markup Language (SBML). SOSlib provides efficient interfaces to well-established methods in theoretical chemistry, biology, and systems theory.

http://www.tbi.univie.ac.at/~raim/odeSolver

4.4.2. Stochastic Simulator

Dizzy is a software for stochastic chemical relations simulation. It provides a model definition, implementation of several stochastic and deterministic algorithms, and a graphical display of a model.

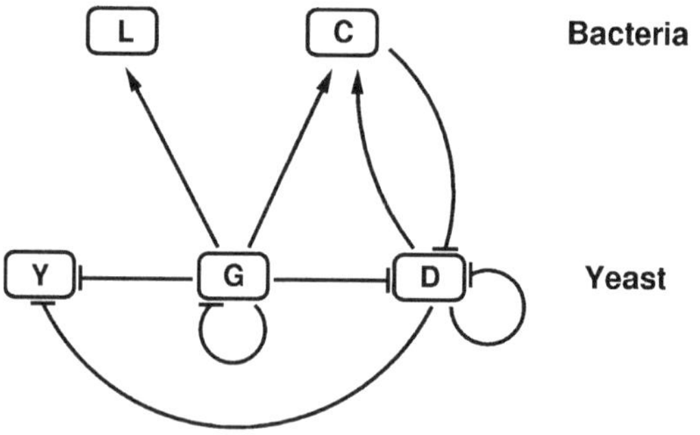

Fig. 8. Interaction among cheese microbial community is reconstructed by using LV-type modeling *(78)*. *Arrows* and *blunt ends* stand for positive and negative interactions, respectively. D *Debaryomyces hansenii*; Y *Yarrowia lipolytica*; G *Geotrichumcandidum*; L *Leucobacter* sp.; C Group including *Arthrobacter arilaitensis*, *Hafnia alvei*, *Corynebacterium casei*, *Brevibacterium aurantiacum*, and *Staphylococcus xylosus*.

It is a standard free software written in Java and is supported on Windows XP, Fedora Core 1 Linux, and Macintosh.

http://magnet.systemsbiology.net/software/Dizzy

Acknowledgments

This work was performed under the auspices of the US Department of Energy by Lawrence Livermore National Laboratory (LLNL) under Contract DE-AC52-07NA27344, and supported by the LLNL Laboratory Directed Research and Development program on grant 06-ERD-061.

References

1. Ram, R. J., Verberkmoes, N. C., Thelen, M. P., Tyson, G. W., Baker, B. J., Blake, R. C., 2nd, Shah, M., Hettich, R. L., and Banfield, J. F. (2005) Community proteomics of a natural microbial biofilm. *Science* 308, 1915–1920.
2. Breitbart, M., Salamon, P., Andresen, B., Mahaffy, J. M., Segall, A. M., Mead, D., Azam, F., and Rohwer, F. (2002) Genomic analysis of uncultured marine viral communities. *Proc. Natl Acad. Sci. USA* 99, 14250–14255.
3. Venter, J. C., Remington, K., Heidelberg, J. F., Halpern, A. L., Rusch, D., Eisen, J. A., Wu, D., Paulsen, I., Nelson, K. E., Nelson, W., Fouts, D. E., Levy, S., Knap, A. H., Lomas, M. W., Nealson, K., White, O., Peterson, J., Hoffman, J., Parsons, R., Baden-Tillson, H., Pfannkoch, C., Rogers, Y. H., and Smith, H. O. (2004) Environmental genome shotgun sequencing of the Sargasso Sea. *Science* 304, 66–74.
4. Tyson, G. W., Chapman, J., Hugenholtz, P., Allen, E. E., Ram, R. J., Richardson, P. M., Solovyev, V. V., Rubin, E. M., Rokhsar, D. S., and Banfield, J. F. (2004) Community structure and metabolism through reconstruction of microbial genomes from the environment. *Nature* 428, 37–43.
5. Turnbaugh, P. J., Ley, R. E., Mahowald, M. A., Magrini, V., Mardis, E. R., and Gordon, J. I. (2006) An obesity-associated gut microbiome with increased capacity for energy harvest. *Nature* 444, 1027–1031.
6. Colwell, R. R., Huq, A., Islam, M. S., Aziz, K. M., Yunus, M., Khan, N. H., Mahmud, A., Sack, R. B., Nair, G. B., Chakraborty, J., Sack, D. A., and Russek-Cohen, E. (2003) Reduction of cholera in Bangladeshi villages by simple filtration. *Proc. Natl Acad. Sci. USA* 100, 1051–1055.
7. Lotka, A. J. (1925) Elements of Physical Biology. Williams & Wilkins, Baltimore, MD.
8. Volterra, V. (1926) Fluctuations in the abundance of a species considered mathematically. *Nature* 118, 558–560.
9. Chassagnole, C., Noisommit-Rizzi, N., Schmid, J. W., Mauch, K., and Reuss, M. (2002) Dynamic modeling of the central carbon metabolism of *Escherichia coli*. *Biotechnol. Bioeng.* 79, 53–73.
10. Maher, A. D., Kuchel, P. W., Ortega, F., de Atauri, P., Centelles, J., and Cascante, M. (2003) Mathematical modelling of the urea cycle. A numerical investigation into substrate channelling. *Eur. J. Biochem.* 270, 3953–3961.
11. Teusink, B., Passarge, J., Reijenga, C. A., Esgalhado, E., van der Weijden, C. C., Schepper, M., Walsh, M. C., Bakker, B. M., van Dam, K., Westerhoff, H. V., and Snoep, J. L. (2000) Can yeast glycolysis be understood in terms of in vitro kinetics of the constituent enzymes? Testing biochemistry. *Eur. J. Biochem.* 267, 5313–5329.
12. Hynne, F., Dano, S., and Sorensen, P. G. (2001) Full-scale model of glycolysis in *Saccharomyces cerevisiae*. *Biophys. Chem.* 94, 121–163.
13. Zhdanov, V. P. and Kasemo, B. (2001) Simulations of oscillatory glycolytic patterns in cells. *Phys. Chem. Chem. Phys.* 3, 3786–3791.
14. Klipp, E. (2007) Modelling dynamic processes in yeast. *Yeast* 24, 943–959.
15. Bakker, B. M., Westerhoff, H. V., Opperdoes, F. R., and Michels, P. A. (2000) Metabolic

control analysis of glycolysis in trypanosomes as an approach to improve selectivity and effectiveness of drugs. *Mol. Biochem. Parasitol.* 106, 1–10.

16. Navid, A. and Ortoleva, P. J. (2004) Simulated complex dynamics of glycolysis in protozoan parasite *Trypanosoma brucei*. *J. Theor. Biol.* 228, 449–458.
17. Smolen, P. (1995) A model for glycolylic oscillations based on skeletal muscle phosphotructokinase kinetics. *J. Theor. Biol.* 174, 137–148.
18. Westermark, P. O. and Lansner, A. (2003) A model of phosphofructokinase and glycolytic oscillations in the pancreatic b-cell. *Biophys. J.* 85, 126–139.
19. Dano, S., Madsen, M. F., Schmidt, H., and Cedersund, G. (2006) Reduction of a biochemical model with preservation of its basic dynamic properties. *FEBS J.* 273, 4862–4877.
20. Chance, B., Hess, B., and Betz, A. (1964) DPNH oscillations in a cell-free extract of *S. carlsbergensis*. *Biochem. Biophys. Res. Commun.* 16, 182–187.
21. Chance, B., Schoener, B., and Elsaesser, S. (1964) Control of the waveform of oscillations of the reduced pyridine nucleotide level in a cell-free extract. *Proc. Natl Acad. Sci. USA* 52, 337–341.
22. Ghosh, A. and Chance, B. (1964) Oscillations of glycolytic intermediates in yeast cells. *Biochem. Biophys. Res. Commun.* 16, 174–181.
23. Markus, M. and Hess, B. (1984) Transitions between oscillatory modes in a glycolytic model system. *Proc. Natl Acad. Sci. USA* 81, 4394–4398.
24. Wolf, J. and Heinrich, R. (1997) Dynamics of two-component biochemical systems in interacting cells; synchronization and desynchronization of oscillations and multiple steady states. *Biosystems* 43, 1–24.
25. Markus, M. and Hess, B. (1985) Input–response relationships in the dynamics of glycolysis. *Arch. Biol. Med. Exp. (Santiago)* 18, 261–271.
26. Markus, M., Kuschmitz, D., and Hess, B. (1984) Chaotic dynamics in yeast. Glycolysis under periodic substrate input flux. *FEBS Lett.* 172, 235–238.
27. Markus, M., Mueller, S. C., and Hess, B. (1985) Observation of entrainment, quasiperiodicity and chaos in glycolysing yeast extracts under periodic glucose input. *Ber. Bunsenges. Phys. Chem.* 89, 651–654.
28. Patnaik, P. R. (2003) Oscillatory metabolism of *Saccharomyces cerevisiae*: An overview of mechanisms and models. *Biotechnol. Adv.* 21, 183–192.
29. Wolf, J. and Heinrich, R. (1997) Dynamics of biochemical oscillators in a large number of interacting cells. *Nonlinear Anal.* 30, 1835–1845.
30. Zhdanov, V. P. and Kasemo, B. (2001) Synchronization of metabolic oscillations: Two cells and ensembles of adsorbed cells. *J. Biol. Phys.* 27, 295–311.
31. Richard, P., Bakker, B. M., Teusink, B., Van Dam, K., and Westerhoff, H. V. (1996) Acetaldehyde mediates the synchronization of sustained glycolytic oscillations in populations of yeast cells. *Eur. J. Biochem.* 235, 238–241.
32. Wolf, J., Passarge, J., Somsen, O. J., Snoep, J. L., Heinrich, R., and Westerhoff, H. V. (2000) Transduction of intracellular and intercellular dynamics in yeast glycolytic oscillations. *Biophys. J.* 78, 1145–1153.
33. Richard, P., Teusink, B., Hemker, M. B., Van Dam, K., and Westerhoff, H. V. (1996) Sustained oscillations in free-energy state and hexose phosphates in yeast. *Yeast* 12, 731–740.
34. Bergmeyer, H. U. (1974) Methods of Enzymatic Analysis. Verlag Chemie, Weinheim.
35. Fuqua, W. C., Winans, S. C., and Greenberg, E. P. (1994) Quorum sensing in bacteria: The LuxR–LuxI family of cell density-responsive transcriptional regulators. *J. Bacteriol.* 176, 269–275.
36. Fuqua, C., Winans, S. C., and Greenberg, E. P. (1996) Census and consensus in bacterial ecosystems: The LuxR–LuxI family of quorum-sensing transcriptional regulators. *Annu. Rev. Microbiol.* 50, 727–751.
37. McFall-Ngai, M. J. and Ruby, E. G. (2000) Developmental biology in marine invertebrate symbioses. *Curr. Opin. Microbiol.* 3, 603–607.
38. Miller, M. B. and Bassler, B. L. (2001) Quorum sensing in bacteria. *Annu. Rev. Microbiol.* 55, 165–199.
39. Hammer, B. K. and Bassler, B. L. (2003) Quorum sensing controls biofilm formation in *Vibrio cholerae*. *Mol. Microbiol.* 50, 101–104.
40. Henke, J. M. and Bassler, B. L. (2004) Quorum sensing regulates type III secretion in *Vibrio harveyi* and *Vibrio parahaemolyticus*. *J. Bacteriol.* 186, 3794–3805.
41. Waters, C. M. and Bassler, B. L. (2005) Quorum sensing: Cell-to-cell communication in bacteria. *Annu. Rev. Cell Dev. Biol.* 21, 319–346.
42. Freeman, J. A., Lilley, B. N., and Bassler, B. L. (2000) A genetic analysis of the functions of LuxN: A two-component hybrid sensor kinase

that regulates quorum sensing in *Vibrio harveyi*. *Mol. Microbiol.* 35, 139–149.

43. Miller, M. B., Skorupski, K., Lenz, D. H., Taylor, R. K., and Bassler, B. L. (2002) Parallel quorum sensing systems converge to regulate virulence in *Vibrio cholerae*. *Cell* 110, 303–314.
44. Mok, K. C., Wingreen, N. S., and Bassler, B. L. (2003) *Vibrio harveyi* quorum sensing: A coincidence detector for two autoinducers controls gene expression. *EMBO J.* 22, 870–881.
45. Henke, J. M. and Bassler, B. L. (2004) Three parallel quorum-sensing systems regulate gene expression in *Vibrio harveyi*. *J. Bacteriol.* 186, 6902–6914.
46. Fuqua, C. and Greenberg, E. P. (2002) Listening in on bacteria: Acyl-homoserine lactone signalling. *Nat. Rev. Mol. Cell Biol.* 3, 685–695.
47. Pesci, E. C., Milbank, J. B., Pearson, J. P., McKnight, S., Kende, A. S., Greenberg, E. P., and Iglewski, B. H. (1999) Quinolone signaling in the cell-to-cell communication system of *Pseudomonas aeruginosa*. *Proc. Natl Acad. Sci. USA* 96, 11229–11234.
48. McKnight, S. L., Iglewski, B. H., and Pesci, E. C. (2000) The *Pseudomonas* quinolone signal regulates rhl quorum sensing in *Pseudomonas aeruginosa*. *J. Bacteriol.* 182, 2702–2708.
49. Pearson, J. P., Van Delden, C., and Iglewski, B. H. (1999) Active eflux and diffusion are involved in transport of *Pseudomonas aeruginosa* cell-to-cell signals. *J. Bacteriol.* 181, 1203–1210.
50. James, S., Nilsson, P., James, G., Kjelleberg, S., and Fagerstrom, T. (2000) Luminescence control in the marine bacterium *Vibrio fischeri*: An analysis of the dynamics of lux regulation. *J. Mol. Biol.* 296, 1127–1137.
51. Eberhard, A., Burlingame, A. L., Eberhard, C., Kenyon, G. L., Nealson, K. H., and Oppenheimer, N. J. (1981) Structural identification of autoinducer of *Photobacterium fischeri* luciferase. *Biochemistry* 20, 2444–2449.
52. Engebrecht, J., Nealson, K., and Silverman, M. (1983) Bacterial bioluminescence: Isolation and genetic analysis of functions from *Vibrio fischeri*. *Cell* 32, 773–781.
53. Engebrecht, J. and Silverman, M. (1984) Identification of genes and gene products necessary for bacterial bioluminescence. *Proc. Natl Acad. Sci. USA* 81, 4154–4158.
54. Fuqua, C., Parsek, M. R., and Greenberg, E. P. (2001) Regulation of gene expression by cell-to-cell communication: Acyl-homoserine lactone quorum sensing. *Annu. Rev. Genet.* 35, 439–468.
55. Mueller, J., Kuttler, C., Hense, B. A., Rothballer, M., and Hartmann, A. (2006) Cell-cell communication by quorum sensing and dimension-reduction. *J. Math. Biol.* 53, 672–702.
56. Kuttler, C. and Hense, B. A. (2008) The interplay of two quorum sensing regulation systems of *Vibrio fischeri*. *J. Theor. Biol.* 251, 167–180.
57. Ward, J. P., King, J. R., Koerber, A. J., Williams, P., Croft, J. M., and Sockett, R. E. (2001) Mathematical modelling of quorum sensing in bacteria. *IMA J. Math. Appl. Med. Biol.* 18, 263–292.
58. Dockery, J. D. and Keener, J. P. (2001) A mathematical model for quorum sensing in *Pseudomonas aeruginosa*. *Bull. Math. Biol.* 63, 95–116.
59. Kreft, J. U., Picioreanu, C., Wimpenny, J. W., and van Loosdrecht, M. C. (2001) Individual-based modelling of biofilms. *Microbiology* 147, 2897–2912.
60. Wimpenny, J. W. T. and Colasanti, R. (1997) A unifying hypothesis for the structure of microbial biofilms based on cellular automaton models. *FEMS Microbiol. Ecol.* 22, 1–16.
61. Pizarro, G., Griffeath, D., and Noguera, D. R. (2001) Quantitative cellular automaton model for biofilms. *J. Environ. Eng.* 127, 782–789.
62. Xavier, J. B. and Foster, K. R. (2007) Cooperation and conflict in microbial biofilms. *Proc. Natl Acad. Sci. USA* 104, 876–881.
63. West, S. A., Griffin, A. S., Gardner, A., and Diggle, S. P. (2006) Social evolution theory for microorganisms. *Nat. Rev. Microbiol.* 4, 597–607.
64. Kreft, J. U. (2004) Biofilms promote altruism. *Microbiology* 150, 2751–2760.
65. Chambless, J. D., Hunt, S. M., and Stewart, P. S. (2006) A three-dimensional computer model of four hypothetical mechanisms protecting biofilms from antimicrobials. *Appl. Environ. Microbiol.* 72, 2005–2013.
66. Hunt, S. M., Hamilton, M. A., Sears, J. T., Harkin, G., and Reno, J. (2003) A computer investigation of chemically mediated detachment in bacterial biofilms. *Microbiology* 149, 1155–1163.
67. Mobilia, M., Georgiev, I. T., and Täuber, U. C. (2007) Phase transitions and spatio-temporal fluctuations in stochastic lattice Lotka–Volterra models. *J. Stat. Phys.* 128, 447–483.
68. May, R. M. (1976) Simple mathematical models with very complicated dynamics. *Nature* 261, 459–467.
69. May, R. M. (1973) Stability and Complexity in Model Ecosystems. Princeton University Press, Princeton, NJ.
70. Kondoh, M. (2003) Foraging adaptation and the relationship between food-web complexity and stability. *Science* 299, 1388–1391.

71. Ackland, G. J. and Gallagher, I. D. (2004) Stabilization of large generalized Lotka–Volterra foodwebs by evolutionary feedback. *Phys. Rev. Lett.* 93, 158701.
72. Murray, J. D. (2002) Mathematical Biology. Springer, New York, NY.
73. Collet, P. and Eckmann, J. P. (1990) Instabilities and Fronts in Extended Systems. Princeton University Press, Princeton, NJ.
74. Kussell, E. and Leibler, S. (2005) Phenotypic diversity, population growth, and information in fluctuating environments. *Science* 309, 2075–2078.
75. Thattai, M. and van Oudenaarden, A. (2004) Stochastic gene expression in fluctuating environments. *Genetics* 167, 523–530.
76. Hoffmann, K. H., Rodriguez-Brito, B., Breitbart, M., Bangor, D., Angly, F., Felts, B., Nulton, J., Rohwer, F., and Salamon, P. (2007) Power law rank-abundance models for marine phage communities. *FEMS Microbiol. Lett.* 273, 224–228.
77. Thingstad, T. F. (2000) Elements of a theory for the mechanisms controlling abundance, diversity, and biogeochemical role of lytic bacterial viruses in aquatic systems. *Limnol. Oceanogr.* 45, 1320–1328.
78. Mounier, J., Monnet, C., Vallaeys, T., Arditi, R., Sarthou, A. S., Helias, A., and Irlinger, F. (2008) Microbial interactions within a cheese microbial community. *Appl. Environ. Microbiol.* 74, 172–181.

INDEX

V

W

X

Y

Z

MIX
Papier aus verantwortungsvollen Quellen
Paper from responsible sources
FSC® C105338

If you have any concerns about our products,
you can contact us on
ProductSafety@springernature.com

In case Publisher is established outside the EU,
the EU authorized representative is:
Springer Nature Customer Service Center GmbH
Europaplatz 3, 69115 Heidelberg, Germany

Printed by Libri Plureos GmbH
in Hamburg, Germany